Topics in

Quaternion Linear Algebra

Princeton Series in Applied Mathematics

Editors

Ingrid Daubechies (Princeton University); Weinan E (Princeton University); Jan Karel Lenstra (Eindhoven University); Endre Süli (University of Oxford)

The Princeton Series in Applied Mathematics publishes high quality advanced texts and monographs in all areas of applied mathematics. Books include those of a theoretical and general nature as well as those dealing with the mathematics of specific applications areas and real-world situations.

A list of titles in this series appears at the back of the book.

Topics in

Quaternion Linear Algebra

Leiba Rodman

Princeton University Press

Princeton and Oxford

Copyright © 2014 by Princeton University Press

Published by Princeton University Press, 41 William Street, Princeton, New Jersey 08540

In the United Kingdom: Princeton University Press, 6 Oxford Street, Woodstock, Oxfordshire OX20 1TW

press.princeton.edu

Library of Congress Cataloging-in-Publication Data
Rodman, L.
    Topics in quaternion linear algebra / Leiba Rodman.
      pages cm. – (Princeton series in applied mathematics)
    Includes bibliographical references and index.
ISBN 978-0-691-16185-3 (hardcover)

1. Algebras, Linear–Textbooks. 2. Quaternions–Textbooks. I. Title.
QA196.R63 2014
512′.5–dc23
                    2013050581

British Library Cataloging-in-Publication Data is available

The publisher would like to acknowledge the author of this volume for providing the camera-ready copy from which this book was printed.

This book has been composed in LaTeX

Printed on acid-free paper ∞

Printed in the United States of America

10 9 8 7 6 5 4 3 2 1

To Ella

Contents

Preface

This is probably the first book devoted entirely to linear algebra and matrix analysis over the skew field of real quaternions.

The book is intended for the primary audience of mathematicians working in the area of linear algebra and matrix analysis, instructors and students of these subjects, mathematicians working in related areas such as operator theory and differential equations, researchers who work in other areas and for whom the book is intended as a reference, and scientists (primarily physicists, chemists, and computer scientists) and engineers who may use the book as a reference as well.

The exposition is accessible to upper undergraduate and graduate students in mathematics, science, and engineering. A background in college linear algebra and a modicum of complex analysis and multivariable calculus will suffice.

I intend to keep up with the use of the book. So, I have a request of the readers: please send remarks, corrections, criticism, etc., concerning the book to me at lxrodm@gmail.com or lxrodm@math.wm.edu.

I thank J. Baez for consultation concerning automorphisms of the division algebra of real quaternions, R. M. Guralnick and R. Pereira for consultations concerning invariants for equivalence of matrices over noncommutative principal ideal domains (in particular polynomials with quaternion coefficients), C.-K. Li and Y.-T. Poon for consultations concerning numerical ranges, F. Zhang for helping out with determinants, and V. Bolotnikov. In the final stages of preparation of the manuscript I took advice from several people whose input is greatly appreciated: P. Lancaster, N. J. Higham, H. Schneider, R. Brualdi, and H. J. Woerdeman. I also thank M. Karow and two anonymous reviewers for careful reading of the manuscript and many helpful suggestions.

Leiba Rodman

Williamsburg, Virginia, September 2013

2000 Mathematics Subject Classification: 15-01, 15-02, 15A21, 15A63, 15A22, 15A60, 15B57, 15A66, 15A99.

Key words: quaternion, quaternion matrix, canonical form, numerical range, similarity, congruence, Smith form, Jordan form, Kronecker form, invariant subspaces, semidefinite subspaces, matrix pencils, hermitian matrix, Hamiltonian matrix, skewhermitian matrix, skew-Hamiltonian matrix, symmetric matrix.

Topics in

Quaternion Linear Algebra

Chapter One

Introduction

Besides the introduction, front matter, back matter, and Appendix (Chapter 15), the book consists of two parts. The first part comprises Chapters 2–7. Here, fundamental properties and constructions of linear algebra are explored in the context of quaternions, such as matrix decompositions, numerical ranges, Jordan and Kronecker canonical forms, canonical forms under congruence, determinants, invariant subspaces, etc. The exposition in the first part is on the level of an upper undergraduate or graduate textbook. The second part comprises Chapters 8–14. Here, the emphasis is on canonical forms of quaternion matrix pencils with symmetries or, what is the same, pairs of matrices with symmetries, and the exposition approaches that of a research monograph. Applications are given to systems of linear differential equations with symmetries, and matrix equations.

The mathematical tools used in the book are easily accessible to undergraduates with a background in linear algebra and rudiments of complex analysis and, on occasion, multivariable calculus. The exposition is largely based on tools of matrix analysis. The author strived to make the book self-contained and inclusive of complete proofs as much as possible, at the same time keeping the size of the book within reasonable limits. However, some compromises were inevitable here. Thus, proofs are often omitted for many linear algebra results that are standard for real and complex matrices, are often presented in textbooks, and are valid for quaternion matrices as well with essentially the same proofs.

The book can be used in a variety of ways. More than 200 exercises are provided, on various levels of difficulty, ranging from routine verification of facts and numerical examples designed to illustrate the results to open-ended questions. The exercises and detailed exposition make the book suitable in teaching as supplementary material for undergraduate courses in linear algebra, as well as for students' independent study or reading courses. For students' benefit, several appendices are included that contain background material used in the main text. The book can serve as a basis for a graduate course named advanced linear algebra, topics in linear algebra, or (for those who want to keep the narrower focus) quaternion linear algebra. For example, one can build a graduate course based on Chapters 2–8 and selections from later chapters.

Open problems presented in the book provide an opportunity to do original research. The open problems are on various levels: open-ended problems that may serve as subject for research by mathematicians and concrete, more-specific problems that are perhaps more suited for undergraduate research work under faculty supervision, honors theses, and the like.

For working mathematicians in both theoretical and applied areas, the book may serve as a reference source. Such areas include, in particular, vector calculus, ordinary and partial differential equations, and boundary value problems (see, e.g., Gürlebeck and Sprössig [60]), and numerical analysis (Bunse-Gerstner et al. [22]). The accessibility and importance of the mathematics should make this book a widely useful work not only for mathematicians, but also for scientists and engineers.

Quaternions have become increasingly useful for practitioners in research, both in theory and applications. For example, a significant number of research papers on quaternions, perhaps even most of them, appear regularly in mathematical physics journals, and quantum mechanics based on quaternion analysis is mainstream physics. In engineering, quaternions are often used in control systems, and in computer science they play a role in computer graphics. Quaternion formalism is also used in studies of molecular symmetry. For practitioners in these areas, the book can serve as a valuable reference tool.

New, previously unpublished results presented in the book with complete proofs will hopefully be useful for experts in linear algebra and matrix analysis. Much of the material appears in a book for the first time; this is true for Chapters 5–14, most of Chapter 4, and a substantial part of Chapter 3.

As far as the author is aware, this is the first book dedicated to systematic exposition of quaternion linear algebra. So far, there are only a few expository papers and chapters in books on the subject (for example, Chapter 1 in Gürlebeck and Sprössig [60], Brieskorn [20], Zhang [164], or Farenick and Pidkowich [38]) as well as algebraic treatises on skew fields (e.g., Cohn [29] or Wan [156]).

It is inevitable that many parts of quaternion linear algebra are not reflected in the book, most notably those parts pertaining to numerical analysis (Bunse-Gerstner et al. [22] and Faßbender et al. [40]). Also, the important classes of orthogonal, unitary, and symplectic quaternion matrices are given only brief exposure.

We now describe briefly the contents of the book chapter by chapter.

Chapter 2 concerns (scalar) quaternions and the basic properties of quaternion algebra, with emphasis on solution of equations such as $axb = c$ and $ax - xb = c$. Description of all automorphisms and antiautomoprhisms of quaternions is given, and representations of quaternions in terms of 2×2 complex matrices and 4×4 real matrices are introduced. These representations will play an important role throughout the book.

Chapter 3 covers basics on the vector space of columns with quaternion components, matrix algebra, and various matrix decomposition. The real and complex representations of quaternions are extended to vectors and matrices. Various matrix decompositions are studied; in particular, Cholesky factorization is proved for matrices that are hermitian with respect to involutions other than the conjugation. A large part of this chapter is devoted to numerical ranges of quaternion matrices with respect to conjugation as well as with respect to other involutions. Finally, a brief exposition is given for the set of quaternion subspaces, understood as a metric space with respect to the metric induced by the gap function.

In a short Chapter 4 we develop diagonal canonical forms and prove inertia theorems for hermitian and skewhermitian matrices with respect to involutions (including the conjugation). We also identify dimensions of subspaces that are neutral or semidefinite relative to a given hermitian matrix and are maximal with respect to this property. The material in Chapters 3 and 4 does not depend on the more involved constructions such as the Jordan form and its proof.

Chapter 5 is a key chapter in the book. Root subspaces of quaternion matrices are introduced and studied. The Jordan form of a quaternion matrix is presented in full detail, including a complete proof. The complex matrix representation plays a crucial role here. Although the standard definition of a determinant is not very useful when applied to quaternion matrices, nevertheless several notions of determinant-like functions for matrices over quaternions have been defined and used in the literature; a few of these are explored in this chapter as well. Several applications of the Jordan form are treated. These include matrix equations of the form $AX - XB = C$, functions of matrices, and stability of systems of differential equations of the form

$$A_\ell x^{(\ell)}(t) + A_{\ell-1} x^{(\ell-1)}(t) + \cdots + A_1 x'(t) + A_0 x(t) = 0, \qquad t \in \mathsf{R},$$

with constant quaternion matrix coefficients $A_\ell, \ldots, A_0$. Stability of an analogous system of difference equations is studied as well.

The main theme of Chapter 6 concerns subspaces that are simultaneously invariant for one matrix and semidefinite (or neutral) with respect to another. Such subspaces show up in many applications, some of them presented later in Chapter 13. For a given invertible plus-matrix A, the main result here asserts that any subspace which is A-invariant and at the same time nonnegative with respect to the underlying indefinite inner product can be extended to an A-invariant subspace which is maximal nonnegative. Analogous results are proved for related classes of matrices, such as unitary and dissipative, as well as in the context of indefinite inner products induced by involutions other than the conjugation.

Chapter 7 treats matrix polynomials with quaternion coefficients. A diagonal form (known as the Smith form) is proved for such polynomials. In contrast to matrix polynomials with real or complex coefficients, a Smith form is generally not unique. For matrix polynomials of first degree, a Kronecker form—the canonical form under strict equivalence—is available, which is presented with a complete proof. Furthermore, a comparison is given for the Kronecker forms of complex or real matrix polynomials with the Kronecker forms of such matrix polynomials under strict equivalence using quaternion matrices.

In Chapters 8, 9, and 10 we develop canonical forms of quaternion matrix pencils $A + tB$ in which the matrices A and B are either hermitian or skewhermitian and their applications. Chapter 8 is concerned with the case when both matrices A and B are hermitian. Full and detailed proofs of the canonical forms under strict equivalence and simultaneous congruence are provided, based on the Kronecker form of the pencil $A + tB$. Several variations of the canonical forms are included as well. Among applications here: criteria for existence of a nontrivial positive semidefinite real linear combination and sufficient conditions for simultaneous diagonalizability of two hermitian matrices under simultaneous congruence. A comparison is made with pencils of real symmetric or complex hermitian matrices. It turns out that two pencils of real symmetric matrices are simultaneously congruent over the reals if and only if they are simultaneously congruent over the quaternions. An analogous statement holds true for two pencils of complex hermitian matrices.

The subject matter of Chapter 9 is concerned mainly with matrix pencils of the form $A + tB$, where one of the matrices A or B is skewhermitian and the other may be hermitian or skewhermitian. Canonical forms of such matrix pencils are given under strict equivalence and under simultaneous congruence, with full detailed proofs, again based on the Kronecker forms. Comparisons with real and complex matrix pencils are presented. In contrast to hermitian matrix pencils, two complex skewhermitian matrix pencils that are simultaneously congruent under quaternions need not be simultaneously congruent under the complex field, although an analogous property is valid for pencils of real skewsymmetric matrices. Similar results hold for real or complex matrix pencils $A+tB$, where A is real symmetric or complex hermitian and B is real skewsymmetric or complex skewhermitian. In each case, we sort out the relationships of simultaneous congruence over the complex field of complex matrix pencils where one matrix is hermitian and the other is skewhermitian versus simultaneous congruence over the skew field of quaternions for such pencils. As an applications we obtain a canonical form for quaternion matrices under (quaternion) congruence.

In Chapter 10 we study matrices (or linear transformations) that are selfadjoint or skewadjoint with respect to a nondegenerate hermitian or skewhermitian inner product. As an application of the canonical forms obtained in Chapters 8 and 9, canonical forms for such matrices are derived. Matrices that are skewadjoint with respect to skewhermitian inner products are known as Hamiltonian matrices; they play a key role in many applications such as linear control systems (see Chapter 14). The canonical forms allow us to study invariant Lagrangian subspaces; in particular, they give criteria for existence of such subspaces. Another application involves boundedness and stable boundedness of linear systems of differential equations with constant coefficients under suitable symmetry requirements.

The development of material in Chapters 11, 12, and 13 is largely parallel to that in Chapters 8, 9, and 10, but with respect to an involution of the quaternions other than the conjugation and with respect to indefinite inner products induced by matrices that are hermitian or skewhermitian with respect to such involutions. Thus, letting ϕ be a fixed involution of the quaternions which is different from the conjugation, the canonical forms (under both strict equivalence and simultaneous ϕ-congruence) of quaternion matrix pencils $A+tB$, where each of A and B is either ϕ-hermitian or ϕ-skewhermitian, are given in Chapters 11 and 12. As before, full and detailed proofs are supplied.

Applications are made in Chapter 14 to various types of matrix equations over quaternions, such as

$$Z^m + \sum_{j=0}^{m-1} A_j Z^j = 0,$$

where $A_0, \ldots, A_{m-1}$ are given $n \times n$ quaternion matrices,

$$ZBZ + ZA - DZ - C = 0,$$

where A, B, C, and D are given quaternion matrices of suitable sizes, and the symmetric version of the latter equation,

$$ZDZ + ZA + A^*Z - C = 0, \qquad (1.0.1)$$

where D and C are assumed to be hermitian. The theory of invariant subspaces of quaternion matrices—and for equation (1.0.1) also of subspaces that are simultaneously invariant and semidefinite—plays a crucial role in study of these matrix equations. Equation (1.0.1) and its solutions, especially hermitian solutions, are important in linear control systems. A brief description of such systems and their relation to equations of the type of (1.0.1) is also provided.

For the readers' benefit, in Chapter 15 we bring several well-known canonical forms for real and complex matrices that are used extensively in the text. No proofs are given; instead we supply references that contain full proofs and further bibliographical information.

1.1   NOTATION AND CONVENTIONS

<u>Numbers, sets, spaces</u>

$A := B$—the expression or item A is defined by the expression or item B

R—the real field

$\lfloor x \rfloor$—the greatest integer not exceeding $x \in \mathsf{R}$

C—the complex field

$\mathfrak{I}(z) = (z - \overline{z})/(2\mathsf{i}) \in \mathsf{R}$—the imaginary part of a complex number z

C_+—closed upper complex half-plane

$D_\epsilon(\lambda) := \{ z \in \mathsf{C}_+ : |z - \lambda| < \epsilon \}$—part of the open circular disk centered at λ with radius ϵ that lies in C_+

$\mathsf{C}_{+,0}$—open upper complex half-plane

H—the skew field of the quaternions

$\mathsf{i}, \mathsf{j}, \mathsf{k}$—the standard quaternion imaginary units

$\mathfrak{R}(x) = x_0$ and $\mathfrak{V}(x) = x_1\mathsf{i} + x_2\mathsf{j} + x_3\mathsf{k}$—the real and the vector part of x, respectively, for $x = x_0 + x_1\mathsf{i} + x_2\mathsf{j} + x_3\mathsf{k} \in \mathsf{H}$, where $x_0, x_1, x_2, x_3 \in \mathsf{R}$

$|x| = \sqrt{x_0^2 + x_1^2 + x_2^2 + x_3^2}$—the length of $x \in \mathsf{H}$

$\mathrm{Inv}\,(\phi) := \{ x \in \mathsf{H} : \phi(x) = x \}$—the set (real vector space) of quaternions invariant under an involution ϕ of H

$\beta(\phi) \in \mathsf{H}$—quaternion with the properties that $\phi(\beta(\phi)) = -\beta(\phi)$ and $|\beta(\phi)| = 1$, where ϕ is an involution of H that is different from the quaternion conjugation; for a given ϕ, the quaternion $\beta(\phi) \in \mathsf{H}$ is unique up to negation

$\mathrm{Con}\,(\alpha) = \{ y^*\alpha y : y \in \mathsf{H} \setminus \{0\} \}$—the congruence orbit of $\alpha \in \mathsf{H}$

$\mathrm{Sim}\,(\alpha) = \{ y^{-1}\alpha y : y \in \mathsf{H} \setminus \{0\} \}$—the similarity orbit of $\alpha \in \mathsf{H}$

$\mathsf{F}^{n \times 1}$—the vector space of n-components columns with components in F, where $\mathsf{F} = \mathsf{R}$, $\mathsf{F} = \mathsf{C}$, or $\mathsf{F} = \mathsf{H}$; $\mathsf{H}^{n \times 1}$ is understood as a right quaternion vector space

$e_j \in \mathsf{H}^{n \times 1}$—the vector with 1 in the jth component and zero elsewhere; n is understood from context

$\langle x, y \rangle := y^*x$, $x, y \in \mathsf{H}^{n \times 1}$—the standard inner product defined on $\mathsf{H}^{n \times 1}$

$\|x\|_\mathsf{H} = \sqrt{\langle x, x \rangle}$ or $\|x\|$—the norm of $x \in \mathsf{H}^{n \times 1}$

$\mathsf{F}^{m \times n}$—the vector space of $m \times n$ matrices with entries in F, where $\mathsf{F} = \mathsf{R}$, $\mathsf{F} = \mathsf{C}$, or $\mathsf{F} = \mathsf{H}$, and $\mathsf{H}^{m \times n}$ is understood as a left quaternion vector space; if $m = n$, then $\mathsf{C}^{n \times n}$ is a complex algebra, whereas $\mathsf{R}^{n \times n}$ and $\mathsf{H}^{n \times n}$ are real algebras

<u>Subspaces</u>

$P_{\mathcal{M}} \in \mathsf{F}^{n \times n}$—the orthogonal projection onto the subspace $\mathcal{M} \subseteq \mathsf{F}^{n \times 1}$; here $\mathsf{F} = \mathsf{R}$, $\mathsf{F} = \mathsf{C}$, or $\mathsf{F} = \mathsf{H}$

$S_{\mathcal{M}} := \{ x \in \mathcal{M} : \|x\| = 1 \}$—the unit sphere of a nonzero subspace $\mathcal{M}$

$A|_{\mathcal{R}}$—restriction of a square-size matrix A to its invariant subspace $\mathcal{R}$ (we represent $A|_{\mathcal{R}}$ as a matrix with respect to some basis in $\mathcal{R}$)

$\mathcal{M} \dotplus \mathcal{N}$—direct sum of subspaces $\mathcal{M}$ and $\mathcal{N}$

$\mathrm{Span}\,\{x_1, \ldots, x_p\}$ or $\mathrm{Span}_\mathsf{H}\,\{x_1, \ldots, x_p\}$—the quaternion subspace spanned by vectors $x_1, \ldots, x_p \in \mathsf{H}^{n \times 1}$

$\mathrm{Span}_\mathsf{R}\,\{x_1, \ldots, x_p\}$—the real subspace spanned by vectors $x_1, \ldots, x_p \in \mathsf{H}^{n \times 1}$

$\dim_\mathsf{H} \mathcal{M}$ or $\dim \mathcal{M}$—the (quaternion) dimension of a quaternion vector space $\mathcal{M}$

Grass_n—the set of all (quaternion) subspaces in $\mathsf{H}^{n\times 1}$

$\theta(\mathcal{M},\mathcal{N}) = \|P_\mathcal{M} - P_\mathcal{N}\|$—the gap between subspaces $\mathcal{M}$ and $\mathcal{N}$

<u>Matrix-related notation</u>

I_n or I (with n understood from context)—$n \times n$ identity matrix

$0_{u\times v}$, abbreviated to 0_u, if $u = v$—the $u \times v$ zero matrix, also 0 (with u and v understood from context)

C-eigenvalues of A—for a square-size complex matrix A, defined as the (complex) roots of the characteristic polynomial of A, and $\sigma_\mathsf{C}(A)$ is the set of all C-eigenvalues; e.g., for $A = \begin{bmatrix} \mathsf{i} & 1 \\ 0 & 2\mathsf{i} \end{bmatrix}$ we have

$$\sigma_\mathsf{C}(A) = \{\mathsf{i}, 2\mathsf{i}\},$$

$\sigma(A) = \{a\mathsf{i} + b\mathsf{j} + c\mathsf{k} : a, b, c \in \mathsf{R}, \quad a^2 + b^2 + c^2 = 1 \quad \text{or} \quad a^2 + b^2 + c^2 = 4\}$

A^T—transposed matrix

A^*—conjugate transposed matrix

$\overline{A}$—the matrix obtained from $A \in \mathsf{C}^{m\times n}$ or $A \in \mathsf{H}^{m\times n}$ by replacing each entry with its complex or quaternion conjugate

$A_\phi \in \mathsf{H}^{n\times m}$, $A = [a_{i,j}]_{i,j=1}^{m,n} \in \mathsf{H}^{m\times n}$—stands for the matrix $[\phi(a_{j,i})]_{j,i=1}^{n,m}$, where ϕ is an involution of H

$\mathrm{Ran}\,A = \{Ax : x \in \mathsf{F}^{n\times 1}\} \subseteq \mathsf{F}^{m\times 1}$—the image or range of $A \in \mathsf{F}^{m\times n}$; here $\mathsf{F} \in \{\mathsf{R}, \mathsf{C}, \mathsf{H}\}$ (understood from context)

$\mathrm{Ker}\,A = \{x \in \mathsf{F}^{n\times 1} : Ax = 0\}$—the kernel of $A \in \mathsf{F}^{m\times n}$; here $\mathsf{F} \in \{\mathsf{R}, \mathsf{C}, \mathsf{H}\}$ (understood from context)

$\|A\|_\mathsf{H}$ or $\|A\|$—the norm of $A \in \mathsf{H}^{m\times n}$; it is taken to be the largest singular value of A

$\mathrm{rank}\,A$—the (quaternion) rank of a matrix $A \in \mathsf{H}^{m\times n}$; if A is real or complex, then $\mathrm{rank}\,A$ coincides with the rank of A as a real or complex matrix

$B \geq C$ or $C \leq B$—for hermitian matrices $B, C \in \mathsf{H}^{n\times n}$, indicates that the difference $B - C$ is positive semidefinite

$\mathrm{In}_+(A)$, $\mathrm{In}_-(A)$, $\mathrm{In}_0(A)$—the number of positive, negative, or zero eigenvalues of a quaternion hermitian matrix A, respectively, counted with algebraic multiplicities

$(\mathrm{In}_+(H), \mathrm{In}_-(H), \mathrm{In}_0(H))$—the $\beta(\phi)$-inertia, or the $\beta(\phi)$-signature, of a ϕ-skewhermitian matrix $H \in \mathsf{H}^{n\times n}$; here ϕ is an involution of H different from the quaternion conjugation

$\mathrm{diag}\,(X_1, \ldots, X_k) = \mathrm{diag}\,(X_j)_{j=1}^k = X_1 \oplus X_2 \oplus \cdots \oplus X_k$—block diagonal matrix with the diagonal blocks $X_1, X_2, \ldots, X_k$ (in that order)

$\mathrm{row}_{j=1,2,\ldots,p}\,X_j = \mathrm{row}\,(X_j)_{j=1}^p = [X_1\ X_2\ \cdots\ X_p]$—block row matrix

$\mathrm{col}_{j=1,2,\ldots,p}\,X_j = \mathrm{col}\,(X_j)_{j=1}^p = \begin{bmatrix} X_1 \\ X_2 \\ \vdots \\ X_p \end{bmatrix}$—block column matrix

$X^{\oplus m}$—$X \in \mathsf{H}^{\delta_1 \times \delta_2}$, the $m\delta_1 \times m\delta_2$ matrix $X \oplus \cdots \oplus X$, where X is repeated m times

1.2 STANDARD MATRICES

In this section we collect matrices in standard forms and fixed notation that will be used throughout the book, sometimes without reference. The subscript in notation for a square-size matrix will always denote the size of the matrix.

I_r or I (with r understood from context)—the $r \times r$ identity matrix

$0_{u \times v}$—often abbreviated to 0_u; if $u = v$ or 0 (with u and v understood from context), the $u \times v$ zero matrix

Jordan blocks:

$$J_m(\lambda) = \begin{bmatrix} \lambda & 1 & 0 & \cdots & 0 \\ 0 & \lambda & 1 & \cdots & 0 \\ \vdots & \vdots & \ddots & \ddots & 0 \\ \vdots & \vdots & & \lambda & 1 \\ 0 & 0 & \cdots & 0 & \lambda \end{bmatrix} \in \mathsf{H}^{m \times m}, \quad \lambda \in \mathsf{H} \tag{1.2.1}$$

real Jordan blocks:

$$J_{2m}(a \pm \mathrm{i}b) = \begin{bmatrix} a & b & 1 & 0 & \cdots & 0 & 0 \\ -b & a & 0 & 1 & \cdots & 0 & 0 \\ 0 & 0 & a & b & \cdots & 0 & 0 \\ 0 & 0 & -b & a & \cdots & \vdots & \vdots \\ \vdots & \vdots & \vdots & \vdots & & 1 & 0 \\ \vdots & \vdots & \vdots & \vdots & & 0 & 1 \\ 0 & 0 & 0 & 0 & \cdots & a & b \\ 0 & 0 & 0 & 0 & \cdots & -b & a \end{bmatrix} \in \mathsf{R}^{2m \times 2m}, \tag{1.2.2}$$

$$a \in \mathsf{R}, \quad b \in \mathsf{R} \setminus \{0\}$$

symmetric matrices:

$$F_m := \begin{bmatrix} 0 & \cdots & \cdots & 0 & 1 \\ \vdots & & & 1 & 0 \\ \vdots & & \iddots & & \vdots \\ 0 & 1 & & & \vdots \\ 1 & 0 & \cdots & \cdots & 0 \end{bmatrix} = F_m^{-1}, \tag{1.2.3}$$

$$G_m := \begin{bmatrix} 0 & \cdots & \cdots & 1 & 0 \\ \vdots & & & 0 & 0 \\ \vdots & & & & \vdots \\ 1 & 0 & & & \vdots \\ 0 & 0 & \cdots & \cdots & 0 \end{bmatrix} = \begin{bmatrix} F_{m-1} & 0 \\ 0 & 0 \end{bmatrix}, \tag{1.2.4}$$

$$\widetilde{G}_m := F_m G_m F_m = \begin{bmatrix} 0 & 0 \\ 0 & F_{m-1} \end{bmatrix}. \tag{1.2.5}$$

We also define

$$\Xi_m(\alpha) := \begin{bmatrix} 0 & 0 & \cdots & 0 & \alpha \\ 0 & 0 & \cdots & -\alpha & 0 \\ \vdots & \vdots & \ddots & \vdots & \vdots \\ 0 & (-1)^{m-2}\alpha & \cdots & 0 & 0 \\ (-1)^{m-1}\alpha & 0 & \cdots & 0 & 0 \end{bmatrix} \in \mathsf{H}^{m\times m}, \qquad (1.2.6)$$

where $\alpha \in \mathsf{H}$, and

$$\begin{aligned} \Phi_m(\beta) &:= \begin{bmatrix} 0 & 0 & \cdots & 0 & 0 & \beta \\ 0 & 0 & \cdots & 0 & -\beta & -1 \\ 0 & 0 & \cdots & \beta & -1 & 0 \\ \vdots & \vdots & \ddots & \vdots & \vdots & \vdots \\ (-1)^{m-1}\beta & -1 & 0 & \cdots & 0 & 0 \end{bmatrix} \\ &= \Xi_m(\beta) - \widetilde{G}_m \in \mathsf{H}^{m\times m}, \end{aligned} \qquad (1.2.7)$$

where $\beta \in \mathsf{H}$. Note that

$$\Xi_m(\alpha) = (-1)^{m-1}(\Xi_m(\alpha))^T, \qquad \alpha \in \mathsf{H};$$

in particular $\Xi_m(\alpha) = (-1)^m(\Xi_m(\alpha))^*$ if and only if the real part of α is zero.

$$Y_{2m} = \begin{bmatrix} 0 & & & & 1 & 0 \\ & & & & 0 & -1 \\ & & 1 & 0 & & \\ & & 0 & -1 & & \\ & \iddots & & & & \\ 1 & 0 & & & & \\ 0 & -1 & & & & 0 \end{bmatrix}. \qquad (1.2.8)$$

Note that Y_{2m} is real symmetric.

Real matrix pencils—

$$Z_{2m}(t, \mu, \nu) := (t + \mu)F_{2m} + \nu Y_{2m} + \begin{bmatrix} F_{2m-2} & 0 \\ 0 & 0_2 \end{bmatrix}, \qquad (1.2.9)$$

where $\mu \in \mathsf{R}$, $\nu \in \mathsf{R} \setminus \{0\}$.

Singular matrix pencils—

$$L_{\varepsilon \times (\varepsilon+1)}(t) = [0_{\varepsilon \times 1} \quad I_\varepsilon] + t[I_\varepsilon \quad 0_{\varepsilon \times 1}] \in \mathsf{H}^{\varepsilon \times (\varepsilon+1)}. \qquad (1.2.10)$$

Here ε is a positive integer; $L_{\varepsilon \times (\varepsilon+1)}(t)$ is of size $\varepsilon \times (\varepsilon + 1)$.

Chapter Two

The algebra of quaternions

In this chapter we introduce the quaternions and their algebra: multiplication, norm, automorphisms and antiautomorphisms, etc. We give matrix representations of various real linear maps associated with quaternion algebra. We also introduce representations of quaternions as real 4×4 matrices and as complex 2×2 matrices.

2.1 BASIC DEFINITIONS AND PROPERTIES

Fix an ordered basis $\{e, i, j, k\}$ in a 4-dimensional real vector space H (we may take $H = R^4$, the vector space of columns consisting of four real components), and introduce multiplication in H by the formulas

$$ei = ie = i, \quad ej = je = j, \quad ek = ke = k,$$

$$i^2 = j^2 = k^2 = -e, \quad e^2 = e, \quad ij = -ji = k, \quad jk = -kj = i, \quad ki = -ik = j,$$

and by the requirement that the multiplication of elements of H is distributive with respect to addition and commutes with scalar multiplication:

$$x(y + z) = xy + xz, \quad (y + z)x = yx + yz, \quad x(\lambda y) = (\lambda x)y = \lambda(xy)$$

for all $x, y, z \in H$ and all $\lambda \in R$.

Definition 2.1.1. The elements of H with the algebraic operations of H as a real vector space and with the multiplication introduced as above are called the (real) *quaternions*.

The letter H stands for William Rowan Hamilton (1805–1865), inventor of quaternions. Clearly, the multiplication in the algebra H is noncommutative.

Proposition 2.1.2. H *is a unital associative algebra with the unity* e:

$$x(yz) = (xy)z, \quad ex = xe = x$$

for all $x, y, z \in H$.

In the sequel we identify the real number λ with the quaternion λe; in particular, 1 stands for $1e$. Also, it is easy to see that the real span of 1 and i is isomorphic (as a subalgebra of H) to C; thus, we identify, when convenient, C with the subalgebra of H spanned (as a real vector space) by 1 and i.

Definition 2.1.3. For a quaternion $x = x_0 + x_1 i + x_2 j + x_3 k$, where $x_0, x_1, x_2, x_3 \in R$, we define $\mathfrak{R}(x) = x_0$, the *real part* of x, and $\mathfrak{V}(x) = x_1 i + x_2 j + x_3 k$, the *vector part* (or *imaginary part*) of x. The *conjugate quaternion* of x is defined by $x_0 - x_1 i - x_2 j - x_3 k = \mathfrak{R}(x) - \mathfrak{V}(x)$ and denoted $\bar{x}$ or x^*. The *norm* of x is $|x| = \sqrt{x^* x} = \sqrt{x_0^2 + x_1^2 + x_2^2 + x_3^2} \in R$. We say that $x \in H$ is a *unit quaternion* if $|x| = 1$.

Some elementary properties of the algebra of quaternions are listed below.

Proposition 2.1.4. *Let $x, y \in \mathsf{H}$. Then:*

1. $x^* x = x x^*$;

2. $|x| = |x^*|$;

3. $|\cdot|$ *is indeed a norm on* H; *in more detail, for all* $x, y \in \mathsf{H}$ *we have:*
$$|x| \geq 0 \quad \text{with equality if and only if } x = 0;$$
$$|x + y| \leq |x| + |y|; \quad |xy| = |yx| = |x| \cdot |y|;$$

4. $\mathsf{j} c \mathsf{j}^* = \mathsf{k} c \mathsf{k}^* = \bar{c}$ *for every* $c \in \mathsf{C}$;

5. $(xy)^* = y^* x^*$;

6. $x = x^*$ *if and only if* $x \in \mathsf{R}$;

7. *if* $a \in \mathsf{H}$, *then* $ax = xa$ *for every* $x \in \mathsf{H}$ *if and only if* $a \in \mathsf{R}$;

8. *every* $x \in \mathsf{H} \setminus \{0\}$ *has an inverse* $x^{-1} = x^* / |x|^2 \in \mathsf{H}$; *in more detail,*
$$x \cdot (x^* / |x|^2) = (x^* / |x|^2) \cdot x = 1;$$

9. $|x^{-1}| = |x|^{-1}$ *for every* $x \in \mathsf{H} \setminus \{0\}$;

10. $x \in \mathsf{H}$ *and* x^* *are solutions of the following quadratic equation with real coefficients:* $t^2 - 2\Re(x)t + |x|^2 = 0$;

11. *Cauchy-Schwarz-type inequality is* $\max\{|\Re(xy)|, |\mathfrak{V}(xy)|\} \leq |x| \cdot |y|$;

12. $\Re(xy) = \Re(yx)$ *for all* $x, y \in \mathsf{H}$;

13. *if* $\Re(x) = 0$, *then* $x^2 = -|x|^2$.

We indicate a proof of $|xy| = |x| \cdot |y|$:
$$|xy|^2 = xy(xy)^* = xyy^* x^* = yy^* x x^* = |y|^2 |x|^2, \quad \text{for all } x, y \in \mathsf{H}.$$

Thus, H is a *division ring*, i.e., a unital ring in which every nonzero element has a multiplicative inverse, and also a 4-dimensional *algebra* over the field of real numbers R.

Note that the multiplication of quaternions with zero real parts can be expressed in terms of the usual inner product and cross product of vectors in R^3, namely, if $x = x_1 \mathsf{i} + x_2 \mathsf{j} + x_3 \mathsf{k}$, $y = y_1 \mathsf{i} + y_2 \mathsf{j} + y_3 \mathsf{k} \in \mathsf{H}$, where $x_\ell, y_\ell \in \mathsf{R}$, then

$$xy = -p_x^T p_y + [\mathsf{i} \ \ \mathsf{j} \ \ \mathsf{k}](p_x \times p_y), \tag{2.1.1}$$

where

$$p_x = [x_1 \ \ x_2 \ \ x_3]^T, \quad p_y = [y_1 \ \ y_2 \ \ y_3]^T \in \mathsf{R}^{3 \times 1}, \tag{2.1.2}$$

and where in the right-hand side of (2.1.1) $\times$ denotes the cross product (also known as vector product) of vectors in $\mathsf{R}^{3 \times 1}$:

$$[x_1 \ \ x_2 \ \ x_3]^T \times [y_1 \ \ y_2 \ \ y_3]^T = (x_2 y_3 - x_3 y_2, -(x_1 y_3 - x_3 y_1), x_1 y_2 - x_2 y_1)^T.$$

The verification of (2.1.1) is straightforward. More generally, let

$$x = x_0 + x_1 \mathsf{i} + x_2 \mathsf{j} + x_3 \mathsf{k}, \quad y = y_0 + y_1 \mathsf{i} + y_2 \mathsf{j} + y_3 \mathsf{k} \in \mathsf{H}, \quad x_\ell, y_\ell \in \mathsf{R},$$

and define p_x, p_y by (2.1.2). Then

$$\Re(xy) = x_0 y_0 - p_x^T p_y, \quad \mathfrak{V}(xy) = x_0 \mathfrak{V}(y) + y_0 \mathfrak{V}(x) + [\mathsf{i} \ \ \mathsf{j} \ \ \mathsf{k}](p_x \times p_y).$$

2.2 REAL LINEAR TRANSFORMATIONS AND EQUATIONS

For fixed $a, b \in \mathsf{H}$, the map $x \mapsto axb$ is obviously a real linear transformation on H. We give the matrix form of this transformation with respect to the (ordered) basis $1, \mathsf{i}, \mathsf{j}, \mathsf{k}$ of H as a real vector space.

Theorem 2.2.1. *Let*

$$a = a_0 + a_1\mathsf{i} + a_2\mathsf{j} + a_3\mathsf{k}, \qquad b = b_0 + b_1\mathsf{i} + b_2\mathsf{j} + b_3\mathsf{k} \in \mathsf{H},$$

where $a_j, b_j \in \mathsf{R}$ for $j = 0, 1, 2, 3$. Let

$$T_{a,b}x = axb, \qquad x \in \mathsf{H}, \tag{2.2.1}$$

be a real linear transformation. Then $T_{a,b}$ is given by the following matrix with respect to the ordered real basis $\{1, \mathsf{i}, \mathsf{j}, \mathsf{k}\}$ in H:

$$\begin{bmatrix} a_0b_0 - a_1b_1 - a_2b_2 - a_3b_3 & -a_0b_1 - a_1b_0 + a_2b_3 - a_3b_2 \\ a_0b_1 + a_1b_0 + a_2b_3 - a_3b_2 & a_0b_0 - a_1b_1 + a_2b_2 + a_3b_3 \\ a_0b_2 - a_1b_3 + a_2b_0 + a_3b_1 & -a_0b_3 - a_1b_2 - a_2b_1 + a_3b_0 \\ a_0b_3 + a_1b_2 - a_2b_1 + a_3b_0 & a_0b_2 - a_1b_3 - a_2b_0 - a_3b_1 \end{bmatrix}$$

$$\begin{bmatrix} -a_0b_2 - a_1b_3 - a_2b_0 + a_3b_1 & -a_0b_3 + a_1b_2 - a_2b_1 - a_3b_0 \\ a_0b_3 - a_1b_2 - a_2b_1 - a_3b_0 & -a_0b_2 - a_1b_3 + a_2b_0 - a_3b_1 \\ a_0b_0 + a_1b_1 - a_2b_2 + a_3b_3 & a_0b_1 - a_1b_0 - a_2b_3 - a_3b_2 \\ -a_0b_1 + a_1b_0 - a_2b_3 - a_3b_2 & a_0b_0 + a_1b_1 + a_2b_2 - a_3b_3 \end{bmatrix}.$$

The proof is obtained by tedious but straightforward computation.

The following particular cases are of interest.

Corollary 2.2.2. *The real linear transformations $T_{1,b}$, $T_{a,1}$, T_{a,a^*}, and $T_{a,a^{-1}}$ (in the latter case it is assumed $a \neq 0$) are given by the following matrices, respectively, with respect to the ordered real basis $\{1, \mathsf{i}, \mathsf{j}, \mathsf{k}\}$ of H and using the notation of Theorem 2.2.1:*

$$\begin{bmatrix} b_0 & -b_1 & -b_2 & -b_3 \\ b_1 & b_0 & b_3 & -b_2 \\ b_2 & -b_3 & b_0 & b_1 \\ b_3 & b_2 & -b_1 & b_0 \end{bmatrix}, \quad \begin{bmatrix} a_0 & -a_1 & -a_2 & -a_3 \\ a_1 & a_0 & -a_3 & a_2 \\ a_2 & a_3 & a_0 & -a_1 \\ a_3 & -a_2 & a_1 & a_0 \end{bmatrix},$$

$$\begin{bmatrix} a_0^2 + a_1^2 + a_2^2 + a_3^2 & 0 \\ 0 & a_0^2 + a_1^2 - a_2^2 - a_3^2 \\ 0 & 2a_0a_3 + 2a_1a_2 \\ 0 & -2a_0a_2 + 2a_1a_3 \end{bmatrix}$$

$$\begin{bmatrix} 0 & 0 \\ -2a_0a_3 + 2a_1a_2 & 2a_0a_2 + 2a_1a_3 \\ a_0^2 - a_1^2 + a_2^2 - a_3^2 & -2a_0a_1 + 2a_2a_3 \\ 2a_0a_1 + 2a_2a_3 & a_0^2 - a_1^2 - a_2^2 + a_3^2 \end{bmatrix}, \tag{2.2.2}$$

and $(a_0^2 + a_1^2 + a_2^2 + a_3^2)^{-1}X$, where X is the matrix (2.2.2).

The statement of Corollary 2.2.2 concerning $T_{a,a^{-1}}$ follows from the observation that $T_{a,a^{-1}} = |a|^{-2}T_{a,a^*}$. Note that the matrices corresponding to $T_{b,1}$, resp. to $T_{1,a}$, are skewsymmetric if and only if $\mathfrak{R}(b) = 0$, resp. $\mathfrak{R}(a) = 0$, whereas the matrices corresponding to T_{a,a^*} and $T_{a,a^{-1}}$ are symmetric if and only if $\mathfrak{R}(a) = 0$ or $\mathfrak{V}(a) = 0$.

Corollary 2.2.3. *Let $a = a_0 + a_1\mathsf{i} + a_2\mathsf{j} + a_3\mathsf{k} \in \mathsf{H}$, $a_0, a_1, a_2, a_3 \in \mathsf{R}$. Then the real linear transformation $T_{1,a} - T_{a,1}$ that maps $x \in \mathsf{H}$ to $xa - ax$ is given by the skewsymmetric matrix*

$$
\begin{bmatrix}
0 & 0 & 0 & 0 \\
0 & 0 & 2a_3 & -2a_2 \\
0 & -2a_3 & 0 & 2a_1 \\
0 & 2a_2 & -2a_1 & 0
\end{bmatrix}
$$

with respect to the ordered real basis $\{1, \mathsf{i}, \mathsf{j}, \mathsf{k}\}$.

Observe that for $a = a_0 + a_1\mathsf{i} + a_2\mathsf{j} + a_3\mathsf{k} \in \mathsf{H} \setminus \{0\}$, where a_0, a_1, a_2, a_3 are real, the matrix

$$
U := \frac{1}{|a|^2}
\begin{bmatrix}
a_0^2 + a_1^2 - a_2^2 - a_3^2 & -2a_0a_3 + 2a_1a_2 & 2a_0a_2 + 2a_1a_3 \\
2a_0a_3 + 2a_1a_2 & a_0^2 - a_1^2 + a_2^2 - a_3^2 & -2a_0a_1 + 2a_2a_3 \\
-2a_0a_2 + 2a_1a_3 & 2a_0a_1 + 2a_2a_3 & a_0^2 - a_1^2 - a_2^2 + a_3^2
\end{bmatrix}
$$

is *orthogonal*, i.e., $U^T U = I$. A straightforward computation will verify this assertion. Moreover, $\det U = 1$. Indeed, the set of all nonzero quaternions is connected, and $\det U$ is a continuous function of the components of a. Therefore, the values of $\det U$ also form a connected set (Theorem 3.10.7). But determinants of real orthogonal matrices can be only 1 or -1. It follows that either $\det U = 1$ for all U, or $\det U = -1$ for all U. Since for $a = 1$ we have $U = I$, the second possibility cannot happen.

We obtain that 1 is an eigenvalue of U, and the corresponding eigenvector is unique up to scaling (apart from the case $U = I$). So, in a suitable orthonormal basis in R^3, the matrix U has the form

$$
U =
\begin{bmatrix}
\cos\mu & -\sin\mu & 0 \\
\sin\mu & \cos\mu & 0 \\
0 & 0 & 1
\end{bmatrix}, \quad 0 \le \mu < 2\pi.
$$

Comparing with Corollary 2.2.2, the following geometric description of the transformation $T_{a,a^{-1}}$ is obtained. In this description, H_0 stands for the real vector space of all quaternions with zero real parts, and orthogonality in H_0 is understood in the sense of the real-valued inner product that has $\mathsf{i}, \mathsf{j}, \mathsf{k}$ as an orthonormal basis.

Corollary 2.2.4. *Let $a \in \mathsf{H} \setminus \{0\}$, and assume $T_{a,a^{-1}} \neq I$. Then $T_{a,a^{-1}}$ maps H_0 onto itself. Moreover, there is a unique (up to scaling) nonzero $x_0 \in \mathsf{H}_0$ such that $T_{a,a^{-1}}x_0 = x_0$, and denoting by $\mathsf{H}_0 \ominus \operatorname{Span}_\mathsf{R}\{x_0\}$ the 2-dimensional plane in H_0 orthogonal to x_0, we have that $T_{a,a^{-1}}$ acts as a rotation through a fixed angle μ, $0 < \mu < 2\pi$, in $\mathsf{H}_0 \ominus \operatorname{Span}_\mathsf{R}\{x_0\}$.*

Definition 2.2.5. *We say that two quaternions x, y are* similar *if $axa^{-1} = y$ for some $a \in \mathsf{H} \setminus \{0\}$ and* congruent *if $axa^* = y$ for some $a \in \mathsf{H} \setminus \{0\}$.*

Clearly, both similarity and congruence are equivalence relations. Denote by

$$
\operatorname{Sim}(x) = \{y \in \mathsf{H} : y \text{ similar to } x\}
$$

and

$$
\operatorname{Con}(x) = \{y \in \mathsf{H} : y \text{ congruent to } x\}
$$

the similarity orbit and the congruence orbit of $x \in \mathsf{H}$, respectively.

We have

$$\mathrm{Con}\,(x) = \bigcup_{\lambda > 0} \{\lambda \mathrm{Sim}\,(x)\}.$$

Indeed, this follows from the formula $a^* = |a|^2 a^{-1}$, $a \in \mathsf{H} \setminus \{0\}$, with $\lambda = |a|^2$.

Theorem 2.2.6. *Fix* $x = x_0 + x_1 \mathsf{i} + x_2 \mathsf{j} + x_3 \mathsf{k} \in \mathsf{H}$, *where* $x_j \in \mathsf{R}$. *The following statements are equivalent for* $y = y_0 + y_1 \mathsf{i} + y_2 \mathsf{j} + y_3 \mathsf{k} \in \mathsf{H}$, $y_j \in \mathsf{R}$:

(1) $y \in \mathrm{Sim}\,(x)$;

(2) $y = axa^*$ *for some unit quaternion* a;

(3) $[y_0 \ \ y_1 \ \ y_2 \ \ y_3]^T = \begin{bmatrix} 1 & 0 \\ 0 & Q \end{bmatrix} [x_0 \ \ x_1 \ \ x_2 \ \ x_3]^T$ *for some* 3×3 *real orthogonal matrix* Q;

(4) $[y_0 \ \ y_1 \ \ y_2 \ \ y_3]^T = \begin{bmatrix} 1 & 0 \\ 0 & Q' \end{bmatrix} [x_0 \ \ x_1 \ \ x_2 \ \ x_3]^T$ *for some* 3×3 *real orthogonal matrix* Q' *having determinant* 1;

(5) $\mathfrak{R}(y) = \mathfrak{R}(x)$ *and* $|\mathfrak{V}(y)| = |\mathfrak{V}(x)|$.

Proof. $(1) \Longrightarrow (2)$: If $y = bxb^{-1}$, $b \in \mathsf{H} \setminus \{0\}$, then (2) holds with $a = b/|b|$.
$(2) \Longrightarrow (4)$: Follows from Corollary 2.2.2 and Ex. 2.7.7.
$(4) \Longrightarrow (3)$: Obvious.
$(3) \Longrightarrow (5)$: Follows from the isometric property of real orthogonal matrices, i.e., $\|Qu\| = \|u\|$ for all real orthogonal $Q \in \mathsf{R}^{n \times n}$ and all $u \in \mathsf{R}^{n \times 1}$.
$(5) \Longrightarrow (1)$: By hypothesis, $y_0 = x_0$ and $y_1^2 + y_2^+ y_3^2 = x_1^2 + x_2^2 + x_3^2$. Consider the equation

$$(z_0 + z_1 \mathsf{i} + z_2 \mathsf{j} + z_3 \mathsf{k})y = x(z_0 + z_1 \mathsf{i} + z_2 \mathsf{j} + z_3 \mathsf{k}) \qquad (2.2.3)$$

with real unknowns z_0, z_1, z_2, z_3. Equating the coefficients of each of $1, \mathsf{i}, \mathsf{j}, \mathsf{k}$ in the left and the right sides of (2.2.3), we see that (2.2.3), after some simple algebra, boils down to the system of equations

$$\begin{bmatrix} 0 & x_1 - y_1 & x_2 - y_2 & x_3 - y_3 \\ -x_1 + y_1 & 0 & x_3 + y_3 & -x_2 - y_2 \\ -x_2 + y_2 & -x_3 - y_3 & 0 & x_1 + y_1 \\ -x_3 + y_3 & x_2 + y_2 & -x_1 - y_1 & 0 \end{bmatrix} \begin{bmatrix} z_0 \\ z_1 \\ z_2 \\ z_3 \end{bmatrix} = 0. \qquad (2.2.4)$$

We claim that the matrix in the left-hand side of (2.2.4), call it X, is singular. Indeed, $X \begin{bmatrix} 0 \\ x_1 + y_1 \\ x_2 + y_2 \\ x_3 + y_3 \end{bmatrix} = 0$, so, unless

$$x_1 + y_1 = x_2 + y_2 = x_3 + y_3 = 0, \qquad (2.2.5)$$

the matrix X is singular. But if (2.2.5) holds, X is easily seen to be singular as well. Thus, (2.2.3) has a nontrivial solution, and (1) follows. $\qquad \square$

Theorem 2.2.6 allows us to express the similarity orbits of quaternions in a more detailed way:

$$\mathrm{Sim}\,(x) = \mathfrak{R}(x) + |\mathfrak{V}(x)|\mathsf{S},$$

where

$$S := \{q \in \mathsf{H} : \mathfrak{R}(q) = 0, \quad |q| = 1\} = \{q \in \mathsf{H} : q^2 = -1\}. \qquad (2.2.6)$$

Geometrically, S is the unit sphere in $\mathsf{R}^{3 \times 1}$.

2.3   THE SYLVESTER EQUATION

Let $a, b \in \mathsf{H}$. In this section we study the *Sylvester equation*

$$ax - xb = y, \qquad x, y \in \mathsf{H},$$

and the corresponding real linear transformation

$$S_{a,b}(x) = ax - xb, \qquad x \in \mathsf{H}. \qquad (2.3.1)$$

In what follows, we make use of the inner product.

Definition 2.3.1. The real-valued *inner product* of two quaternions is defined by

$$\backslash x_0 + x_1 \mathsf{i} + x_2 \mathsf{j} + x_3 \mathsf{k}, y_0 + y_1 \mathsf{i} + y_2 \mathsf{j} + y_3 \mathsf{k}/ := x_0 y_0 + x_1 y_1 + x_2 y_2 + x_3 y_3, \quad x_\ell, y_\ell \in \mathsf{R}.$$

Note that $\backslash x, x/ = |x|^2$ for every $x \in \mathsf{H}$. Also, for $x, y \in \mathsf{H}$ with zero real parts, we have $\backslash x, xy/ = \backslash y, xy/ = 0$, as can be easily verified using formula (2.1.1).

For given $a, b \in \mathsf{H}$, define the quaternions $\{x_+, y_+, x_-, y_-\}$ as follows:

(i) If $\mathfrak{V}(a)$ and $\mathfrak{V}(b)$ are linearly independent over R, we set

$$\begin{aligned} x_\pm &= (\pm |\mathfrak{V}(a)||\mathfrak{V}(b)| - (\mathfrak{V}(a))(\mathfrak{V}(b)))/n_\pm, \\ y_\pm &= (|\mathfrak{V}(a)|\mathfrak{V}(b) \pm |\mathfrak{V}(b)|\mathfrak{V}(a))/n_\pm, \end{aligned} \qquad (2.3.2)$$

where

$$n_\pm = \sqrt{2|\mathfrak{V}(a)| \, |\mathfrak{V}(b)| \, (|\mathfrak{V}(a)||\mathfrak{V}(b)| \pm \backslash \mathfrak{V}(a), \mathfrak{V}(b)/)}.$$

Note that in view of the Cauchy-Schwarz inequality applied to $\mathfrak{V}(a)$ and $\mathfrak{V}(b)$ (interpreted as vectors in R^3) and the linear independence of $\mathfrak{V}(a)$ and $\mathfrak{V}(b)$, we have $|\mathfrak{V}(a)||\mathfrak{V}(b)| \pm \backslash \mathfrak{V}(a), \mathfrak{V}(b)/ > 0$.

(ii) Suppose $\mathfrak{V}(a)$ and $\mathfrak{V}(b)$ are linearly dependent over R. Then there exists $q \in \mathsf{H}$ with $\mathfrak{R}(q) = 0$, $|q| = 1$, $\mathfrak{V}(a) = |\mathfrak{V}(a)|q$, and $\mathfrak{V}(b) = |\mathfrak{V}(b)|q$ or $\mathfrak{V}(b) = -|\mathfrak{V}(b)|q$. Let $\widehat{q} \in \mathbb{H}$ be such that $\mathfrak{R}(\widehat{q}) = 0$, $|\widehat{q}| = 1$ and $\langle q, \widehat{q} \rangle = 0$. If $\mathfrak{V}(b) = |\mathfrak{V}(b)|q$, we define

$$x_+ = 1, \quad y_+ = q, \quad x_- = \widehat{q}, \quad y_- = q\widehat{q}.$$

If $\mathfrak{V}(b) = -|\mathfrak{V}(b)|q$, we define

$$x_+ = \widehat{q}, \quad y_+ = q\widehat{q}, \quad x_- = 1, \quad y_- = q.$$

Note that q and $\widehat{q}$ are not unique (for given a and b); more precisely, q is unique if $\mathfrak{V}(a) \neq 0$ and is unique up to negation if $\mathfrak{V}(b) \neq \mathfrak{V}(a) = 0$.

Furthermore, we define the subspaces

$$\mathcal{V}_{a,b}^+ = \mathrm{Span}_{\mathsf{R}} \{x_+, y_+\}, \qquad \mathcal{V}_{a,b}^- = \mathrm{Span}_{\mathsf{R}} \{x_-, y_-\}.$$

The subspaces $\mathcal{V}_{a,b}^{\pm}$ are uniquely determined by a and b (unless $\mathfrak{V}(a) = \mathfrak{V}(b) = 0$), see Ex. 2.7.25. Moreover, we have

$$\mathcal{V}_{a,b}^+ = \mathcal{V}_{a,b^*}^-, \quad \mathcal{V}_{a,b}^- = V_{a,b^*}^+$$

if at least one of a and b is nonreal.

Finally, we set

$$Q(\alpha, \beta) := \begin{bmatrix} \alpha & -\beta \\ \beta & \alpha \end{bmatrix}, \qquad \alpha, \beta \in \mathsf{R}.$$

Note the easily observed equality $\|Q(\alpha, \beta)\xi\| = \sqrt{\alpha^2 + \beta^2}\,\|\xi\|$ for all $\xi \in \mathsf{R}^{2 \times 1}$, where we have used the euclidean norm in $\mathsf{R}^{2 \times 1}$.

The main result of this section gives explicitly the orthonormal basis that reduces all three linear transformations $T_{a,1}$, $T_{1,b}$ (defined in (2.2.1)) and $S_{a,b}$ to a real Jordan form.

Theorem 2.3.2. (a) *The vectors x_+, y_+, x_-, y_- form an orthonormal basis (with respect to $\langle \cdot, \cdot \rangle$) of* H.

(b) *The equalities*

$$\begin{bmatrix} T_{a,1}(x_+) & T_{a,1}(y_+) & T_{a,1}(x_-) & T_{a,1}(y_-) \end{bmatrix} = \begin{bmatrix} x_+ & y_+ & x_- & y_- \end{bmatrix}$$
$$\cdot\, (Q(\mathfrak{R}(a), |\mathfrak{V}(a)|) \oplus Q(\mathfrak{R}(a), |\mathfrak{V}(a)|))$$

and

$$\begin{bmatrix} T_{1,b}(x_+) & T_{1,b}(y_+) & T_{1,b}(x_-) & T_{1,b}(y_-) \end{bmatrix} = \begin{bmatrix} x_+ & y_+ & x_- & y_- \end{bmatrix}$$
$$\cdot\, (Q(\mathfrak{R}(b), |\mathfrak{V}(b)|) \oplus Q(\mathfrak{R}(b), -|\mathfrak{V}(b)|))$$

hold true.

(c) *The subspaces $\mathcal{V}_{a,b}^+$ and $\mathcal{V}_{a,b}^-$ are both $T_{a,1}$- and $T_{1,b}$-invariant.*

(d) *The equality*

$$\begin{bmatrix} S_{a,b}(x_+) & S_{a,b}(y_+) & S_{a,b}(x_-) & S_{a,b}(y_-) \end{bmatrix} = \begin{bmatrix} x_+ & y_+ & x_- & y_- \end{bmatrix}$$
$$\cdot\, (Q(\mathfrak{R}(a) - \mathfrak{R}(b), |\mathfrak{V}(a)| - |\mathfrak{V}(b)|) \oplus Q(\mathfrak{R}(a) - \mathfrak{R}(b), |\mathfrak{V}(a)| + |\mathfrak{V}(b)|))$$

holds true.

Proof. Part (a) is established by a straightforward but tedious computation. Parts (c) and (d) are immediate from (b) as $S_{a,b} = T_{a,1} - T_{1,b}$. The verification of (b) is straightforward (Proposition 2.1.4(13) is used repeatedly). $\qquad\square$

We can read off many important properties of $S_{a,b}$ from Theorem 2.3.2, such as the following.

Theorem 2.3.3. *Let $a, b \in \mathsf{H}$, and $S_{a,b}$ defined by (2.3.1). Then:*

(1) *the four singular values of $S_{a,b}$ are*

$$\sigma_1 = \sigma_2 = \sqrt{(\mathfrak{R}(a) - \mathfrak{R}(b))^2 + (|\mathfrak{V}(a)| + |\mathfrak{V}(b)|)^2},$$
$$\sigma_3 = \sigma_4 = \sqrt{(\mathfrak{R}(a) - \mathfrak{R}(b))^2 + (|\mathfrak{V}(a)| - |\mathfrak{V}(b)|)^2};$$

moreover, $|S_{a,b}(x)| = \sigma_4|x|$ for $x \in \mathcal{V}_{a,b}^+$, and $|S_{a,b}(x)| = \sigma_1|x|$ for $x \in \mathcal{V}_{a,b}^-$;

(2) $S_{a,b}$ *is singular if and only if* $\Re(a) = \Re(b)$ *and* $|\mathfrak{V}(a)| = |\mathfrak{V}(b)|$. *If these conditions hold and* $a, b \notin \mathsf{R}$, *then*

$$\operatorname{Ker} S_{a,b} = \mathcal{V}_{a,b}^+ = \mathcal{V}_{a,b^*}^- = \operatorname{Ran} S_{a,b^*}, \quad \operatorname{Ran} S_{a,b} = \mathcal{V}_{a,b}^- = \mathcal{V}_{a,b^*}^+ = \operatorname{Ker} S_{a,b^*};$$

(3) $S_{a,b}$ *has a real eigenvalue (which then is* $\Re(a) - \Re(b)$*) if and only if* $|\mathfrak{V}(a)| = |\mathfrak{V}(b)|$ *and the associated eigenspace is* $\mathcal{V}_{a,b}^+$;

(4) *the centralizer of* $a \in \mathsf{H}$ *is*

$$\operatorname{Cen}(a) := \{x \in \mathsf{H} : ax = xa\} = \operatorname{Ker} S_{a,a};$$

we have

$$\operatorname{Cen}(a) = \begin{cases} \mathsf{H} & \text{if } a \in \mathsf{R}, \\ \mathcal{V}_{a,a}^+ = \operatorname{Span}_\mathsf{R}\{1, a\} & \text{otherwise.} \end{cases}$$

In the case a and b are similar (by Theorem 2.2.6 this happens if and only if $\Re(a) = \Re(b)$ and $|\mathfrak{V}(a)| = |\mathfrak{V}(b)|$), the kernel and image of $S_{a,b}$ have alternative descriptions.

Theorem 2.3.4. *Assume* $a, b \in \mathsf{H}\backslash\mathsf{R}$ *are similar, so that* $b = z^{-1}az$, $z \in \mathsf{H}\backslash\{0\}$. *Then:*

(a) $\operatorname{Ran} S_{a,b} = \operatorname{Ker} S_{a,b^*}$. *In other words, the equation* $ax - xb = y$ *has a solution* x *if and only if* $ay = yb^*$;

(b) $\operatorname{Ker} S_{a,b} = \operatorname{Cen}(a)\,z = \operatorname{Span}_\mathsf{R}\{z, az\}$.

Proof. Part (a) follows from Theorem 2.3.3(2). Part (b) is a consequence of the identity

$$ax - x(z^{-1}az) = (a(xz^{-1}) - (xz^{-1})a)z.$$

$\square$

We conclude this section with formulas for the unique solution of the Sylvester equation, provided $S_{a,b}$ is invertible. For $a, b \in \mathsf{H}$, define

$$f_1(a, b) = b^2 - 2\Re(a)b + |a|^2, \quad f_2(a, b) = a^2 - 2\Re(b)a + |b|^2.$$

Proposition 2.3.5. *The following statements are equivalent:*

(1) $f_1(a, b) = 0$;

(2) $f_2(a, b) = 0$;

(3) a *and* b *are similar;*

(4) $\Re(a) = \Re(b)$ *and* $|a| = |b|$.

Proof. Equivalence of (3) and (4) follows from equivalence of (1) and (5) in Theorem 2.2.6. The implications (3) $\Rightarrow$ (1) and (3) $\Rightarrow$ (2) follow easily from Proposition 2.1.4(10). Suppose (1) holds true. Subtracting the equality $b^2 - 2\Re(b)b + |b|^2 = 0$ from $f_1(a, b) = 0$, we obtain $2(\Re(a) - \Re(b))b = |a|^2 - |b|^2$. If $\Re(a) \neq \Re(b)$, then b must be real. Subtracting $a^2 - 2\Re(a)a + |a|^2 = 0$ from $f_1(a, b) = 0$ yields $a = b$. If $\Re(a) = \Re(b)$, then $|a|^2 - |b|^2 = 0$, and (4) follows. Analogously, one proves that (2) implies (4). $\square$

Theorem 2.3.6. *If $S_{a,b}$ is nonsingular, then the unique solution to the equation $S_{a,b}(x) = y$ satisfies*

$$x = a^*y(f_1(a,b))^{-1} - y(f_1(a,b))^{-1}b = a(f_2(a,b))^{-1}y - (f_2(a,b))^{-1}yb^*.$$

The proof follows from the equalities

$$S_{a,b}(a^*z - zb) = zf_1(a,b), \qquad S_{a,b}(az - zb^*) = f_2(a,b)z,$$

for all $z \in \mathsf{H}$, which can be verified without difficulty.

2.4 AUTOMORPHISMS AND INVOLUTIONS

Definition 2.4.1. An ordered triple of quaternions (q_1, q_2, q_3) is said to be a *units triple* if

$$q_1^2 = q_2^2 = q_3^2 = -1, \qquad q_1q_2 = -q_2q_1 = q_3,$$
$$q_2q_3 = -q_3q_2 = q_1, \qquad q_3q_1 = -q_1q_3 = q_2. \tag{2.4.1}$$

For example, $\{\mathsf{i}, \mathsf{j}, \mathsf{k}\}$ is a units triple.

Proposition 2.4.2. *An ordered triple (q_1, q_2, q_3), $q_j \in \mathsf{H}$, is a units triple if and only if there exists a 3×3 real orthogonal matrix $P = [p_{\alpha,\beta}]_{\alpha,\beta=1}^3$ with determinant 1 such that*

$$q_\alpha = p_{1,\alpha}\mathsf{i} + p_{2,\alpha}\mathsf{j} + p_{3,\alpha}\mathsf{k}, \quad \alpha = 1, 2, 3. \tag{2.4.2}$$

Proof. A straightforward computation verifies that $x \in \mathsf{H}$ satisfies $x^2 = -1$ if and only if

$$x = a_1\mathsf{i} + a_2\mathsf{j} + a_3\mathsf{k},$$

where $a_1, a_2, a_3 \in \mathsf{R}$ and $a_1^2 + a_2^2 + a_3^2 = 1$. Thus, we may assume that q_α are given by (2.4.2) with the vectors $p_\alpha := (p_{1,\alpha}, p_{2,\alpha}, p_{3,\alpha})^T \in \mathsf{R}^{3\times 1}$ having euclidean norm 1, for $\alpha = 1, 2, 3$. Next, in view of (2.1.1) we have

$$q_uq_v = -p_u^Tp_v + \begin{bmatrix} \mathsf{i} & \mathsf{j} & \mathsf{k} \end{bmatrix}(p_u \times p_v), \quad u, v \in \{1, 2, 3\}. \tag{2.4.3}$$

The result of Proposition 2.4.2 now follows easily. $\square$

In particular, for every units triple (q_1, q_2, q_3) the quaternions $1, q_1, q_2, q_3$ form a basis of the real vector space H.

Next, we consider endomorphisms and antiendomorphisms of quaternions.

Definition 2.4.3. A map $\phi : \mathsf{H} \longrightarrow \mathsf{H}$ is called an *endomorphism*, resp. an *antiendomorphism*, if $\phi(xy) = \phi(x)\phi(y)$, resp., $\phi(xy) = \phi(y)\phi(x)$ for all $x, y \in \mathsf{H}$, and $\phi(x + y) = \phi(x) + \phi(y)$ for all $x, y \in \mathsf{H}$. An antiendomorphism ϕ is called an *involution* if $\phi(\phi(x)) = x$ for every $x \in \mathsf{H}$.

An involution is necessarily one-to-one and onto.

Theorem 2.4.4. *Let ϕ be an endomorphism or an antiendomorphism of H. Assume that ϕ does not map H into zero. Then ϕ is one-to-one and onto H; thus, ϕ is in fact an automorphism or an antiautomorphism. Moreover, ϕ is real linear, and representing ϕ as a 4×4 real matrix with respect to the basis $\{1, \mathsf{i}, \mathsf{j}, \mathsf{k}\}$, we have:*

(a) ϕ *is an automorphism if and only if*

$$\phi = \begin{bmatrix} 1 & 0 \\ 0 & T \end{bmatrix}, \tag{2.4.4}$$

where T is a 3×3 real orthogonal matrix of determinant 1;

(b) ϕ *is an antiautomorphism if and only if ϕ has the form* (2.4.4)*, where T is a 3×3 real orthogonal matrix of determinant* -1;

(c) ϕ *is an involution if and only if*

$$\phi = \begin{bmatrix} 1 & 0 \\ 0 & T \end{bmatrix},$$

where either $T = -I_3$ or T is a 3×3 real orthogonal symmetric matrix with eigenvalues 1, 1, -1.

Proof. Clearly, ϕ is one-to-one (indeed, if $\phi(x) = 0$ for some nonzero x, then for all $y \in \mathsf{H}$ we have $\phi(y) = \phi(yx^{-1})\phi(x) = 0$, a contradiction to the hypotheses of Theorem 2.4.4). Also, $\phi(x) = x$ for every real rational x.

We will use an observation which can be easily checked by straightforward algebra: *if $x \in \mathsf{H}$ is nonreal, then the commutant of x, namely, the set of all $y \in \mathsf{H}$ such that $xy = yx$ coincides with the set of all quaternions of the form $a + bx$, where $a, b \in \mathsf{R}$.*

Next, we prove that ϕ maps reals into reals. Arguing by contradiction, assume that $\phi(x)$ is nonreal for some real x. Since $xy = yx$ for every $y \in \mathsf{H}$, we have that $\phi(\mathsf{H})$ is contained in the commutant of $\phi(x)$, i.e., by the above observation,

$$\phi(\mathsf{H}) \subseteq \mathsf{R} + \mathsf{R}\phi(x).$$

However, the set $\mathsf{R} + \mathsf{R}\phi(x)$ contains only two square roots of -1, namely,

$$\pm \frac{\mathfrak{V}(\phi(x))}{|\mathfrak{V}(\phi(x))|}.$$

On the other hand, H contains a continuum of square roots of -1. Since ϕ maps square roots of -1 onto square roots of -1, ϕ cannot be one-to-one, a contradiction. Thus, ϕ maps reals into reals, and the restriction of ϕ to R is a nonzero endomorphism of the field of real numbers. Now R has no nontrivial endomorphisms (indeed, any nonzero endomorphism of R fixes every rational number, and since only nonnegative real numbers have real square roots, any such endomorphism is also order preserving, and these properties easily imply that any such endomorphism must be the identity). Therefore, we must have $\phi(x) = x$ for all $x \in \mathsf{R}$. Now, clearly, ϕ is a real linear map. Representing ϕ as a 4×4 matrix with respect to the basis $\{1, \mathsf{i}, \mathsf{j}, \mathsf{k}\}$, we obtain the result of Part (a) from Proposition 2.4.2. For Part (b), note that if ϕ is an antiautomorphism, then a composition of ϕ with any fixed antiautomorphism is an automorphism. Taking a composition of ϕ with the antiautomorphism of standard conjugation $\mathsf{i} \longrightarrow -\mathsf{i}, \mathsf{j} \longrightarrow -\mathsf{j}, \mathsf{k} \longrightarrow -\mathsf{k}$, we see by Part (a) that the composition has the form (2.4.4) with T a real orthogonal matrix of determinant 1. Since the standard conjugation has the form (2.4.4) with $T = -I$, we obtain the result of Part (b). Finally, clearly ϕ is involutory if and only if the matrix T of (2.4.4) has eigenvalues ± 1, and (c) follows at once from (a) and (b). $\qquad\square$

Definition 2.4.5. If the former case of (c) holds true, then ϕ is the standard conjugation, and we say that ϕ is *standard*. If the latter case of (c) holds true, we say that these involutions are *nonstandard*.

Thus, the nonstandard involutions are parameterized by 1-dimensional real subspaces (representing eigenvectors of T corresponding to the eigenvalue -1) in $\mathbf{R}^3$. In other words, the set of nonstandard involutions can be identified (as a topological space) with the 2-dimensional real projective space.

Here is another useful property of nonstandard involutions.

Lemma 2.4.6. *Let ϕ_1 and ϕ_2 be two distinct nonstandard involutions. Then for any $\alpha \in \mathsf{H}$, the equality $\phi_1(\alpha) = \phi_2(\alpha)$ holds if and only if $\phi_1(\alpha) = \phi_2(\alpha) = \alpha$.*

Proof. The "if" part is obvious. To prove the "only if" part, let T_1 and T_2 be the 3×3 real orthogonal matrices with eigenvalues $1, 1, -1$ such that

$$\phi_j = \begin{bmatrix} 1 & 0 \\ 0 & T_j \end{bmatrix}, \quad j = 1, 2,$$

as in Theorem 2.4.4(c). Then the "only if" part amounts to the following: if $T_1 \neq T_2$ and $T_1 x = T_2 x$ for some $x \in \mathbf{R}^{3 \times 1}$, then $T_1 x = T_2 x = x$. Considering the orthogonal complement of a common eigenvector of T_1 and T_2 corresponding to the eigenvalue 1, the proof reduces to the statement that if $\widehat{T}_1 \neq \widehat{T}_2$ are 2×2 real orthogonal symmetric matrices with determinants -1, then $\det(\widehat{T}_1 - \widehat{T}_2) \neq 0$. This can be verified by elementary matrix manipulations, taking $\widehat{T}_j$ in the form $\begin{bmatrix} \cos \tau_j & \sin \tau_j \\ -\sin \tau_j & -\cos \tau_j \end{bmatrix}$, where $0 \leq \tau_j < 2\pi$, for $j = 1, 2$. $\square$

Theorem 2.4.4 allows us to prove easily, using elementary linear algebra, the following well-known fact.

Proposition 2.4.7. *Every automorphism of H is inner—i.e., if $\phi : \mathsf{H} \to \mathsf{H}$ is an automorphism, then there exists $\alpha \in \mathsf{H} \setminus \{0\}$ such that*

$$\phi(x) = \alpha^{-1} x \alpha, \qquad \text{for all } x \in \mathsf{H}. \tag{2.4.5}$$

Proof. An elementary (but tedious) calculation shows that for

$$\alpha = a + b\mathsf{i} + c\mathsf{j} + d\mathsf{k} \in \mathsf{H} \setminus \{0\},$$

the 4×4 matrix representing the R-linear transformation $x \mapsto \alpha^{-1} x \alpha$ in the standard basis $\{1, \mathsf{i}, \mathsf{j}, \mathsf{k}\}$ is equal to

$$\begin{bmatrix} 1 & 0 \\ 0 & S \end{bmatrix},$$

where

$$S := \frac{1}{|\alpha|^2} \left(\begin{bmatrix} b \\ c \\ d \end{bmatrix} \begin{bmatrix} b & c & d \end{bmatrix} + \begin{bmatrix} a & d & -c \\ -d & a & b \\ c & -b & a \end{bmatrix}^2 \right) \tag{2.4.6}$$

(cf. Corollary 2.2.2). In view of Theorem 2.4.4, the matrix S is orthogonal with determinant 1, because the transformation $x \mapsto \alpha^{-1} x \alpha$ is an automorphism. Conversely, every 3×3 real orthogonal matrix T with determinant 1 has the form of the right-hand side of (2.4.6): Indeed, if $T = I$, choose $a \neq 0$ and $b = c = d = 0$. If $T \neq I$, choose $(b, c, d)^T$ as a unit length eigenvector of T corresponding to the eigenvalue 1. Observe that $(b, c, d)^T$ is also an eigenvector of S corresponding to the eigenvalue 1. Since the trace of S is equal to

$$1 + \frac{2}{a^2 + b^2 + c^2 + d^2}(a^2 - b^2 - c^2 - d^2),$$

it remains to choose a so that

$$\frac{a^2 - b^2 - c^2 - d^2}{a^2 + b^2 + c^2 + d^2}$$

coincides with the real part of those eigenvalues of T which are different from 1. The choice of $a \in \mathsf{R}$ is always possible because the real part of the eigenvalues of T other than 1 is between -1 and 1 and is not equal to 1. $\qquad\square$

One can write down concrete formulas for the automorphism ϕ given by (2.4.5). Namely, without loss of generality, we can assume that $\alpha \in \mathsf{H} \setminus \{0\}$ has the form

$$\alpha = \cos\left(\frac{\theta}{2}\right) - \sin\left(\frac{\theta}{2}\right) q_3, \quad \text{for some} \quad \theta, \quad 0 \leq \theta < 2\pi,$$

where (q_1, q_2, q_3) is a suitable units triple. Indeed, if α is real, we take $\theta = 0$, and if $\alpha \notin \mathsf{R}$, we take $q_3 = \mathfrak{V}(\alpha)/|\mathfrak{V}(\alpha)|$ and, in both cases, divide α by $|\alpha|$. Then

$$\begin{aligned}
\phi(1) &= 1, & \phi(q_1) &= \cos(\theta)q_1 + \sin(\theta)q_2, \\
\phi(q_2) &= -\sin(\theta)q_1 + \cos(\theta)q_2, & \phi(q_3) &= q_3.
\end{aligned}$$

These formulas can be verified by direct computation. One also verifies that in terms of formula (2.4.4), the automorphism ϕ is given with

$$T = P \cdot \begin{bmatrix} \cos(\theta) & -\sin(\theta) & 0 \\ \sin(\theta) & \cos(\theta) & 0 \\ 0 & 0 & 1 \end{bmatrix} \cdot P^T, \tag{2.4.7}$$

where the real orthogonal matrix $P \in \mathsf{R}^{3 \times 3}$ is defined by

$$[q_1 \ q_2 \ q_3] = [\mathsf{i} \ \mathsf{j} \ \mathsf{k}]P.$$

Since every antiautomorphism of H is a composition of a fixed antiautomorphism (such as the conjugation) and a suitable automorphism, Proposition 2.4.7 that every antiautomorphism ϕ of H has the form $\phi(x) = \beta^{-1}x^*\beta$ for some $\beta \in \mathsf{H}$ with $|\beta| = 1$. One easily verifies that ϕ is an involution if and only if β^2 is real. Also, ϕ is nonstandard if and only if $\beta^2 = -1$, and then $\phi(\beta) = -\beta$. It follows from Theorem 2.4.4(c) that for every nonstandard involution ϕ, there is a unique (up to negation) $\beta \in \mathsf{H}$ such that $\phi(\beta) = -\beta$ and $\beta^2 = -1$. We select one of the two quaternions β with these properties and denote $\beta(\phi) = \beta$. Letting (q_1, q_2, q_3) be a units triple with $q_1 = \beta(\phi)$, we see that for every $x_0, x_1, x_2, x_3 \in \mathsf{R}$,

$$\phi(x_0 + x_1q_1 + x_2q_2 + x_3q_3) = x_0 - x_1q_1 + x_2q_2 + x_3q_3. \tag{2.4.8}$$

In particular, $\phi(x) = -x$, $x \in \mathsf{H}$, if and only if $x \in \mathsf{R}\beta$. We denote by $\mathrm{Inv}\,(\phi)$ the set of quaternions left invariant by a nonstandard involution ϕ:

$$\mathrm{Inv}\,(\phi) := \{x \in \mathsf{H} : \phi(x) = x\} = \mathrm{Span}_{\mathsf{R}}\,\{1, q_2, q_3\}.$$

2.5 QUADRATIC MAPS

For a fixed involution ϕ consider quadratic maps of the form $x \mapsto \phi(x)\alpha x$, where $\alpha \in \mathsf{H} \setminus \{0\}$ is such that either $\phi(\alpha) = \alpha$ or $\phi(\alpha) = -\alpha$. (We exclude the trivial case $\alpha = 0$.) It is useful to find information about ranges of these maps. For example, if $\phi(\alpha) = \alpha$, is it true that the range of the quadratic map coincides with the set of quaternions that are fixed by ϕ?

We use the notation $\mathrm{Quad}_\phi(\alpha)$ for the map $x \mapsto \phi(x)\alpha x$, $x \in \mathsf{H}$.

If ϕ is a nonstandard involution, then there exists a unique (up to negation) $\beta \in \mathsf{H}$ such that $\phi(\beta) = -\beta$ and $|\beta| = 1$; note that β has zero real part. When working with nonstandard involutions ϕ, we often fix one such β and write $\beta(\phi)$ for β.

Theorem 2.5.1.   (a) *If ϕ is a nonstandard involution, then*

$$\phi(x)\beta(\phi)x = \beta(\phi)|x|^2, \qquad \text{for every } x \in \mathsf{H}. \tag{2.5.1}$$

In particular, the range of $\mathrm{Quad}_\phi(\alpha)$ is the half-line

$$\{a\alpha \ : \ a \in \mathsf{R}, \ a \geq 0\} \tag{2.5.2}$$

for every $\alpha \neq 0$ such that $\phi(\alpha) = -\alpha$.

(b) *If ϕ is a nonstandard involution, then for every $\alpha \neq 0$ such that $\phi(\alpha) = \alpha$, the range of $\mathrm{Quad}_\phi(\alpha)$ coincides with the set (actually, a real subspace of H) $\mathrm{Inv}\,(\phi) := \{x \in \mathsf{H} \ : \ \phi(x) = x\}$; moreover, for every $\lambda \in \mathrm{Inv}\,(\phi)$ there exists $x \in \mathrm{Inv}\,(\phi)$ such that $\phi(x)\alpha x = \lambda$.*

(c) *If ϕ is the conjugation, then for every $\alpha \neq 0$ such that $\phi(\alpha) = -\alpha$ (in other words, $\mathfrak{R}(\alpha) = 0$, $\mathfrak{V}(\alpha) \neq 0$), the range of $\mathrm{Quad}_\phi(\alpha)$ coincides with the set (real subspace of H) of quaternions with zero real parts.*

(d) *If ϕ is the conjugation, then for every $\alpha \neq 0$ such that $\phi(\alpha) = \alpha$ (in other words, $\mathfrak{R}(\alpha) \neq 0$, $\mathfrak{V}(\alpha) = 0$), we have $\phi(x) = \alpha|x|^2$ for all $x \in \mathsf{H}$. In particular, the range of $\mathrm{Quad}_\phi(\alpha)$ is the half-line $\{x \in \mathsf{R} : x \geq 0\}$ if $\alpha > 0$ and the half-line $\{x \in \mathsf{R} : x \leq 0\}$ if $\alpha < 0$.*

Proof. The verification of (2.5.1) is straightforward, and Part (d) follows immediately from the basic properties of Proposition 2.1.4. Part (c) follows from the description of the conjugate orbit of α given in Ex. 2.7.24; indeed,

$$\mathrm{Quad}_\phi(\alpha) = \mathrm{Con}\,(\alpha) \cup \{0\}$$

if ϕ is the conjugation. Part (b) will be proved later. $\square$

For a proof of Theorem 2.5.1(b), we investigate the square root function on quaternions, which is of independent interest. In the next theorem, we use the notation

$$\mathsf{R}_- := \{x \in \mathsf{H} \ : \ \mathfrak{R}(x) \leq 0, \ \mathfrak{V}(x) = 0\},$$

and $\sqrt{\cdot}$ stands for the nonnegative square root of a nonnegative number.

Theorem 2.5.2.   (a) *Let $\lambda \in \mathsf{R}_-$. Then $x^2 = \lambda$ if and only if $x = \sqrt{|\lambda|}q$ for some $q \in \mathsf{H}$ with $\mathfrak{R}(q) = 0$, $|q| = 1$.*

(b) *Let $\lambda \in H \setminus R_-$. Then $x^2 = \lambda$ if and only if $x = x_\lambda$ or $x = -x_\lambda$, where*

$$x_\lambda := \frac{|\lambda| + \lambda}{\sqrt{2(|\lambda| + \Re(\lambda))}}.$$

If $\lambda \notin R$ then x_λ can be written in the form

$$x_\lambda = \sqrt{\frac{|\lambda| + \Re(\lambda)}{2}} + \sqrt{\frac{|\lambda| - \Re(\lambda)}{2}} \frac{\mathfrak{V}(\lambda)}{|\mathfrak{V}(\lambda)|}.$$

Since x_λ in Part (b) is the unique square root of λ with positive real part, we introduce the notation

$$\sqrt{\lambda} := x_\lambda, \qquad \text{for all } \lambda \in H \setminus R_-. \tag{2.5.3}$$

Proof. We leave aside the trivial case $\lambda = 0$. Part (a) follows from the fact (easily proved by straightforward verification using the representation $x = x_0 + x_1 i + x_2 j + x_3 k \in H$, where $x_0, x_1, x_2, x_3 \in R$) that $x^2 = -1$ if and only if $\Re(x) = 0$ and $|x| = 1$.

Part (b). The equation $x^2 = \lambda$ implies

$$\lambda = x^2 = (2\Re(x) - x^*)x = 2\Re(x)x - |x|^2 = 2\Re(x)x - |\lambda|.$$

Hence, $|\lambda| + \lambda = 2\Re(x)x$, and (taking the real part in this equality) $|\lambda| + \Re(\lambda) = 2(\Re(x))^2$. If $\lambda \notin R_-$ then $\Re(\lambda) + |\lambda| > 0$, and it follows that

$$x = \frac{|\lambda| + \lambda}{2\Re(x)} = \frac{|\lambda| + \lambda}{2\sqrt{(|\lambda| + \Re(\lambda))/2}} = \frac{|\lambda| + \lambda}{\sqrt{2(|\lambda| + \Re(\lambda))}}.$$

Thus,

$$\Re(x) = \frac{|\lambda| + \Re(\lambda)}{\sqrt{2(\Re(\lambda) + |\lambda|)}}$$

and, if $\mathfrak{V}(\lambda) \neq 0$, then

$$\mathfrak{V}(x) = \frac{\mathfrak{V}(\lambda)}{\sqrt{2(|\lambda| + \Re(\lambda))}} = \frac{\Re(\lambda)\sqrt{|\lambda| - \Re(\lambda)}}{\sqrt{2(|\lambda|^2 - \Re(\lambda)^2)}} = \sqrt{(|\lambda| - \Re(\lambda))/2} \cdot \frac{\mathfrak{V}(\lambda)}{|\mathfrak{V}(\lambda)|},$$

and the theorem is proved. $\qquad\qquad\square$

Following from Theorem 2.5.2, the function $\lambda \mapsto \sqrt{\lambda}$ is continuous (even real analytic) on $H \setminus \{0\}$. However, there is no continuous square root function on H as the following corollary shows.

Corollary 2.5.3. *Let S be a nonempty and connected subset of $H \setminus R_-$, and let $u : S \to H$ be a function such that $u(\lambda)^2 = \lambda$ for all $\lambda \in S$.*

(a) *The function $u(\cdot)$ is continuous if and only if either $u(\lambda) = \sqrt{\lambda}$ for all $\lambda \in S$ or $u(\lambda) = -\sqrt{\lambda}$ for all $\lambda \in S$.*

(b) *Suppose that the function $u(\cdot)$ is continuous. Let $\mu \in R_-$, $\mu \neq 0$. Suppose further that there is a sequence $\lambda_k \in S \setminus R$ such that $\lim_{k \to \infty} \lambda_k = \mu$ and $\lim_{k \to \infty} \mathfrak{V}(\lambda_k)/|\mathfrak{V}(\lambda_k)|$ does not exist. Then the function $u(\cdot)$ admits no continuous extension to $S \cup \{\mu\}$.*

Proof. Part (a): For $\lambda \notin \mathsf{R}_-$ the equation $u(\lambda)^2 = \lambda$ implies $u(\lambda) = s(\lambda)\sqrt{\lambda}$, where $s(\lambda) \in \{-1, 1\}$. Continuity of u and connectedness of S yield that $s(\lambda)$ is independent of λ.

Part (b): Without loss of generality, we may suppose $u(\lambda) = \sqrt{\lambda}$. Thus, for a sequence λ_k as in (b), we have

$$u(\lambda_k) = \underbrace{\sqrt{\frac{|\lambda_k| + \mathfrak{R}(\lambda_k)}{2}}}_{=a_k} + \underbrace{\sqrt{\frac{|\lambda_k| - \mathfrak{R}(\lambda_k)}{2}}}_{=b_k} \frac{\mathfrak{V}(\lambda_k)}{|\mathfrak{V}(\lambda_k)|}.$$

As k tends to infinity, a_k tends to 0 and b_k tends to $\sqrt{|\mu|}$, while the sequence $(\mathfrak{V}\lambda_k)/|\mathfrak{V}(\lambda_k)|$ does not converge. Thus each extension of $u(\cdot)$ to $S \cup \{\mu\}$ is discontinuous at μ. □

Proof of Theorem 2.5.1(b). Clearly, the range of $\mathrm{Quad}_\phi(\alpha)$ is contained in $\mathrm{Inv}(\phi)$. We show that for all $\lambda \in \mathrm{Inv}(\phi)$, the equation $x_\phi \alpha x = \lambda$ has a solution $x \in \mathrm{Inv}(\phi)$. Suppose first that $\alpha\lambda \notin \mathsf{R}_-$. Then

$$x = \alpha^{-1}\sqrt{\alpha\lambda} = \alpha^{-1}\frac{|\alpha\lambda| + \alpha\lambda}{\sqrt{2(|\alpha\lambda| + \mathfrak{R}(\alpha\lambda))}} = \frac{\alpha^* \frac{|\lambda|}{|\alpha|} + \lambda}{\sqrt{2(|\alpha\lambda| + \mathfrak{R}(\alpha\lambda))}}$$

is a solution. To see this, observe that

$$x \in \mathrm{Span}_{\mathsf{R}}\{\alpha^*, \lambda\} \subseteq \mathrm{Inv}(\phi).$$

Thus,

$$x_\phi \alpha x = x\alpha x = \alpha^{-1}(\alpha x)^2 = \lambda.$$

Now suppose that $\alpha\lambda \in \mathsf{R}_-$. Then for any $q \in \mathsf{H}$ with $\mathfrak{R}(q) = 0$ and $|q| = 1$ the quaternion $x = \alpha^{-1}\sqrt{|\alpha\lambda|}\,q$ satisfies $(\alpha x)^2 = \alpha\lambda$. Hence, $x\alpha x = \lambda$. If $\alpha \notin \mathsf{R}$, choose $q = \mathfrak{V}(\alpha)/|\mathfrak{V}(\alpha)|$. Then $x \in \mathrm{Inv}(\phi)$. If $\alpha \in \mathsf{R}$, then $x \in \mathrm{Inv}(\phi)$ for any $q \in \mathrm{Inv}(\phi)$. □

2.6 REAL AND COMPLEX MATRIX REPRESENTATIONS

In the sequel it will be often useful to represent quaternions as 4×4 real matrices or 2×2 complex matrices. These representations are described as follows.

Define the map

$$\chi : \mathsf{H} \to \mathsf{R}^{4 \times 4}, \quad \chi(a_0 + a_1\mathsf{i} + a_2\mathsf{j} + a_3\mathsf{k}) = \begin{bmatrix} a_0 & -a_1 & a_3 & -a_2 \\ a_1 & a_0 & -a_2 & -a_3 \\ -a_3 & a_2 & a_0 & -a_1 \\ a_2 & a_3 & a_1 & a_0 \end{bmatrix},$$

where $a_0, a_1, a_2, a_3 \in \mathsf{R}$.

Proposition 2.6.1. *The map χ is a unital (i.e., $\chi(I) = I$) isomorphism of H onto the algebra of all 4×4 real matrices of the form $\lambda I + S$, where $\lambda \in \mathsf{R}$ and $S \in \mathsf{R}^{4 \times 4}$ is a skew symmetric matrix of the form*

$$S = \left\{ \begin{bmatrix} 0 & -a_1 & a_3 & -a_2 \\ a_1 & 0 & -a_2 & -a_3 \\ -a_3 & a_2 & 0 & -a_1 \\ a_2 & a_3 & a_1 & 0 \end{bmatrix} : a_1, a_2, a_3 \in \mathsf{R} \right\}. \tag{2.6.1}$$

Proposition 2.6.1 can be proved by straightforward verification that $\chi(x+y) = \chi(x) + \chi(y)$, $\chi(xy) = \chi(x)\chi(y)$ for all $x, y \in \mathsf{H}$, $\chi(1) = I$, and χ is one-to-one and onto map on the indicated algebra of 4×4 real matrices (Ex. 2.7.10).

For $\alpha = a_0 + ia_1 + ja_2 + ka_3 \in \mathsf{H}$, $a_0, a_1, a_2, a_3 \in \mathsf{R}$, define

$$\omega(\alpha) = \begin{bmatrix} a_0 + ia_1 & a_2 + ia_3 \\ -a_2 + ia_3 & a_0 - ia_1 \end{bmatrix} \in \mathsf{C}^{2 \times 2}.$$

Proposition 2.6.2. *The map ω is a unital isomorphism of H onto the algebra of all 2×2 complex matrices of the form $\begin{bmatrix} z & u \\ -\overline{u} & \overline{z} \end{bmatrix}$, where $z, u \in \mathsf{C}$ are arbitrary.*

The proof is again by straightforward verification of the required properties (Ex. 2.7.11).

The ordered triple of matrices

$$-i\omega(\mathsf{k}) = \begin{bmatrix} 0 & 1 \\ 1 & 0 \end{bmatrix}, \quad -i\omega(\mathsf{j}) = \begin{bmatrix} 0 & -i \\ i & 0 \end{bmatrix}, \quad -i\omega(\mathsf{i}) = \begin{bmatrix} 1 & 0 \\ 0 & -1 \end{bmatrix}$$

is known as *Pauli spin matrices*, of importance of studies of spin in quantum mechanics. Note that the Pauli spin matrices are hermitian and unitary. Together with I_2, the Pauli spin matrices form a basis in the real vector space of 2×2 hermitian matrices.

2.7   EXERCISES

Ex. 2.7.1. Verify Proposition 2.1.2.

Ex. 2.7.2. Verify Proposition 2.1.4.

Ex. 2.7.3. Prove that $|x + y| = |x| + |y|$, $x, y \in \mathsf{H}$, holds if and only if either at least one of x, y is zero or $x \neq 0$ and $y \neq 0$ are positive real multiples of each other.

Ex. 2.7.4. Solve the following equations:

(1) $x^4 + 1 = 0$, $x \in \mathsf{H}$;

(2) $x^m + 1 = 0$, $x \in \mathsf{H}$, where m is a fixed even positive integer.

Hint: For a fixed $q \in \mathsf{H}$, with $\Re(q) = 0$ and $|q| = 1$, consider solutions in $\mathrm{Span}_{\mathsf{R}}\{1, q\}$ which is isomorphic to C.

Ex. 2.7.5. Verify the following equality for $x \in \mathsf{H}$:

$$x^2 = |\Re(x)|^2 - |\mathfrak{V}(x)|^2 + 2\Re(x)\mathfrak{V}(x).$$

Ex. 2.7.6. Let $a \in \mathsf{H}$ be such that $ax = xa$ for every $x \in \mathsf{H}_0$, where H_0 is a real 3-dimensional subspace of H. Show that $a \in \mathsf{R}$. Show by example that this statement is generally not true if H_0 is taken to be a real 2-dimensional subspace of H.

Ex. 2.7.7. Verify that the matrix $Y := (a_0^2 + a_1^2 + a_2^2 + a_3^2)^{-1}X$ of Corollary 2.2.2 is orthogonal and has determinant 1. Hint: The determinant of a real orthogonal matrix can be only ± 1, and the case that the determinant is equal to -1 is impossible here.

Ex. 2.7.8. Prove that H has no divisors of zero: If $x_1, \ldots, x_k \in \mathsf{H}$ are such that $x_1 \cdots x_k = 0$, then at least one of the x_j's is equal to zero.

Ex. 2.7.9. Find the kernel and the range of the real linear transformation $x \mapsto xa - ax$, $x \in \mathsf{H}$, where $a \in \mathsf{H} \setminus \{0\}$ is fixed.

Ex. 2.7.10. Prove Proposition 2.6.1.

Ex. 2.7.11. Prove Proposition 2.6.2.

Ex. 2.7.12. Show that any two involutions ϕ_1 and ϕ_2 of H different from the conjugation are similar: There exists $\alpha \in \mathsf{H} \setminus \{0\}$ (which depends on ϕ_1 and ϕ_2) such that $\phi_2(x) = \alpha^{-1}\phi_1(x)\alpha$, $x \in \mathsf{H}$.

Ex. 2.7.13. Let (q_1, q_2, q_3) be a units triple. Show that $xq_1 = -q_1x$, $x \in \mathsf{H}$ if and only if x is a real linear combination of q_2 and q_3.

Ex. 2.7.14. Let α and β be two quaternions with zero real parts. Prove that $\alpha\beta = \beta^*\alpha$ if and only if α and β are *orthogonal*, i.e., writing

$$\alpha = \mathsf{i}\alpha_1 + \mathsf{j}\alpha_2 + \mathsf{k}\alpha_3, \quad \beta = \mathsf{i}\beta_1 + \mathsf{j}\beta_2 + \mathsf{k}\beta_3, \quad \alpha_\ell, \beta_\ell \in \mathsf{R} \text{ for } \ell = 1, 2, 3,$$

we have

$$\alpha_1\beta_1 + \alpha_2\beta_2 + \alpha_3\beta_3 = 0. \tag{2.7.1}$$

Ex. 2.7.15. Let u, v and u', v' be two pairs of unit length quaternions with zero real parts. Prove that there exists an automorphism ϕ of H such that $\phi(u) = u'$, $\phi(v) = v'$ if and only if the angle θ ($0 \le \theta \le \pi$) between u and v is equal to that between u' and v'. Is ϕ unique?

Ex. 2.7.16. Let u, v and u', v' be as in Ex. 2.7.15. Prove that there exists an antiautomorphism ψ of H such that $\psi(u) = u'$, $\psi(v) = v'$ if and only if the angle between u and v is equal to that between u' and v'. Is ψ unique?

Ex. 2.7.17. Identify all pairs (x, y) of quaternions in the set $\{1 + \mathsf{i}, 1 + 2\mathsf{j}, 2 + 2\mathsf{j}, 1 + 3\mathsf{k}\}$ that are similar, resp. congruent, to each other. Find $a \in \mathsf{H}$ such that $axa^{-1} = y$ in case x and y are similar or $axa^* = y$ in case x and y are congruent.

Ex. 2.7.18. Find:

(a) all $x \in \mathsf{H}$ such that $\mathsf{i}x = x\mathsf{j}$;

(b) all $y \in \mathsf{H}$ such that $2\mathsf{i}y = y(\mathsf{j} + \mathsf{k})$;

(c) all $z \in \mathsf{H}$ such that $3\mathsf{i}z = \sqrt{3}z(\mathsf{i} + \mathsf{j} + \mathsf{k})$.

Ex. 2.7.19. Show that all solutions z of the equation $z^2 = p + \mathsf{i}q$, where $p, q \in \mathsf{R}$ and $q \neq 0$ and $z \in \mathsf{H}$ is to be found, belong to $\mathrm{Span}_\mathsf{R}\{1, \mathsf{i}\}$.

Ex. 2.7.20. Prove Bohr's inequality:

$$|z + w|^2 \le p|z|^2 + q|w|^2, \qquad z, w \in \mathsf{H}, \tag{2.7.2}$$

where p, q are positive real numbers such that $(1/p) + (1/q) = 1$, with the equality holding in (2.7.2) if and only if $w = (p - 1)z$.

Ex. 2.7.21. Show that every quaternion can be written in infinitely many ways as product of two quaternions with zero real parts. Hint: Show that for all $x \in \mathrm{Span}_\mathsf{R}\{1, \mathsf{i}\}$, $x_1 \in \mathrm{Span}_\mathsf{R}\{\mathsf{j}, \mathsf{k}\}$ with $x_1 \neq 0$, there exists a $y \in \mathrm{Span}_\mathsf{R}\{\mathsf{j}, \mathsf{k}\}$ such that $x = x_1 y$.

Ex. 2.7.22. Let $S := \{x \in \mathsf{H} : |x| = 1\}$ be the set of quaternions of norm one.

(1) Show that S is a (multiplicative) subgroup of $\mathsf{H} \setminus \{0\}$, i.e., $x, y \in S$ implies $xy \in S$ and $x^{-1} \in S$.

(2) Prove that the function $f : S \times S \to S$ defined by $f(x, y) = y^{-1}x^{-1}yx$ maps onto S. Hint: Use the result of Ex. 2.7.21.

Ex. 2.7.23. Define S as in Ex. 2.7.22.

(a) Show that every automorphism or antiautomorphism of H maps S onto itself.

(b) Show that every automorphism ϕ of H has at least four fixed points x in S, i.e., $x \in S$ such that $\phi(x) = x$. Hint: Every 3×3 real orthogonal matrix with determinant one has eigenvalue 1.

Ex. 2.7.24. Verify the following formula for the congruence orbit of $x \in \mathsf{H}$:

$$\mathrm{Con}\,(x) = \{(\Re(x))r + \mathsf{S}|\mathfrak{V}x|r : r > 0\},$$

where S is defined in (2.2.6).

Ex. 2.7.25. Verify that if at least one of $a, b \in \mathsf{H}$ is nonreal, then the subspaces $\mathcal{V}_{a,b}^{\pm}$ (defined in Section 2.3) are uniquely determined by a and b.

Ex. 2.7.26. Verify formula (2.4.7).

2.8   NOTES

Most of the material in this chapter is standard (except Section 2.3), although some of it, e.g., Theorem 2.2.1, is difficult to locate in the literature. The representation of quaternion multiplication as rotations in 3-dimensional real vector space, as stated in Corollary 2.2.4, as well as connection with cross products (2.1.1), is a key to many applications in geometry, mechanics, and engineering, which is written up in many books; see, e.g., Ward [157]. For applications in feedback regulator problems and stability analysis, see, e.g., Wie et al. [159].

Proposition 2.1.4 contains basic elementary properties of H; a large set of such properties is found in Zhang [164]. The result of Theorem 2.3.3(2), as well as that of Theorem 2.3.4(a), is found in Johnson [74]. Ex. 2.7.21 is taken from Koecher and Remmert [81].

The material of Section 2.3 is taken from Karow [79]. Theorem 2.5.2 and Corollary 2.5.3, as well as the proof of Theorem 2.5.1(b), were suggested by Karow [80].

For some material on antiautomorphisms of quaternions, see von Randow [127].

We mention the following "fundamental theorem of algebra" concerning polynomial equations in quaternions:

Theorem 2.8.1. *Let $f(x)$ be a function of the form*

$$f(x) = a_0 x a_1 x a_2 \cdots a_{n-1} x a_n + g(x), \qquad x \in \mathsf{H},$$

where $a_0, a_1, \ldots a_n \in \mathsf{H} \setminus \{0\}$, and $g(x)$ is a finite sum of monomials of the form $b_0 x b_1 x \cdots b_{k-1} x b_k$ with $k < n$. Then there exists $x_0 \in \mathsf{H}$ such that $f(x_0) = 0$.

Theorem 2.8.1 was proved by Eilenberg and Niven [36]. The complete proof is based on topological methods and is beyond the scope of this book.

Quaternions were a starting point in many important developments in modern algebra: octonions, division algebras, and Clifford algebras, to name a few. While we cannot here go in depth into any of these developments, some references for further reading are provided: Conway and Smith [30], Lounesto [103], and Cohn [29]. In particular, note that the division ring of quaternions is a special case of a finite dimensional central simple algebra over a field. Many well-known results for such algebras apply to H, for example, the Skolem-Noether theorem that asserts (in a basic formulation) that every automorphism of such algebras is inner (cf. Proposition 2.4.7); see, e.g., Berhuy and Oggier [14] or Farb and Dennis [37] for more details.

Rotations transforms in 3- and 4-dimensional real vector space via quaternions (cf. Corollary 2.2.4) and related topics are studied in depth in many books, e.g., Ward [157], Vince [153], and Kuipers [84].

For remarks and further references concerning historical developments of quaternions, consult Koecher and Remmert [81] or Vince [153].

Chapter Three

Vector spaces and matrices: Basic theory

We introduce the basic structures in the quaternion vector space $H^{n \times 1}$ and in quaternion matrix algebras, including various types of matrix decompositions and factorizations. In particular, representations of quaternion matrix algebras in terms of real and complex matrix algebra are developed. Numerical ranges, joint numerical ranges, and their convexity properties are emphasized. We also provide an Appendix containing a few basic facts on analysis of sets and continuous multivariable functions that are used in the book.

3.1 FINITE DIMENSIONAL QUATERNION VECTOR SPACES

Consider $H^{n \times 1}$, the set of all n-component columns with quaternion components, as a right quaternion vector space; in other words, besides the standard addition, we consider multiplication on the right by quaternions: For $v = \begin{bmatrix} v_1 \\ \vdots \\ v_n \end{bmatrix} \in H^{n \times 1}$, $v_j \in H$, and $\alpha \in H$, we let

$$v\alpha = \begin{bmatrix} v_1 \alpha \\ \vdots \\ v_n \alpha \end{bmatrix} \in H^{n \times 1}.$$

We will use standard linear algebra concepts and properties of vectors and subspaces of $H^{n \times 1}$, such as linear independence, spanning set (of a subspace), basis, dimension, direct sum, etc.; they work in exactly the same way as for finite dimensional vector spaces over (commutative) fields. See, e.g., Hungerford [69] or Wan [156].

We denote by $\dim(\mathcal{M})$, or $\dim_H(\mathcal{M})$ if the quaternion nature of $\mathcal{M}$ is to be emphasized, the dimension (understood in the sense of right quaternion vector spaces) of a subspace $\mathcal{M}$ of $H^{n \times 1}$. The subspace spanned by $v_1, \ldots, v_p \in H^{n \times 1}$ is denoted

$$\text{Span}_H \{v_1, \ldots, v_p\} := \{v_1 \alpha_1 + \cdots + v_p \alpha_p : \alpha_1, \ldots, \alpha_p \in H\}.$$

As an example, we prove the following *replacement theorem*:

Theorem 3.1.1. *Let $u_1, \ldots, u_s$ be a linearly independent subset of*

$$\text{Span}_H \{v_1, \ldots, v_p\}, \quad \text{where} \quad v_1, \ldots, v_p \in H^{n \times 1}.$$

Then there exist s elements $v_{i_1}, \ldots, v_{i_s}$, $1 \leq i_1 < i_2 < \cdots < i_s \leq r$, such that upon replacing $v_{i_1}, \ldots, v_{i_s}$ with $u_1, \ldots, u_s$, respectively, in $v_1, \ldots, v_p$, a spanning set for $\text{Span}_H \{v_1, \ldots, v_p\}$ is obtained.

Proof. We use induction on s. Suppose $s = 1$. Then $u_1 \neq 0$ and $u_1 = v_1\alpha_1 + v_2\alpha_2 + \cdots + v_p\alpha_p$ for some $\alpha_1, \ldots, \alpha_p \in \mathsf{H}$. Not all the α_j's are zeros; if $\alpha_{i_1} \neq 0$, then it is easy to see that replacement of v_{i_1} by u_1 results in a spanning set for $\mathrm{Span}_{\mathsf{H}}\{v_1, \ldots, v_p\}$.

Assume now that the theorem holds for $s - 1$. By this assumption, there exist $v_{i_1}, \ldots, v_{i_{s-1}}$ such that

$$u_1, \ldots, u_{s-1}, v_{i_s}, \ldots, v_{i_r}$$

is a spanning set for $\mathrm{Span}_{\mathsf{H}}\{v_1, \ldots, v_p\}$; here $\{i_1, \ldots, i_s, \ldots, i_r\}$ is a permutation of $\{1, 2, \ldots, r\}$. Since

$$u_s \in \mathrm{Span}_{\mathsf{H}}\{v_1, \ldots, v_p\} = \mathrm{Span}_{\mathsf{H}}\{u_1, \ldots, u_{s-1}, v_{i_s}, \ldots, v_{i_r}\},$$

we have

$$u_s = u_1\alpha_1 + \cdots + u_{s-1}\alpha_{s-1} + v_{i_s}\alpha_s + \cdots + v_{i_r}\alpha_r, \quad \text{where} \quad \alpha_1, \ldots, \alpha_r \in \mathsf{H}. \quad (3.1.1)$$

Since $u_1, \ldots, u_s$ are linearly independent, there is a nonzero quaternion among $\alpha_s, \ldots, \alpha_r$. We can adjust the permutation $\{i_1, \ldots, i_s, \ldots, i_r\}$ so that $\alpha_s \neq 0$. Then, by replacing v_{i_s} with u_s, we get a set such that

$$\mathrm{Span}_{\mathsf{H}}\{u_1, \ldots, u_{s-1}, u_s, v_{i_{s+1}}, \ldots, v_{i_r}\}$$

$$= \mathrm{Span}_{\mathsf{H}}\{u_1, \ldots, u_{s-1}, v_{i_s}, v_{i_{s+1}}, \ldots, v_{i_r}\}. \quad (3.1.2)$$

Indeed, inclusion $\subseteq$ in (3.1.2) holds in view of (3.1.1), and since (3.1.1) can be solved for v_{i_s}:

$$v_{i_s} = -u_1\alpha_1\alpha_s^{-1} - \cdots - u_{s-1}\alpha_{s-1}\alpha_s^{-1} + u_s\alpha_s^{-1} - v_{i_{s+1}}\alpha_{i_{s+1}}\alpha_s^{-1} - \cdots - v_{i_r}\alpha_r\alpha_s^{-1};$$

the converse inclusion holds as well. □

Definition 3.1.2. For $v = \begin{bmatrix} v_1 \\ \vdots \\ v_n \end{bmatrix} \in \mathsf{H}^{n \times 1}$, define the *adjoint* as the n-component row $v^* = [v_1^* \ v_2^* \ \ldots \ v_n^*]$.

The vector space $\mathsf{H}^{n \times 1}$ is endowed with the quaternion-valued inner product $\langle u, v \rangle = v^* u$, $u, v \in \mathsf{H}^{n \times 1}$. Observe the following properties of $\langle \cdot, \cdot \rangle$:

$$\langle u_1\alpha_1 + u_2\alpha_2, v \rangle = \langle u_1, v \rangle\alpha_1 + \langle u_2, v \rangle\alpha_2, \quad u_1, u_2, v \in \mathsf{H}^{n \times 1}, \quad \alpha_1, \alpha_2 \in \mathsf{H};$$
$$\langle u, v_1\alpha_1 + v_2\alpha_2 \rangle = \alpha_1^*\langle u, v_1 \rangle + \alpha_2^*\langle u, v_2 \rangle, \quad u, v_1, v_2 \in \mathsf{H}^{n \times 1}, \quad \alpha_1, \alpha_2 \in \mathsf{H};$$
$$\langle u, v \rangle = \langle v, u \rangle^*, \quad u, v \in \mathsf{H}^{n \times 1};$$

$\langle u, u \rangle \geq 0$ for all $u \in \mathsf{H}^{n \times 1}$, with equality only if $u = 0$.

Definition 3.1.3. We say that $u, v \in \mathsf{H}^{n \times 1}$ are *orthogonal* if $\langle u, v \rangle = 0$.

A p-tuple $\{v_1, \ldots, v_p\}$, where $v_1, \ldots, v_p \in \mathsf{H}^{n \times 1}$, is said to be *orthogonal* if $\langle v_i, v_j \rangle = 0$ for $i \neq j$, and *orthonormal* if it is orthogonal and $\langle v_i, v_i \rangle = 1$ for $i = 1, 2, \ldots, p$.

Standard concepts and results related to orthogonality, in particular, the Gram-Schmidt algorithm, apply in this context, with essentially the same proofs (see, e.g., Farenick and Pidkowich [38]; note that the definition of the inner product in Farenick and Pidkowich is slightly different). Thus, every nonzero subspace of $\mathsf{H}^{n \times 1}$ has an orthonormal basis.

We use $\|u\|_{\mathsf{H}} := \sqrt{\langle u, u \rangle}$, also denoted $\|u\|$, as the norm on $\mathsf{H}^{n \times 1}$. Analogous norms on real and complex vector spaces are used: $\|u\|_{\mathsf{F}} = \sqrt{u^* u}$ for $u \in \mathsf{F}^{n \times 1}$, where $\mathsf{F} = \mathsf{R}$ or $\mathsf{F} = \mathsf{C}$.

Fact 3.1.4. *Let* $\mathsf{F} = \mathsf{R}$ *or* $\mathsf{F} = \mathsf{C}$. *It is easy to see that vectors* $v_1, \ldots, v_p \in \mathsf{F}^{n \times 1}$ *are linearly independent over* F *if and only if they are linearly independent over* H. *Thus,* $\dim_{\mathsf{F}} \operatorname{Span} \{v_1, \ldots, v_p\}$ *(as a subspace of* $\mathsf{F}^{n \times 1}$*) is equal to* $\dim_{\mathsf{H}} \operatorname{Span} \{v_1, \ldots, v_p\}$ *(as a subspace of* $\mathsf{H}^{n \times 1}$*).*

3.2 MATRIX ALGEBRA

Denote by $\mathsf{H}^{m \times n}$ the set of all $m \times n$ matrices with entries in H, considered as a *left* quaternion vector space, with the standard matrix addition and multiplication: $AB \in \mathsf{H}^{p \times m}$, where $A \in \mathsf{H}^{p \times n}$, $B \in \mathsf{H}^{n \times m}$. The multiplication of quaternion matrices, when defined, is associative. The setting of the left quaternion vector space allows us to interpret $A \in \mathsf{H}^{m \times n}$ as a linear transformation $\mathsf{H}^{m \times 1} \to \mathsf{H}^{n \times 1}$, which acts by the standard matrix-vector multiplication. (There is an ambiguity of notation here: $\mathsf{H}^{n \times 1}$ is a right vector space, but $\mathsf{H}^{n \times 1}$ is a left vector space when considered as the set of linear transformations $\mathsf{H} \to \mathsf{H}^{n \times 1}$. It will be clear from context which interpretation applies in every particular instance.)

Definition 3.2.1. Define the *adjoint matrix* : $A^* = [a_{j,i}^*]_{i=1,j=1}^{n,m} \in \mathsf{H}^{m \times n}$ for $A = [a_{i,j}]_{i=1,j=1}^{n,m} \in \mathsf{H}^{n \times m}$, where $a_{i,j} \in \mathsf{H}$.

For example,

$$\begin{bmatrix} 1 & i & j \\ k & i+j & 1+k \end{bmatrix}^* = \begin{bmatrix} 1 & -k \\ -i & -i-j \\ -j & 1-k \end{bmatrix}.$$

Note the following algebraic properties:

(a) $(\alpha A + \beta B)^* = A^* \alpha^* + B^* \beta^*$, for all $\alpha, \beta \in \mathsf{H}$, $A, B \in \mathsf{H}^{m \times n}$.

(b) $(A\alpha + B\beta)^* = \alpha^* A^* + \beta^* B^*$, for all $\alpha, \beta \in \mathsf{H}$, $A, B \in \mathsf{H}^{m \times n}$.

(c) $(AB)^* = B^* A^*$, for all $A \in \mathsf{H}^{m \times n}$, $B \in \mathsf{H}^{n \times p}$.

(d) $(A^*)^* = A$, for all $A \in \mathsf{H}^{m \times n}$.

(e) If $A \in \mathsf{H}^{n \times n}$ is invertible, then $(A^*)^{-1} = (A^{-1})^*$.

In particular, the adjoint operation is real linear. The classes of matrices familiar in the matrix analysis of real and complex matrices are introduced in the same way for quaternion matrices:

Definition 3.2.2. A matrix $A \in \mathsf{H}^{n \times n}$ is called *hermitian, positive definite, positive semidefinite, skewhermitian, invertible (nonsingular), unitary,* or *normal* if $A = A^*$, $x^* A x$ is real and positive for all $x \in \mathsf{H}^{n \times 1} \setminus \{0\}$, $x^* A x$ is real and nonnegative for all $x \in \mathsf{H}^{n \times 1}$ and $A = -A^*$, there exists $A^{-1} \in \mathsf{H}^{n \times n}$ such that $A^{-1} A = A A^{-1} = I$, A is invertible and $A^{-1} = A^*$, or $A A^* = A^* A$, respectively.

We shall see later—Proposition 3.5.1—that $x^* A x$ is real for all $x \in \mathsf{H}^{n \times 1}$ if and only if A is hermitian.

Definition 3.2.3. The *rank* of $A \in \mathsf{H}^{m \times n}$ is defined as the (quaternion) dimension of
$$\operatorname{Ran}(A) := \{Ax \ : \ x \in \mathsf{H}^{m \times 1}\},$$
the range of A; by convention, rank$(0) = 0$.

Several elementary properties of the rank function that will be useful in the sequel are listed in the next proposition. For $A \in \mathsf{H}^{m \times n}$, we denote by $A_\phi \in \mathsf{H}^{n \times m}$ the matrix obtained from the transposed matrix A^T by applying an involution ϕ entrywise.

Proposition 3.2.4. *Let $A \in \mathsf{H}^{m \times n}$. Then:*

(1) *if $S \in \mathsf{H}^{m \times n}$, $T \in \mathsf{H}^{m \times n}$ are invertible, then* rank$(SAT) = $ rank A;

(2) rank $A^* = $ rank A;

(3) *if ϕ is a nonstandard involution, then* rank $A_\phi = $ rank A.

See Section 3.6 for the algebraic properties of the map $A \ \mapsto \ A_\phi$.

Many matrix decompositions that are standard in real and complex matrix analysis remain valid for quaternion matrices, with essentially the same proofs. In particular, the following holds.

Proposition 3.2.5. *Let $A \in \mathsf{H}^{m \times n}$. Then:*

(a) $A = X A_0$, *where $X \in \mathsf{H}^{m \times m}$ is invertible and $A_0 \in \mathsf{H}^{m \times n}$ is a row reduced echelon form; moreover, A_0 is unique, i.e., uniquely determined by A;*

(b) $A = A_0' Y$, *where $Y \in \mathsf{H}^{n \times n}$ is invertible and $A_0' \in \mathsf{H}^{m \times n}$ is a column reduced echelon form; moreover, A_0' is unique;*

(c) *(QR factorization) if $m = n$, then $A = QR$, where Q is unitary and R is upper triangular with nonnegative diagonal elements; moreover, if A is invertible, then Q and R are unique;*

(d) *(polar decomposition) if $m = n$, then $A = RU$, where R is positive semidefinite and U is unitary; moreover, if A is invertible, then R and U are unique;*

(e) *(rank decompositions) if* rank$(A) = k \neq 0$, *then $A = BC$, where $B \in \mathsf{H}^{m \times k}$, $C \in \mathsf{H}^{k \times n}$; also,*
$$A = \widetilde{B} \begin{bmatrix} I_k & 0 \\ 0 & 0_{(m-k) \times (n-k)} \end{bmatrix} \widetilde{C},$$
where $\widetilde{B} \in \mathsf{H}^{m \times m}$, $\widetilde{C} \in \mathsf{H}^{n \times n}$ are invertible;

(f) *(singular value decomposition) if $A \neq 0$, then there exist unitary $U \in \mathsf{H}^{m \times m}$, $V \in \mathsf{H}^{n \times n}$, and real positive numbers $a_1 \geq a_2 \geq \cdots \geq a_k$, where $k = $ rank(A), such that*
$$A = U \begin{bmatrix} \operatorname{diag}(a_1, \ldots, a_k) & 0 \\ 0 & 0_{(m-k) \times (n-k)} \end{bmatrix} V;$$
moreover, the a_j's are unique.

For example, rank decompositions can be easily deduced from (a) and (b).

Definition 3.2.6. The a_j's of Proposition 3.2.5(f) are called the *singular values* of A; by convention, the singular values of the zero matrix are zeros.

We also note the *Cholesky decomposition* (Proposition 3.2.7(b) and (c) below), again proved analogously to the familiar cases of real and complex matrices:

Proposition 3.2.7. *The following statements are equivalent for* $A \in \mathsf{H}^{n \times n}$, $A \neq 0$:

(a) *A is positive semidefinite*;

(b) $A = L^*L$ *for some lower triangular* $L \in \mathsf{H}^{n \times n}$ *with nonnegative elements on the diagonal; moreover, L is unique*;

(c) $A = U^*U$ *for some upper triangular* $U \in \mathsf{H}^{n \times n}$ *with nonnegative elements on the diagonal; moreover, U is unique*;

(d) $A = B^*B$ *for some* $B \in \mathsf{H}^{k \times n}$, *where* $k = \operatorname{rank}(A)$.

For matrices, we use the operator norm throughout:

$$\|A\|_{\mathsf{H}} = \max\{\|Au\|_{\mathsf{H}} \; : \; u \in \mathsf{H}^{n \times 1}, \;\; \|u\|_{\mathsf{H}} = 1\},$$

where $A \in \mathsf{H}^{m \times n}$. It is immediate to see that $\|\cdot\|_{\mathsf{H}}$ is indeed a multiplicative norm on matrices:

$$\|A + B\|_{\mathsf{H}} \leq \|A\|_{\mathsf{H}} + \|B\|_{\mathsf{H}}, \quad \|\alpha A\|_{\mathsf{H}} = \|A\|_{\mathsf{H}} \, |\alpha|, \quad \|AC\|_{\mathsf{H}} \leq \|A\|_{\mathsf{H}} \cdot \|C\|_{\mathsf{H}},$$

and $\|A\|_{\mathsf{H}} \geq 0$ with $\|A\|_{\mathsf{H}} = 0$ only if $A = 0$; for all $A, B \in \mathsf{H}^{m \times n}$, all $C \in \mathsf{H}^{n \times k}$, and all $\alpha \in \mathsf{H}$.

Note that $\|A\|_{\mathsf{H}}$ coincides with the largest singular value of A. Analogous norms on real and complex matrices are used:

$$\|A\|_{\mathsf{F}} = \max\{\|Au\|_{\mathsf{F}} \; : \; u \in \mathsf{F}^{n \times 1}, \;\; \|u\|_{\mathsf{F}} = 1\}$$

for $A \in \mathsf{F}^{m \times n}$, where $\mathsf{F} = \mathsf{R}$ or $\mathsf{F} = \mathsf{C}$. Since for $A \in \mathsf{F}^{m \times n}$ we have $\|A\|_{\mathsf{F}} = \|A\|_{\mathsf{H}}$ (indeed, both $\|A\|_{\mathsf{F}}$ and $\|A\|_{\mathsf{H}}$ are equal to the largest singular value of A), the $\|\cdot\|$ notation for vector and matrix norms will be used unambiguously.

For future reference, we record the following characterization of linear independence.

Proposition 3.2.8. *Let* $A \in \mathsf{H}^{m \times n}$. *Then the columns of A are linearly independent (as elements of a right quaternion vector space) if and only A is left-invertible, i.e., $BA = I$ for some* $B \in \mathsf{H}^{n \times m}$.

Analogously, the rows of A are linearly independent (as elements of a left quaternion vector space) if and only if A is right-invertible.

A proof follows from a rank decomposition for A (see also Ex. 3.11.4).

3.3   REAL MATRIX REPRESENTATION OF QUATERNIONS

We extend to matrices the standard representation of quaternions as 4×4 real matrices of Proposition 2.6.1. Define the map

$$\chi : \mathsf{H} \to \mathsf{R}^{4\times4}, \qquad \chi(a_0 + a_1\mathsf{i} + a_2\mathsf{j} + a_3\mathsf{k}) = \begin{bmatrix} a_0 & -a_1 & a_3 & -a_2 \\ a_1 & a_0 & -a_2 & -a_3 \\ -a_3 & a_2 & a_0 & -a_1 \\ a_2 & a_3 & a_1 & a_0 \end{bmatrix},$$

where $a_0, a_1, a_2, a_3 \in \mathsf{R}$, and its matrix extension

$$\chi_{m,n} : \mathsf{H}^{m\times n} \to \mathsf{R}^{4m\times4n}, \qquad \chi_{m,n}\left([x_{i,j}]_{i,j=1}^{m,n}\right) = [\chi(x_{i,j})]_{i,j=1}^{m,n},$$

where $x_{i,j} \in \mathsf{H}$. The following properties are routinely verified:

(i) $\chi_{n,n}$ is an isomorphism of the real algebra $\mathsf{H}^{n\times n}$ onto the real subalgebra

$$\left\{ [z_{i,j}]_{i,j=1}^{n} : z_{i,j} \in \{\lambda I_4 + S : \lambda \in \mathsf{R}, \quad S \text{ has the form } (2.6.1)\} \right\}$$

of $\mathsf{R}^{4n\times4n}$, and $\chi_{n,n}(I) = I$.

(ii) If $X \in \mathsf{H}^{m\times n}$, $Y \in \mathsf{H}^{n\times p}$, then $\chi_{m,p}(XY) = \chi_{m,n}(X)\chi_{n,p}(Y)$.

(iii) If $X, Y \in \mathsf{H}^{m\times n}$ and $s, t \in \mathsf{R}$, then

$$\chi_{m,n}(sX + tY) = s\chi_{m,n}(X) + t\chi_{m,n}(Y).$$

(iv) $\chi_{n,m}(X^*) = (\chi_{m,n}(X))^T, \quad$ for all $X \in \mathsf{H}^{m\times n}$.

(v) There exist positive constants $c_{m,n}$, $C_{m,n}$ such that

$$c_{m,n}\|\chi_{m,n}(X)\|_{\mathsf{R}} \le \|X\|_{\mathsf{H}} \le C_{m,n}\|\chi_{m,n}(X)\|_{\mathsf{R}}$$

for every $X \in \mathsf{H}^{m\times n}$.

In fact, $c_{m,n}$ can be taken to be the minimal value of the nonzero continuous function $\|X\|_{\mathsf{H}}$ on the compact set

$$\{X \in \mathsf{H}^{m\times n} : \|\chi_{m.n}(X)\|_{\mathsf{R}} = 1\}$$

(cf. Theorem 3.10.5); an analogous statement is valid for $1/C_{m,n}$.

Often, we will abbreviate $\chi_{m,n}$ to χ (with m, n understood from context). The following properties of the map χ are useful.

Proposition 3.3.1. *A matrix $A \in \mathsf{H}^{n\times n}$ is hermitian, positive definite, positive semidefinite, skewhermitian, unitary, or normal if and only if $\chi(A)$ is symmetric, positive definite, positive semidefinite, skewsymmetric, orthogonal, or normal, respectively.*

The proof is left as Ex. 3.11.5. To prove that if A is positive semidefinite, then so is $\chi(A)$, use the Cholesky decomposition for A and property (ii) of χ.

Note that automorphisms, resp., antiautomorphisms, of H translate into unitary similarity, resp., unitary similarity followed by transposition, in terms of the map χ, a fact that will be useful in the sequel.

Proposition 3.3.2. *Let ϕ be an automorphism, resp. antiautomorphism, of* H, *and let $\phi(A)$ be the matrix obtained from $A \in \mathsf{H}^{m \times n}$, resp. $A^T \in \mathsf{H}^{n \times m}$, by applying ϕ entrywise. Then there exist real orthogonal matrices $U_{\phi,n} \in \mathsf{R}^{4n \times 4n}$ (independent of A) such that*

$$\chi(\phi(A)) = U_{\phi,m}^T (\chi(A)) U_{\phi,n} \quad \forall \ \ A \in \mathsf{H}^{m \times n},$$

resp.

$$\chi(\phi(A)) = U_{\phi,n}^T (\chi(A))^T U_{\phi,m} \quad \forall \ \ A \in \mathsf{H}^{m \times n}.$$

Proof. We prove only the part for automorphisms. Use the fact that every automorphism of H is inner (Proposition 2.4.7): there exists $\alpha_0 \in \mathsf{H}$ with $|\alpha_0| = 1$ such that $\phi(\alpha) = \alpha_0^{-1} \alpha \alpha_0$ for every $\alpha \in \mathsf{H}$. Then let

$$U_{\phi,n} = \underbrace{\chi(\alpha_0) \oplus \chi(\alpha_0) \oplus \cdots \oplus \chi(\alpha_0)}_{n \ \text{times}}, \quad U_{\phi,m} = \underbrace{\chi(\alpha_0) \oplus \chi(\alpha_0) \oplus \cdots \oplus \chi(\alpha_0)}_{m \ \text{times}}.$$

$\square$

Proposition 3.3.3. *Let $u_1, \ldots, u_p \in \mathsf{H}^{n \times 1}$. Then $u_1, \ldots, u_p$ are linearly independent (over* H) *if and only if the columns of $\chi_{n,p}([u_1 \ \cdots \ u_p])$ are linearly independent (over* R).

Moreover, $u_1, \ldots, u_p$ is an orthonormal, resp. orthogonal, set (over H) *if and only if the columns of $\chi_{n,p}([u_1 \ \cdots \ u_p])$ form an orthonormal, resp. orthogonal, set (over* R).

Proof. Assume $\chi_{n,p}([u_1 \ \cdots \ u_p])$ are linearly independent. Let $\alpha_1, \ldots, \alpha_p \in \mathsf{H}$ be such that

$$[u_1 \ \cdots \ u_p] \begin{bmatrix} \alpha_1 \\ \vdots \\ \alpha_p \end{bmatrix} = 0.$$

Applying χ to this equality and using the linear independence of $\chi_{n,p}[u_1 \ \cdots \ u_p]$, we obtain that every column of $\chi_{p,1}\left(\begin{bmatrix} \alpha_1 \\ \vdots \\ \alpha_p \end{bmatrix}\right)$ is zero. The definition of χ now implies that $\alpha_j = 0$, $j = 1, 2, \ldots, p$.

Assume $u_1, \ldots, u_p$ are linearly independent. Let

$$\chi_{n,p}([u_1 \ \cdots \ u_p]) \begin{bmatrix} a_1 \\ \vdots \\ a_p \end{bmatrix} = 0 \tag{3.3.1}$$

for some $a_1, \ldots, a_p \in \mathsf{R}^{4 \times 1}$, and, in turn,

$$a_j = \begin{bmatrix} a_{j0} \\ a_{j1} \\ -a_{j3} \\ a_{j2} \end{bmatrix}, \quad a_{j\ell} \in \mathsf{R}, \ \ \ell = 0, 1, 2, 3.$$

Write

$$u_j = \begin{bmatrix} u_{j1} \\ \vdots \\ u_{jn} \end{bmatrix}, \qquad u_{jk} \in \mathsf{H}, \qquad (3.3.2)$$

and, in turn,

$$u_{jk} = u_{jk}^{(0)} + u_{jk}^{(1)}\mathsf{i} + u_{jk}^{(2)}\mathsf{j} + u_{jk}^{(3)}\mathsf{k}, \quad u_{jk}^{(\ell)} \in \mathsf{R}, \quad \ell = 0, 1, 2, 3. \qquad (3.3.3)$$

It will be convenient to use the following notation for block row matrices:

$$\text{row}_{j=1,2,\dots,p} X_j = [X_1 \ X_2 \ \dots \ X_p]. \qquad (3.3.4)$$

Equation (3.3.1) now can be rewritten in the form

$$\text{row}_{j=1,2,\dots,p} \begin{bmatrix} u_{jk}^{(0)} & -u_{jk}^{(1)} & u_{jk}^{(3)} & -u_{jk}^{(2)} \\ u_{jk}^{(1)} & u_{jk}^{(0)} & -u_{jk}^{(2)} & -u_{jk}^{(3)} \\ -u_{jk}^{(3)} & u_{jk}^{(2)} & u_{jk}^{(0)} & -u_{jk}^{(1)} \\ u_{jk}^{(2)} & u_{jk}^{(3)} & u_{jk}^{(1)} & u_{jk}^{(0)} \end{bmatrix} \cdot \begin{bmatrix} a_{10} \\ a_{11} \\ -a_{13} \\ a_{12} \\ a_{20} \\ a_{21} \\ -a_{23} \\ a_{22} \\ \vdots \\ a_{p0} \\ a_{p1} \\ -a_{p3} \\ a_{p2} \end{bmatrix} = 0, \qquad (3.3.5)$$

for $k = 1, 2, \dots, n$. Using (3.3.5), a straightforward computation verifies that the second, third, and forth columns in the matrix

$$\text{row}_{j=1,2,\dots,p} \begin{bmatrix} u_{jk}^{(0)} & -u_{jk}^{(1)} & u_{jk}^{(3)} & -u_{jk}^{(2)} \\ u_{jk}^{(1)} & u_{jk}^{(0)} & -u_{jk}^{(2)} & -u_{jk}^{(3)} \\ -u_{jk}^{(3)} & u_{jk}^{(2)} & u_{jk}^{(0)} & -u_{jk}^{(1)} \\ u_{jk}^{(2)} & u_{jk}^{(3)} & u_{jk}^{(1)} & u_{jk}^{(0)} \end{bmatrix} \cdot \begin{bmatrix} \chi(a_{10} + a_{11}\mathsf{i} + a_{12}\mathsf{j} + a_{13}\mathsf{k}) \\ \vdots \\ \chi(a_{p0} + a_{p1}\mathsf{i} + a_{p2}\mathsf{j} + a_{p3}\mathsf{k}) \end{bmatrix}$$

are zero. Together with (3.3.5) this implies

$$\sum_{j=1}^{p} u_j(a_{j0} + a_{j1}\mathsf{i} + a_{j2}\mathsf{j} + a_{j3}\mathsf{k}) = 0,$$

and the linear independence of $u_1, \dots, u_p$ yields $a_{j\ell} = 0$ for all j and ℓ, as required.

The statement of Proposition (3.3.3) concerning orthonormal and orthogonal sets follows easily from properties (ii) and (iv) of χ, taking advantage of the orthogonality of the columns (and rows) of $\chi(x)$, for any $x \in \mathsf{H}$. □

If $u_1, \dots, u_p$ of Proposition 3.3.3 form a basis of a subspace $\mathcal{M} \subseteq \mathsf{H}^{n \times 1}$, then the columns of $\chi([u_1 \ \dots \ u_p])$ form a basis of a subspace, denoted $\chi(\mathcal{M})$, of $\mathsf{R}^{4n \times 1}$.

Note that $\chi(\mathcal{M})$ is independent of the choice of the basis $u_1, \ldots, u_p$. Indeed, if $u_1, \ldots, u_p$ and $u_1', \ldots, u_p'$ are two bases for $\mathcal{M}$, then

$$[u_1' \ \cdots \ u_p'] = [u_1 \ \cdots \ u_p] \cdot S \qquad (3.3.6)$$

for some invertible matrix $S \in \mathsf{H}^{p \times p}$. Applying the map χ to (3.3.6), we see the columns of $\chi([u_1 \ \cdots \ u_p])$ and those of $\chi([u_1' \ \cdots \ u_p'])$ span the same subspace in $\mathsf{R}^{4n \times 1}$.

3.4   COMPLEX MATRIX REPRESENTATION OF QUATERNIONS

For $\alpha = a_0 + i a_1 + j a_2 + k a_3 \in \mathsf{H}$, $a_0, a_1, a_2, a_3 \in \mathsf{R}$, we define the map ω as in Proposition 2.6.2,

$$\omega(\alpha) = \begin{bmatrix} a_0 + i a_1 & a_2 + i a_3 \\ -a_2 + i a_3 & a_0 - i a_1 \end{bmatrix} \in \mathsf{C}^{2 \times 2},$$

and extend ω entrywise to a map

$$\omega_{m,n} : \mathsf{H}^{m \times n} \rightarrow \mathsf{C}^{2m \times 2n}, \quad \omega_{m,n}\left([x_{i,j}]_{i,j=1}^{m,n}\right) = [\omega(x_{i,j})]_{i,j=1}^{m,n}, \qquad (3.4.1)$$

where $x_{i,j} \in \mathsf{H}$. Analogously to the map χ, we have:

(i) $\omega_{n,n}$ is a unital isomorphism of the real algebra $\mathsf{H}^{n \times n}$ onto the real subalgebra

$$\left\{ [z_{i,j}]_{i,j=1}^{n} \ : \ z_{i,j} \in \left\{ \begin{bmatrix} u & v \\ -\overline{v} & \overline{u} \end{bmatrix}, \ u, v \in \mathsf{C} \right\} \right\}$$

of $\mathsf{C}^{2n \times 2n}$;

(ii) if $X \in \mathsf{H}^{m \times n}$, $Y \in \mathsf{H}^{n \times p}$, then $\omega_{m,p}(XY) = \omega_{m,n}(X)\omega_{n,p}(Y)$;

(iii) if $X, Y \in \mathsf{H}^{m \times n}$ and $s, t \in \mathsf{R}$, then

$$\omega_{m,n}(sX + tY) = s\omega_{m,n}(X) + t\omega_{m,n}(Y);$$

(iv) $\omega_{n,m}(X^*) = (\omega_{m,n}(X))^*$,   for all $X \in \mathsf{H}^{m \times n}$;

(v) there exist positive constants $k_{m,n}$, $K_{m,n}$ such that

$$k_{m,n}\|\omega_{m,n}(X)\|_{\mathsf{C}} \leq \|X\|_{\mathsf{H}} \leq K_{m,n}\|\omega_{m,n}(X)\|_{\mathsf{C}}$$

for every $X \in \mathsf{H}^{m \times n}$.

Often, we will abbreviate $\omega_{m,n}$ to ω (with m, n understood from the context).

Proposition 3.4.1. *A matrix $A \in \mathsf{H}^{n \times n}$ is hermitian, positive definite, positive semidefinite, skewhermitian, unitary, or normal if and only if $\omega(A)$ is such.*

See Ex. 3.11.6.

Proposition 3.4.2. *Let $u_1, \ldots, u_p \in \mathsf{H}^{n \times 1}$. Then $u_1, \ldots, u_p$ are linearly independent (over H) if and only if the columns of $\omega_{n,p}([u_1 \ \cdots \ u_p])$ are linearly independent (over C).*

Moreover, $u_1, \ldots, u_p$ is an orthonormal, resp. orthogonal, set (over H) if and only if the columns of $\omega_{n,p}([u_1 \ \cdots \ u_p])$ form an orthonormal, resp. orthogonal, set (over C).

Proof. The "if" part is proved as in the proof of Proposition 3.3.3. Assume $u_1, \ldots, u_p$ are linearly independent. Let

$$\omega_{n,p}\left(\begin{bmatrix} u_1 & \cdots & u_p \end{bmatrix}\right) \begin{bmatrix} a_1 \\ \vdots \\ a_p \end{bmatrix} = 0 \tag{3.4.2}$$

for some $a_1, \ldots, a_p \in \mathsf{C}^{2 \times 1}$, and, in turn,

$$a_j = \begin{bmatrix} a_{j0} \\ a_{j1} \end{bmatrix}, \quad a_{j0}, a_{j1} \in \mathsf{C}.$$

Partition u_j as in (3.3.2), (3.3.3). Then (using the notation (3.3.4)) we have

$$\mathrm{row}_{j=1,2,\ldots,p} \begin{bmatrix} u_{jk}^{(0)} + i u_{jk}^{(1)} & u_{jk}^{(2)} + i u_{jk}^{(3)} \\ -u_{jk}^{(2)} + i u_{jk}^{(3)} & u_{jk}^{(0)} - i u_{jk}^{(1)} \end{bmatrix} \cdot \begin{bmatrix} a_{10} \\ a_{11} \\ a_{20} \\ a_{21} \\ \vdots \\ a_{p0} \\ a_{p1} \end{bmatrix} = 0, \tag{3.4.3}$$

for $k = 1, 2, \ldots, n$. A straightforward computation shows that (3.4.3) also yields

$$\mathrm{row}_{j=1,2,\ldots,p} \begin{bmatrix} u_{jk}^{(0)} + i u_{jk}^{(1)} & u_{jk}^{(2)} + i u_{jk}^{(3)} \\ -u_{jk}^{(2)} + i u_{jk}^{(3)} & u_{jk}^{(0)} - i u_{jk}^{(1)} \end{bmatrix} \cdot \begin{bmatrix} -\overline{a_{11}} \\ \overline{a_{10}} \\ \vdots \\ -\overline{a_{p1}} \\ \overline{a_{p0}} \end{bmatrix} = 0,$$

for $k = 1, 2, \ldots, n$, and

$$\mathrm{row}_{j=1,2,\ldots,p} \begin{bmatrix} u_{jk}^{(0)} + i u_{jk}^{(1)} & u_{jk}^{(2)} + i u_{jk}^{(3)} \\ -u_{jk}^{(2)} + i u_{jk}^{(3)} & u_{jk}^{(0)} - i u_{jk}^{(1)} \end{bmatrix} \cdot \begin{bmatrix} \omega(a_{10} - \mathsf{j}a_{11}) \\ \vdots \\ \omega(a_{p0} - \mathsf{j}a_{p1}) \end{bmatrix} = 0,$$

which, in turn, gives

$$\sum_{j=1}^{p} u_j (a_{j0} - \mathsf{j}a_{j1}) = 0.$$

By the linear independence of $u_1, \ldots, u_p$ we obtain $a_j = 0$, $j = 1, \ldots, p$, as required.

The statement of Proposition (3.4.2) concerning orthonormal and orthogonal sets follows easily from properties (ii) and (iv) of ω, taking advantage of the orthogonality of the columns (and rows) of $\omega(x)$, for any $x \in \mathsf{H}$.   □

If $\mathcal{M}$ is a subspace of $\mathsf{H}^{n \times 1}$, we let $\omega(\mathcal{M})$ be the subspace of $\mathsf{C}^{2n \times 1}$ spanned by the columns of $\omega(\begin{bmatrix} u_1 & \cdots & u_p \end{bmatrix})$, where $u_1, \ldots, u_p$ is a basis for $\mathcal{M}$. Analogously to $\chi(\mathcal{M})$, one verifies that $\omega(\mathcal{M})$ is independent of the choice of a basis for $\mathcal{M}$.

3.5   NUMERICAL RANGES WITH RESPECT TO CONJUGATION

In this section we will work with the standard involution, i.e., the conjugation. Let $A \in \mathsf{H}^{n \times n}$. Consider the quadratic form

$$\langle Ax, x \rangle := x^* A x, \quad x \in \mathsf{H}^{n \times 1}.$$

We have the *polarization identity* (3.5.1), which can be verified by a straightforward computation, as follows. Let (q_1, q_2, q_3) be a units triple of quaternions. Using Proposition 2.4.2, for $A \in \mathsf{H}^{n \times n}$ we then have

$$\sum_{j=1}^{3} \left(\sum_{k=0}^{3} q_j^k (x + y q_j^k)^* A (x + y q_j^k) \right) = 12\, y^* A x + 8\, \mathfrak{V} \left(x^* A y \right)$$

$$\text{for all} \quad x, y \in \mathsf{H}^{n \times 1}. \quad (3.5.1)$$

Proposition 3.5.1. *Let $A \in \mathsf{H}^{n \times n}$. Then:*

(1) $x^* A x = 0$ *for all* $x \in \mathsf{H}^{n \times 1}$ *if and only if* $A = 0$;

(2) $x^* A x \in \mathsf{R}$ *for all* $x \in \mathsf{H}^{n \times 1}$ *if and only if* $A = A^*$;

(3) $\mathfrak{R} \left(x^* A x \right) = 0$ *for all* $x \in \mathsf{H}^{n \times 1}$ *if and only if* $A = -A^*$.

Proof. The "if" parts are obvious. For the "only if" part of (1), assuming $x^* A x = 0$ for all $x \in \mathsf{H}^{n \times 1}$, (3.5.1) gives

$$\mathfrak{R}(y^* A x) = 0 \quad \text{and} \quad \mathfrak{V} \left(x^* A y \right) = -\frac{3}{2} \mathfrak{V} \left(y^* A x \right) \quad \text{for all} \quad x, y \in \mathsf{H}^{n \times 1}.$$

Iterating the second equality we obtain

$$\mathfrak{V} \left(x^* A y \right) = -\frac{3}{2} \mathfrak{V} \left(y^* A x \right) = \frac{9}{4} \mathfrak{V} \left(x^* A y \right), \quad \text{for all} \quad x, y \in \mathsf{H}^{n \times 1};$$

hence, $\mathfrak{V} \left(x^* A y \right) = 0$ for all $x, y \in \mathsf{H}^{n \times 1}$, and $A = 0$ follows.

The "only if" part of (2), resp. (3), follows by applying (1) to the matrix $A - A^*$, resp. $A + A^*$. □

Definition 3.5.2. The set

$$W_*^{\mathsf{H}}(A) := \left\{ x^* A x \; : \; x^* x = 1, \quad x \in \mathsf{H}^{n \times 1} \right\} \subset \mathsf{H}$$

is known as the (quaternion) *numerical range* of $A \in \mathsf{H}^{n \times n}$ with respect to the conjugation.

Clearly, $W_*^{\mathsf{H}}(A)$ is compact and connected (Theorem 3.10.7). Also, if $y_1 \in W_*^{\mathsf{H}}(A)$ and $y_2 \in \mathsf{H}$ is such that $\mathfrak{R} y_2 = \mathfrak{R} y_1$ and $|\mathfrak{V} y_2| = |\mathfrak{V} y_1|$, then also $y_2 \in W_*^{\mathsf{H}}(A)$. Indeed, by Theorem 2.2.6 we have $y_2 = a^* y_1 a$ for some $a \in \mathsf{H}$, $|a| = 1$, so if $y_1 = x^* A x$ for some $x \in \mathsf{H}^{n \times 1}$, $x^* x = 1$, then $y_2 = (xa)^* A(xa)$. Proposition 3.5.1 yields the following.

Corollary 3.5.3. *For $A \in \mathsf{H}^{n \times n}$, we have $W_*^{\mathsf{H}}(A) = \{0\}$, resp. $W_*^{\mathsf{H}}(A) \subset \mathsf{R}$ or $\mathfrak{R}(W_*^{\mathsf{H}}(A)) = \{0\}$, if and only if $A = 0$, resp. $A = A^*$ or $A = -A^*$.*

Indeed, the "if" part is obvious. Assume $W_*^{\mathsf{H}}(A) = \{0\}$. Then, by scaling $x \in \mathsf{H}^{n \times 1}$, i.e., multiplying x on the right by a nonzero quaternion, we obtain that $x^* A x = 0$ for every nonzero $x \in \mathsf{H}$. Thus, $x^* A x = 0$ for all $x \in \mathsf{H}^{n \times 1}$, and it remains to apply Proposition 3.5.3(1) to complete the proof of the first part of Corollary 3.5.3. The proof of other parts of the corollary follows the same pattern.

Some algebraic properties of the numerical ranges are worth noticing.

Proposition 3.5.4. *For $A \in \mathsf{H}^{n \times n}$, unitary $U \in \mathsf{H}^{n \times n}$, and real a, we have*

$$W_*^{\mathsf{H}}(U^* A U) = W_*^{\mathsf{H}}(A), \quad W_*^{\mathsf{H}}(A + aI) = a + W_*^{\mathsf{H}}(A), \quad W_*^{\mathsf{H}}(aA) = a W_*^{\mathsf{H}}(A).$$

In contrast to the classical convexity property (the Toeplitz-Hausdorff theorem) of numerical ranges of complex matrices

$$W_*^{\mathsf{C}}(B) := \left\{ x^* B x : x^* x = 1, \quad x \in \mathsf{C}^{n \times 1} \right\} \subset \mathsf{C},$$

where $B \in \mathsf{C}^{n \times n}$ (Theorem 3.5.7 below), the quaternion numerical ranges are generally nonconvex:

Example 3.5.5. Let

$$A = \begin{bmatrix} \lambda & 0 \\ 0 & A_0 \end{bmatrix}, \qquad n \geq 2,$$

where $\lambda \in \mathsf{H} \setminus \{0\}$ has zero real part, and $A_0 \in \mathsf{H}^{(n-1) \times (n-1)}$ is hermitian and either positive or negative definite. Clearly, $\lambda \in W_*^{\mathsf{H}}(A)$, and, therefore, also $-\lambda \in W_*^{\mathsf{H}}(A)$ (because $-\lambda$ is congruent to λ). But one easily checks that $0 = \frac{1}{2}\lambda + \frac{1}{2}(-\lambda)$ does not belong to $W_*^{\mathsf{H}}(A)$. Indeed, if we had

$$x^* \lambda x + y^* A_0 y = 0, \quad \text{for some} \quad x \in \mathsf{H}, \quad y \in \mathsf{H}^{(n-1) \times 1},$$

then, since $\mathfrak{R}(x^* \lambda x) = 0$ and $\mathfrak{V}(y^* A_0 y) = 0$, we must have

$$x^* \lambda x = y^* A_0 y = 0,$$

which yields $x = 0$ and $y = 0$. $\qquad\square$

Nevertheless, some convexity-like properties of quaternion numerical ranges turn out to be valid. For example, it is proved by So and Thompson [148] that the intersection of $W_*^{\mathsf{H}}(A)$ with the closed upper half-plane of the complex plane (or, for that matter, with the closed upper half-plane of any 2-dimensional plane in $\mathsf{R}^4 = \mathrm{Span}_{\mathsf{R}}\{1, \mathsf{i}, \mathsf{j}, \mathsf{k}\}$ that contains the reals) is convex, for any $A \in \mathsf{H}^{n \times n}$.

We now introduce joint numerical ranges.

Definition 3.5.6. For $\mathsf{F} \in \{\mathsf{R}, \mathsf{C}, \mathsf{H}\}$ and for a p-tuple of hermitian (real symmetric in the case $\mathsf{F} = \mathsf{R}$) matrices $A_1, \ldots, A_p \in \mathsf{F}^{n \times n}$, the F-*joint numerical range* is defined by

$$W J_*^{\mathsf{F}}(A_1, \ldots, A_p) := \left\{ (x^* A_1 x, \ldots, x^* A_p x) \in \mathsf{R}^p : x^* x = 1, \quad x \in \mathsf{F}^{n \times 1} \right\} \subset \mathsf{R}^p.$$

Here $x^* = x^T$ if $\mathsf{F} = \mathsf{R}$.

Since $W J_*^{\mathsf{F}}(A_1, \ldots, A_p)$, $\mathsf{F} \in \{\mathsf{R}, \mathsf{C}, \mathsf{H}\}$, are ranges of continuous functions defined on the compact sets $\{x \in \mathsf{F}^{n \times 1} : x^* x = 1\}$, it follows that $W J_*^{\mathsf{F}}(A_1, \ldots, A_p)$ themselves are compact.

The following is a basic convexity result for numerical ranges.

Theorem 3.5.7. *Let* F *be one of* R, C, *or* H, *and assume that* $n \neq 2$ *in the case* $\mathsf{F} = \mathsf{R}$. *Then the set* $WJ_*^{\mathsf{F}}(A, B)$ *is convex for every pair of hermitian (symmetric in the case* $\mathsf{F} = \mathsf{R}$) *matrices* $A, B \in \mathsf{F}^{n \times n}$.

The case $\mathsf{F} = \mathsf{C}$ is just a reformulation of the Toeplitz-Hausdorff theorem.

Proof. The case $n = 1$ is trivial, so assume $n \geq 2$ (and $n \geq 3$ in the real case). We prove first the following statement: *if* $(0,0) \notin WJ_*^{\mathsf{F}}(A, B)$, *then for any* $(a,b) \in \mathsf{R}^2$, (a,b) *and* $(-a,-b)$ *cannot both belong to* $WJ_*^{\mathsf{F}}(A, B)$.

Arguing by contradiction, suppose

$$(a, b) = (u_1^* A u_1, u_1^* B u_1), \quad (-a, -b) = (u_2^* A u_2, u_2^* B u_2)$$

for some $u_1, u_2 \in \mathsf{F}^{n \times 1}$, $\|u_1\| = \|u_2\| = 1$. Obviously, u_1 and u_2 are linearly independent over F. Because of our assumptions on n, there exists $u_3 \in \mathsf{F}^{n \times 1}$ such that u_1, u_2, u_3 are linearly independent over R. Now, for any $(x, y, z) \in \mathsf{R}^3$ we have

$$(u_1 x + u_2 y + u_3 z)^* A (u_1 x + u_2 y + u_3 z)$$

$$= a(x^2 - y^2) + a_1 xy + a_2 xz + a_3 yz + a_4 z^2$$

for some real numbers a_1, a_2, a_3, a_4 and

$$(u_1 x + u_2 y + u_3 z)^* B (u_1 x + u_2 y + u_3 z)$$

$$= b(x^2 - y^2) + b_1 xy + b_2 xz + b_3 yz + b_4 z^2$$

for some real numbers b_1, b_2, b_3, b_4. For example, $b_1 = u_2^* B u_1 + u_1^* B u_2$. It turns out that the system

$$\begin{aligned} a(x^2 - y^2) + a_1 xy + a_2 xz + a_3 yz + a_4 z^2 &= 0, \\ b(x^2 - y^2) + b_1 xy + b_2 xz + b_3 yz + b_4 z^2 &= 0 \end{aligned} \tag{3.5.2}$$

has a nontrivial solution $(x, y, z) \in \mathsf{R}^3$. Let us verify this claim. If $ab_1 - ba_1 = 0$, set $z = 0$, and the resulting homogeneous linear system in the variables $x^2 - y^2$ and xy has a nontrivial solution, leading to a nontrivial solution $(x, y, 0)$ of (3.5.2). If $ab_1 - ba_1 \neq 0$, then set $z = 1$. Since $ab_1 - ba_1 \neq 0$, the system resulting from (3.5.2) is equivalent to

$$x^2 - y^2 + a_2' x + a_3' y + a_4' = 0, \qquad xy + b_2' x + b_3' y + b_4' = 0, \tag{3.5.3}$$

for some real numbers $a_2', a_3', a_4', b_2', b_3', b_4'$. Making the change of variables $x \mapsto x + a_2'/2$, $y \mapsto y - a_3'/2$, we may (and do) assume without loss of generality, that $a_2' = a_3' = 0$. Solve the second equation in (3.5.3) for y (assuming $x \neq -b_3'$), to obtain $y = (-b_4' - b_2' x)(x + b_3')^{-1}$, and substitute in the first equation, which results in

$$(x + b_3')^2 x^2 - (b_4' + b_2' x)^2 + a_4'(x + b_3')^2 = 0. \tag{3.5.4}$$

Unless $b_4' - b_2' b_3' = 0$, the value of the quartic polynomial in the left-hand side of (3.5.4) at $x = -b_3'$ is negative; hence, the polynomial has a real root different from $-b_3'$, leading to a nontrivial solution (x, y, z) of (3.5.2). In the remaining case, when $b_4' - b_2' b_3' = 0$, the second equation in (3.5.3) takes the form $(x + b_3')(y + b_2') = 0$; thus, $x = -b_3'$ or $y = -b_2'$. If $x = -b_3'$, then the first equation in (3.5.3) boils down to

$$y^2 = (b_3')^2 + a_4', \tag{3.5.5}$$

and if $y = -b'_2$, then the first equation in (3.5.3) is

$$x^2 = (b'_2)^2 - a'_4. \qquad (3.5.6)$$

Clearly, at least one of (3.5.5) and (3.5.6) has a real solution. We have shown that the system (3.5.2) has a nontrivial solution $(x_0, y_0, z_0) \in \mathsf{R}^3$. But now

$$(u_1 x_0 + u_2 y_0 + u_3 z_0)^* A (u_1 x_0 + u_2 y_0 + u_3 z_0)$$

$$= (u_1 x_0 + u_2 y_0 + u_3 z_0)^* B (u_1 x_0 + u_2 y_0 + u_3 z_0) = 0;$$

thus, $(0, 0) \in WJ_*^\mathsf{F}(A, B)$, a contradiction to our hypothesis. This proves the statement.

Now the proof of convexity of $WJ_*^\mathsf{F}(A, B)$ can be easily completed. Assuming the contrary, suppose $(a, b), (a', b') \in WJ_*^\mathsf{F}(A, B)$ but

$$(\alpha_0 a + (1 - \alpha_0) a', \alpha_0 b + (1 - \alpha_0) b') \notin WJ_*^\mathsf{F}(A, B)$$

for some real α_0, $0 < \alpha_0 < 1$. Let $(\alpha', \alpha'') \subseteq (0, 1)$ be the maximal open interval that contains α_0 and is such that

$$(\alpha a + (1 - \alpha) a', \alpha b + (1 - \alpha) b') \notin WJ_*^\mathsf{F}(A, B) \qquad \text{for all } \alpha \in (\alpha', \alpha'').$$

(The existence of such an interval (α', α'') is guaranteed in view of the openness of the complement of the numerical range $WJ_*^\mathsf{F}(A, B)$.) Note that

$$(\alpha' a + (1 - \alpha') a', \alpha' b + (1 - \alpha') b'),$$

$$(\alpha'' a + (1 - \alpha'') a', \alpha'' b + (1 - \alpha'') b') \in WJ_*^\mathsf{F}(A, B) \qquad (3.5.7)$$

because of compactness of $WJ_*^\mathsf{F}(A, B)$. Set

$$\widetilde{A} = A - \left[\frac{1}{2}(\alpha' + \alpha'') a + (1 - \frac{1}{2}(\alpha' + \alpha'')) a' \right] I,$$

$$\widetilde{B} = B - \left[\frac{1}{2}(\alpha' + \alpha'') b + (1 - \frac{1}{2}(\alpha' + \alpha'')) b' \right] I.$$

Then $(0, 0) \notin WJ_*^\mathsf{F}(\widetilde{A}, \widetilde{B})$ but, setting $z := \frac{1}{2}\alpha' - \frac{1}{2}\alpha''$, we have as a consequence of (3.5.7),

$$\pm(za - za', zb - zb') \in WJ_*^\mathsf{F}(\widetilde{A}, \widetilde{B}),$$

a contradiction to the statement proved above. $\square$

The result of Theorem 3.5.7 fails for $\mathsf{F} = \mathsf{R}$ and $n = 2$ (Ex. 3.11.14).

This theorem has been extended and generalized in many ways, and the literature on this subject is voluminous. We indicate here only a few facts.

Theorem 3.5.8.   (1) *If $n \neq 2$ and $B_1, B_2, B_3 \in \mathsf{C}^{n \times n}$ are hermitian, then the joint numerical range $WJ_*^\mathsf{C}(B_1, B_2, B_3)$ is convex.*

(2) *If $n \neq 2$ and $A_1, \ldots, A_5 \in \mathsf{H}^{n \times n}$ are hermitian, then $WJ_*^\mathsf{H}(A_1, \ldots, A_5)$ is convex.*

(3) *If $A_1, \ldots, A_4 \in \mathsf{H}^{n \times n}$ are hermitian, then $WJ_*^\mathsf{H}(A_1, \ldots, A_4)$ is convex.*

Part (1) is proved by Au-Yeung and Tsing [12, 11], parts (2) and (3) are proved by Au-Yeung and Poon [10], and a far-reaching, more-general result is proved by Poon [120].

Using Theorem 3.5.8, we give a criterion for convexity of joint numerical ranges of five 2×2 hermitian quaternion matrices.

Corollary 3.5.9. *Let* $A_1, \ldots, A_5 \in \mathsf{H}^{2 \times 2}$ *be hermitian matrices. Then the joint numerical range* $W J_*^{\mathsf{H}}(A_1, \ldots, A_5)$ *is convex if and only if the 6-tuple of matrices* $\{A_1, \ldots, A_5, I_2\}$ *is linearly dependent over the reals.*

Proof. First consider a particular situation. Let

$$B_1 = \begin{bmatrix} 1 & 0 \\ 0 & -1 \end{bmatrix}, \quad B_2 = \begin{bmatrix} 0 & 1 \\ 1 & 0 \end{bmatrix}, \quad B_3 = \begin{bmatrix} 0 & \mathsf{i} \\ -\mathsf{i} & 0 \end{bmatrix},$$

$$B_4 = \begin{bmatrix} 0 & \mathsf{j} \\ -\mathsf{j} & 0 \end{bmatrix}, \quad B_5 = \begin{bmatrix} 0 & \mathsf{k} \\ -\mathsf{k} & 0 \end{bmatrix}.$$

We claim that $W J_*^{\mathsf{H}}(B_1, \ldots, B_5, I_2) \subset \mathsf{R}^6$ is not convex. Since

$$W J_*^{\mathsf{H}}(B_1, \ldots, B_5, I_2) = (W J_*^{\mathsf{H}}(B_1, \ldots, B_5), 1),$$

it suffices to show that $W J_*^{\mathsf{H}}(B_1, \ldots, B_5)$ is not convex. Indeed,

$$e_j^* B_1 e_j = \pm 1, \quad e_j^* B_2 e_j = \cdots = e_j^* B_5 e_j = 0, \quad j = 1, 2,$$

which shows that

$$(1, 0, 0, 0, 0), (-1, 0, 0, 0, 0) \in W J_*^{\mathsf{H}}(B_1, \ldots, B_5).$$

On the other hand, $0 \notin W J_*^{\mathsf{H}}(B_1, \ldots, B_5)$. To establish that, we will prove that the system of equations

$$x^* B_k x = 0 \quad \text{for} \quad k = 1, 2, 3, 4, 5, \qquad x \in \mathsf{H}^{2 \times 1}, \tag{3.5.8}$$

has only the trivial solution $x = 0$. Write $x = \begin{bmatrix} y \\ z \end{bmatrix}$, where $y, z \in \mathsf{H}$. Since the assumption that $y = 0$ yields only the trivial solution of (3.5.8), we may assume $y \neq 0$, and then by replacing x with xy^{-1}, we may further assume that $y = 1$. But then (3.5.8) gives

$$z^* \mathsf{i} = \mathsf{i} z, \quad z^* \mathsf{j} = \mathsf{j} z, \qquad z^* \mathsf{k} = \mathsf{k} z. \tag{3.5.9}$$

It is easy to see from (3.5.9) that z must be real. Now

$$x^* B_1 x = 1 - z^2 = 0, \qquad x^* B_2 x = 2z = 0$$

leads to a contradiction.

If the 6-tuple $\{A_1, \ldots, A_5, I_2\}$ is linearly independent over the reals, then (because the real vector space of 2×2 quaternion hermitian matrices is 6-dimensional) we have

$$\begin{bmatrix} A_1 \\ \vdots \\ A_5 \\ I_2 \end{bmatrix} = C \begin{bmatrix} B_1 \\ \vdots \\ B_5 \\ I_2 \end{bmatrix}$$

for some real invertible 6×6 matrix C. Thus,

$$W J_*^{\mathsf{H}}(A_1, \ldots, A_5, I_2) = C\, W J_*^{\mathsf{H}}(B_1, \ldots, B_5, I_2),$$

and, since $W J_*^{\mathsf{H}}(B_1, \ldots, B_5, I_2)$ is not convex, $W J_*^{\mathsf{H}}(A_1, \ldots, A_5, I_2)$ is not convex either, and the nonconvexity of $W J_*^{\mathsf{H}}(A_1, \ldots, A_5)$ follows.

Conversely, assume that $A_1, \ldots, A_5, I_2$ are linearly dependent over the reals. If $A_1, \ldots, A_5$ are linearly dependent, say

$$A_5 = a_1 A_1 + \cdots + a_4 A_4, \qquad \text{where } a_1, a_2, a_3, a_4 \in \mathsf{R};$$

then

$$W J_*^{\mathsf{H}}(A_1, \ldots, A_5) = \left[\begin{array}{c} I_4 \\ \left[\begin{array}{cccc} a_1 & a_2 & a_3 & a_4 \end{array}\right] \end{array} \right] W J_*^{\mathsf{H}}(A_1, A_2, A_3, A_4),$$

and so the convexity of $W J_*^{\mathsf{H}}(A_1, \ldots, A_5)$ follows from that of $W J_*^{\mathsf{H}}(A_1, A_2, A_3, A_4)$, by Theorem 3.5.8. If $A_1, \ldots, A_5$ are linearly independent, then, for some index $j \in \{1, 2, 3, 4, 5\}$, the matrix A_j is a real linear combination of

$$A_1, \ldots, A_{j-1}, A_{j+1}, \ldots, A_5, I_2,$$

and as before, the convexity of $W J_*^{\mathsf{H}}(A_1, \ldots, A_5)$ follows from that of the joint numerical range $W J_*^{\mathsf{H}}(A_1, \ldots, A_{j-1}, A_{j+1}, \ldots, A_5, I_2)$. $\square$

In a different direction, we have another convexity property.

Theorem 3.5.10. *Let S be any 2-dimensional subspace in $\mathsf{R}^4 = \mathrm{Span}_{\mathsf{R}}\{1, \mathsf{i}, \mathsf{j}, \mathsf{k}\}$ that contains the reals, and let P_S be the orthogonal projection on S. Then, for any $A \in \mathsf{H}^{n \times n}$, the set $P_S\left(W_*^{\mathsf{H}}(A)\right)$ is convex.*

Proof. Let (q_1, q_2, q_3) be a units triple and such that $1, q_1$ form an orthonormal basis in S and q_2, q_3 orthogonal to S (one can take for q_1 any of the two unit quaternions in S with zero real part). Write $x \in \mathsf{H}^{n \times 1}$ in the form

$$x = y + q_2 z, \quad y, z \in \mathsf{R}^{n \times 1} + q_1 \mathsf{R}^{n \times 1}$$

and $A \in \mathsf{H}^{n \times n}$ in the form

$$A = B + q_2 C, \quad B, C \in \mathsf{R}^{n \times n} + q_1 \mathsf{R}^{n \times n}.$$

Then for $x \in \mathsf{H}^{n \times 1}$, such that $x^* x = 1$, we have $|y|^2 + |z|^2 = 1$ and write

$$x^* A x = (y + q_2 z)^* (B + q_2 C)(y + q_2 z) = Q_1 + Q_2,$$

where

$$Q_1 \in \mathrm{Span}_{\mathsf{R}}\{1, q_1\}, \qquad Q_2 \in \mathrm{Span}_{\mathsf{R}}\{q_2, q_3\}.$$

A calculation shows that

$$Q_1 = \left[\begin{array}{cc} y^* & z^* \end{array}\right] \left[\begin{array}{cc} B & -\overline{C} \\ C & B^* \end{array}\right] \left[\begin{array}{c} y \\ z \end{array}\right],$$

where $\overline{C}$ is obtained from C by replacing each entry with its conjugate. Thus, $P_S(W_*(A))$ coincides with the complex numerical range of the complex matrix $\left[\begin{array}{cc} B & -\overline{C} \\ C & B^* \end{array}\right]$ (if we identify $a + q_1 b$, $a, b \in \mathsf{R}$, with the compex number $a + ib$) and, therefore, is convex by Theorem 3.5.7. $\square$

3.6   MATRIX DECOMPOSITIONS: NONSTANDARD INVOLUTIONS

In this section, we fix a nonstandard involution ϕ. By analogy with conjugation, for $A \in \mathsf{H}^{m \times n}$, we denote by A_ϕ the $n \times m$ matrix obtained by applying ϕ entrywise to the *transposed* matrix A^T. For example, if ϕ is such that $\phi(\mathsf{i}) = -\mathsf{i}$, $\phi(\mathsf{j}) = \mathsf{j}$, $\phi(\mathsf{k}) = \mathsf{k}$, then

$$\begin{bmatrix} 1 & 2 + \mathsf{i} \\ 3 + \mathsf{j} & 4 + \mathsf{k} \\ \mathsf{i} + \mathsf{j} & \mathsf{i} + \mathsf{k} \end{bmatrix}_\phi = \begin{bmatrix} 1 & 3 + \mathsf{j} & -\mathsf{i} + \mathsf{j} \\ 2 - \mathsf{i} & 4 + \mathsf{k} & -\mathsf{i} + \mathsf{k} \end{bmatrix}.$$

Note the following algebraic properties:

(a) $(\alpha A + \beta B)_\phi = A_\phi \phi(\alpha) + B_\phi \phi(\beta)$, $\alpha, \beta \in \mathsf{H}$, $A, B \in \mathsf{H}^{m \times n}$.

(b) $(A\alpha + B\beta)_\phi = \phi(\alpha) A_\phi + \phi(\beta) B_\phi$, $\alpha, \beta \in \mathsf{H}$, $A, B \in \mathsf{H}^{m \times n}$.

(c) $(AB)_\phi = B_\phi A_\phi$, $A \in \mathsf{H}^{m \times n}$, $B \in \mathsf{H}^{n \times p}$.

(d) $(A_\phi)_\phi = A$, $A \in \mathsf{H}^{m \times n}$.

(e) If $A \in \mathsf{H}^{n \times n}$ is invertible, then $(A_\phi)^{-1} = (A^{-1})_\phi$.

The standard classes of matrices with respect to ϕ are introduced in a familiar way.

Definition 3.6.1. $A \in \mathsf{H}^{n \times n}$ is said to be ϕ-*hermitian*, ϕ-*skewhermitian*, ϕ-*unitary*, or ϕ-*normal* if $A = A_\phi$, $A = -A_\phi$, A is invertible and $A^{-1} = A_\phi$, or $AA_\phi = A_\phi A$, respectively.

Note that the sets of ϕ-hermitian and ϕ-skewhermitian matrices form real vector subspaces of $\mathsf{H}^{n \times n}$ (for a fixed ϕ), whereas the set of ϕ-unitaries is a multiplicative group: if U, V are ϕ-unitaries, then so are UV and U^{-1}.

Definition 3.6.2. By analogy with the standard inner product $\langle \cdot, \cdot \rangle$ in $\mathsf{H}^{n \times 1}$, we introduce the quaternion-valued ϕ-*inner product* $\langle u, v \rangle^\phi := v_\phi u$, for $u, v \in \mathsf{H}^{n \times 1}$.

The ϕ-inner product obeys linearity and symmetry properties,

$$\begin{aligned} \langle u_1 \alpha_1 + u_2 \alpha_2, v \rangle^\phi &= \langle u_1, v \rangle^\phi \alpha_1 + \langle u_2, v \rangle^\phi \alpha_2, \quad u_1, u_2, v \in \mathsf{H}^{n \times 1}, \\ &\quad \alpha_1, \alpha_2 \in \mathsf{H}, \\ \langle u, v_1 \alpha_1 + v_2 \alpha_2 \rangle^\phi &= (\alpha_1)_\phi \langle u, v_1 \rangle^\phi + (\alpha_2)_\phi \langle u, v_2 \rangle^\phi, \quad u, v_1, v_2 \in \mathsf{H}^{n \times 1}, \\ &\quad \alpha_1, \alpha_2 \in \mathsf{H}, \\ \langle u, v \rangle^\phi &= ((\langle v, u \rangle^\phi)_\phi, \quad u, v \in \mathsf{H}^{n \times 1}, \end{aligned}$$

but not the positive definiteness property: $\langle u, u \rangle^\phi$ need not be nonnegative, or even real. Note that

$$\langle u, Av \rangle^\phi = \langle A_\phi u, v \rangle^\phi \quad \text{for all } u \in \mathsf{H}^{n \times 1}, \ v \in \mathsf{H}^{m \times 1}, A \in \mathsf{H}^{m \times n}.$$

We say that $u, v \in \mathsf{H}^{n \times 1}$ are ϕ-*orthogonal* if $\langle u, v \rangle^\phi = 0$.

Definition 3.6.3. For a set $Z \subseteq \mathsf{H}^{n \times 1}$ define the ϕ-*orthogonal companion*

$$Z^{\perp_\phi} := \{x \in \mathsf{H}^{n \times 1} : \langle x, u \rangle^\phi = 0 \text{ for all } u \in Z\}.$$

We list some basic properties of ϕ-orthogonal companions.

Proposition 3.6.4. *Let $Z \subseteq \mathsf{H}^{n \times 1}$. Then:*

(a) *$Z^{\perp \phi}$ is a (quaternion) subspace in $\mathsf{H}^{n \times 1}$;*

(b) *if Z is a subspace in $\mathsf{H}^{n \times 1}$, then $\dim Z + \dim Z^{\perp \phi} = n$;*

(c) *$((Z)^{\perp \phi})^{\perp \phi} \supseteq \mathrm{Span}_{\mathsf{H}} \{Z\}$, and if Z is a subspace, then $((Z)^{\perp \phi})^{\perp \phi} = Z$;*

(d) *if Z is a subspace, then $Z^{\perp \phi}$ is a direct complement of Z in $\mathsf{H}^{n \times 1}$ if and only if Z does not contain a nonzero vector which is ϕ-orthogonal to Z.*

Proof. We prove in detail only part (b). Let $\{f_1, \ldots, f_d\}$ be a basis for Z (we leave aside the trivial case when $Z = \{0\}$), and choose $f_{d+1}, \ldots, f_n$ so that $\{f_1, f_2, \ldots, f_n\}$ is a basis for $\mathsf{H}^{n \times 1}$. Then the matrix $X := [f_1 \ f_2 \ \cdots \ f_n]$ is invertible. Partition the inverse $X^{-1} = \begin{bmatrix} Y_1 \\ Y_2 \end{bmatrix}$, where $Y_1 \in \mathsf{H}^{d \times n}$ and $Y_2 \in \mathsf{H}^{(n-d) \times n}$. The equality $X^{-1} X = I$ yields $Y_2 \cdot [f_1 \ \cdots \ f_d] = 0$. So the $(n-d)$-dimensional subspace spanned by the linearly independent columns of $(Y_2)_\phi$, call this subspace $\mathcal{N}$, is contained in $Z^{\perp \phi}$. If the subspace $Z^{\perp \phi}$ were strictly larger than $\mathcal{N}$, then we would find a nonzero vector in $Z^{\perp \phi}$ which is a linear combination of the columns of $(Y_1)_\phi$. Thus, there would exist $\alpha_1, \ldots, \alpha_d \in \mathsf{H}$, not all zeros, such that

$$[\alpha_1 \ \alpha_2 \ \cdots \ \alpha_d] Y_1 \, u = 0 \quad \text{for all } u \in Z,$$

a contradiction with the equality $Y_1 \cdot [f_1 \ f_2 \ \cdots \ f_d] = I$. $\qquad \square$

The analogues of QR factorization, polar decomposition, and singular value decomposition (using ϕ-unitaries instead of unitaries) in the presence of a nonstandard involution generally fail. Indeed, it may happen that $x_\phi x = 0$ for a nonzero $x \in \mathsf{H}^{n \times 1}$; hence, the Gram-Schmidt orthogonalization procedure is not always available. However, a version of Cholesky factorization remains valid.

Theorem 3.6.5. (a) *A nonzero matrix $A = [a_{i,j}]_{i,j=1}^n \in \mathsf{H}^{n \times n}$, $a_{i,j} \in \mathsf{H}$, is ϕ-hermitian if and only if it admits a factorization $A = B_\phi B$ for some $B \in \mathsf{H}^{k \times n}$. Here k is the rank of A.*

(b) *If the principal submatrices $[a_{i,j}]_{i,j=1}^s$, $s = 1, 2, \ldots, \mathrm{rank}\,(A)$, are invertible, then B can be taken upper triangular in $A = B_\phi B$, and if the principal submatrices $[a_{i,j}]_{i,j=n-s}^n$, $s = 0, 1, \ldots, \mathrm{rank}\,(A) - 1$, are invertible, then B can be taken lower triangular.*

Proof. Part (a). We need to prove the nontrivial part "only if." Use induction on n. The case $n = 1$ is taken care of by Theorem 2.5.1(b).

Consider now the general case. We claim, replacing A if necessary by $C_\phi AC$ for a suitable invertible matrix C, that it can be assumed $a_{1,1} \neq 0$. Indeed, if $a_{j,j} \neq 0$ for some index j, just simultaneously interchange rows 1 and j and columns 1 and j. And if all diagonal entries of A are zeros, but $a_{i,j} \neq 0$ for some indices $i \neq j$, then the following transformation in rows and columns i and j yields a ϕ-hermitian matrix with a nonzero diagonal:

$$\begin{bmatrix} 0 & a_{i,j} \\ (a_{i,j})_\phi & 0 \end{bmatrix} \mapsto \begin{bmatrix} 1 & (a_{i,j})_\phi^{-1} \\ 0 & 1 \end{bmatrix} \begin{bmatrix} 0 & a_{i,j} \\ (a_{i,j})_\phi & 0 \end{bmatrix} \begin{bmatrix} 1 & 0 \\ a_{i,j}^{-1} & 1 \end{bmatrix}$$

$$= \begin{bmatrix} 2 & a_{i,j} \\ (a_{i,j})_\phi & 0 \end{bmatrix}.$$

Next, using Theorem 2.5.1(b), we can assume that actually $a_{1,1} = 1$. Now it is easy to see that simultaneous row and column replacement operations yield a ϕ-hermitian matrix of the form $\begin{bmatrix} 1 & 0_{1 \times n-1} \\ 0_{n-1 \times 1} & A' \end{bmatrix}$, where $A' = A'_\phi \in \mathsf{H}^{(n-1) \times (n-1)}$, and an application of the induction hypothesis completes the proof of (a).

For Part (b), we use again induction on n. Say the matrices $[a_{i,j}]_{i,j=1}^s$, $s = 1, 2, \ldots, \operatorname{rank}(A)$, are invertible. We distinguish two cases: (1) $\operatorname{rank}(A) = n$, i.e., A is invertible (cf. Ex. 3.11.7); (2) $\operatorname{rank}(A) < n$. Consider case (1). By the induction hypothesis, we have

$$[a_{i,j}]_{i,j=1}^{n-1} = (B_1)_\phi B_1$$

for some upper triangular, necessarily invertible (cf. Ex. 3.11.8) matrix $B_1 \in \mathsf{H}^{(n-1) \times (n-1)}$. We seek B in the form

$$B = \begin{bmatrix} B_1 & B_2 \\ 0 & B_3 \end{bmatrix}, \qquad B_2 \in \mathsf{H}^{(n-1) \times 1}, \quad B_3 \in \mathsf{H};$$

then the equality $A = B_\phi B$ amounts to the following:

$$\begin{bmatrix} a_{1,n} \\ a_{2,n} \\ \vdots \\ a_{n-1,n} \end{bmatrix} = (B_1)_\phi B_2, \qquad (a_{n,n})_\phi = (B_2)_\phi B_2 + (B_3)_\phi B_3. \tag{3.6.1}$$

The first equation in (3.6.1) can be solved for B_2 in view of the invertibility of B_1; then the second equation in (3.6.1) can be solved for B_3 in view of Theorem 2.5.1(b).

Now let $k := \operatorname{rank}(A) < n$. Partition: $A = \begin{bmatrix} A_1 & A_2 \\ (A_2)_\phi & A_3 \end{bmatrix}$, where $A_1 = (A_1)_\phi \in \mathsf{H}^{k \times k}$, $A_3 = (A_3)_\phi \in \mathsf{H}^{(n-k) \times (n-k)}$. Since $\operatorname{rank}(A) = k$ and the principal submatrix $[a_{i,j}]_{i,j=1}^k$ is invertible, the last $n-k$ columns of A are linearly dependent on its first k columns, in other words, we have

$$A \begin{bmatrix} X \\ I_{n-k} \end{bmatrix} = 0 \tag{3.6.2}$$

for some $X \in \mathsf{H}^{k \times (n-k)}$. Note that (by (3.6.2))

$$\widetilde{A} := \begin{bmatrix} I_k & 0 \\ X_\phi & I_{n-k} \end{bmatrix} A \begin{bmatrix} I_k & X \\ 0 & I_{n-k} \end{bmatrix} = \begin{bmatrix} A_1 & 0 \\ 0 & 0 \end{bmatrix},$$

and apply the induction hypothesis to $\widetilde{A}$.

The case when the submatrices $[a_{i,j}]_{i,j=n-s}^n$, $s = 0, 1, \ldots, \operatorname{rank}(A) - 1$ are invertible, is treated analogously. $\qquad \square$

Denote by $\tau_\phi : \mathsf{H}^{n \times 1} \rightarrow \mathsf{H}$ the map $\tau_\phi(x) = x_\phi x$, $x \in \mathsf{H}^{n \times 1}$ (we suppress in the notation the dependence of τ_ϕ on n). The map τ_ϕ has a local right inverse at nonzero vectors.

Proposition 3.6.6. *For every $x \in \mathsf{H}^{n \times 1} \setminus \{0\}$, there exist an open neighborhood $\mathcal{U}_x$ of x in $\mathsf{H}^{n \times 1}$, an open neighborhood $\mathcal{V}_x$ of $\tau_\phi(x)$ in $\mathrm{Inv}\,(\phi)$ such that τ_ϕ maps $\mathcal{U}_x$ onto $\mathcal{V}_x$, and a continuously differentiable map*

$$\mathsf{u}_x : \tau_\phi(\mathcal{U}_x) = \mathcal{V}_x \rightarrow \mathcal{U}_x$$

such that $\lambda = (\mathsf{u}_x(\lambda))_\phi\, \mathsf{u}_x(\lambda)$ for every $\lambda \in \mathcal{V}_x$, and $\mathsf{u}_x(\tau_\phi(x)) = x$.

Proof. Consider τ_ϕ as a map onto $\mathrm{Inv}\,(\phi)$ (cf. Theorem 2.5.1(b)). Identify

$$\mathsf{H}^{n \times 1} = (\mathrm{Span}_{\mathsf{R}}\, \{1, q_1, q_2, q_3\})^n$$

and $\mathrm{Inv}\,(\phi) = \mathrm{Span}_{\mathsf{R}}\, \{1, q_2, q_3\}$ with R^{4n} and R^3, respectively, where (q_1, q_2, q_3) is a suitable units triple. Then τ_ϕ becomes a map (for which we use the same notation) $\tau_\phi : \mathsf{R}^{4n \times 1} \rightarrow \mathsf{R}^{3 \times 1}$ given by

$$\tau_\phi \left(\mathrm{col}_{j=1,2,\ldots,n} \begin{bmatrix} a_j \\ b_j \\ c_j \\ d_j \end{bmatrix} \right) = \sum_{j=1}^{n} \begin{bmatrix} a_j^2 + b_j^2 - c_j^2 - d_j^2 \\ 2a_j c_j + 2b_j d_j \\ 2a_j d_j - 2b_j c_j \end{bmatrix},$$

where $a_j, b_j, c_j, d_j \in \mathsf{R}$ for $j = 1, 2, \ldots, n$.

The Jacobian matrix of τ_ϕ is

$$\mathrm{Jac} = 2\, \mathrm{row}_{j=1,2,\ldots,n} \left(\begin{bmatrix} a_j & b_j & -c_j & -d_j \\ c_j & d_j & a_j & b_j \\ d_j & -c_j & -b_j & a_j \end{bmatrix} \right),$$

which has full rank (unless $a_j = b_j = c_j = d_j = 0$ for all j); indeed, if $a_j \neq 0$ for some j, then the 3×3 submatrix formed by the first, third, and fourth columns of $\begin{bmatrix} a_j & b_j & -c_j & -d_j \\ c_j & d_j & a_j & b_j \\ d_j & -c_j & -b_j & a_j \end{bmatrix}$ is invertible, and if $a_j = 0$ then the determinants of the three other 3×3 submatrices of $\begin{bmatrix} a_j & b_j & -c_j & -d_j \\ c_j & d_j & a_j & b_j \\ d_j & -c_j & -b_j & a_j \end{bmatrix}$ are $b_j(b_j^2 + c_j^2 + d_j^2)$, $c_j(b_j^2 + c_j^2 + d_j^2)$, and $d_j(b_j^2 + c_j^2 + d_j^2)$. Now we use the implicit function theorem of multivariable calculus (which is presented in many textbooks; see, for instance, Kaplan [77], Edwards [35], or Fleming [44]) to complete the proof. □

3.7 NUMERICAL RANGES WITH RESPECT TO NONSTANDARD INVOLUTIONS

In this section we fix a nonstandard involution ϕ. Let (q_1, q_2, q_3) be a units triple such that $\phi(q_1) = -q_1$, $\phi(q_2) = q_2$, $\phi(q_3) = q_3$. A straightforward but tedious calculation shows that for any $A \in \mathsf{H}^{n \times n}$, we have the polarization identity

$$\sum_{j=1}^{3} \left(\sum_{k=0}^{3} q_j^k (x + y q_j^k)_\phi A(x + y q_j^k) \right) = 4\, y_\phi A x + 8\, \mathfrak{V}\,(x_\phi A y)$$

$$\text{for all} \qquad x, y \in \mathsf{H}^{n \times 1}. \quad (3.7.1)$$

Thus, using (3.7.1), analogous to Proposition 3.5.1, the following result is obtained.

Proposition 3.7.1. *Let $A \in \mathsf{H}^{n \times n}$. Then:*

(1) $x_\phi A x = 0$ *for all $x \in \mathsf{H}^{n \times 1}$ if and only if $A = 0$;*

(2) $x_\phi A x \in \operatorname{Span}_{\mathsf{R}} \{1, q_2, q_3\}$ *for all $x \in \mathsf{H}^{n \times 1}$ if and only if $A = A_\phi$;*

(3) $x_\phi A x \in \operatorname{Span}_{\mathsf{R}} \{q_1\}$ *for all $x \in \mathsf{H}^{n \times 1}$ if and only if $A = -A_\phi$.*

Since $x_\phi x$ can take on any value in $\operatorname{Inv}(\phi)$, it makes sense to introduce numerical ranges of a matrix $A \in \mathsf{H}^{n \times n}$ with respect to ϕ in the following way, for a fixed $\alpha \in \operatorname{Inv}(\phi)$:

$$W_\phi^{(\alpha)}(A) := \{x_\phi A x \ : \ x_\phi x = \alpha, \quad x \in \mathsf{H}^{n \times 1}\}.$$

Writing $\alpha = \gamma_\phi \gamma$ for some $\gamma \in \mathsf{H}$ (Theorem 2.5.1(b)) we see that

$$W_\phi^{(\alpha)}(A) = \gamma_\phi W_\phi^{(1)}(A) \gamma, \tag{3.7.2}$$

assuming $\alpha \neq 0$. Thus, we can focus on two particular numerical ranges for a given $A \in \mathsf{H}^{n \times n}$: $W_\phi^{(1)}(A)$ and $W_\phi^{(0)}(A)$. To avoid trivialities, in the latter case $n \geq 2$ will be assumed. Note some elementary algebraic properties.

Proposition 3.7.2. (a) *If $A, U \in \mathsf{H}^{n \times n}$, where U is ϕ-unitary, and if $a \in \mathsf{R}$, then*

$$W_\phi^{(\alpha)}(U_\phi A U) = W_\phi^{(\alpha)}(A), \qquad W_\phi^{(\alpha)}(A + aI) = W_\phi^{(\alpha)}(A) + a\alpha,$$

$$W_\phi^{(\alpha)}(aA) = a \, W_\phi^{(\alpha)}(A)$$

for every $\alpha \in \operatorname{Inv}(\phi)$.

(b) *If, for a fixed $\alpha \in \operatorname{Inv}(\phi)$, the numerical ranges of matrices $A_1, \ldots, A_p \in \mathsf{H}^{n \times n}$ with respect to ϕ are contained in the same real subspace $\mathcal{V} \subseteq \mathsf{H}$, then the numerical ranges of all real linear combinations of $A_1, \ldots, A_p$ with respect to ϕ are also contained in $\mathcal{V}$.*

The proof is left to the reader as Ex. 3.11.10.

Proposition 3.7.1 yields the following information about the numerical ranges.

Proposition 3.7.3. *Let $A \in \mathsf{H}^{n \times n}$. Then:*

(1) $W_\phi^{(\alpha)}(A) = \{0\}$ *for some (equivalently, for all) $\alpha \in \operatorname{Inv}(\phi) \setminus \{0\}$ if and only if $A = 0$;*

(2) $W_\phi^{(\alpha)}(A) \subseteq \operatorname{Inv}(\phi)$ *for some (equivalently, for all) $\alpha \in \operatorname{Inv}(\phi) \setminus \{0\}$ if and only if $A = A_\phi$;*

(3) $W_\phi^{(\alpha)}(A) \subseteq \operatorname{Span}_{\mathsf{R}}\{q_1\}$ *for some (equivalently, for all) $\alpha \in \operatorname{Inv}(\phi) \setminus \{0\}$ if and only if $A = -A_\phi$.*

Proof. Note that (1) follows from (2) and (3).

We prove Part (2). The "if" statement is trivial. Conversely, assume $W_\phi^{(\alpha)}(A) \subseteq \operatorname{Inv}(\phi)$ for some $\alpha \in \operatorname{Inv}(\phi)$, $\alpha \neq 0$. Then (cf. (3.7.2)) $W_\phi^{(\alpha)}(A) \subseteq \operatorname{Inv}(\phi)$ for every nonzero $\alpha \in \operatorname{Inv}(\phi)$. In other words, $x_\phi A x \in \operatorname{Inv}(\phi)$ for all $x \in \mathsf{H}^{n \times 1}$ such that $x_\phi x \neq 0$. Since the set of all $x \in \mathsf{H}^{n \times 1}$ such that $x_\phi x \neq 0$ is dense in $\mathsf{H}^{n \times 1}$ (indeed, its complement

$$\{x = \operatorname{col}(x_j)_{j=1}^n \in \mathsf{H}^{n \times 1} \ : \ x_1, \ldots, x_n \in \mathsf{H} \text{ and } x_\phi x = 0\}$$

can be represented as common zeros of a set of polynomial equations in $4n$ real variables, the coefficients of the x_j's as real linear combinations of $1, q_1, q_2, q_3$), we obtain $x_\phi A x \in \text{Inv}(\phi)$ for all $x \in \mathsf{H}^{n \times 1}$. Now use Proposition 3.7.1.

Part (3) is proved analogously. $\square$

The case $\alpha = 0$ is excluded in Proposition 3.7.3. In fact, the result is generally not valid for $W_\phi^{(0)}(A)$, as the following example shows.

Example 3.7.4. Let $A = \begin{bmatrix} q_1 & 0 \\ 0 & -q_1 \end{bmatrix}$. We claim that $W_\phi^{(0)}(A) = \{0\}$. Indeed, for $x = \begin{bmatrix} b \\ c \end{bmatrix} \in \mathsf{H}^{2 \times 1}$, where $b, c \in \mathsf{H}$, the condition $x_\phi x = 0$ amounts to $b_\phi b + c_\phi c = 0$, which, in turn, implies $|b| = |c|$ (because ϕ is an isometry on H). On the other hand, by Theorem 2.5.1(a) (with $\beta(\phi) = q_1$) we have $b_\phi q_1 b = q_1 |b|^2$; hence,

$$x_\phi A x = b_\phi q_1 b - c_\phi q_1 c = q_1(|b|^2 - |c|^2),$$

which is equal to zero as long as $|b| = |c|$. $\square$

A suitably revised analogue of Proposition 3.7.3, however, is valid for $\alpha = 0$; recall that $\phi(q_1) = -q_1$, $q_1^2 = -1$.

Theorem 3.7.5. *Let $A \in \mathsf{H}^{n \times n}$, $n \geq 2$.*

(1) $W_\phi^{(0)}(A) = \{0\}$ *if and only if either $n = 2$ and A has the form*

$$A = \begin{bmatrix} a_0 + a_1 q_1 & a_2 + a_3 q_1 \\ -a_2 + a_3 q_1 & a_0 - a_1 q_1 \end{bmatrix} \tag{3.7.3}$$

for some $a_0, a_1, a_2, a_3 \in \mathsf{R}$, or $n \geq 3$ and $A = aI$ for some real a.

(2) $W_\phi^{(0)}(A) \subseteq \text{Inv}(\phi)$ *if and only if $n \geq 3$ and A is ϕ-hermitian or $n = 2$ and A has the form*

$$A = \begin{bmatrix} a_1 q_1 & a_2 + a_3 q_1 \\ -a_2 + a_3 q_1 & -a_1 q_1 \end{bmatrix} + B, \tag{3.7.4}$$

for some $a_1, a_2, a_3 \in \mathsf{R}$ and some ϕ-hermitian B.

(3) $W_\phi^{(0)}(A) \subseteq \text{Span}_{\mathsf{R}}\{q_1\}$ *if and only if A has the form $A = aI + B$, where $a \in \mathsf{R}$ and B is ϕ-skewhermitian.*

The proof of Theorem 3.7.5 is rather long and is, therefore, relegated to the next section.

Note that the numerical ranges and joint numerical ranges with respect to conjugation are connected (Ex. 3.11.13); everywhere in the book connectivity is understood in the sense of pathwise connectivity. It turns out that connectivity holds also for numerical ranges with respect to a nonstandard involution.

Theorem 3.7.6. *For all nonstandard involutions ϕ and all $\alpha \in \text{Inv}(\phi)$, the numerical range $W_\phi^{(\alpha)}(A)$ is connected for every $A \in \mathsf{H}^{n \times n}$.*

We need a lemma for the proof of Theorem 3.7.6.

Lemma 3.7.7. *The set $\mathcal{S}_n^{(\alpha)}$ is connected for every $\alpha \in \text{Inv}(\phi)$.*

Proof. The connectivity of $\mathcal{S}_n^{(0)}$ is obvious: the line segment $\{tx : 0 \leq t \leq 1\}$ connects any $x \in \mathcal{S}_n^{(0)}$ to zero within $\mathcal{S}_n^{(0)}$. The case $\alpha \neq 0$ is reduced to $\alpha = 1$ by virtue of (3.7.2). So, it suffices to prove that $\mathcal{S}_n^{(1)}$ is connected.

First, we verify that $\mathcal{S}_1^{(1)}$ is connected. Note that $\alpha = a + q_1 b + q_2 c + q_3 d$, where $a, b, c, d \in \mathsf{R}$, satisfies $\alpha_\phi \alpha = 1$ if and only if

$$a^2 + b^2 - c^2 - d^2 = 1, \qquad ac + bd = 0, \qquad ad - bc = 0,$$

which boils down to

$$c = d = 0, \qquad a^2 + b^2 = 1,$$

and this is a connected set.

More generally, we have: for every $\alpha \in \mathrm{Inv}\,(\phi)$, the set

$$\mathcal{S}(\alpha) := \{\gamma \in \mathsf{H} : \gamma_\phi \alpha \gamma = \alpha\} \tag{3.7.5}$$

is connected. Indeed, for $\alpha = 0$ this is trivial, and for $\alpha \neq 0$ write $\alpha = \lambda_\phi \lambda$ by Theorem 2.5.1(b) for some $\lambda \in \mathsf{H} \setminus \{0\}$ and note that $\gamma \in \mathcal{S}(\alpha)$ if and only if $\lambda \gamma \lambda^{-1} \in \mathcal{S}_1^{(1)}$.

Consider now the case when $n \geq 2$. First note that for every $x \in \mathcal{S}_n^{(1)}$, the set

$$x\mathsf{H} \cap \mathcal{S}_n^{(1)} = \{x\alpha : \alpha \in \mathcal{S}_1^{(1)}\}$$

is connected. Suppose now that $x, y \in \mathcal{S}_n^{(1)}$ are linearly independent. We construct a continuous curve γ in $\mathcal{S}_n^{(1)}$ that connects x and y. Let the curves

$$\gamma_0 : [0,1] \to \mathsf{H}^{n \times 1} \quad \text{and} \quad \lambda : [0,1] \to \mathsf{H}$$

be defined by

$$\begin{aligned} \gamma_0(t) &= (1-t)x + ty, \\ \lambda(t) &= \gamma_0(t)_\phi \gamma_0(t) = (1-t)^2 + t^2 + t(1-t)((x_\phi y) + (x_\phi y)_\phi) \in \mathrm{Inv}\,(\phi). \end{aligned}$$

Suppose $\Re(x_\phi y) \geq 0$. Then $\Re(\lambda(t)) > 0$ for all $t \in [0,1]$. Thus, $\sqrt{\lambda(t)}$ (see (2.5.3) for the definition) is well defined, is nonzero, and depends continuously on t (Corollary 2.5.3). We have

$$(\sqrt{\lambda(t)})^{-1} \in \mathrm{Span}_{\mathsf{R}} \{1, \lambda(t)^*\} \subseteq \mathrm{Inv}\,(\phi).$$

The curve $t \mapsto \gamma(t) := \gamma_0(t)(\sqrt{\lambda(t)})^{-1}$ is continuous and satisfies $\gamma(0) = x$, $\gamma(1) = y$, $\gamma(t)_\phi \gamma(t) = 1$ for all $t \in [0,1]$. Suppose $\Re(x_\phi y) < 0$. Then choose a continuous curve in $x\mathsf{H} \cap \mathcal{S}_n^{(1)}$ that connects x with $-x$ and apply the construction of γ for $-x$ and y instead of x and y.

Finally, if $x, y \in \mathcal{S}_n^{(1)}$ are linearly dependent, then $x, y \in x\mathsf{H} \cap \mathcal{S}_n^{(1)}$, and, therefore, x and y are connected within $x\mathsf{H} \cap \mathcal{S}_n^{(1)}$. $\qquad \square$

Since $x_\phi A x$ (for a fixed $A \in \mathsf{H}^{n \times n}$) is a continuous function of $x \in \mathsf{H}^{n \times 1}$, the connectivity of $W_\phi^{(\alpha)}(A)$ follows immediately from Lemma 3.7.7.

Since the sets $\mathcal{S}_n^{(\alpha)}$, $\alpha = 0$ or $\alpha = 1$, are unbounded (if $n > 1$) we cannot expect that the numerical ranges $W_\phi^{(\alpha)}(A)$ be bounded, for a general $A \in \mathsf{H}^{n \times n}$.

Open Problem 3.7.8. *Characterize those matrices $A \in \mathsf{H}^{n \times n}$ for which the numerical range $W_\phi^{(\alpha)}(A)$ is bounded.*

We have the answer for the case $\alpha = 0$.

Theorem 3.7.9. *The numerical range $W_\phi^{(0)}(A)$, where $A \in \mathsf{H}^{n \times n}$, $n \geq 2$, is bounded if and only if $n \geq 3$ and $A = aI$ for some real a, or $n = 2$ and A has the form* (3.7.3).

Proof. Since the vectors $x \in \mathsf{H}^{n \times 1}$ satisfying $x_\phi x = 0$ can be arbitrarily scaled without losing this property, it follows that $W_\phi^{(0)}(A)$ is bounded if and only if $W_\phi^{(0)}(A) = \{0\}$. In view of Theorem 3.7.5 we are done. $\square$

By analogy with the joint numerical ranges with respect to the conjugation, we define the joint numerical ranges with respect to a nonstandard involution ϕ.

Definition 3.7.10. Fix $\alpha \in \mathrm{Inv}\,(\phi)$, and for a p-tuple of ϕ-hermitian matrices $A_1, \ldots, A_p \in \mathsf{H}^{n \times n}$, let

$$WJ_\phi^{(\alpha)}(A_1, \ldots, A_p) := \{(x_\phi A_1 x, \ldots, x_\phi A_p x) \in (\mathrm{Inv}\,(\phi))^p : x_\phi x = \alpha, \quad x \in \mathsf{H}^{n \times 1}\},$$

be the *joint ϕ-numerical range* of $A_1, \ldots, A_p$.

Open Problem 3.7.11. *Study geometric properties of joint ϕ-numerical ranges versus algebraic properties of the constituent matrices.*

For ϕ-skewhermitian matrices we consider still another version of joint numerical ranges and a result on their convexity.

We start with a result on convexity of numerical ranges.

Definition 3.7.12. Let there be given a p-tuple of ϕ-skewhermitian $n \times n$ quaternion matrices $(A_1, \ldots, A_p)$; thus, $A_j = -(A_j)_\phi$, $j = 1, 2, \ldots, p$. Define the *joint ϕ-numerical range*

$$WJ_\phi(A_1, \ldots, A_p) := \{(x_\phi A_1 x, x_\phi A_2 x, \ldots, x_\phi A_p x) : x \in \mathsf{H}^{n \times 1},$$
$$\|x\| = 1\} \subseteq \mathsf{H}^p.$$

Since $\phi(x_\phi A x) = -x_\phi A x$, we clearly have that

$$WJ_\phi(A_1, \ldots, A_p) \subseteq \{(y_1 \beta(\phi), \ldots, y_p \beta(\phi)) : y_1, y_2, \ldots, y_p \in \mathsf{R}\}. \tag{3.7.6}$$

Theorem 3.7.13. (1) *If $n \neq 2$, then the joint ϕ-numerical range*

$$WJ_\phi(A_1, A_2, A_3, A_4, A_5)$$

is convex for every 5-tuple of ϕ-skewhermitian matrices $A_1, \ldots, A_5$.

(2) *If $n = 2$, then the joint ϕ-numerical range $WJ_\phi(A_1, A_2, A_3, A_4)$ is convex for every 4-tuple of ϕ-skewhermitian matrices $A_1, \ldots, A_4$.*

(3) *If $n = 2$, then the joint ϕ-numerical range $WJ_\phi(A_1, A_2, A_3, A_4, A_5)$ is convex for a 5-tuple of ϕ-skewhermitian matrices $A_1, \ldots, A_5$ if and only if the 6-tuple of ϕ-skewhermitian matrices $(A_1, \ldots, A_5, \beta(\phi)I)$ is linearly dependent over the reals.*

Proof. It follows from (2.4.8) that

$$x_\phi A x = \beta(\phi)^{-1} x^* (\beta \phi A) x \quad \text{for all } x \in \mathsf{H}^{n \times 1}, \quad A \in \mathsf{H}^{n \times n}.$$

Therefore, A is ϕ-skewhermitian if and only if $\beta(\phi)A$ is hermitian (see Proposition 3.5.1(2)). Now the parts (1) and (3) follow by applying Theorem 3.5.7 and Corollary 3.5.8 to the matrices $\beta(\phi)A_j$ (cf. formula (3.7.6)). Part (2) follows from part (3) by taking $A_5 = 0$. $\qquad\square$

Another proof of Theorem 3.7.13 (in the case $p = 2$) can be obtained by reduction to the real joint numerical range, as follows. Write

$$\begin{aligned} x &= x_1 + x_2 \mathsf{i} + x_3 \mathsf{j} + x_4 \mathsf{k} \in \mathsf{H}^{n \times 1}, \quad x_1, x_2, x_3, x_4 \in \mathsf{R}^{n \times 1}, \\ A &= A_1 + A_2 \mathsf{i} + A_3 \mathsf{j} + A_4 \mathsf{k} \in \mathsf{H}^{n \times n}, \quad A_1, A_2, A_3, A_4 \in \mathsf{R}^{n \times n}. \end{aligned}$$

Then we have $x_\phi A x = a\beta(\phi)$, where the real number a is represented as a bilinear form of

$$\widetilde{x} := [x_1 \quad x_2 \quad x_3 \quad x_4]^T \in \mathsf{R}^{4n \times 1}.$$

In other words, there exists a real symmetric $4n \times 4n$ matrix $\widetilde{A}$ such that

$$a = \widetilde{x}^T \widetilde{A} \widetilde{x}, \quad \widetilde{x} \in \mathsf{R}^{4n \times 1}.$$

Note also that $\|x\| = \|\widetilde{x}\|$. Now, clearly,

$$WJ_\phi(A, B) = WJ_*^{\mathsf{R}}(\widetilde{A}, \widetilde{B})\beta(\phi),$$

where

$$WJ_*^{\mathsf{R}}(X, Y) := \{(x^T X x, x^T Y x) : x \in \mathsf{R}^{4n \times 1}, \ \|x\| = 1\}$$

is the *real joint numerical range* of the pair (X, Y) of two real symmetric $4n \times 4n$ matrices X and Y. By Theorem 3.5.7 the result follows. $\qquad\square$

3.8 PROOF OF THEOREM 3.7.5

We start with preliminary results. As in the preceding section, ϕ is a nonstandard involution, and (q_1, q_2, q_3) is a units triple such that $\phi(q_1) = -q_1$, $\phi(q_2) = q_2$, $\phi(q_3) = q_3$.

Lemma 3.8.1. *Let $A = -A_\phi \in \mathsf{H}^{2 \times 2}$ be such that*

$$W_\phi^{(0)}(A) = \{0\}. \tag{3.8.1}$$

Then A has the form

$$A = \begin{bmatrix} a_1 q_1 & a_2 + a_3 q_1 \\ -a_2 + a_3 q_1 & -a_1 q_1 \end{bmatrix} \tag{3.8.2}$$

for some $a_1, a_2, a_3 \in \mathsf{R}$.

Conversely, if A has the form (3.8.2), then (3.8.1) holds.

Proof. Write

$$A = \begin{bmatrix} a_1q_1 & a_2 + a_3q_1 + a_4q_2 + a_5q_3 \\ -a_2 + a_3q_1 - a_4q_2 - a_5q_3 & a_6q_1 \end{bmatrix}$$

where $a_1, a_2, a_3, a_4, a_5, a_6 \in \mathsf{R}$. A computation shows that for $b = b_1 + b_2q_1 + b_3q_2 + b_4q_3 \in \mathsf{H}$, where $b_1, b_2, b_3, b_4 \in \mathsf{R}$, we have

$$\begin{bmatrix} 1 & b_\phi \end{bmatrix} A \begin{bmatrix} 1 \\ b \end{bmatrix}$$

$$= q_1(a_1 + a_6(b_1^2 + b_2^2 + b_3^2 + b_4^2) + 2b_1a_3 + 2b_2a_2 - 2b_3a_5 + 2b_4a_4). \tag{3.8.3}$$

On the other hand, the condition $\begin{bmatrix} 1 & b_\phi \end{bmatrix} \cdot \begin{bmatrix} 1 \\ b \end{bmatrix} = 0$ boils down to the system of equalities

$$b_1^2 + b_2^2 - b_3^2 - b_4^2 = -1, \quad b_1b_3 + b_2b_4 = 0, \quad b_1b_4 - b_2b_3 = 0. \tag{3.8.4}$$

It is easy to see (3.8.4) is equivalent to the system

$$b_1 = b_2 = 0, \qquad b_3^2 + b_4^2 = 1. \tag{3.8.5}$$

Indeed, if at least one of b_1 and b_2 is nonzero, then the second and the third equations in (3.8.4) yield only the trivial solution $b_3 = b_4 = 0$, which gives a contradiction when substituted in the first equation in (3.8.4). Now the hypothesis (3.8.1) implies

$$b_3^2 + b_4^2 = 1 \implies a_1 + a_6 - 2b_3a_5 + 2b_4a_4 = 0. \tag{3.8.6}$$

It is easy to see that (3.8.6) is possible only if $a_4 = a_5 = 0$ and $a_1 + a_6 = 0$.

For the converse statement, we have to prove, in view of Proposition 3.7.2(b), only that

$$W_\phi^{(0)}(A_1) = W_\phi^{(0)}(A_2) = W_\phi^{(0)}(A_3) = \{0\} \tag{3.8.7}$$

for

$$A_1 = \begin{bmatrix} q_1 & 0 \\ 0 & -q_1 \end{bmatrix}, \quad A_2 = \begin{bmatrix} 0 & 1 \\ -1 & 0 \end{bmatrix}, \quad A_3 = \begin{bmatrix} 0 & q_1 \\ q_1 & 0 \end{bmatrix}.$$

For A_1 see Example 3.7.4. For A_2 note that for $x := \begin{bmatrix} c \\ b \end{bmatrix} \in \mathsf{H}^{2\times 1}$ such that $x_\phi x = 0$ and $c \neq 0$, the equalities (3.8.5) are valid, where b_1, b_2, b_3, b_4 are the real coefficients of the expansion of bc^{-1} as a linear combination of $1, q_1, q_2, q_3$, whereas

$$\begin{bmatrix} c_\phi & b_\phi \end{bmatrix} A \begin{bmatrix} c \\ b \end{bmatrix} = c_\phi \begin{bmatrix} 1 & (c_\phi)^{-1}b_\phi \end{bmatrix} A \begin{bmatrix} 1 \\ bc^{-1} \end{bmatrix} c = c_\phi(2q_1b_2)c = 0,$$

as $b_2 = 0$. If $x_\phi x = 0$ and $c = 0$, then $x = 0$ and $x_\phi Ax = 0$ trivially holds. This proves (3.8.7) for A_2. An analogous proof works for A_3 (cf. Ex. 3.11.16). $\qquad\square$

Lemma 3.8.2. *Let $A = -A_\phi \in \mathsf{H}^{n\times n}$, $n \geq 3$, be such that (3.8.1) holds true. Then*

$$A \in \mathsf{R}^{n\times n} + q_1\mathsf{R}^{n\times n} \tag{3.8.8}$$

and A has zero diagonal.

Proof. We apply Lemma 3.8.1 to the 2×2 principal submatrices of A. Then (3.8.8) follows from (3.8.2). Also, if $a_{i,i}$ and $a_{j,j}$ are the ith and the jth ($i \neq j$) diagonal entries of A, then $a_{i,i} + a_{j,j} = 0$. Since $n \geq 3$, this is possible only if all diagonal entries of A are zeros. $\qquad \square$

Lemma 3.8.3. *If $A = -A_\phi \in \mathsf{H}^{n \times n}$, where $n \geq 3$, and if (3.8.1) holds true, then $A = 0$.*

Proof. We identify $\mathsf{R} + q_1 \mathsf{R}$ with the field of complex numbers, with q_1 playing the role of the imaginary unit and ϕ acting as complex conjugation. By Lemma 3.8.2, $A \in \mathsf{C}^{n \times n}$ is skewhermitian under this identification. It is well known that (complex) hermitian matrices—and, hence, also (complex) skewhermitian matrices—can be diagonalized by (complex) unitary similarity; in other words, there exists a ϕ-unitary $U \in \mathsf{R} + q_1 \mathsf{R}$ such that $B := U_\phi A U$ is diagonal. Obviously, B satisfies the hypotheses of Lemma 3.8.2 (cf. Proposition 3.7.2), so B must have zero diagonal; hence, $B = 0$ and $A = 0$. $\qquad \square$

Lemma 3.8.4. *If $A = A_\phi \in \mathsf{H}^{2 \times 2}$ is such that $W_\phi^{(0)}(A) = \{0\}$, then $A = aI$ for some real a.*

Proof. Write

$$A = \left[\begin{array}{cc} a_1 + a_2 q_2 + a_3 q_3 & a_4 + a_5 q_1 + a_6 q_2 + a_7 q_3 \\ a_4 - a_5 q_1 + a_6 q_2 + a_7 q_3 & a_8 + a_9 q_2 + a_{10} q_3 \end{array} \right],$$

$$a_1, a_2, \ldots, a_{10} \in \mathsf{R}.$$

We compute $x_\phi A x$, where $x = \left[\begin{array}{c} 1 \\ b_3 q_2 + b_4 q_3 \end{array} \right]$, and b_3, b_4 are real numbers such that $b_3^2 + b_4^2 = 1$ but otherwise arbitrary. By (3.8.5), we have $x_\phi x = 0$. Now

$$\begin{aligned} x_\phi A x & = \ [1 \ \ b_3 q_2 + b_4 q_3] \\ & \quad \cdot \left[\begin{array}{cc} a_1 + a_2 q_2 + a_3 q_3 & a_4 + a_5 q_1 + a_6 q_2 + a_7 q_3 \\ a_4 - a_5 q_1 + a_6 q_2 + a_7 q_3 & a_8 + a_9 q_2 + a_{10} q_3 \end{array} \right] \\ & \quad \cdot \left[\begin{array}{c} 1 \\ b_3 q_2 + b_4 q_3 \end{array} \right], \end{aligned}$$

which, after simple straightforward algebra, turns out to be equal to

$$\begin{aligned} a_1 - 2 a_6 b_3 - 2 a_7 b_4 - a_8 \ & + \ \ q_2 (a_2 + 2 a_4 b_3 - 2 a_5 b_4 + a_9 (b_4^2 - b_3^2) - 2 a_{10} b_3 b_4) \\ & + \ \ q_3 (a_3 - 2 a_4 b_4 + 2 a_5 b_3 + a_{10} (b_3^2 - b_4^2) - 2 a_9 b_3 b_4). \end{aligned}$$

Thus, the hypothesis that $W_\phi^{(0)}(A) = \{0\}$ amounts to equalities

$$a_1 - 2 a_6 b_3 - 2 a_7 b_4 - a_8 \ = \ 0; \qquad (3.8.9)$$
$$a_2 + 2 a_4 b_3 - 2 a_5 b_4 + a_9 (b_4^2 - b_3^2) - 2 a_{10} b_3 b_4 \ = \ 0; \qquad (3.8.10)$$
$$a_3 - 2 a_4 b_4 + 2 a_5 b_3 + a_{10} (b_3^2 - b_4^2) - 2 a_9 b_3 b_4 \ = \ 0 \qquad (3.8.11)$$

for every pair $b_3, b_4 \in \mathsf{R}$ such that $b_3^2 + b_4^2 = 1$. It is easy to see that (3.8.9) holds for every such b_3, b_4 if and only if $a_6 = a_7 = 0$, $a_1 = a_8$. It will be convenient to represent $b_3 = \sin\psi$, $b_4 = \cos\psi$, where $\psi \in \mathsf{R}$. Then (3.8.10) and (3.8.11) take the form

$$a_2 + 2a_4 \sin\psi - 2a_5 \cos\psi + a_9 \cos(2\psi) - a_{10}\sin(2\psi) = 0 \qquad (3.8.12)$$

and

$$a_3 - 2a_4 \cos\psi + 2a_5 \sin\psi - a_{10}\cos(2\psi) - a_9\sin(2\psi) = 0, \qquad (3.8.13)$$

respectively. Denoting the left-hand side of (3.8.12) by $f(\psi)$, a function of ψ, we see that

$$\limsup_{k\to\infty} |f^{(k)}(0)| = \infty$$

(unless both a_9 and a_{10} are equal to zero); on the other hand, $f(\psi)$ is identically zero by virtue of (3.8.12); therefore, so are all the derivatives of $f(\psi)$. So, we must have $a_9 = a_{10} = 0$. Now we obtain, from (3.8.12) and (3.8.13), that a_2, a_3, a_4, a_5 are also all zeros. $\qquad \square$

Lemma 3.8.5. *If $A = A_\phi \in \mathsf{H}^{n\times n}$, $n \geq 2$, is such that $W_\phi^{(0)}(A) = \{0\}$, then $A = aI$ for some real a.*

Proof. Apply Lemma 3.8.4 to 2×2 principal submatrices of A. $\qquad \square$

Proof of Theorem 3.7.5. As in Proposition 3.7.3, (1) is a consequence of (2) and (3).

Proof of Part (2), the "if" part: The case $n \geq 3$ is obvious, and the case $n = 2$ follows from Lemma 3.8.1 (cf. Proposition 3.7.2(b)).

Proof of Part (2), the "only if" part. Assume

$$W_\phi^{(0)}(A) \subseteq \mathrm{Inv}\,(\phi) \qquad (3.8.14)$$

Write

$$A = \frac{1}{2}(A + A_\phi) + \frac{1}{2}(A - A_\phi), \qquad (3.8.15)$$

and note that

$$W_\phi^{(0)}\left(\frac{1}{2}(A - A_\phi)\right) \subseteq W_\phi^{(0)}(A) - W_\phi^{(0)}\left(\frac{1}{2}(A + A_\phi)\right). \qquad (3.8.16)$$

Since $\frac{1}{2}(A+A_\phi)$ is ϕ-hermitian, the numerical range $W_\phi^{(0)}(\frac{1}{2}(A+A_\phi))$ is contained in $\mathrm{Inv}\,(\phi)$. In view of (3.8.14), the right-hand side of (3.8.16) is contained in $\mathrm{Inv}\,(\phi)$. But $\frac{1}{2}(A - A_\phi)$ is ϕ-skewhermitian; therefore, $W_\phi^{(0)}(\frac{1}{2}(A - A_\phi))$ is contained in $\mathrm{Span}_\mathsf{R}\,\{q_1\}$. As

$$\mathrm{Inv}\,(\phi) \,\cap\, \mathrm{Span}_\mathsf{R}\,\{q_1\} = \{0\},$$

we obtain that $W_\phi^{(0)}(\frac{1}{2}(A - A_\phi)) = \{0\}$. If $n = 2$, we are done by Lemma 3.8.1 (with $\frac{1}{2}(A - A_\phi)$ playing the role of A), and if $n \geq 3$, then $\frac{1}{2}(A - A_\phi) = 0$ by Lemma 3.8.3, and we are done again.

Proof of Part (3). The "if" part being obvious (cf. Proposition 3.7.2(b)), we focus on the "only if" part. Assume $W_\phi^{(0)}(A) \subseteq \mathrm{Span}_\mathsf{R}\{q_1\}$. Write A in the form (3.8.15); then

$$W_\phi^{(0)}\left(\frac{1}{2}(A + A_\phi)\right) \subseteq W_\phi^{(0)}(A) - W_\phi^{(0)}\left(\frac{1}{2}(A - A_\phi)\right). \qquad (3.8.17)$$

It follows that

$$W_\phi^{(0)}\left(\frac{1}{2}(A + A_\phi)\right) \subseteq \mathrm{Span}_{\mathsf{R}}\{q_1\}.$$

Since $\frac{1}{2}(A + A_\phi)$ is skewhermitian, we obtain $W_\phi^{(0)}(\frac{1}{2}(A + A_\phi)) = \{0\}$. Now $\frac{1}{2}(A + A_\phi) = aI$ for some real a by Lemma 3.8.5, and so A has the form as claimed in (3). $\qquad\square$

3.9   THE METRIC SPACE OF SUBSPACES

Let $\mathcal{M}, \mathcal{N}$ be (quaternion) subspaces in $\mathsf{H}^{n\times 1}$. A standard measure of the closeness of these two subspaces is the *gap*.

Definition 3.9.1. The gap between $\mathcal{M}$ and $\mathcal{N}$ is defined by

$$\theta(\mathcal{M}, \mathcal{N}) := \|P_{\mathcal{M}} - P_{\mathcal{N}}\|,$$

where we denote by $P_{\mathcal{X}} \in \mathsf{H}^{n\times n}$ the orthogonal projection on the subspace $\mathcal{X} \subseteq \mathsf{H}^{n\times 1}$.

Note that $P_{\mathcal{X}}$ is characterized by the properties that $\mathrm{Ran}\, P_{\mathcal{X}} = \mathcal{X}$, $P_{\mathcal{X}} x = x$ for every $x \in \mathcal{X}$, and $\mathrm{Ker}\, P_{\mathcal{X}} = \mathcal{X}^\perp$, the orthogonal complement to $\mathcal{X}$. Also, $P_{\mathcal{X}}^2 = P_{\mathcal{X}} = P_{\mathcal{X}}^*$ and $\|P_{\mathcal{X}}\| = 1$ (unless $\mathcal{X} = \{0\}$, in which case $P_{\mathcal{X}} = 0$). If $\{v_1, \ldots, v_p\}$ is an orthonormal basis for X, then

$$P_{\mathcal{X}} = [u_1 \ u_2 \ \ldots \ u_p][u_1 \ u_2 \ \ldots \ u_p]^*. \qquad (3.9.1)$$

Denote by Grass_n the set of all (quaternion) subspaces in $\mathsf{H}^{n\times 1}$ (the Grassmannian). The gap function is easily seen to be a *metric* on Grass_n; i.e., for all $\mathcal{M}, \mathcal{N}, \mathcal{Q} \in \mathrm{Grass}_n$ we have:

(1) $\theta(\mathcal{M}, \mathcal{N}) = \theta(\mathcal{N}, \mathcal{M})$;

(2) $\theta(\mathcal{M}, \mathcal{N}) \geq 0$, and $\theta(\mathcal{M}, \mathcal{N}) = 0$ if and only if $\mathcal{M} = \mathcal{N}$;

(3) triangle inequality: $\theta(\mathcal{M}, \mathcal{N}) \leq \theta(\mathcal{M}, \mathcal{Q}) + \theta(\mathcal{Q}, \mathcal{N})$.

The convergence in Grass_n is naturally defined in terms of the gap: a sequence $\{\mathcal{X}_m\}_{m=1}^\infty$, $\mathcal{X}_m \in \mathrm{Grass}_n$, is said to converge to $\mathcal{Y} \in \mathrm{Grass}_n$ if

$$\lim_{m \to \infty} \theta(\mathcal{X}_m, \mathcal{Y}) = 0.$$

This notion of convergence can be expressed in terms of vectors that belong to the subspaces in question (Theorem 3.9.4 below) and turns Grass_n into a compact metric space; compactness is understood in the sense that every sequence contains a subsequence that converges to an element of Grass_n (cf. Theorem 3.10.2 and Remark 3.10.3). A basic theory of metric spaces is found in many textbooks on real analysis; see, e.g., Wade [154] or Bruckner et al. [21].

Theorem 3.9.2. *Every sequence* $\{\mathcal{X}_m\}_{m=1}^\infty$, *where* $\mathcal{X}_m \in \mathrm{Grass}_n$, *contains a subsequence that converges to an element of* Grass_n.

Proof. It will suffice to prove that, for every fixed integer k, $1 \le k \le n-1$, the set $\mathrm{Grass}_{n,k}$ of all k-dimensional subspaces of $\mathsf{H}^{n \times 1}$ is compact, with respect to the gap metric.

To this end consider the set $\mathcal{O}_{n,k}$ of all (ordered) orthonormal k-tuples in $\mathsf{H}^{n \times 1}$:

$$\mathcal{O}_{n,k} \quad := \quad \{(u_1, \ldots, u_k) \, : \, \langle u_j, u_i \rangle = 1 \text{ if } i = j;$$
$$\langle u_j, u_i \rangle = 0 \text{ if } i \ne j; \;\; u_1, \ldots, u_k \in \mathsf{H}^{n \times 1}\}.$$

Let

$$\mathcal{S}_{n,k} := \{(x_1, \ldots, x_k) \, : \, \|x_1\|_{\mathsf{H}}^2 + \cdots + \|x_k\|_{\mathsf{H}}^2 = 1, \quad x_1, \ldots, x_k \in \mathsf{H}^{n \times 1}\}.$$

Then $\mathcal{S}_{n,k}$ can be identified (as a normed space) with the unit sphere in $\mathsf{R}^{4nk \times 1}$; therefore, $\mathcal{S}_{n,k}$ is compact. It is easy to see that $\mathcal{O}_{n,k}$ is a closed subset of $\mathcal{S}_{n,k}$, and consequently $\mathcal{O}_{n,k}$ is compact as well (Theorem 3.10.2).

Define a map $\Omega_{n,k} : \mathcal{O}_{n,k} \to \mathrm{Grass}_{n,k}$ by

$$\Omega_{n,k}((u_1, \ldots, u_k)) = \mathrm{Span}_{\mathsf{H}} \{u_1, \ldots, u_k\}, \quad (u_1, \ldots, u_k) \in \mathcal{O}_{n,k}.$$

It turns out that $\Omega_{n,k}$ is continuous; moreover, it has the Lipschitz property (see (3.9.6) below) with respect to the gap metric in $\mathrm{Grass}_{n,k}$ and with respect to the standard metric $\mho$ induced by the norm in $\mathsf{R}^{4nk \times 1}$:

$$\mho((u_1, \ldots, u_k), (v_1, \ldots, v_k)) \quad := \quad \sqrt{\sum_{j=1}^{k} \|u_j - v_j\|^2},$$
$$(u_1, \ldots, u_k), (v_1, \ldots, v_k) \in \mathcal{O}_{n,k}.$$

For the reader's convenience, we present a proof of this fact (adapted from Gohberg et al. [54, Section 13.4, 55, Section S.4]).

We need some preparation for the proof. Let $\mathcal{L} \in \mathrm{Grass}_{n,k}$, and let $v := (v_1, \ldots, v_k)$ be an orthonormal basis for $\mathcal{L}$. Pick some $u = (u_1, \ldots, u_k) \in \mathcal{O}_{n,k}$, and let $\mathcal{M} := \Omega_{n,k}(u)$. Then

$$\|(P_\mathcal{M} - P_\mathcal{L})v_i\| \;=\; \|P_\mathcal{M} v_i - v_i\| \;\le\; \|P_\mathcal{M}(v_i - u_i)\| + \|u_i - v_i\|$$
$$\le \;\; \|P_\mathcal{M}\| \, \|v_i - u_i\| + \|u_i - v_i\| \;\le\; 2\mho(u, v)$$

for $i = 1, 2, \ldots, k$. Thus, for $x = \sum_{i=1}^{k} v_i \alpha_i \in \mathcal{L}$, $\alpha_i \in \mathsf{H}$, we have

$$\|(P_\mathcal{M} - P_\mathcal{L})x\| \le 2 \sum_{j=1}^{k} |\alpha_j| \mho(u, v). \tag{3.9.2}$$

Assuming $\|x\| = 1$, it follows that $\sum_{j=1}^{k} |\alpha_j|^2 = \|x\|^2 = 1$; so, $|\alpha_j| \le 1$ for $j = 1, 2, \ldots, k$, and (3.9.2) gives

$$\|(P_\mathcal{M} - P_\mathcal{L})|_\mathcal{L}\| \le 2k\mho(u, v). \tag{3.9.3}$$

Now fix some $y \in \mathcal{L}^\perp$, $\|y\| = 1$. We wish to evaluate $P_\mathcal{M} y$. For every $x \in \mathcal{L}$, write

$$\langle x, P_\mathcal{M} y \rangle = \langle (P_\mathcal{M} - P_\mathcal{L})x, y \rangle + \langle x, y \rangle = \langle (P_\mathcal{M} - P_\mathcal{L})x, y \rangle;$$

hence,

$$|\langle x, P_{\mathcal{M}}y \rangle| \leq 2k\|x\|\mho(u,v) \tag{3.9.4}$$

by (3.9.3). On the other hand, if we write

$$P_{\mathcal{M}}y = u_1\alpha_1 + \cdots + u_k\alpha_k, \qquad \alpha_1, \ldots, \alpha_k \in \mathsf{H},$$

then for every $z \in \mathcal{L}^\perp$:

$$\langle z, P_{\mathcal{M}}y \rangle = \langle z, \sum_{i=1}^k (u_i - v_i)\alpha_i \rangle + \langle z, \sum_{i=1}^k v_i\alpha_i \rangle = \langle z, \sum_{i=1}^k (u_i - v_i)\alpha_i \rangle$$

and

$$|\langle z, P_{\mathcal{M}}y \rangle| \leq \|z\| \, \|\sum_{i=1}^k (u_i - v_i)\alpha_i\| \leq \|z\| \, k \left(\max_{1 \leq i \leq k} (|\alpha_i| \, \|u_i - v_i\|) \right).$$

But $\|y\| = 1$ implies $\sum_{i=1}^k |\alpha_i|^2 \leq 1$, so $\max_{1 \leq i \leq k}\{|\alpha_1|, \ldots, |\alpha_k|\} \leq 1$. Hence,

$$|\langle z, P_{\mathcal{M}}y \rangle| \leq \|z\| \, k \, \mho(u,v),$$

and combining this inequality with (3.9.4), it is found that $|\langle t, P_{\mathcal{M}}y \rangle| \leq 3 \, k \, \mho(u,v)$ for every $t \in \mathsf{H}^{n \times 1}$ with $\|t\| = 1$. Thus,

$$\|P_{\mathcal{M}}y\| \leq 3 \, k \, \mho(u,v). \tag{3.9.5}$$

Now continuity of the map $\Omega_{n,k}$ can be easily proven. Indeed, pick an $x \in \mathsf{H}^{n \times 1}$ with $\|x\| = 1$. Then, using (3.9.3) and (3.9.5), we have

$$\|(P_{\mathcal{M}} - P_{\mathcal{L}})x\| \leq \|(P_{\mathcal{M}} - P_{\mathcal{L}})P_{\mathcal{L}}x\| + \|P_{\mathcal{M}}(x - P_{\mathcal{L}}x)\| \leq 5k\mho(u,v),$$

so

$$\theta(\mathcal{M}, \mathcal{L}) = \|P_{\mathcal{M}} - P_{\mathcal{L}}\| \leq 5k\mho(u,v), \tag{3.9.6}$$

and the continuity of $\Omega_{n,k}$ follows.

Since the continuous map $\Omega_{n,k}$ is onto and $\mathcal{O}_{n,k}$ is compact, we obtain that $\mathrm{Grass}_{n,k}$ is compact as well (Theorem 3.10.5). $\qquad \square$

It follows from Theorem 3.9.2 that Grass_n, as well as each set $\mathrm{Grass}_{n,k}$, for $k = 1, 2, \ldots, n-1$, are complete, i.e., every Cauchy sequence of elements of the set converges within the set.

The following property connecting the gap between subspaces and their dimensions is very useful.

Theorem 3.9.3. *If* $\mathcal{X}, \mathcal{Y} \in \mathrm{Grass}_n$ *and* $\theta(\mathcal{X}, \mathcal{Y}) < 1$, *then* $\dim \mathcal{X} = \dim \mathcal{Y}$.

Proof. We have

$$(I - (P_{\mathcal{Y}} - P_{\mathcal{X}}))P_{\mathcal{Y}} = P_{\mathcal{X}}(I - (P_{\mathcal{X}} - P_{\mathcal{Y}})). \tag{3.9.7}$$

Since $\|P_{\mathcal{Y}} - P_{\mathcal{X}}\| < 1$, the matrices $(I - (P_{\mathcal{Y}} - P_{\mathcal{X}}))$ and $(I - (P_{\mathcal{X}} - P_{\mathcal{Y}}))$ are invertible, with the inverses given by the converging series $\sum_{j=0}^\infty (P_{\mathcal{Y}} - P_{\mathcal{X}})^j$ and $\sum_{j=0}^\infty (P_{\mathcal{X}} - P_{\mathcal{Y}})^j$, respectively, and hence, (3.9.7) gives $\mathrm{rank}\, P_{\mathcal{Y}} = \mathrm{rank}\, P_{\mathcal{X}}$. It remains to observe that $\mathrm{rank}\, P_{\mathcal{Y}} = \dim \mathcal{Y}$, and similarly for $\mathcal{X}$. $\qquad \square$

Theorem 3.9.4. *Assume* $\lim_{m \to \infty} \mathcal{X}_m = \mathcal{Y}$, *where* $\mathcal{X}_m, \mathcal{Y} \in \mathrm{Grass}_n$. *Then* $\mathcal{Y}$ *consists of exactly those vectors* $y \in \mathsf{H}^{n \times 1}$ *for which there exists a sequence* $\{x_m\}_{m=1}^{\infty}$ *such that* $x_m \in \mathcal{X}_m$ *for* $m = 1, 2, \ldots$ *and* $\lim_{m \to \infty} x_m = y$.

Proof. We ignore the trivial case $\mathcal{Y} = \{0\}$.

Suppose $y \in \mathcal{Y}$. Let $x_m := P_{\mathcal{X}_m} y$. Then, clearly, $x_m \in \mathcal{X}_m$, and

$$\|x_m - y\| = \|P_{\mathcal{X}_m} y - P_{\mathcal{Y}} y\| \leq \|P_{\mathcal{X}_m} - P_{\mathcal{Y}}\| \cdot \|y\| \to 0$$

as $m \to \infty$.

Conversely, suppose $y \in \mathsf{H}^{n \times 1}$ is such that $x_m \in \mathcal{X}_m$ for $m = 1, 2, \ldots$ and $\lim_{m \to \infty} x_m = y$ for some sequence $\{x_m\}_{m=1}^{\infty}$. Then

$$
\begin{aligned}
\|P_{\mathcal{Y}} y - y\| &\leq \|P_{\mathcal{Y}} y - P_{\mathcal{X}_m} y\| + \|P_{\mathcal{X}_m} y - P_{\mathcal{X}_m} x_m\| + \|x_m - y\| \\
&\leq \theta(\mathcal{Y}, \mathcal{X}_m) \|y\| + \|P_{\mathcal{X}_m}\| \|y - x_m\| + \|x_m - y\| \\
&\leq \theta(\mathcal{Y}, \mathcal{X}_m) \|y\| + 2\|x_m - y\|,
\end{aligned}
\tag{3.9.8}
$$

where the equality $\|P_{\mathcal{X}_m}\| = 1$ was used. The right-hand side of (3.9.8) tends to zero as $m \to \infty$. Thus, $P_{\mathcal{Y}} y = y$, and $y \in \mathcal{Y}$, as required. $\qquad \square$

3.10 APPENDIX: MULTIVARIABLE REAL ANALYSIS

For the reader's convenience we collect here a few basic facts on analysis of sets and continuous multivariable functions that are used in the main text. All this material, and much more, can be found in many textbooks on real analysis; see, e.g., Wade [154].

Definition 3.10.1. A set $E \subseteq \mathsf{R}^n$ is called *open* if for every $x \in E$ there exists $\delta > 0$ such that the ball $\{y \in \mathsf{R}^n : \|y - x\|_{\mathsf{R}} < \delta\}$ is contained in E. The complements (in R^m) of open sets are *closed* sets. An *open cover* of a set $E \subseteq \mathsf{R}^n$ is, by definition, a collection $\{V_i\}_{i \in K}$ of open sets $V_i \subseteq \mathsf{R}^n$ indexed by the index set K such that $E \subseteq \cup_{i \in K} V_i$. A set $E \subseteq \mathsf{R}^n$ is said to be *compact* if every open cover of E contains a finite subcover: if $E \subseteq \cup_{i \in K} V_i$, V_i are open, then there exists a finite subset $K_0 \subseteq K$ such that $E \subseteq \cup_{i \in K_0} V_i$.

Theorem 3.10.2. *The following statements are equivalent for a set* $E \subseteq \mathsf{R}^n$:

(a) E *is compact*;

(b) E *is closed and bounded*;

(c) E *is sequentially compact, i.e., every sequence* $\{x_m\}_{m=1}^{\infty}$, *where* $x_m \in E$ *for* $m = 1, 2, \ldots$, *contains a subsequence that converges to an element of* E.

This result is known as the Heine-Borel theorem.

Remark 3.10.3. Although the Heine-Borel theorem generally fails for subsets of metric spaces (in general, only the implications (a) $\Rightarrow$ (c) $\Rightarrow$ (b) are valid), it does hold true for subsets of the metric space Grass_n with the gap metric. Indeed, Grass_n can be identified with the closed set of all orthogonal projections in $\mathsf{H}^{n \times n}$, with the norm $\| \cdot \|$. Furthermore, $\mathsf{H}^{n \times n}$ with the norm $\| \cdot \|$ can be identified (by considering the four real components of every entry in an $n \times n$ quaternion matrix) with $\mathsf{R}^{4n^2 \times 1}$ but with a norm $\| \cdot \|_d$ which is different from $\| \cdot \|_{\mathsf{R}}$. However, any two norms in $\mathsf{R}^{4n^2 \times 1}$ are equivalent, a well-known fact proved in many linear algebra textbooks (see, e.g., Lancaster and Tismenetsky [94, Section 10.1] or Horn and Johnson [62, Section 5.4.4]); thus, there exist positive constants m and M such that

$$m\|x\|_d \leq \|x\|_\mathsf{R} \leq M\|x\|_d, \quad \text{for all} \ \ x \in \mathsf{R}^{4n^2 \times 1}.$$

It is now not difficult to see that Theorem 3.10.2 holds for $\mathsf{R}^{4n^2 \times 1}$ with the norm $\| \cdot \|_d$ and, hence, also for Grass_n.

Definition 3.10.4. A function $f : E \to \mathsf{R}^m$, where E is a subset of R^n is said to be *continuous* if for every given $x \in E$ and $\varepsilon > 0$, there exists $\delta > 0$ such that $\|f(y) - f(x)\|_\mathsf{R} < \varepsilon$ as soon as $y \in E$ and $\|y - x\|_\mathsf{R} < \delta$.

Equivalently, a function $f, : E \to \mathsf{R}^m$ is continuous if for every open set $V \subseteq \mathsf{R}^m$, the preimage $f^{-1}(V) := \{x \in E : f(x) \in V\}$ is relatively open in E; in other words, $\{x \in E : f(x) \in V\}$ is the intersection of some open set in R^n with E. Also, f is continuous if and only if the preimage $f^{-1}(U)$ of every closed set $U \subseteq \mathsf{R}^m$ is relatively closed in E.

Theorem 3.10.5. *Let $E \subseteq \mathsf{R}^n$ be a compact set and $f : E \to \mathsf{R}^m$ a continuous function. Then:*

(1) *the range of f,*

$$\mathrm{Ran}\,(f) := \{f(x) : x \in E\},$$

is a compact set of R^m;

(2) *the infimum and supremum of the norm of f are attained; in other words, there exist $x_0, y_0 \in E$ such that*

$$\|f(x_0)\|_\mathsf{R} = \inf_{x \in E} \|f(x)\|_\mathsf{R}, \qquad \|f(y_0)\|_\mathsf{R} = \sup_{x \in E} \|f(x)\|_\mathsf{R}.$$

We note in passing that in general the range of a continuous function defined on an open (resp. closed) set need not be open (resp. closed).

Finally. we consider connected sets.

Definition 3.10.6. A set $E \subseteq \mathsf{R}^n$ is said to be (pathwise) *connected* if for any two points $x, y \in E$, there is a continuous function $f : [0, 1] \to E$ such that $f(0) = x$ and $f(1) = y$.

(This definition is slightly different from the definition of connectedness given in Wade [154].)

Theorem 3.10.7. *If E is connected and $f : E \to \mathsf{R}^m$ is a continuous function, then the range of f is connected as well.*

Proof. Let $a = f(x)$, $b = f(y)$, where $x, y \in E$, be two arbitrary elements of the range of f. Since E is connected, there is a continuous function $g : [0, 1] \to E$ such that $g(0) = x$, $g(1) = y$. Then the composition $f \circ g$ is a continuous function from $[0, 1]$ into the range of f such that $(f \circ g)(0) = a$, $(f \circ g)(1) = b$. $\qquad\square$

3.11   EXERCISES

Ex. 3.11.1. Verify that $\| \cdot \|_H$ is indeed a norm on $H^{n \times 1}$:

$$\|u + v\|_H \leq \|u\|_H + \|v\|_H; \quad \|u\alpha\|_H = \|u\|_H \, |\alpha|; \quad \|u\|_H \geq 0$$

for all $u, v \in H^{n \times 1}$ and all $\alpha \in H$, and the equality $\|u\|_H = 0$ holds only for $u = 0$.

Ex. 3.11.2. Let $A \in H^{n \times n}$. Show that if there exists $B \in H^{n \times n}$ such that $BA = I$, then A is invertible and $B = A^{-1}$.

Ex. 3.11.3. Repeat Ex. 3.11.2, with $BA = I$ replaced by $AB = I$.

Ex. 3.11.4. The *row rank* of a matrix $A \in H^{m \times n}$ is defined as the dimension of the (left quaternion) subspace spanned by the rows of A. Show that the row rank of A coincides with $\operatorname{rank}(A)$. Hint: Use the rank decomposition for A.

Ex. 3.11.5. Prove Proposition 3.3.1.

Ex. 3.11.6. Prove Proposition 3.4.1.

Ex. 3.11.7. Show that the following three statements are equivalent for $A \in H^{n \times n}$:

(1) A is invertible;

(2) $\operatorname{Ker}(A) = \{0\}$;

(3) $\operatorname{Ran}(A) = H^{n \times 1}$.

Ex. 3.11.8. Prove that if $A_1, A_2 \in H^{n \times n}$ and the product $A_1 A_2$ is invertible, then A_1 and A_2 are separately invertible.

Ex. 3.11.9. Let $A = \begin{bmatrix} 0 & 1 \\ 1 & 0 \end{bmatrix}$, and let ϕ be a nonstandard involution.

(a) Prove that A does not admit a decomposition $A = B_\phi B$ with either upper triangular or lower triangular $B \in H^{2 \times 2}$.

(b) Find $B \in H^{2 \times 2}$ such that $A = B_\phi B$.

Ex. 3.11.10. Prove Proposition 3.7.2.

Ex. 3.11.11. Prove parts (2) and (3) of Proposition 3.2.4.

Ex. 3.11.12. Verify formula (3.9.1).

Ex. 3.11.13. Show that the numerical ranges $W_*^H(A)$, $A \in H^{n \times n}$ and joint numerical ranges $WJ_*^H(A_1, \ldots, A_p)$, where $A_1, \ldots, A_p \in H^{n \times n}$ are hermitian, are connected sets. Hint: These are ranges of the connected set (unit sphere) $\{x \in H^{n \times 1} : x^* x = 1\}$ under continuous functions.

Ex. 3.11.14. Find $WJ_*^R(A, B)$ for

$$A = \begin{bmatrix} 1 & 0 \\ 0 & -1 \end{bmatrix}, \qquad B = \begin{bmatrix} 0 & 1 \\ 1 & 0 \end{bmatrix},$$

and verify that $WJ_*^R(A, B)$ is not convex.

Ex. 3.11.15. Provide details for the proof of Proposition 3.3.2 in case ϕ is an antiautomorphism.

Ex. 3.11.16. Prove that if ϕ is a nonstandard involution, and $q_1 \in \mathsf{H}$ is such that $\phi(q_1) = -q_1$, then

$$W_\phi^{(0)} \left(\begin{bmatrix} 0 & q_1 \\ q_1 & 0 \end{bmatrix} \right) = \{0\}.$$

Ex. 3.11.17. Show that a block upper triangular matrix $B = [B_{i,j}]_{i,j=1}^k$, where $B_{i,j} \in \mathsf{H}^{n_i \times n_j}$ for $i, j = 1, 2, \ldots, k$ and $B_{i,j} = 0$ if $i > j$, is invertible if and only if the diagonal blocks $B_{1,1}, \ldots, B_{k,k}$ are invertible. Prove the analogous result for block lower triangular matrices.

Ex. 3.11.18. Let $A \in \mathsf{H}^{n \times n}$ be a matrix which is block triangular with respect to the "other" diagonal, and the "other" diagonal blocks are of square sizes

$$A = [A_{i,j}]_{i,j=1}^k,$$

where $A_{i,j} \in \mathsf{H}^{n_{k+1-i} \times n_j}$ for $i, j = 1, 2, \ldots, k$, and $A_{i,j} = 0$ if $i + j < k + 1$. Show that A is invertible if and only if all "other" diagonal blocks $A_{k,1}, A_{k-1,2}, \ldots, A_{1,k}$ are invertible.

Ex. 3.11.19. Let $\mathsf{F} \in \{\mathsf{R}, \mathsf{C}, \mathsf{H}\}$ and let $(A_1, \ldots, A_p)$ be a p-tuple of hermitian matrices in $\mathsf{F}^{n \times n}$. Define the *joint numerical cone* of $(A_1, \ldots, A_p)$ by

$$CJ_*^{\mathsf{F}}(A_1, \ldots, A_p) := \{(x^* A_1 x, \ldots, x^* A_p x) \,:\, x \in \mathsf{F}^{n \times 1} \subseteq \mathsf{R}^p.$$

Show that if $WJ_*^{\mathsf{F}}(A_1, \ldots, A_p)$ is convex, then so is $CJ_*^{\mathsf{F}}(A_1, \ldots, A_p)$.

Ex. 3.11.20. Give example of a situation when $CJ_*^{\mathsf{F}}(A_1, \ldots, A_p)$ is convex but $WJ_*^{\mathsf{F}}(A_1, \ldots, A_p)$ is not.

Ex. 3.11.21. Given a real, resp. complex, skewsymmetric matrix X, show that

$$B^T X B = \text{diag} \left(\begin{bmatrix} 0 & 1 \\ -1 & 0 \end{bmatrix}, \ldots, \begin{bmatrix} 0 & 1 \\ -1 & 0 \end{bmatrix}, 0, \ldots, 0 \right)$$

for some invertible real, resp. complex, matrix B.

Hint: Use induction on n. If the first row of X has a nonzero entry, by simultaneous column and row interchange and column and row scaling, bring the matrix X to the form where the top left 2×2 corner in the matrix is $\begin{bmatrix} 0 & 1 \\ -1 & 0 \end{bmatrix}$. Then, by simultaneous row and column replacements, reduce the proof to the case when

$$X = \begin{bmatrix} 0 & 1 \\ -1 & 0 \end{bmatrix} \oplus X_0,$$

where X_0 is again skewsymmetric. Now use the induction.

Conclude that a real or complex skewsymmetric matrix has even rank.

Ex. 3.11.22. Show that the skewsymmetric hermitian quaternion matrix $A = \begin{bmatrix} 0 & i & j \\ -i & 0 & k \\ -j & -k & 0 \end{bmatrix}$ has rank 3—in other words, is invertible. Find A^{-1}. Hint: $A^2 = 2I - A$.

Ex. 3.11.23. Show that if $A \in \mathsf{R}^{2 \times 2}$, then $W_*^{\mathsf{H}}(A)$ is convex.
Hint: We may assume $A \neq 0$. By using transformations of types

$$A \mapsto U^T A U, \quad A \mapsto \alpha A, \quad A \mapsto A + \lambda I,$$

where A is a 2×2 real matrix, U is a 2×2 real orthogonal matrix, and λ, α are nonzero real numbers, we reduce the problem to one of the following two situations:
(1) $A = \begin{bmatrix} 0 & 1 \\ -1 & 0 \end{bmatrix}$; (2) $A = \begin{bmatrix} 1 & b \\ -b & 0 \end{bmatrix}$, where b is real. In case (1) holds true, show that

$$W_*^{\mathsf{H}}(A) = \{z \in \mathsf{H} : \mathfrak{R}(z) = 0, \quad |\mathfrak{V}(z)| \leq 1\},$$

which is convex. In case (2) holds true, show that

$$W_*^{\mathsf{H}}(A) = \{1 - \alpha + bz : 0 \leq \alpha \leq 1, \quad \mathfrak{R}(z) = 0, \quad |\mathfrak{V}(z)| \leq \sqrt{\alpha - \alpha^2}\},$$

which is convex as well (to verify convexity use the fact that the function $\sqrt{\alpha - \alpha^2}$ is convex on the interval $\alpha \in [0, 1]$).

3.12   NOTES

Real and complex matrix representations of quaternions are well known and found in many sources. Propositions 3.3.3 and 3.4.2 are taken from Rodman [137]. Theorem 3.1.1 and its proof are from Wan [156]. Theorem 3.6.5(b) is in the spirit of a well-known result on LU decompositions of real and complex matrices; see, e.g., Golub and Van Loan [56]. Details on many matrix decompositions of Proposition 3.2.5 are found in Farenick and Pidkowich [38]; see also Loring [102]. Applications to signal processing are found in Le Bihan and Mars [97].

Example 3.5.5 (with $A_0 = I$) is due to Au-Yeung [9].

The case $\mathsf{F} = \mathsf{R}$ of Theorem 3.5.7 was proved in Brickman [19]. The proof of Theorem 3.5.7 is adapted from the papers by Au-Yeung [7, 8]. More information about convexity of quaternion numerical ranges (in certain situations) is found in Au-Yeung [9].

The results of Sections 3.7 and 3.8 seem to be new.

Many properties of the gap function that are well known in the context of real and complex matrices and vector spaces also remain valid for quaternion matrices and vector spaces, with essentially the same proofs; see, e.g., Gohberg et al. [54, Chapter 13] and Rodman [137] for more details. The proof of Theorem 3.9.4 follows the exposition in Section 13.4 of Gohberg et al. [54].

Chapter Four

Symmetric matrices and congruence

In this chapter we develop canonical forms of hermitian and skewhermitian matrices under congruence, in the contexts of the conjugation and of nonstandard involutions. These canonical forms allow us to identify maximal neutral and maximal semidefinite subspaces, with respect to a given hermitian or skewhermitian matrix, in terms of their dimensions.

4.1 CANONICAL FORMS UNDER CONGRUENCE

Definition 4.1.1. Two matrices $A, B \in \mathsf{H}^{n \times n}$ are said to be *congruent* if $A = S^* B S$ for some invertible $S \in \mathsf{H}^{n \times n}$. If ϕ is a nonstandard involution, then we say that $A, B \in \mathsf{H}^{n \times n}$ are *ϕ-congruent* if $A = S_\phi B S$ for some invertible S.

Clearly, congruence and ϕ-congruence are equivalence relations. Also, if A and B are congruent and A is hermitian or skewhermitian, then so is B. Analogously, if A and B are ϕ-congruent and A is ϕ-hermitian or ϕ-skewhermitian, then so is B.

To formulate the result on canonical forms of ϕ-skewhermitian matrices under ϕ-congruence, we make use of a unit quaternion β such that $\beta_\phi = -\beta$. Note that the quaternion β with these properties is unique up to negation for a given nonstandard involution ϕ. In what follows, we fix one such β and denote it $\beta(\phi)$. Note that $\beta^{-1} = -\beta$.

Theorem 4.1.2. *Let ϕ be a nonstandard involution.*
(a) *For every ϕ-hermitian $H \in \mathsf{H}^{n \times n}$, there exists an invertible S such that*

$$S_\phi H S = \begin{bmatrix} I_r & 0 \\ 0 & 0 \end{bmatrix} \tag{4.1.1}$$

for a nonnegative integer $r \leq n$. Moreover, r is uniquely determined by H.
(b) *For every ϕ-skewhermitian matrix $H \in \mathsf{H}^{n \times n}$, there exists an invertible $S \in \mathsf{H}^{n \times n}$ such that*

$$S_\phi H S = \begin{bmatrix} \beta I_p & 0 & 0 \\ 0 & -\beta I_q & 0 \\ 0 & 0 & 0_{n-p-q} \end{bmatrix}, \qquad \beta = \beta(\phi). \tag{4.1.2}$$

Moreover, the integers p and q are uniquely determined by H (for a fixed $\beta(\phi)$).

Proof. Part (a). To prove existence of S, we use induction on n and argue similarly to the proof of Theorem 3.6.5, part (a). The uniqueness of r is evident; in fact, $r = \operatorname{rank} H$ (cf. Proposition 3.2.4).

Part (b). We show the existence of S first. Again, we use induction on n. The case $n = 1$ is trivial. Now consider the general case. We may assume $H \neq 0$. If $H = [h_{i,j}]_{i,j=1}^n$ has a nonzero diagonal entry, say $h_{1,1} \neq 0$, then by scaling we may assume $h_{1,1} = \pm\beta$; then by simultaneous row and column replacement operations, we reduce A to a block diagonal form:

$$\begin{bmatrix} 1 & 0 & \dots & 0 \\ (h_{1,2})_\phi(\mp\beta) & 1 & \dots & 0 \\ \vdots & \vdots & \ddots & \vdots \\ (h_{1,n})_\phi(\mp\beta) & 0 & \dots & 1 \end{bmatrix} \begin{bmatrix} \pm\beta & h_{1,2} & h_{1,3} & \dots & h_{1,n} \\ -(h_{1,2})_\phi & h_{2,2} & h_{2,3} & \dots & h_{2,n} \\ \vdots & \vdots & \vdots & \dots & \vdots \\ -(h_{1,n})_\phi & h_{n,2} & h_{n,3} & \dots & h_{n,n} \end{bmatrix}$$

$$\cdot \begin{bmatrix} 1 & \pm\beta h_{1,2} & \pm\beta h_{1,3} & \dots & \pm\beta h_{1,n} \\ 0 & 1 & 0 & \dots & 0 \\ \vdots & \vdots & \ddots & \dots & \vdots \\ 0 & 0 & 0 & \dots & 1 \end{bmatrix} = \begin{bmatrix} \pm\beta & 0_{1\times(n-1)} \\ 0_{(n-1)\times 1} & * \end{bmatrix}.$$

(* stands for entries of no immediate interest.) It remains to use the induction hypothesis. If H has zero diagonal, then a simultaneous row and column replacement yields a ϕ-skewhermitian matrix with nonzero diagonal, thereby reducing the proof to the case already considered. Say $h_{1,2} \neq 0$; then, with $b = \beta((h_{1,2})_\phi)^{-1}$, we have

$$\begin{bmatrix} 1 & b & 0 \\ 0 & 1 & 0 \\ 0 & 0 & I_{n-2} \end{bmatrix} H \begin{bmatrix} 1 & 0 & 0 \\ b_\phi & 1 & 0 \\ 0 & 0 & I_{n-2} \end{bmatrix} = \begin{bmatrix} -2\beta & * \\ * & * \end{bmatrix}.$$

It remains to prove the uniqueness of p and q. First note that $p + q = \operatorname{rank} H$ is uniquely determined by H. Assume that (4.1.2) holds, and let

$$\mathcal{V} \quad := \quad \operatorname{Span}_{\mathsf{H}}\{Se_1, \dots, Se_p, Se_{p+q+1}, \dots, Se_n\},$$
$$\mathcal{W} \quad := \quad \operatorname{Span}_{\mathsf{H}}\{Se_{p+1}, \dots, Se_{p+q}\},$$

where $e_1, \dots, e_n$ are the standard unit coordinate vectors in $\mathsf{H}^{n\times 1}$. From (2.5.1) it follows that $v_\phi H v \beta^{-1} \geq 0$ for all $v \in \mathcal{V}$ and $w_\phi H w \beta^{-1} < 0$ for all nonzero vectors $w \in \mathcal{W}$. It is now clear that q is the maximal dimension of a subspace $\mathcal{M} \subseteq \mathsf{H}^{n\times 1}$ with the property that $x_\phi H x \beta^{-1} < 0$ for all nonzero $x \in \mathcal{M}$ is equal to q. Indeed, if there were such a subspace $\mathcal{M}$ of dimension larger than q, then it would have a nonzero intersection with $\mathcal{V}$, leading to a contradiction. Thus, q and, therefore, also p, are uniquely determined by H. $\square$

Definition 4.1.3. We say that the ordered triple of nonnegative integers

$$(\operatorname{In}_+(H), \operatorname{In}_-(H), \operatorname{In}_0(H)) := (p, q, n - p - q)$$

is the $\beta(\phi)$-*inertia*, or the $\beta(\phi)$-*signature*, of a ϕ-skewhermitian matrix $H \in \mathsf{H}^{n\times n}$, as in Theorem 4.1.2(b). The matrix H is said to be $\beta(\phi)$-*positive definite*, resp. $\beta(\phi)$-*positive semidefinite*, if $\operatorname{In}_+(H) = n$, resp. $\operatorname{In}_+(H) + \operatorname{In}_0(H) = n$. Analogously, $\beta(\phi)$-*negative definite* and $\beta(\phi)$-*negative semidefinite* ϕ-skewhermitian matrices are defined.

We record the following corollary of Theorem 4.1.2 to be used later.

Corollary 4.1.4. *Let ϕ be a nonstandard involution. Fix $\beta(\phi) \in \mathsf{H}$ such that $\phi(\beta(\phi)) = -\beta(\phi)$, $|\beta(\phi)| = 1$. Then for every nonzero $\alpha_0 \in \operatorname{Inv}(\phi)$, there exists $\omega \in \mathsf{H}$, $|\omega| = 1$, with the properties that $\phi(\omega)\alpha_0\omega$ is real positive, and, hence,*

$$\phi(\omega)\alpha_0\omega = |\alpha_0| \quad \text{and} \quad \phi(\omega)\beta(\phi)\omega = \beta(\phi).$$

Proof. By Theorem 4.1.2(a) there exists $\omega \in \mathsf{H}$, $|\omega| = 1$, such that $\phi(\omega)\alpha_0\omega$ is real positive. The quaternion $\phi(\omega)\beta(\phi)\omega$ is ϕ-skewhermitian. The condition $|\omega| = 1$ implies that $\phi(\omega)\beta(\phi)\omega = \pm\beta(\phi)$. In fact, we must have sign $+$. Indeed, the continuous map $\mathcal{F}$ defined by

$$\omega \in \mathsf{H} \setminus \{0\} \quad \mapsto \quad \beta(\phi)^{-1}\phi(\omega)\beta(\phi)\omega \in \mathsf{R} \setminus \{0\}$$

is defined on a connected set $\mathsf{H} \setminus \{0\}$; therefore, the range (or image) of $\mathcal{F}$ must be also connected, and so $\operatorname{Ran}\mathcal{F} \subseteq (0,\infty)$ or $\operatorname{Ran}\mathcal{F} \subseteq (-\infty,0)$. But the latter possibility is excluded because $\mathcal{F}(1) = 1$. $\qquad\square$

Definition 4.1.5. For a nonstandard involution ϕ, a ϕ-skewhermitian $H \in \mathsf{H}^{n\times n}$ is said to have *balanced inertia* if $\operatorname{In}_+(H) = \operatorname{In}_-(H)$.

Note that the property of having balanced inertia is independent of the choice of $\beta(\phi)$, as $\operatorname{In}_\pm(H)$ with respect to $-\beta(\phi)$ coincides with $\operatorname{In}_\mp(H)$ with respect to $\beta(\phi)$.

For the (quaternion) conjugation, a theorem similar to Theorem 4.1.2 holds, but with the inertia present in the case of hermitian matrices, in complete analogy with the standard results of complex linear algebra.

Theorem 4.1.6. (a) *For every hermitian $H \in \mathsf{H}^{n\times n}$, there exists an invertible S such that*

$$S^*HS = \begin{bmatrix} I_p & 0 & 0 \\ 0 & -I_q & 0 \\ 0 & 0 & 0_{n-p-q} \end{bmatrix}. \tag{4.1.3}$$

Moreover, the integers p and q are uniquely determined by H.

(b) *For every skewhermitian matrix $H \in \mathsf{H}^{n\times n}$, there exists an invertible $S \in \mathsf{H}^{n\times n}$ such that*

$$S^*HS = \begin{bmatrix} iI_r & 0 \\ 0 & 0 \end{bmatrix}.$$

Moreover, the integer r is uniquely determined by H.

The quaternion i in Theorem 4.1.6 can be replaced by any nonzero quaternion with zero real part; indeed, this can be achieved by diagonal scaling

$$S^*HS \mapsto \operatorname{diag}(a_j^*)_{j=1}^n S^*HS\operatorname{diag}(a_j)_{j=1}^n, \qquad a_1,\ldots,a_n \in \mathsf{H} \setminus \{0\};$$

cf. Theorem 2.2.6(2)$\Leftrightarrow$ (5).

The proof of Theorem 4.1.6(a) follows the same pattern as that of Theorem 4.1.2(b) and is, therefore, omitted. For the proof of part (b), argue as in the proof of Theorem 3.6.5(a), using the statement in part (c) of Theorem 2.5.1. Alternatively, Theorem 4.1.6 may be obtained from Theorem 4.1.2 using the next proposition (Ex. 4.6.18). We state the proposition in the form suitable for later references.

Proposition 4.1.7. *Let ϕ be a nonstandard involution.*

(a) *A matrix $H \in \mathsf{H}^{n\times n}$ is hermitian, resp. skewhermitian, if and only if $\beta(\phi)H$ is ϕ-skewhermitian, resp. ϕ-hermitian; equivalently $H\beta(\phi)$ is ϕ-skewhermitian, resp. ϕ-hermitian.*

(b) *If $X \in \mathsf{H}^{m\times n}$, $Y \in \mathsf{H}^{p\times r}$, $S \in \mathsf{H}^{m\times p}$, $T \in \mathsf{H}^{n\times r}$, then the equality $Y = S^*XT$ holds true if and only if the equality*

$$\beta(\phi)Y = S_\phi(\beta(\phi)X)T \tag{4.1.4}$$

does.

Proof. Part (a): Assume $H = [h_{i,j}]_{i,j=1}^n = \pm H^*$, where $h_{i,j} \in \mathsf{H}$. Using the easily verifiable equalities

$$h_{i,j}(-\beta(\phi)) = -\beta(\phi)h_{i,j}^*, \quad i,j = 1,2,\ldots,n,$$

we obtain

$$(\beta(\phi)h_{i,j})_\phi = -\beta(\phi)h_{i,j}^* = \mp\beta(\phi)h_{j,i}, \quad i,j = 1,2,\ldots,n;$$

therefore, $(\beta(\phi)H)_\phi = \mp\beta(\phi)H$. Reversing the argument, the "if" direction of part (a) follows. Finally, a straightforward algebra shows that $\beta(\phi)H$ is ϕ-skewhermitian or ϕ-hermitian if and only if $H\beta(\phi)$ is such.

Part (b): Letting the lowercase letter (with indices) denote the entries of a matrix given by the corresponding capital letter, equation $Y = S^*XT$ amounts to

$$y_{i,j} = \sum_{k,\ell} s_{k,i}^* x_{k,\ell} t_{\ell,j}, \qquad i = 1,2,\ldots,p, \quad j = 1,2,\ldots,r, \tag{4.1.5}$$

whereas (4.1.4) amounts to

$$\beta(\phi)y_{i,j} = \sum_{k,\ell} (s_{k,i})_\phi \beta(\phi) x_{k,\ell} t_{\ell,j}, \qquad i = 1,2,\ldots,p, \quad j = 1,2,\ldots,r. \tag{4.1.6}$$

Comparing (4.1.5) and (4.1.6), we see that statement (b) boils down to the equality $s_\phi\beta(\phi) = \beta(\phi)s^*$, $s \in \mathsf{H}$, which can be easily verified (cf. (2.4.8)). $\qquad\square$

Definition 4.1.8. We say that the ordered triple of nonnegative integers

$$(\mathrm{In}_+(H), \mathrm{In}_-(H), \mathrm{In}_0(H)) := (p, q, n - p - q),$$

as in Theorem 4.1.6, is the *inertia* of a hermitian matrix $H \in \mathsf{H}^{n\times n}$. The *signature* of a hermitian matrix $H \in \mathsf{H}^{n\times n}$ is defined as the difference between the number of positive eigenvalues, counted with multiplicities, and the number of negative eigenvalues, counted with multiplicities.

In terms of the inertia, the signature of H is equal to $\mathrm{In}_+(H) - \mathrm{In}_-(H)$.

Definition 4.1.9. The matrix H is said to be *positive definite*, resp. *positive semidefinite, negative definite, negative semidefinite*, if $\mathrm{In}_+(H) = n$, resp. $\mathrm{In}_+(H) + \mathrm{In}_0(H) = n$, $\mathrm{In}_-(H) = n$, $\mathrm{In}_-(H) + \mathrm{In}_0(H) = n$, in complete analogy with the corresponding classes of real symmetric and complex hermitian matrices.

Note that this definition of positive (semi)definiteness is equivalent to the definition of positive (semi)definiteness given in Section 3.2.

Theorems 4.1.2 and 4.1.6 lead to the *inertia theorem*.

Theorem 4.1.10. *Two (quaternion) hermitian matrices are congruent if and only if they have the same inertia. For a given nonstandard involution ϕ, two ϕ-skewhermitian matrices are ϕ-congruent if and only if they have the same $\beta(\phi)$-inertia.*

For ϕ-skewhermitian matrices with balanced inertia and having even rank, an alternative canonical form is available.

Theorem 4.1.11. *Let ϕ be an involution of H. Assume either* (a) *or* (b) *holds:*

(a) ϕ is nonstandard, and $H \in \mathsf{H}^{n \times n}$ is a ϕ-skewhermitian matrix with balanced inertia;

(b) ϕ is the conjugation, and $H \in \mathsf{H}^{n \times n}$ is a skewhermitian matrix of even rank.

Then there exists an invertible S such that

$$S_\phi H S = \begin{bmatrix} 0 & I_{k/2} & 0 \\ -I_{k/2} & 0 & 0 \\ 0 & 0 & 0 \end{bmatrix}, \tag{4.1.7}$$

where k is the rank of H (note that the rank of H is even in the case (a) as well).

Conversely, if equality (4.1.7) holds for some invertible S, then either (a) or (b) is valid.

Proof. If (a) holds, then the result follows from the canonical form of Theorem 4.1.2(b) and the equality

$$\begin{bmatrix} 1 & -1 \\ \frac{1}{2}\beta(\phi) & \frac{1}{2}\beta(\phi) \end{bmatrix} \begin{bmatrix} \beta(\phi) & 0 \\ 0 & -\beta(\phi) \end{bmatrix} \begin{bmatrix} 1 & -\frac{1}{2}\beta(\phi) \\ -1 & -\frac{1}{2}\beta(\phi) \end{bmatrix} = \begin{bmatrix} 0 & 1 \\ -1 & 0 \end{bmatrix}. \tag{4.1.8}$$

If (b) holds, then the result follows from the canonical form of Theorem 4.1.6(b) and the equality

$$S^* \begin{bmatrix} \mathsf{i} & 0 \\ 0 & \mathsf{i} \end{bmatrix} S = \begin{bmatrix} 0 & -1 \\ 1 & 0 \end{bmatrix}, \tag{4.1.9}$$

where

$$S = \frac{1}{\sqrt{2}} \begin{bmatrix} -\mathsf{j} & -\mathsf{k} \\ 1 & \mathsf{i} \end{bmatrix}.$$

Note that $S^{-1} = S^*$.

The converse statement follows from the equalities (4.1.8) and (4.1.9) and the uniqueness of the canonical form of A, as given in Theorems 4.1.2 and 4.1.6. $\square$

It is easy to see that the matrix in the right-hand side of (4.1.7) can be transformed by congruence (or ϕ-congruence) to the form

$$\mathrm{diag}\left(\underbrace{\begin{bmatrix} 0 & 1 \\ -1 & 0 \end{bmatrix}, \dots, \begin{bmatrix} 0 & 1 \\ -1 & 0 \end{bmatrix}}_{k/2 \text{ times}}, 0, \dots, 0 \right).$$

Sometimes we are interested in diagonalization under congruence of matrices that are ϕ-hermitian or ϕ-skewhermitian, not necessarily in the canonical form under congruence. It turns out that this can be achieved using matrices S from a restricted set. We denote by $Q_n(k, j)$ the matrix obtained from I_n by interchanging the kth and jth rows (or columns) and by $Q_n(k, j_\lambda)$ the matrix obtained from I_n by adding λ times the jth column to the kth column. Here $j \neq k$, $j, k = 1, 2, \dots, n$, and $\lambda \in \mathsf{H}$. Further, denote by $\mathcal{G}_n$ the multiplicative group generated by all matrices $Q_n(k, j)$ and $Q_n(k, j_\lambda)$ (for a fixed n); in other words, $\mathcal{G}_n$ consists of all products of matrices of the form $Q_n(k, j)$, $Q_n(k, j_\lambda)$. Note that $Q_n(k, j)$ is its own inverse and $Q_n(k, j_\lambda)^{-1} = Q_n(k, j_{-\lambda})$.

Theorem 4.1.12. *Let ϕ be the conjugation or a nonstandard involution, and let $H = \pm H_\phi \in \mathsf{H}^{n \times n}$. Then there exists $S \in \mathcal{G}_n$ such that $S_\phi H S$ is diagonal.*

The proof follows the pattern of the proof of Theorem 4.1.2, Part (b). We omit details.

4.2 NEUTRAL AND SEMIDEFINITE SUBSPACES

Let ϕ be an involution; thus, either ϕ is nonstandard or ϕ is the conjugation. To keep a uniform notation, let us agree that X_ϕ, $X \in \mathsf{H}^{m \times n}$, where ϕ being the conjugation simply means X^*.

Definition 4.2.1. Given a ϕ-hermitian or ϕ-skewhermitian matrix $H \in \mathsf{H}^{n \times n}$, in other words, $H_\phi = \pm H$, a subspace $\mathcal{M} \subseteq \mathsf{H}^{n \times 1}$ is said to be (H, ϕ)-*neutral*, or (H, ϕ)-*isotropic* if $y_\phi H x = 0$ for all $x, y \in \mathcal{M}$ or, equivalently, if $x_\phi H x = 0$ for all $x \in \mathcal{M}$.

The equivalence of these two definitions follows from the polarization identities (3.5.1) and (3.7.1); cf. the proof of Proposition 3.5.1.

In the cases when ϕ is nonstandard and $H \in \mathsf{H}^{n \times n}$ is ϕ-skewhermitian or ϕ is the conjugation and H is hermitian, we introduce also semidefinite and definite subspaces.

Definition 4.2.2. Let H be hermitian and ϕ be the conjugation. A subspace $\mathcal{M} \subseteq \mathsf{H}^{n \times n}$ is said to be $(H, {}^*)$-*nonnegative*, resp. $(H, {}^*)$-*positive*, $(H, {}^*)$-*nonpositive*, $(H, {}^*)$-*negative*, if $x^* H x \geq 0$ holds for all $x \in \mathcal{M}$, resp. $x^* H x > 0$ holds for all nonzero $x \in \mathcal{M}$, $x^* H x \leq 0$ holds for all $x \in \mathcal{M}$, $x^* H x < 0$ holds for all nonzero $x \in \mathcal{M}$. If ϕ is a nonstandard involution and $H \in \mathsf{H}^{n \times n}$ is ϕ-skewhermitian, then we say that a subspace $\mathcal{M} \subseteq \mathsf{H}^{n \times n}$ is (H, ϕ)-*nonnegative*, resp. (H, ϕ)-*positive*, (H, ϕ)-*nonpositive*, (H, ϕ)-*negative*, if $(\beta(\phi))^{-1} x_\phi H x \geq 0$ holds for all $x \in \mathcal{M}$, resp. $(\beta(\phi))^{-1} x_\phi H x > 0$ holds for all nonzero $x \in \mathcal{M}$, $(\beta(\phi))^{-1} x_\phi H x \leq 0$ holds for all $x \in \mathcal{M}$, $(\beta(\phi))^{-1} x_\phi H x < 0$ holds for all nonzero $x \in \mathcal{M}$.

Note that $x_\phi H x$ in the above definition is a real multiple of $\beta(\phi)$, so

$$(\beta(\phi))^{-1} x_\phi H x = x_\phi H x (\beta(\phi))^{-1}$$

is a real number.

We use the notation $\mathcal{S}_0(H)$, $\mathcal{S}_{\geq 0}(H)$, $\mathcal{S}_{>0}(H)$, $\mathcal{S}_{\leq 0}(H)$, and $\mathcal{S}_{<0}(H)$ to designate the classes of all (H, ϕ)-neutral, all (H, ϕ)-nonnegative, all (H, ϕ)-positive, all (H, ϕ)-nonpositive, and all (H, ϕ)-negative subspaces, respectively. In this notation, we suppress the dependence of $\mathcal{S}_\mu(H)$, $\mu \in \{0, \geq 0, > 0, \leq 0, < 0\}$, on ϕ; it will be clear from context whether ϕ is the conjugation or a nonstandard involution and whether H is ϕ-hermitian or ϕ-skewhermitian.

Observe that by default (nonexistence of nonzero vectors) the zero subspace is simultaneously (H, ϕ)-positive and (H, ϕ)-negative (here H is hermitian and ϕ is the conjugation, or ϕ is a nonstandard involution and H is ϕ-skewhermitian).

Clearly, for a fixed ϕ and H, if a subspace $\mathcal{V} \subseteq \mathsf{H}^{n \times 1}$ belongs to one of the classes $\mathcal{S}_{\geq 0}(H)$, $\mathcal{S}_{>0}(H)$, $\mathcal{S}_{\leq 0}(H)$, $\mathcal{S}_{<0}(H)$, $\mathcal{S}_0(H)$, then so does every subspace of $\mathcal{V}$. On the other hand, the sum of two subspaces in $\mathcal{S}_{\geq 0}(H)$ does not necessarily belong to $\mathcal{S}_{\geq 0}(H)$; this observation applies also to the other four classes $\mathcal{S}_{>0}(H)$, $\mathcal{S}_{\leq 0}(H)$, $\mathcal{S}_{<0}(H)$, $\mathcal{S}_0(H)$.

The following example illustrates this phenomenon.

Example 4.2.3. (a) Let

$$H = H^* = \begin{bmatrix} 1 & a \\ a & 1 \end{bmatrix} \in \mathsf{H}^{2 \times 2},$$

where a is real and $|a| > 1$. Then $\text{Span}_\mathsf{H}\{e_1\}, \text{Span}_\mathsf{H}\{e_2\} \in \mathcal{S}_{>0}(H)$, but

$$\text{Span}_\mathsf{H}\{e_1\} + \text{Span}_\mathsf{H}\{e_2\} = \mathsf{H}^{2\times 1} \notin \mathcal{S}_{>0}(H),$$

because H is not positive definite.

(b) Let

$$H = \begin{bmatrix} 0 & 1 \\ 1 & 0 \end{bmatrix} \in \mathsf{H}^{2\times 2}.$$

Then $\text{Span}_\mathsf{H}\{e_1\}, \text{Span}_\mathsf{H}\{e_2\} \in \mathcal{S}_0(H)$, but

$$\text{Span}_\mathsf{H}\{e_1\} + \text{Span}_\mathsf{H}\{e_2\} = \mathsf{H}^{2\times 1},$$

which is not in $\mathcal{S}_0(H)$. $\qquad\qquad\qquad\qquad\qquad\qquad\qquad\qquad\qquad$ □

Let ϕ be the conjugation and $H = H^*$, or let ϕ be a nonstandard involution and $H = -H_\phi$.

Definition 4.2.4. A subspace $\mathcal{M} \in \mathcal{S}_{>0}(H)$ is said to be *maximal (H, ϕ)-nonnegative* if there is no larger subspace in the set $\mathcal{S}_{>0}(H)$, in other words, if $\mathcal{N} \in \mathcal{S}_{>0}(H)$ and $\mathcal{N} \supseteq \mathcal{M}$ implies $\mathcal{N} = \mathcal{M}$. Analogously, maximal (H, ϕ)-nonpositive, maximal (H, ϕ)-neutral, maximal (H, ϕ)-positive, and maximal (H, ϕ)-negative subspaces are defined.

In Example 4.2.3(a), the subspaces $\text{Span}_\mathsf{H}\{e_1\}$ and $\text{Span}_\mathsf{H}\{e_2\}$ are maximal (H, ϕ)-nonnegative, ϕ being the conjugation. In Example 4.2.3(b), $\text{Span}_\mathsf{H}\{e_1\}$ and $\text{Span}_\mathsf{H}\{e_2\}$ are maximal (H, ϕ)-neutral.

Also, for ϕ the conjugation and $H = -H^*$ or ϕ a nonstandard involution and $H = H_\phi$, we define maximal (H, ϕ)-neutral subspaces.

In this section we will be concerned with characterization of maximal (by inclusion) neutral and semidefinite subspaces in terms of dimensions and inertia (when inertia is defined). First, a simple observation showing that all the classes of subspaces introduced above in this section respect transformations of congruence.

Proposition 4.2.5. *Let ϕ be an involution (conjugation or nonstandard).*
(a) *Let $H = \pm H_\phi \in \mathsf{H}^{n\times n}$. Then for every invertible $T \in \mathsf{H}^{n\times n}$, we have*

$$\mathcal{S}_0(T_\phi H T) = T^{-1}\left(\mathcal{S}_0(H)\right).$$

(b) *Let $H \in \mathsf{H}^{n\times n}$ be hermitian if ϕ is the conjugation or ϕ-skewhermitian if ϕ is a nonstandard involution. Then for every invertible $T \in \mathsf{H}^{n\times n}$, we have*

$$\mathcal{S}_\mu(T_\phi H T) = T^{-1}\left(\mathcal{S}_\mu(H)\right), \qquad\qquad (4.2.1)$$

where μ is one of the symbols $\{0, \geq 0, > 0, \leq 0, < 0\}$.

The proof is immediate. Consider, for instance, the case $\mu = 0$. If a subspace $\mathcal{M}$ is such that $x_\phi H x = 0$ for all $x \in \mathcal{M}$, then for any invertible $T \in \mathsf{H}^{n\times n}$ we have $(T^{-1}x)_\phi T_\phi H T \cdot (T^{-1}x) = 0$, so the set of vectors of the form $T^{-1}x$, $x \in \mathcal{M}$, forms a $(T_\phi H T, \phi)$-neutral subspace. This shows the $\supseteq$ inclusion in (4.2.1). The converse inclusion follows by reversing the roles of H and $T_\phi H T$.

It turns out that the maximal elements in each of the sets $\mathcal{S}_{\geq 0}(H)$, $\mathcal{S}_{>0}(H)$, $\mathcal{S}_0(H)$, $\mathcal{S}_{\leq 0}(H)$, $\mathcal{S}_{<0}(H)$ (when defined) have the same dimension. In the next theorem we state this property formally and identify the dimensions of the maximal elements in terms of inertia.

Theorem 4.2.6. *Assume that either ϕ is the conjugation and $H \in \mathsf{H}^{n \times n}$ is hermitian or ϕ is a nonstandard involution and $H \in \mathsf{H}^{n \times n}$ is ϕ-skewhermitian. Let $(\mathrm{In}_+(H), \mathrm{In}_-(H), \mathrm{In}_0(H))$ be the inertia of H in the former case and the $\beta(\phi)$-inertia of H in the latter case. Then:*

(a) *$\mathcal{M} \in \mathcal{S}_{\geq 0}(H)$ is maximal (H, ϕ)-nonnegative if and only if $\dim \mathcal{M} = \mathrm{In}_+(H) + \mathrm{In}_0(H)$;*

(b) *$\mathcal{M} \in \mathcal{S}_{>0}(H)$ is maximal (H, ϕ)-positive if and only if $\dim \mathcal{M} = \mathrm{In}_+(H)$;*

(c) *$\mathcal{M} \in \mathcal{S}_{\leq 0}(H)$ is maximal (H, ϕ)-nonpositive if and only if $\dim \mathcal{M} = \mathrm{In}_-(H) + \mathrm{In}_0(H)$;*

(d) *$\mathcal{M} \in \mathcal{S}_{<0}(H)$ is maximal (H, ϕ)-negative if and only if $\dim \mathcal{M} = \mathrm{In}_-(H)$;*

(e) *$\mathcal{M} \in \mathcal{S}_0(H)$ is maximal (H, ϕ)-neutral if and only if*

$$\dim \mathcal{M} = \min(\mathrm{In}_+(H) + \mathrm{In}_0(H), \mathrm{In}_-(H) + \mathrm{In}_0(H)). \qquad (4.2.2)$$

It follows from Theorem 4.2.6 that in each of the cases (a)–(e), there exists a subspace in the indicated class $\mathcal{S}_\mu(H)$, $\mu \in \{\geq 0, > 0, \leq 0, < 0, 0\}$, having the dimension equal to $\mathrm{In}_+(H) + \mathrm{In}_0(H)$, $\mathrm{In}_+(H)$, $\mathrm{In}_-(H) + \mathrm{In}_0(H)$, $\mathrm{In}_-(H)$, or the right-hand side of (4.2.2), respectively.

The lengthy proof of Theorem 4.2.6 is given in Section 4.3.

If inertia is absent, maximal neutral subspaces can be characterized by dimension expressed in terms of the rank.

Theorem 4.2.7. *Let $H = -H_\phi$ and ϕ be the conjugation or $H = H_\phi$ and ϕ be a nonstandard involution. Let $\mathcal{M}$ be an (H, ϕ)-neutral subspace. Then $\mathcal{M}$ is maximal (by inclusion) (H, ϕ)-neutral subspace if and only if*

$$\dim \mathcal{M} = n - \frac{\lceil \mathrm{rank}\, H \rceil}{2},$$

where $\lceil x \rceil$ stands for the least even integer greater than or equal to x.

In particular, there exists an (H, ϕ)-neutral subspace of dimension $n - \lceil \mathrm{rank}\, H \rceil / 2$.

The proof of Theorem 4.2.7 will be given in Section 4.4.

We indicate a simple but very useful corollary of Theorem 4.2.6.

Corollary 4.2.8. *Under the notation and hypotheses of Theorem 4.2.6, suppose that $H \in \mathsf{H}^{n \times n}$ has the form*

$$H = H_1 \oplus H_2 \oplus \cdots \oplus H_s, \quad \text{where} \quad H_j \in \mathsf{H}^{n_j \times n_j}, \quad \text{for} \quad j = 1, 2, \ldots, s. \qquad (4.2.3)$$

Let there be given a subspace $\mathcal{M}$ of $\mathsf{H}^{n \times 1}$ of the form

$$\mathcal{M} = \begin{bmatrix} \mathcal{M}_1 \\ \mathcal{M}_2 \\ \vdots \\ \mathcal{M}_s \end{bmatrix} := \left\{ \begin{bmatrix} x_1 \\ x_2 \\ \vdots \\ x_s \end{bmatrix} : x_1 \in \mathcal{M}_1, \ldots, x_s \in \mathcal{M}_s \right\}, \qquad (4.2.4)$$

where $\mathcal{M}_j$ is a fixed subspace of $\mathsf{H}^{n_j \times 1}$. Then:

(1) $\mathcal{M}$ is (H,ϕ)-nonnegative if and only if each $\mathcal{M}_j$ is (H_j,ϕ)-nonnegative;

(2) $\mathcal{M}$ is maximal (H,ϕ)-nonnegative if and only if each $\mathcal{M}_j$ is maximal (H_j,ϕ)-nonnegative;

(3) $\mathcal{M}$ is (H,ϕ)-negative if and only if each $\mathcal{M}_j$ is (H_j,ϕ)-negative;

(4) $\mathcal{M}$ is maximal (H,ϕ)-negative if and only if each $\mathcal{M}_j$ is maximal (H_j,ϕ)-negative;

(5) $\mathcal{M}$ is (H,ϕ)-neutral if and only if each $\mathcal{M}_j$ is (H_j,ϕ)-neutral.

Remark 4.2.9. The statement obtained from Corollary 4.2.8 by replacing negative with positive everywhere in (1)–(4) is valid as well and can be verified by applying Corollary 4.2.8 to $-H$.

Remark 4.2.10. Part (5) holds true also for H and $\mathcal{M}$ given by (4.2.3) and (4.2.4), respectively, where $H = -H_\phi$ and ϕ is the conjugation or $H = H_\phi$ and ϕ is a nonstandard involution.

Remark 4.2.11. If each $\mathcal{M}_j$ is maximal (H_j,ϕ)-neutral, then generally it does not follow that $\mathcal{M}$ is maximal (H,ϕ)-neutral. For example, if $H = 1\oplus(-1) \in \mathsf{H}^{2\times2}$, then $\mathcal{M}_1 = \mathcal{M}_2 = \{0\}$ is maximal 1-neutral as well as maximal -1-neutral, but $\mathcal{M} = \begin{bmatrix} \mathcal{M}_1 \\ \mathcal{M}_2 \end{bmatrix} = \{0\}$ is not maximal H-neutral.

Proof. Parts (1), (3), and (5) follow without difficulty from the definitions of (H,ϕ)-nonnegative and (H,ϕ)-negative subspaces. Parts (2) and (4) follow from Theorem 4.2.6, the easily verified property that the inertia is additive with respect to direct sum of matrices:

$$\mathrm{In}_\mu(H) = \mathrm{In}_\mu(H_1) + \cdots + \mathrm{In}_\mu(H_s), \qquad (4.2.5)$$

where $\mu \in \{0,+,-\}$, and the additive property of dimensions:

$$\dim_\mathsf{H} \mathcal{M} = \sum_{j=1}^{s} \dim_\mathsf{H} \mathcal{M}_j.$$

$\square$

4.3   PROOF OF THEOREM 4.2.6

First, observe that it suffices to prove statements (a), (b), and (e); statements (c) and (d) follow by applying (a) and (b) to $-H$ and using

$$\mathcal{S}_{\geq0}(-H) = \mathcal{S}_{\leq0}(H), \quad \mathcal{S}_{>0}(-H) = \mathcal{S}_{<0}(H).$$

Step 1. We exhibit (H,ϕ)-nonnegative, (H,ϕ)-positive, and (H,ϕ)-neutral subspaces of dimensions $\mathrm{In}_+(H) + \mathrm{In}_0(H)$, $\mathrm{In}_+(H)$, and

$$\min(\mathrm{In}_+(H) + \mathrm{In}_0(H), \mathrm{In}_-(H) + \mathrm{In}_0(H)),$$

respectively. To this end, in view of Proposition 4.2.5 and Theorems 4.1.2 and 4.1.6, we can assume that

$$H = \begin{bmatrix} \delta I_p & 0 & 0 \\ 0 & -\delta I_q & 0 \\ 0 & 0 & 0_{n-p-q} \end{bmatrix}, \qquad (4.3.1)$$

where $\delta = \beta(\phi)$ if ϕ is a nonstandard involution, and $\delta = 1$ if ϕ is the conjugation. Now, clearly,

$$\mathrm{Span}_\mathsf{H}\{e_1, \ldots, e_p, e_{p+q+1}, \ldots, e_n\}, \quad \mathrm{Span}_\mathsf{H}\{e_1, \ldots, e_p\}, \quad \text{and}$$

$$\mathrm{Span}_\mathsf{H}\{e_1 + e_{p+1}, e_2 + e_{p+2}, \ldots, e_{\min\{p,q\}} + e_{p+\min\{p,q\}}, e_{p+q+1}, \ldots, e_n\}$$

are (H, ϕ)-nonnegative, (H, ϕ)-positive, and (H, ϕ)-neutral subspaces, respectively, of the requisite dimensions.

Step 2. Proof of the "if" parts.

Arguing by contradiction, suppose an (H, ϕ)-nonnegative subspace $\mathcal{M}$ has dimension $\mathrm{In}_+(H) + \mathrm{In}_0(H)$, but it is not maximal (H, ϕ)-nonnegative. Let $\mathcal{M}'$ be an (H, ϕ)-nonnegative subspace strictly larger than $\mathcal{M}$. By Step 1 (applied to $-H$ rather than H), there exists an (H, ϕ)-negative subspace $\mathcal{N}$ of dimension $\mathrm{In}\,(H)$. Since $\dim \mathcal{M}' + \dim \mathcal{N} > n$, there is a nonzero $x \in \mathcal{M}' \cap \mathcal{N}$. Then $\delta^{-1}x_\phi H x \geq 0$ as $x \in \mathcal{M}'$, and $\delta^{-1}x_\phi H x < 0$ as $x \in \mathcal{N}$, a contradiction. The "if" part of (b) is proved analogously. For the "if" part of (e), assume without loss of generality, that $\mathrm{In}_+(H) \leq \mathrm{In}_-(H)$ (otherwise consider $-H$ in place of H; note that $\mathcal{S}_0(-H) = \mathcal{S}_0(H)$). Now argue by contradiction as above.

In Steps 3, 4, and 5 below we suppose that ϕ is the conjugation and $H \in \mathsf{H}^{n \times n}$ is hermitian.

Step 3. Proof of the "only if" part of (b).

Let $\mathcal{M}_0 \in \mathcal{S}_{>0}(H)$, and assume that $\dim \mathcal{M}_0 < \mathrm{In}_+(H)$. We have to prove that there exists $\mathcal{M} \in \mathcal{S}_{>0}(H)$ such that $\mathcal{M}_0$ is properly contained in $\mathcal{M}$. Letting $d = \dim \mathcal{M}_0$, let $f_1, \ldots, f_n \in \mathsf{H}^{n \times 1}$ be an orthonormal basis of $\mathsf{H}^{n \times 1}$ in which the first d vectors $f_1, \ldots, f_d$ form an orthonormal basis of $\mathcal{M}_0$. (Such $f_1, \ldots, f_n$ exist because the Gram-Schmidt orthonormalization process is available in $\mathsf{H}^{n \times 1}$ with respect to the conjugation.) Write H with respect to the basis $\{f_1, \ldots, f_n\}$; in other words, replace H with U^*HU and $\mathcal{M}$ with $U^*\mathcal{M}$, where $U = [f_1 f_2 \ldots f_n]$ is the unitary matrix formed by $f_1, \ldots, f_n$. In view of Proposition 4.2.5, we can assume without loss of generality, that $\mathcal{M}_0 = \mathrm{Span}_\mathsf{H}\{e_1, \ldots, e_d\}$. Partition

$$H = \begin{bmatrix} H_{11} & H_{12} \\ H_{12}^* & H_{22} \end{bmatrix},$$

where H_{11} is $d \times d$ and H_{22} is $(n-d) \times (n-d)$. Because of our assumption that $\mathcal{M}_0 \in \mathcal{S}_{>0}(H)$, we have that H_{11} is positive definite (see Ex. 4.6.5). In particular, H_{11} is invertible.

The equality

$$\begin{bmatrix} I & 0 \\ -H_{12}^* H_{11}^{-1} & I \end{bmatrix} \begin{bmatrix} H_{11} & H_{12} \\ H_{12}^* & H_{22} \end{bmatrix} \begin{bmatrix} I & -H_{11}^{-1}H_{12} \\ 0 & I \end{bmatrix}$$

$$= \begin{bmatrix} H_{11} & 0 \\ 0 & H_{22} - H_{12}^* H_{11}^{-1} H_{12} \end{bmatrix}$$

and Proposition 4.2.5 show that we can (and do) further assume without loss of generality, that $H_{12} = 0$. Since $d < \mathrm{In}_+(H)$ and $\mathrm{In}_+(H) = d + \mathrm{In}_+(H_{22})$ (Ex.

4.6.6), we have $\mathrm{In}_+ (H_{22}) > 0$, and so there is a nonzero $v \in \mathsf{H}^{(n-d)\times 1}$ such that $v^* H_{22} v > 0$. Then the subspace $\mathcal{M}_0 + \mathrm{Span} \begin{bmatrix} 0_d \\ v \end{bmatrix}$ is strictly larger than $\mathcal{M}_0$ and is easily seen to be (H, ϕ)-positive. This completes the proof of Step 3.

Step 4. Proof of the "only if" part of (a).

Let $\mathcal{M} \in \mathcal{S}_{\geq 0}(H)$, and assume that $\dim \mathcal{M} < \mathrm{In}_+ (H) + \mathrm{In}_0 (H)$. Then for every positive integer m, we have $\mathcal{M} \in \mathcal{S}_{>0} (H + (1/m)I)$. As in Step 1, without loss of generality, it will be assumed that H has the form (4.3.1) with $\delta = 1$. Evidently,

$$\mathrm{In}_+ \left(H + \frac{1}{m} I \right) = \mathrm{In}_+ (H) + \mathrm{In}_0 (H) \qquad (m > 1).$$

By Step 3, there exists a subspace $\mathcal{M}_m \in \mathcal{S}_{>0} (H + (1/m)I)$ such that $\mathcal{M}_m \supseteq \mathcal{M}$ and $\dim \mathcal{M}_m = \dim \mathcal{M} + 1$. By Theorem 3.9.2, the sequence $\{\mathcal{M}_m\}_{m=1}^\infty$ has a convergent subsequence $\{\mathcal{M}_{m_k}\}_{k=1}^\infty$. Denote by $\mathcal{M}'$ the limit of $\mathcal{M}_{m_k}$ (as $k \to \infty$). Then $\dim \mathcal{M}' = \dim \mathcal{M} + 1$ (Theorem 3.9.3). Also, $\mathcal{M}' \in \mathcal{S}_{\geq 0}(H)$. Indeed, by Theorem 3.9.4 every vector $x \in \mathcal{M}'$ is the limit of some sequence $\{x_{m_k}\}_{k=1}^\infty$, where $x_{m_k} \in \mathcal{M}_{m_k}$ $(k = 1, 2, \ldots)$. Now

$$\langle Hx, x \rangle = \lim_{k \to \infty} \left\langle \left(H + \frac{1}{m} I \right) x_{m_k}, x_{m_k} \right\rangle \geq 0,$$

because $\left\langle \left(H + \frac{1}{m}I \right) x_{m_k}, x_{m_k} \right\rangle \geq 0$ for every k. Finally, $\mathcal{M}' \supseteq \mathcal{M}$ because $\mathcal{M}_{m_k} \supseteq \mathcal{M}$ for $k = 1, 2, \ldots$. This completes the proof of Step 4.

Step 5. Proof of the "only if" part of (e).

Let be given $\mathcal{M}_0 \in \mathcal{S}_0(H)$ having dimension

$$d < \min(\mathrm{In}_+ (H) + \mathrm{In}_0 (H), \ \mathrm{In}_- (H) + \mathrm{In}_0 (H)).$$

We will prove that there exists $\mathcal{M} \in \mathcal{S}_0(H)$ such that $\mathcal{M} \supseteq \mathcal{M}_0$ and $\dim \mathcal{M} = d + 1$. Since $\mathcal{S}_0(H) \subseteq \mathcal{S}_{\geq 0}(H)$, by part (a) there exists $\mathcal{M}_+ \supseteq \mathcal{M}_0$ such that $\dim \mathcal{M}_+ = d + 1$ and $\mathcal{M}_+ \in \mathcal{S}_{\geq 0}(H)$. Analogously, there exists $\mathcal{M}_- \in \mathcal{S}_{\leq 0}(H)$ such that $\mathcal{M}_- \supseteq \mathcal{M}_0$ and $\dim \mathcal{M}_- = d + 1$. Select $f_+ \in \mathcal{M}_+ \setminus \mathcal{M}_0$ and $f_- \in \mathcal{M}_- \setminus \mathcal{M}_0$. We claim that $\langle Hf_+, g \rangle = 0$ for every $g \in \mathcal{M}_0$. Indeed, for every real λ and $g \in \mathcal{M}_0$ we have

$$0 \leq \langle H(\lambda f_+ + g), \lambda f_+ + g \rangle = \lambda^2 \langle Hf_+, f_+ \rangle + \lambda \cdot 2\Re(\langle Hf_+, g \rangle).$$

Since this inequality holds for all real λ and $\langle Hf_+, f_+ \rangle \geq 0$, we must have the equality $\Re(\langle Hf_+, g \rangle) = 0$. Replacing here g by $g\alpha$, $\alpha \in \mathsf{H}$, it follows that $\Re(\alpha^* \langle Hf_+, g \rangle) = 0$ for all $\alpha \in \mathsf{H}$, which is possible only if $\langle Hf_+, g \rangle = 0$. Analogously, $\langle Hf_-, g \rangle = 0$ for every $g \in \mathcal{M}_0$.

We can (and do) assume that $\mathcal{M}_- \neq \mathcal{M}_+$ (otherwise, $\mathcal{M}_- = \mathcal{M}_+$ is H-neutral of dimension $d + 1$, and we are done). For $0 \leq \alpha \leq 1$, consider the subspace

$$\mathcal{M}_\alpha := \mathrm{Span}_\mathsf{H} \{\alpha f_+ + (1 - \alpha f_- \} + \mathcal{M}_0.$$

We have $f_+ \notin \mathcal{M}_0, f_- \notin \mathcal{M}_0$ and $\alpha f_+ + (1 - \alpha)f_- \notin \mathcal{M}_0$ for $0 < \alpha < 1$ (otherwise, we would have $\mathcal{M}_- = \mathcal{M}_+$). Thus, $\dim \mathcal{M}_\alpha = d + 1$ for every $\alpha \in [0, 1]$. Since

$\langle Hf_-, g \rangle = \langle Hf_+, g \rangle = 0$ for every $g \in \mathcal{M}_0$, we also have $\langle H(\alpha f_+ + (1-\alpha)f_-), g \rangle = 0$ for every $g \in \mathcal{M}_0$. Since $\langle Hf_+, f_+ \rangle \geq 0 \geq \langle Hf_-, f_- \rangle$ and the function

$$z(\alpha) := \langle H(\alpha f_+ + (1-\alpha)f_-), \alpha f_+ + (1-\alpha)f_- \rangle$$

is a continuous real-valued function of $\alpha \in [0,1]$, there is $\alpha_0 \in [0,1]$ such that $z(\alpha_0) = 0$. Then the subspace $\mathcal{M}_{\alpha_0}$ is H-neutral, has dimension $d+1$, and contains $\mathcal{M}_0$.

Step 6. Proof of Theorem 4.2.6 for a nonstandard ϕ.

Using Proposition 4.2.5 and Theorem 4.1.2, we may assume that

$$H = \mathrm{diag}\,(\beta(\phi)I_p, -\beta(\phi)I_q, 0_{n-q-p}).$$

Let

$$\widehat{H} := \mathrm{diag}\,(I_p, -I_q, 0_{n-q-p}) = (\beta(\phi))^{-1} H.$$

A straightforward computation using a units triple (q_1, q_2, q_3) with $q_1 = \beta(\phi)$ (so that $(q_2)_\phi = q_2$, $(q_3)_\phi = q_3$) shows that

$$\beta(\phi)x_\phi = x^* \beta(\phi), \quad \text{for all } x \in \mathsf{H}^{n \times 1}.$$

Thus

$$\beta(\phi)^{-1} x_\phi H x = x^* \widehat{H} x, \quad \text{for all } x \in \mathsf{H}^{n \times 1}.$$

It follows that a subspace $\mathcal{M} \subseteq \mathsf{H}^{n \times 1}$ is (H, ϕ)-nonnegative or (H, ϕ)-positive, etc., if and only if $\mathcal{M}$ is $(\widehat{H}, \phi')$-nonnegative or $(\widehat{H}, \phi')$-positive, etc.; here ϕ' is the conjugation. We have reduced the proof to the case when the involution is the conjugation and H is a hermitian matrix.

This concludes the proof of Theorem 4.2.6. $\qquad\square$

4.4 PROOF OF THEOREM 4.2.7

This section is devoted to the proof of Theorem 4.2.7.

It will be convenient to have a simple lemma first.

Lemma 4.4.1. *If ϕ is a nonstandard involution, then every subspace of $\mathsf{H}^{n \times 1}$ of dimension at least two contains a nonzero ϕ-selforthogonal vector x, i.e., $x_\phi x = 0$.*

Proof. It suffices to consider the two-dimensional subspace $\mathcal{M}$ spanned by vectors $y_1, y_2 \in \mathsf{H}^{n \times 1}$. If $(y_1)_\phi y_1 = 0$ or $(y_2)_\phi y_2 = 0$, we are done: take $x = y_1$ or $x = y_2$, as the case may be. Otherwise, consider the ϕ-orthogonal companion $(\mathrm{Span}_{\mathsf{H}} \{y_1\})^{\perp \phi}$ of $\mathrm{Span}_{\mathsf{H}} \{y_1\}$. By Proposition 3.6.4(b),

$$\mathcal{M} \cap (\mathrm{Span}_{\mathsf{H}} \{y_1\})^{\perp \phi} = \mathrm{Span}_{\mathsf{H}} \{z\}$$

for some nonzero $z \in \mathcal{M}$. If $z_\phi z = 0$, we are done again. Otherwise, we seek x in the form $x = y_1 + z\alpha$ for a suitable $\alpha \in \mathsf{H}$. The condition $x_\phi x = 0$ can be rewritten as

$$(y_1)_\phi y_1 + \alpha_\phi (z_\phi z)\alpha = 0. \tag{4.4.1}$$

Recall that $(y_1)_\phi y_1 \neq 0$, $z_\phi z \neq 0$ by our assumptions. By Theorem 2.5.1(b), we find $\gamma_1, \gamma_2 \in \mathsf{H} \setminus \{0\}$ such that $(\gamma_1)_\phi \gamma_1 = -(y_1)_\phi y_1$, $(\gamma_2)_\phi \gamma_2 = z_\phi z$. Now take $\alpha = \gamma_2^{-1} \gamma_1$ to satisfy (4.4.1). $\qquad\square$

Proof of Theorem 4.2.7. We divide the proof into steps.

Step 1. Proof of the "if" part.

Let $\mathcal{M}$ be (H, ϕ)-neutral of dimension k. Letting $P \in \mathsf{H}^{n \times n}$ be an invertible matrix whose first k columns form a basis for $\mathcal{M}$, we obtain

$$P_\phi H P = \begin{bmatrix} 0_{k \times k} & H_1 \\ \pm (H_1)_\phi & H_2 \end{bmatrix} \tag{4.4.2}$$

for some $H_2 \in \mathsf{H}^{(n-k) \times (n-k)}$, $H_1 \in \mathsf{H}^{k \times (n-k)}$. From (4.4.2), we have

$$\operatorname{rank} H = \operatorname{rank} (P_\phi H P) \leq \operatorname{rank} (H_1)_\phi + \operatorname{rank} \begin{bmatrix} H_1 \\ H_2 \end{bmatrix} \leq 2(n - k),$$

or

$$k \leq n - \frac{\operatorname{rank} H}{2},$$

which implies (since k is an integer) $k \leq n - \lceil \operatorname{rank} H \rceil / 2$. This shows that every (H, ϕ)-neutral subspace of dimension $n - \lceil \operatorname{rank} H \rceil / 2$ is maximal, which proves the "if" part of the theorem.

Such subspaces exist. Indeed, to construct an (H, ϕ)-neutral subspace of dimension $n - \lceil \operatorname{rank} H \rceil / 2$, in the case ϕ is nonstandard, without loss of generality, we may assume in view of Theorem 4.1.2 that $H = I_p \oplus 0_{n-p}$, $p = \operatorname{rank} H$. It is easy to see that the construction boils down to constructing a one-dimensional (I_2, ϕ)-neutral subspace. One such subspace is $\operatorname{Span}_\mathsf{H} \begin{bmatrix} 1 \\ \beta(\phi) \end{bmatrix}$. If ϕ is the conjugation, then in view of Theorem 4.1.6 we may assume $H = \mathfrak{i} I_r \oplus 0_{n-r}$, $r = \operatorname{rank} H$. In this case, $\operatorname{Span}_\mathsf{H} \begin{bmatrix} 1 \\ \mathfrak{j} \end{bmatrix}$ is a one-dimensional $(\mathfrak{i} I_2, \phi)$-neutral subspace.

Step 2. Proof of the "only if" part in the case $H = I$, ϕ a nonstandard involution.

Let $\mathcal{M}$ be an (I, ϕ)-neutral subspace of dimension $d \leq n - 1 - \lceil n \rceil / 2$. We need to show that there exists an (I, ϕ)-neutral subspace that is strictly larger than $\mathcal{M}$. By Proposition 3.6.4, the ϕ-orthogonal companion $\mathcal{M}^{\perp \phi}$ has dimension at least $1 + \lceil n \rceil / 2$. Since

$$1 + \frac{\lceil n \rceil}{2} \geq \left(n - 1 - \frac{\lceil n \rceil}{2} \right) + 2,$$

there exists a two-dimensional subspace in $\mathcal{M}^{\perp \phi}$, call it $\mathcal{N}$, such that $\mathcal{N} \cap \mathcal{M} = \{0\}$. By Lemma 4.4.1, we find a nonzero ϕ-selforthogonal vector $x \in \mathcal{N}$. Then it is easy to see that the subspace spanned by $\mathcal{M}$ and x is strictly larger than $\mathcal{M}$ and is (I, ϕ)-neutral.

Step 3. Proof of the "only if" part in the case $H = H_\phi$, ϕ a nonstandard involution.

As in Step 2, given an (H, ϕ)-neutral subspace $\mathcal{M}$ of dimension d with

$$d \leq n - 1 - \frac{\lceil \operatorname{rank} H \rceil}{2}, \tag{4.4.3}$$

we need to show that there exists an (H, ϕ)-neutral subspace which is strictly larger than $\mathcal{M}$. By Proposition 4.2.5 and Theorem 4.1.2, we may (and do) assume that

$H = \begin{bmatrix} I_\ell & 0 \\ 0 & 0 \end{bmatrix}$, where $\ell = \operatorname{rank} H$. In view of the already proved part of Theorem 4.2.7 in Step 1, we suppose $\ell < n$.

If $\mathcal{M}$ does not contain $\operatorname{Span}_{\mathsf{H}} \{e_{\ell+1}, \ldots, e_n\}$, then the (H, ϕ)-neutral subspace

$$\mathcal{M} + \operatorname{Span}_{\mathsf{H}} \{e_{\ell+1}, \ldots, e_n\}$$

is strictly larger than $\mathcal{M}$. We suppose, therefore, that

$$\mathcal{M} \supseteq \operatorname{Span}_{\mathsf{H}} \{e_{\ell+1}, \ldots, e_n\}. \tag{4.4.4}$$

In particular, (4.4.4) entails the assumption that $d \geq n - \ell$. Choose a basis for $\mathcal{M}$ of the form

$$\{f_1, \ldots, f_{d-n+\ell}, e_{\ell+1}, \ldots, e_n\},$$

where

$$f_1 = \begin{bmatrix} f_1' \\ 0 \end{bmatrix}, f_2 = \begin{bmatrix} f_2' \\ 0 \end{bmatrix}, \ldots, f_{d-n+\ell} = \begin{bmatrix} f_{d-n+\ell}' \\ 0 \end{bmatrix} \in \operatorname{Span}_{\mathsf{H}} \{e_1, \ldots, e_\ell\},$$

with $f_1', f_2', \ldots, f_{d-n+\ell}' \in \mathsf{H}^{\ell \times 1}$. Define the subspace

$$\mathcal{M}' = \operatorname{Span}_{\mathsf{H}} \{f_1', f_2', \ldots, f_{d-n+\ell}'\} \subseteq \mathsf{H}^{\ell \times 1}.$$

Clearly $\mathcal{M}'$ is (I_ℓ, ϕ)-neutral and has dimension $d - n + \ell$. Condition (4.4.3) implies

$$d - n + \ell \leq \ell - \frac{[\ell]}{2} - 1;$$

thus, by the result of Step 2, $\mathcal{M}'$ is not a maximal (I_ℓ, ϕ)-neutral subspace in $\mathsf{H}^{\ell \times 1}$. Now, if $\mathcal{M}''$ is an (I_ℓ, ϕ)-neutral subspace strictly larger than $\mathcal{M}'$ and if $g_1, \ldots, g_s$ is a basis for $\mathcal{M}''$, then

$$\operatorname{Span}_{\mathsf{H}} \left\{ \begin{bmatrix} g_1 \\ 0 \end{bmatrix}, \begin{bmatrix} g_2 \\ 0 \end{bmatrix}, \ldots, \begin{bmatrix} g_s \\ 0 \end{bmatrix} \right\} + \operatorname{Span}_{\mathsf{H}} \{e_1, \ldots, e_\ell\}$$

is an (H, ϕ)-neutral subspace strictly larger than $\mathcal{M}$.

It remains to prove the "only if" part in the case $H = -H_\phi$, ϕ the conjugation. This follows from the already proven part of Theorem 4.2.7 by using Proposition 4.1.7. Indeed, letting ϕ be any nonstandard involution, we see that the matrix $\beta(\phi)H$ is ϕ-hermitian. Moreover, Proposition 4.1.7(b) shows, by taking $X = H$ and S and T vectors in a subspace $\mathcal{M}$, that $\mathcal{M}$ is $(H, {}^*)$-neutral if and only if it is $(\beta(\phi)H, \phi)$-neutral.

Alternatively, the "only if" part in the case $H = -H^*$ can be proved analogously to Steps 2 and 3, using the following lemma in place of Lemma 4.4.1.

Lemma 4.4.2. *Let $G \in \mathsf{H}^{n \times n}$ be skewhermitian. Then every subspace $\mathcal{M}$ of $\mathsf{H}^{n \times 1}$ of dimension at least two contains a nonzero vector x which is selforthogonal in the sense of G, i.e., $x^* G x = 0$.*

Proof. We may assume that the dimension of $\mathcal{M}$ is two. Let $y_1, \ldots, y_n$ be an orthonormal basis in $\mathsf{H}^{n \times 1}$ such that y_1, y_2 is a basis for $\mathcal{M}$. Replacing $\mathcal{M}$ with

.

$S(\mathcal{M})$, where $S = \begin{bmatrix} y_1^* \\ y_2^* \\ \vdots \\ y_n^* \end{bmatrix}$, and replacing H with $(S^*)^{-1}HS^{-1}$, we can (and do)

assume that $\mathcal{M} = \text{Span}_{\mathsf{H}}\{e_1, e_2\}$. Using Theorem 4.1.6 we may further assume that the 2×2 upper-left corner of G has one of the three forms (1) $\begin{bmatrix} i & 0 \\ 0 & i \end{bmatrix}$; (2) $\begin{bmatrix} i & 0 \\ 0 & 0 \end{bmatrix}$; (3) 0_2. If (2) or (3) occurs, we take $x = e_2$. If (1) occurs, we take $x = e_1 + je_2$. In all cases, the equality $x^*Gx = 0$ is easily verified. $\square$

This completes the proof of Theorem 4.2.7.

4.5 REPRESENTATION OF SEMIDEFINITE SUBSPACES

In this section, either ϕ is the conjugation and $H = H^* \in \mathsf{H}^{n \times n}$ or ϕ is a nonstandard involution and $H = -H_\phi$. We present here a representation of semidefinite subspaces with respect to ϕ which will be useful later on.

Lemma 4.5.1. *Let* $\mathcal{M} \subseteq \mathsf{H}^{n \times 1}$ *be an* (H, ϕ)*-nonnegative subspace of dimension* d*. Then there is a nonsingular* $P \in \mathsf{H}^{n \times n}$ *such that* $P^{-1}\mathcal{M} = \text{Span}_{\mathsf{H}}\{e_1, \ldots, e_d\}$ *and* $P_\phi HP$ *has the form*

$$P_\phi HP = \begin{bmatrix} 0_{d''} & 0 & 0 & 0 & 0 \\ 0 & 0 & 0 & I_{d'} & 0 \\ 0 & 0 & \spadesuit I_{d_2} & 0 & 0 \\ 0 & \delta I_{d'} & 0 & 0 & 0 \\ 0 & 0 & 0 & 0 & H_0 \end{bmatrix}, \qquad (4.5.1)$$

where $d_2 + d' + d'' = d$ *and where* $\spadesuit = 1$, $\delta = 1$ *if* $\phi = *$ *and* $\spadesuit = \beta(\phi)$, $\delta = -1$ *if* ϕ *is nonstandard.*

Note that H_0 in (4.5.1) is necessarily hermitian (if $\delta = 1$) or ϕ-skewhermitian (if $\delta = -1$). The cases when some of d_2, d', d'' are zeros are not excluded.

Proof. The result follows from properties analogous to those of *skewly linked* subspaces in the context of complex hermitian matrices; see, e.g., Mehl et al. [110], Iohvidov et al. [70], Bolshakov et al. [16]. For the case when H is hermitian and invertible, Lemma 4.5.1 was proved in Alpay et al. [4]. We follow the approach of Alpay et al.

Applying a transformation $H \mapsto S_\phi HS$ for some invertible S, we may assume that $\mathcal{M} = \text{Span}_{\mathsf{H}}\{e_1, \ldots, e_d\}$. Partition

$$H = \begin{bmatrix} H_1 & H_2 \\ \delta(H_2)_\phi & H_3 \end{bmatrix},$$

where $H_1 \in \mathsf{H}^{d \times d}$. Using one of the canonical forms (4.1.2) or (4.1.3), we may partition further

$$H = \begin{bmatrix} H_{11} & H_{12} & H_{13} & H_{14} \\ (H_{12})_\phi & H_{22} & H_{23} & H_{24} \\ \delta(H_{13})_\phi & \delta(H_{23})_\phi & H_{33} & H_{34} \\ \delta(H_{14})_\phi & \delta(H_{24})_\phi & \delta(H_{34})_\phi & H_{44} \end{bmatrix},$$

where

$$\begin{bmatrix} H_{11} & H_{12} \\ \delta(H_{12})_\phi & H_{22} \end{bmatrix} \in \mathsf{H}^{d \times d} \tag{4.5.2}$$

and where it is assumed $H_{11} = 0$, $H_{12} = 0$, and $H_{22} = \spadesuit I_{d_2}$. Using the rank decomposition for quaternion matrices (Proposition 3.2.5), we have

$$Q_1 \begin{bmatrix} H_{13} & H_{14} \end{bmatrix} Q_2 = \begin{bmatrix} 0 & 0 \\ I_{d_1} & 0 \end{bmatrix}$$

for some invertible (quaternion) matrices Q_1 and Q_2. Setting $P_1 = (Q_1)_\phi \oplus I_{d_2} \oplus Q_2$, we obtain that $(P_1)_\phi H P_1$ has the form

$$(P_1)_\phi H P_1 = \begin{bmatrix} 0 & 0 & 0 & 0 & 0 \\ 0 & 0 & 0 & I_{d_1} & 0 \\ 0 & 0 & \spadesuit I_{d_2} & \widehat{H}_{23} & \widehat{H}_{24} \\ 0 & \delta I_{d_1} & \delta(\widehat{H}_{23})_\phi & \widehat{H}_{33} & \widehat{H}_{34} \\ 0 & 0 & \delta(\widehat{H}_{24})_\phi & \delta(\widehat{H}_{34})_\phi & \widehat{H}_{44} \end{bmatrix}, \quad P_1^{-1} \mathcal{M} = \mathcal{M}.$$

Furthermore, setting

$$P_2 = P_1 \begin{bmatrix} I & 0 & 0 & 0 & 0 \\ 0 & I_{d_1} & 0 & \frac{1}{2}(\widehat{H}_{23}^* \widehat{H}_{23} - \widehat{H}_{23}) & 0 \\ 0 & 0 & I_{d_2} & -\widehat{H}_{23} & -\widehat{H}_{24} \\ 0 & 0 & 0 & I_{d_1} & 0 \\ 0 & 0 & 0 & 0 & I \end{bmatrix},$$

we obtain that $P_2^{-1} \mathcal{M} = \mathcal{M}$ and

$$(P_2)_\phi H P_2 = \begin{bmatrix} 0 & 0 & 0 & 0 & 0 \\ 0 & 0 & 0 & I_{d_1} & 0 \\ 0 & 0 & \spadesuit I_{d_2} & 0 & 0 \\ 0 & \delta I_{d_1} & 0 & 0 & \widetilde{H}_{34} \\ 0 & 0 & 0 & \delta(\widetilde{H}_{34})_\phi & \widetilde{H}_{44} \end{bmatrix},$$

and one more obvious transformation $(P_2)_\phi H P_2 \mapsto (P_3)_\phi (P_2)_\phi H P_2 P_3$ for a suitable invertible P_3 completes the proof. $\qquad \square$

Analogous result holds for (H, ϕ)-nonpositive matrices. The only change to be made is replacement of $\spadesuit I_{d_2}$ in (4.5.1) with $-\spadesuit I_{d_2}$.

Corollary 4.5.2. *Assume the hypotheses and notation of Lemma 4.5.1. Then:*

(1) *the subspace $\mathcal{M}$ is maximal (H, ϕ)-nonnegative if and only if H_0 is negative definite (when ϕ is the conjugation) or $\beta(\phi)H_0$ is negative definite (when ϕ is nonstandard);*

(2) *$\mathcal{M}$ is (H, ϕ)-neutral if and only if $d_2 = 0$;*

(3) *suppose ϕ is the conjugation; then $\mathcal{M}$ is maximal (H, ϕ)-neutral if and only if $d_2 = 0$ and H_0 is definite (positive or negative);*

(4) *suppose ϕ is the nonstandard; then $\mathcal{M}$ is maximal (H, ϕ)-neutral if and only if $d_2 = 0$ and $\beta(\phi)H_0$ is definite (positive or negative).*

We leave the proof as an exercise (Ex. 4.6.19).

4.6 EXERCISES

Ex. 4.6.1. Provide details for the proof of Theorem 4.1.6.

Ex. 4.6.2. Provide details for the proof of Theorem 4.1.12.

Ex. 4.6.3. Show that the ϕ-skewhermitian matrix $\begin{bmatrix} 0 & z \\ -z_\phi & 0 \end{bmatrix}$ has balanced inertia for every $z \in \mathsf{H}$.

Ex. 4.6.4. For a nonstandard involution ϕ and a ϕ-skewhermitian $H \in \mathsf{H}^{n \times n}$, show that H is $\beta(\phi)$-positive semidefinite, $\beta(\phi)$-positive definite, $\beta(\phi)$-negative semidefinite, or $\beta(\phi)$-negative definite if and only if $(\beta(\phi))^{-1}x_\phi Hx \geq 0$ for all $x \in \mathsf{H}^{n \times 1}$, $(\beta(\phi))^{-1}x_\phi Hx > 0$ for all $x \in \mathsf{H}^{n \times 1} \setminus \{0\}$, $(\beta(\phi))^{-1}x_\phi Hx \leq 0$ for all $x \in \mathsf{H}^{n \times 1}$, or $(\beta(\phi))^{-1}x_\phi Hx < 0$ for all $x \in \mathsf{H}^{n \times 1} \setminus \{0\}$, respectively.

Ex. 4.6.5. State and prove results analogous to Ex. 4.6.4 for classes of matrices with definiteness properties with respect to the conjugation.

Ex. 4.6.6. Prove that inertia and $\beta(\phi)$-inertia are additive with respect to block diagonal decompositions: If $H = \mathrm{diag}\,(H_1, \ldots, H_k) \in \mathsf{H}^{n \times n}$ is hermitian (if ϕ is the conjugation) or ϕ-skewhermitian (if ϕ is a nonstandard involution), then $H_1, \ldots, H_k$ are hermitian or ϕ-skewhermitian as the case may be, and

$$\mathrm{In}_\mu\,(H) = \sum_{j=1}^{k} \mathrm{In}_\mu\,(H_j), \quad \text{for } \mu = +, -, 0.$$

Ex. 4.6.7. For a fixed $H \in \mathsf{H}^{n \times n}$ and ϕ, where either H is hermitian and ϕ the conjugation, or H is ϕ-skewhermitian and ϕ a nonstandard involution, prove that the sets $\mathcal{S}_0(H)$, $\mathcal{S}_{\geq 0}(H)$, and $\mathcal{S}_{\leq 0}(H)$ are closed in the gap metric topology of Grass_n.

Ex. 4.6.8. For H and ϕ as in Ex. 4.6.7, prove that the sets $\mathcal{S}_{>0}(H)$ and $\mathcal{S}_{<0}(H)$ are open in the gap metric topology of Grass_n. Hint: Show that the complements of the sets in question are closed in Grass_n.

Ex. 4.6.9. Repeat Ex. 4.6.7 and Ex. 4.6.8, but use the sets of maximal (by inclusion) elements in each of $\mathcal{S}_0(H)$, $\mathcal{S}_{\geq 0}(H)$, $\mathcal{S}_{\leq 0}(H)$, $\mathcal{S}_{>0}(H)$ and $\mathcal{S}_{<0}(H)$ (rather than using $\mathcal{S}_\mu(H)$, $\mu \in \{0, \geq 0, \leq 0, > 0, < 0\}$, as in Ex. 4.6.7 and Ex. 4.6.8).

Ex. 4.6.10. Prove that if $A \in \mathsf{H}^{n \times n}$ is positive semidefinite, then for $x \in \mathsf{H}^{n \times 1}$ the equality $x^* Ax = 0$ is equivalent to $Ax = 0$. Hint: Use Theorem 4.1.6.

Ex. 4.6.11. Find all (H_j, ϕ)-neutral subspaces, $j = 1, 2$, for the following two matrices:

$$H_1 = \begin{bmatrix} 0 & \beta(\phi) \\ \beta(\phi) & 0 \end{bmatrix}, \qquad H_2 = \begin{bmatrix} 0 & 0 & \beta(\phi) \\ 0 & \beta(\phi) & 0 \\ \beta(\phi) & 0 & 0 \end{bmatrix}.$$

Hint: For H_2, consider $\mathrm{Span}_\mathsf{H} \begin{bmatrix} 1 \\ x \\ y \end{bmatrix}$, $x, y \in \mathsf{H}$.

Ex. 4.6.12. Find the canonical form of Theorem 4.1.6 for the following hermitian matrices in $\mathsf{H}^{n \times n}$:

(a)
$$\begin{bmatrix} 1 & \alpha_1 & \alpha_2 & \ldots & \alpha_{n-1} \\ \alpha_1^* & 1 & 0 & \ldots & 0 \\ \alpha_2^* & 0 & 1 & \ldots & 0 \\ \vdots & \vdots & \vdots & \ddots & \vdots \\ \alpha_{n-1}^* & 0 & 0 & \ldots & 1 \end{bmatrix}, \qquad \alpha_1, \ldots, \alpha_{n-1} \in \mathsf{H},$$

(the canonical form depends on $\alpha_1, \ldots, \alpha_{n-1}$);

(b)
$$\begin{bmatrix} 0 & 0 & \ldots & 0 & \alpha_1 \\ 0 & 0 & \ldots & \alpha_2 & 0 \\ \vdots & \vdots & \ddots & \vdots & \vdots \\ \alpha_n & 0 & \ldots & 0 & 0 \end{bmatrix}, \qquad \alpha_1, \ldots, \alpha_n \in \mathsf{H},$$

where $\alpha_j = \alpha_{n+1-j}^*$ for $j = 1, 2, \ldots, n$;

(c) $\left(\begin{bmatrix} 0 & \alpha_1 \\ \alpha_1^* & 0 \end{bmatrix} \right) \oplus \cdots \oplus \left(\begin{bmatrix} 0 & \alpha_k \\ \alpha_k^* & 0 \end{bmatrix} \right), \qquad \alpha_1, \ldots, \alpha_k \in \mathsf{H},$

where n is even and $k = n/2$.

Ex. 4.6.13. Find the canonical form of Theorem 4.1.2 for the following ϕ-skewhermitian matrices in $\mathsf{H}^{n \times n}$:

(a)
$$\begin{bmatrix} \beta(\phi) & \alpha_1 & \alpha_2 & \ldots & \alpha_{n-1} \\ -\phi(\alpha_1) & \beta(\phi) & 0 & \ldots & 0 \\ -\phi(\alpha_2) & 0 & \beta(\phi) & \ldots & 0 \\ \vdots & \vdots & \vdots & \ddots & \vdots \\ -\phi(\alpha_{n-1}) & 0 & 0 & \ldots & \beta(\phi) \end{bmatrix}, \qquad \alpha_1, \ldots, \alpha_{n-1} \in \mathsf{H};$$

(b)
$$\begin{bmatrix} 0 & 0 & \ldots & 0 & \alpha_1 \\ 0 & 0 & \ldots & \alpha_2 & 0 \\ \vdots & \vdots & \ddots & \vdots & \vdots \\ \alpha_n & 0 & \ldots & 0 & 0 \end{bmatrix}, \qquad \alpha_1, \ldots, \alpha_n \in \mathsf{H},$$

where $\alpha_j = -\phi(\alpha_{n+1-j})$ for $j = 1, 2, \ldots, n$;

(c) $\left(\begin{bmatrix} 0 & \alpha_1 \\ -\phi(\alpha_1) & 0 \end{bmatrix} \right) \oplus \cdots \oplus \left(\begin{bmatrix} 0 & \alpha_k \\ -\phi(\alpha_k) & 0 \end{bmatrix} \right), \qquad \alpha_1, \ldots, \alpha_k \in \mathsf{H},$

where n is even and $k = n/2$.

Ex. 4.6.14. Let ϕ be a nonstandard involution. Find all subspaces in the following classes: maximal (H, ϕ)-nonnegative, maximal (H, ϕ)-nonpositive, maximal (H, ϕ)-neutral, for each of the following ϕ-skewhermitian matrices $H \in \mathsf{H}^{3 \times 3}$:

(a) $\quad H = \begin{bmatrix} \beta(\phi) & 0 & 0 \\ 0 & \beta(\phi) & 0 \\ 0 & 0 & \beta(\phi) \end{bmatrix};$

(b) $\quad H = \begin{bmatrix} \beta(\phi) & 0 & \alpha \\ 0 & -\beta(\phi) & 0 \\ -\phi(\alpha) & 0 & \beta(\phi) \end{bmatrix};$

(c) $\quad H = \begin{bmatrix} 0 & 0 & \beta(\phi) \\ 0 & \beta(\phi) & 0 \\ \beta(\phi) & 0 & 0 \end{bmatrix}.$

Ex. 4.6.15. For each of the following ϕ-hermitian matrices $H \in \mathsf{H}^{3 \times 3}$, find all (H, ϕ)-neutral subspaces:

$$\text{(a)} \qquad H = \begin{bmatrix} 1 & \alpha_1 & \alpha_2 \\ \phi(\alpha_1) & 1 & 0 \\ \phi(\alpha_2) & 0 & 1 \end{bmatrix}, \quad \alpha_1, \alpha_2 \in \mathsf{H};$$

$$\text{(b)} \qquad H = \begin{bmatrix} 0 & 0 & \alpha_1 \\ 0 & \alpha_2 & 0 \\ \phi(\alpha_1) & 0 & 1 \end{bmatrix}, \quad \alpha_1, \alpha_2 \in \mathsf{H}, \quad \phi(\alpha_2) = \alpha_2.$$

Ex. 4.6.16. Show that a hermitian matrix $\begin{bmatrix} a & b \\ b^* & c \end{bmatrix} \in \mathsf{H}^{2 \times 2}$ is positive definite if and only if $a > 0$, $c > 0$, and $ac > |b|^2$.

Ex. 4.6.17. Show that a hermitian matrix $\begin{bmatrix} a & b \\ b^* & c \end{bmatrix} \in \mathsf{H}^{2 \times 2}$ is positive semidefinite if and only if $a \geq 0$, $c \geq 0$, and $ac \geq |b|^2$.

Ex. 4.6.18. Deduce Theorem 4.1.6 from Theorem 4.1.2 using Proposition 4.1.7. Hint: Take $T = S$ in Proposition 4.1.7.

Ex. 4.6.19. Prove Lemma 4.5.2. Hint: Use Theorem 4.2.6 to identify the dimensions of maximal (H, ϕ)-nonnegative and maximal (H, ϕ)-neutral subspaces.

4.7   NOTES

Results of Theorems 4.1.6(a) and Theorem 4.1.11, as well as the inertia theorem, are standard for real and complex matrices. For quaternion matrices, Theorems 4.1.2, 4.1.6, and 4.1.11 with complete proofs are found in Brieskorn [20], for example.

Theorem 4.2.6 is known in the real and complex cases; see, e.g., Alpay et al. [4] (the case of invertible H) and Mehl et al. [110]. The proof of Theorem 4.2.6 given in Section 4.2 follows the same approach as in Alpay et al.

Theorem 4.2.7 holds in the context of real and complex matrices, with essentially the same proof.

Chapter Five

Invariant subspaces and Jordan form

We start the chapter by introducing the notion of root subspaces for quaternion matrices. These are basic invariant subspaces, and we prove in particular that they enjoy the Lipschitz property with respect to perturbations of the matrix. Another important class of invariant subspaces are the 1-dimensional ones, i.e., generated by eigenvectors. Existence of quaternion eigenvalues and eigenvectors is proved, which leads to the Schur triangularization theorem (in the context of quaternion matrices) and its many consequences familiar for real and complex matrices. Jordan canonical form for quaternion matrices is stated and proved (both the existence and uniqueness parts) in full detail. We also discuss in this chapter various concepts of determinants for square-size quaternion matrices. Several applications of the Jordan form are given, including functions of matrices and boundedness properties of systems of differential and difference equations with constant quaternion coefficients.

5.1 ROOT SUBSPACES

Definition 5.1.1. A (quaternion) subspace $\mathcal{M} \subseteq \mathsf{H}^{n \times 1}$ is said to be *invariant* for a matrix $A \in \mathsf{H}^{n \times n}$ if $A\mathcal{M} \subseteq \mathcal{M}$.

Clearly, $\{0\}$ and $\mathsf{H}^{n \times 1}$ are A-invariant for any $A \in \mathsf{H}^{n \times n}$; these are the *trivial* invariant subspaces. In this section we consider a class of A-invariant subspaces called root subspaces.

If $p(t) = \sum_{j=0}^{m} p_j t^j$ is a polynomial of the independent variable t with real coefficients $p_0, \ldots, p_m$ and $A \in \mathsf{H}^{n \times n}$, then $p(A)$ is naturally defined as $p(A) = \sum_{j=0}^{m} p_j A^j \in \mathsf{H}^{n \times n}$. For a fixed A, the map $p(t) \mapsto p(A)$ is a unital algebra homomorphism:

$$(p + q)(A) = p(A) + q(A); \quad (pq)(A) = p(A)q(A);$$

$$(\alpha p)(A) = \alpha p(A); \quad 1(A) = I_n \tag{5.1.1}$$

for all polynomials p, q with real coefficients and all $\alpha \in \mathsf{R}$. Also,

$$p(S^{-1}AS) = S^{-1}p(A)S, \quad \text{for all invertible } S \in \mathsf{H}^{n \times n}.$$

Fix $A \in \mathsf{H}^{n \times n}$. Since $\mathsf{H}^{n \times n}$ is finite dimensional (as a real vector space), the powers of A, $\{A^j\}_{j=0}^{\infty}$, cannot be linearly independent, so, $p(A) = 0$ for some nonconstant polynomial $p(t)$ with real coefficients. Denote by $p^{(A)}(t)$ the monic (i.e., with leading coefficient 1) real polynomial of minimal degree such that $p(A) = 0$. The division algorithm for polynomials, together with the algebraic properties (5.1.1), shows that $p^{(A)}(t)$ is unique and that $p^{(A)}(t)$ divides any polynomial $q(t)$ with real coefficients such that $q(A) = 0$.

Definition 5.1.2. The polynomial $p^{(A)}(t)$ is called the *minimal polynomial* of A.

Write

$$p^{(A)}(t) = p_1(t)^{m_1} \cdots p_k(t)^{m_k}, \tag{5.1.2}$$

where the $p_j(t)$'s are distinct monic irreducible real polynomials (i.e., of the form $t - a$, a real, or of the form $t^2 + pt + q$, where $p, q \in \mathsf{R}$, with no real roots), and the m_j's are positive integers.

Definition 5.1.3. The subspaces

$$\mathcal{M}_j := \{u \,:\, u \in \mathsf{H}^{n \times 1}, \quad p_j(A)^{m_j} u = 0\}, \quad j = 1, 2, \ldots, k,$$

are called the *root subspaces* of A.

Since $A p_j(A)^{m_j} = p_j(A)^{m_j} A$, the root subspaces of A are A-invariant. Root subspaces form "building blocks" for invariant subspaces.

Proposition 5.1.4. *Let* $A \in \mathsf{H}^{n \times n}$. *Then:*

(a) *the root subspaces decompose* $\mathsf{H}^{n \times 1}$ *into a direct sum:*

$$\mathsf{H}^{n \times 1} = \mathcal{M}_1 \dotplus \mathcal{M}_2 \dotplus \cdots \dotplus \mathcal{M}_k; \tag{5.1.3}$$

(b) *for every* A-*invariant subspace* $\mathcal{M}$,

$$\mathcal{M} = (\mathcal{M} \cap \mathcal{M}_1) \dotplus \cdots \dotplus (\mathcal{M} \cap \mathcal{M}_k). \tag{5.1.4}$$

Proof. (a) follows from (b) (take $\mathcal{M} = \mathsf{H}^{n \times 1}$ in (b)).
To prove (b), we use (5.1.2) and observe that the polynomials

$$q_\ell(t) := \prod_{\{j \,:\, j \neq \ell\}} p_j(t)^{m_j}, \quad \ell = 1, 2, \ldots, k,$$

have no common real or complex roots. We now use a well-known property of polynomial algebra, which can be proved by repeated applications of division with remainder and is exposed in many texts; see, e.g., Herstein [61] or Artin [6]. Namely, for any finite collection of polynomials $y_1(t), \ldots, y_k(t)$ with coefficients in a field, there exist polynomials $z_1(t), \ldots, z_k(t)$ with coefficients in the same field such that

$$y_1(t)z_1(t) + \cdots + y_k(t)z_k(t) = \text{greatest common divisor of } \{y_1(t), \ldots, y_k(t)\}.$$

Applying this fact to $q_1(t), \ldots, q_k(t)$, we see that there exist polynomials with real coefficients $r_1(t), \ldots, r_k(t)$ such that

$$q_1(t)r_1(t) + \cdots + q_k(t)r_k(t) \equiv 1.$$

Thus, for every $x \in \mathcal{M}$ we have

$$x = q_1(A)r_1(A)x + \cdots + q_k(A)r_k(A)x.$$

Clearly, $q_j(A)r_j(A)x \in \mathcal{M} \cap \mathcal{M}_j$ for $j = 1, 2, \ldots, k$, and the inclusion $\subseteq$ in (5.1.4) follows. The converse inclusion is obvious.

It remains to prove that the $\mathcal{M}_j$'s form a direct sum. Assume that $x_1 + \cdots + x_k = 0$ for some $x_1 \in \mathcal{M}_1, \ldots, x_k \in \mathcal{M}_k$. We have:

$$p_1(A)^{m_1} x_1 = 0, \qquad q_1(A)(x_2 + \cdots + x_k) = 0. \tag{5.1.5}$$

Since the polynomials $p_1(t)^{m_1}$ and $q_1(t)$ are relatively prime, there exist polynomials with real coefficients $s_1(t)$ and $s_2(t)$ such that

$$s_1(t)p_1(t)^{m_1} + s_2(t)q_1(t) \equiv 1.$$

Now

$$
\begin{aligned}
x_1 &= (s_1(A)p_1(A)^{m_1} + s_2(A)q_1(A))x_1 \\
&= s_1(A)p_1(A)^{m_1}x_1 - s_2(A)q_1(A)(x_2 + \cdots + x_k) = 0
\end{aligned}
$$

in view of (5.1.5). Analogously, $x_j = 0$ for $j = 2, 3, \ldots, k$. $\square$

We note that the root subspaces are also given by

$$\mathcal{M}_\ell = \mathrm{Ran}\left(\prod_{\{j \,:\, j \neq \ell\}} p_j(A)^{m_j} \right), \quad \ell = 1, 2, \ldots, k, \tag{5.1.6}$$

where the minimal polynomial of A is (5.1.2). To verify this, note that the right-hand side of (5.1.6) is an A-invariant subspace, and apply Proposition 5.1.6(b) to it.

For a given $A \in \mathsf{H}^{n \times n}$, we will identify a sum $\mathcal{M}$ of root subspaces of A by the set of (complex and real) roots of the minimal real polynomial of the restriction $A|_{\mathcal{M}}$ of A to $\mathcal{M}$ (the restriction $A|_{\mathcal{M}}$ is understood as the matrix of the H-linear transformation $A_{\mathcal{M}} : \mathcal{M} \to \mathcal{M}$ with respect to some basis in $\mathcal{M}$). Note that this set is necessarily closed under complex conjugation. Thus, for example, the sum $\mathcal{M}$ of root subspaces of A corresponding to the set $\{1, \mathrm{i}, -\mathrm{i}\}$ is identified by the property that the minimal real polynomial of $A_{\mathcal{M}}$ is of the form $(x-1)^{m_1}(x^2+1)^{m_2}$ for some integers m_1, m_2.

Open Problem 5.1.5. *Describe the set of A-invariant subspaces in terms of analytic manifolds.*

For complex invariant subspaces of complex matrices, such a description was obtained by Shayman [144].

In view of Proposition 5.1.4, for the open problem 5.1.5 it suffices to consider the case when A has only one root subspace.

5.2   ROOT SUBSPACES AND MATRIX REPRESENTATIONS

In this section, we study the transformation of sums of root subspaces of quaternion matrices under real and complex representations developed in Sections 3.3 and 3.4.

At this point, we need to recall the concepts of root subspaces for real and complex matrices which are familiar in real and complex linear algebra. The construction of root subspaces for a matrix $A \in \mathsf{R}^{n \times n}$ is completely analogous to the construction for quaternion matrices given in Section 5.1, but using $R^{n \times 1}$ rather than $\mathsf{H}^{n \times 1}$; thus the root subspaces of A are (real) subspaces of $\mathsf{R}^{n \times 1}$. For a complex matrix $A \in \mathsf{C}^{n \times n}$ the minimal polynomial $p^{(A)}(t)$ is a polynomial with *complex* coefficients, and we use the factorization

$$p^{(A)}(t) = (t - a_1)^{m_1} (t - a_2)^{m_2} \cdots (t - a_k)^{m_k},$$

where $a_1, \ldots, a_k$ are all the distinct complex roots of $p^{(A)}(t)$. The root subspaces of A are defined as

$$\{u \,:\, u \in \mathsf{C}^{n \times 1}, \ (A - a_j)^{m_j} u = 0\}, \qquad \text{for} \ \ j = 1, 2, \ldots, k.$$

They are (complex) subspaces of $\mathsf{C}^{n \times 1}$.

Analogously to the quaternion matrices in Section 5.1, a sum of root subspaces $\mathcal{M}$ of a real, resp. complex, matrix will be identified by the roots of the real, resp. complex, minimal polynomial of the restriction of the matrix to $\mathcal{M}$.

Definition 5.2.1. For a real or complex matrix $A \in \mathsf{F}^{n \times n}$, $\mathsf{F} = \mathsf{C}$ or $\mathsf{F} = \mathsf{R}$, the (real and complex) roots of the minimal polynomial of A will be called the C-*eigenvalues* of A.

They are precisely the real and complex eigenvalues of A, to be distinguished from quaternion eigenvalues of quaternion matrices that will be introduced later on in this chapter.

Proposition 5.2.2. *Let $A \in \mathsf{H}^{n \times n}$, and let $v_1, \ldots, v_r$ be a basis for the sum of root subspaces of A corresponding to a set $T \subseteq \mathsf{C}$. Then the columns of the matrix $[\chi(v_1) \ \chi(v_2) \ \ldots \ \chi(v_r)]$ form a basis of the sum of root subspaces of $\chi(A)$ corresponding to the same set T.*

Proof. We have

$$A [v_1 \ v_2 \ \ldots \ v_r] = [v_1 \ v_2 \ \ldots \ v_r] X,$$

for some $X \in \mathsf{H}^{n \times n}$ with minimal real polynomial $p^{(X)}(x)$ having all its roots in T. Applying χ and using the real algebras homomorphism property of χ, we have

$$\chi(A) [\chi(v_1) \ \chi(v_2) \ \ldots \ \chi(v_r)] = [\chi(v_1) \ \chi(v_2) \ \ldots \ \chi(v_r)] \chi(X)$$

and $p(\chi(X)) = 0$. By Proposition 3.3.3 we have

$$\dim_{\mathsf{R}} \ (\text{column space of } [\chi(v_1) \ \chi(v_2) \ \ldots \ \chi(v_r)]) = 4r. \tag{5.2.1}$$

Write

$$p^{(A)}(t) = p(t)q(t),$$

where $p(t)$ and $q(t)$ are relatively prime monic real polynomials. Analogously to the above, we obtain the following properties:

(a)
$$\dim_{\mathsf{R}} \ (\text{column space of } [\chi(u_1) \ \chi(u_2) \ \ldots \ \chi(u_s)]) = 4s, \tag{5.2.2}$$

where $u_1, \ldots, u_s$ is a basis for the sum of root subspaces of A corresponding to the roots of $q(t)$;

(b) the column space of $[\chi(u_1) \ \chi(u_2) \ \ldots \ \chi(u_s)]$ is $\chi(A)$-invariant;

(c) the restriction of $\chi(A)$ to this column space is annihilated by $q(t)$.

However, by (5.1.3) we have $r + s = n$, so $4r + 4s = 4n$, the size of $\chi(A)$. It now follows from (5.2.1) and (5.2.2) that the columns of $[\chi(v_1)\ \chi(v_2)\ \ldots\ \chi(v_r)]$, resp. of $[\chi(u_1)\ \chi(u_2)\ \ldots\ \chi(u_s)]$, actually form a basis in the sum of root subspaces of $\chi(A)$ corresponding to the roots of $p(t)$, resp. of $q(t)$. $\qquad\qquad\square$

Corollary 5.2.3. *For $A \in \mathsf{H}^{n \times n}$, the set of the roots of $p^{(A)}(t)$ coincides with the set of C-eigenvalues of $\chi(A)$.*

Proof. Proposition 5.2.2 shows that the roots of $p^{(A)}(t)$ are C-eigenvalues of $\chi(A)$. Conversely, by the homomorphism property of χ, we have $p^{(A)}(\chi(A)) = 0$. Hence, the minimal polynomial of $\chi(A)$ divides $p^{(A)}(t)$. Since the C-eigenvalues of $\chi(A)$ are precisely the roots of the minimal polynomial of $\chi(A)$, it follows that the C-eigenvalues of $\chi(A)$ are roots of $p^{(A)}(t)$. $\qquad\qquad\square$

Proposition 5.2.4. *Let $A \in \mathsf{H}^{n \times n}$, and let $v_1, \ldots, v_r$ be a basis for the sum of root subspaces of A corresponding to a set $T \subseteq \mathsf{C}$. Then the columns of the matrix $[\omega(v_1)\ \omega(v_2)\ \ldots\ \omega(v_r)]$ form a basis of the sum of root subspaces of $\omega(A)$ corresponding to the same set T.*

Proof. Note that the set T is necessarily closed under complex conjugation. The proof is analogous to that of Proposition 5.2.2, using the algebraic properties of the map ω and Proposition 3.4.2. $\qquad\qquad\square$

As an application of the results of this section, we show that sums of root subspaces behave like Lipschitz functions under changes of the underlying matrix A. This statement is part of the next theorem. The notation $\overline{X}$ stands for the complex conjugate of a set $X \subseteq \mathsf{C}$.

Theorem 5.2.5. (a) *The roots of the minimal polynomial of a matrix depend continuously on the matrix:*

Fix $A \in \mathsf{H}^{n \times n}$, and let $\lambda_1, \ldots, \lambda_s$ be all the distinct roots of $p^{(A)}(t)$ in the closed upper complex half-plane C_+. Then for every $\epsilon > 0$, there exists $\delta > 0$ such that if $B \in \mathsf{H}^{n \times n}$ satisfies $\|B - A\| < \delta$, then the roots of $p^{(B)}(t)$ in C_+ are contained in the union

$$\cup_{j=1}^s \{z \in \mathsf{C}_+\ :\ |z - \lambda_j| < \epsilon\}.$$

(b) *Sums of root subspaces are Lipschitz functions of the underlying matrix, in the following sense.*

Given $A \in \mathsf{H}^{n \times n}$ and $\lambda_1, \ldots, \lambda_s$ as in part (a), there exist $\delta_0, K_0 > 0$ such that for every $B \in \mathsf{H}^{n \times n}$ satisfying $\|B - A\| < \delta_0$, it holds that if T is any nonempty subset of $\lambda_1, \ldots, \lambda_s$ and T' is the set of all roots of $p^{(B)}(t)$ contained in

$$\mathcal{U} := \cup_{j \in T}\{z \in \mathsf{C}_+\ :\ |z - \lambda_j| < \delta_0\},$$

then the sum of root subspaces $\mathcal{M}'$ of B corresponding to $T' \cup \overline{T'}$ and the sum of root subspaces $\mathcal{M}$ of A corresponding to $T \cup \overline{T}$ satisfy the inequality

$$\theta(\mathcal{M}, \mathcal{M}') \leq K_0\, \|B - A\|.$$

Here $\theta(\cdot, \cdot)$ is the gap function.

Proof. Part (a). The corresponding property for *real* or *complex* matrices is well known and follows from the continuous dependence of the roots of the polynomial equation $\det(tI - Y) = 0$ (i.e., the eigenvalues of Y) on its coefficients; here $Y \in \mathsf{R}^{n \times n}$ or $Y \in \mathsf{C}^{n \times n}$; see Theorem 5.15.1 in the Appendix. Then (a) follows by using the map χ to reduce the proof to the real matrices and by taking advantage of Corollary 5.2.3 and of property (v) of χ.

Part (b). Let $v_1, \ldots, v_r$ be an orthonormal basis for $\mathcal{M}$. By Proposition 5.2.2, the columns of $[\chi(v_1,) \; \chi(v_2) \; \ldots \; \chi(v_r)]$ form a basis for a sum of root subspaces of $\chi(A)$ corresponding to the eigenvalues in $T \cup \overline{T}$. By Proposition 3.3.3, the columns of $[\chi(v_1) \; \ldots \; \chi(v_r)]$ form an *orthonormal* basis in

$$\widetilde{\mathcal{M}} := \text{column space of } [\chi(v_1) \; \chi(v_2) \; \ldots \; \chi(v_r)] \in \mathsf{R}^{4n \times 1}.$$

We now use the corresponding Lipschitz property result for sums of root subspaces for real matrices (see, e.g., Theorem 15.9.7 in Gohberg et al. [54]). Thus, there exists $\delta_0' > 0$ such that every $B' \in \mathsf{R}^{4n \times 4n}$ with $\|B' - \chi(A)\| \leq \delta_0'$ has an invariant subspace $\widetilde{\mathcal{M}}' \subseteq \mathsf{R}^{4n \times 1}$ such that

$$\|P_{\widetilde{\mathcal{M}}} - P_{\widetilde{\mathcal{M}}'}\| \leq K_0' \|B' - \chi(A)\|, \tag{5.2.3}$$

where the positive constant K_0' is independent of B'.

At this point we need a closer examination of the subspace $\widetilde{\mathcal{M}}'$. It turns out that $\widetilde{\mathcal{M}}'$ is necessarily *the sum* $\mathcal{N}_{\mathcal{U} \cup \overline{\mathcal{U}}}'$ *of root subspaces of B' corresponding to the eigenvalues contained in $\mathcal{U} \cup \overline{\mathcal{U}}$,* at least for δ_0' sufficiently small. Let us provide a proof of this statement. Denoting by $\mathcal{N}' \subseteq \mathsf{R}^{4n \times 1}$ the sum of root subspaces of B' other than those in $\mathcal{U} \cup \overline{\mathcal{U}}$ and arguing by contradiction, suppose that $\widetilde{\mathcal{M}}' \cap \mathcal{N}' \neq \{0\}$ for some B' arbitrarily close to $\chi(A)$. Select a unit length eigenvector $w' \in \widetilde{\mathcal{M}}' \cap \mathcal{N}'$ of B' corresponding to a real eigenvalue μ', or a pair of vectors $u', v' \in \widetilde{\mathcal{M}}' \cap \mathcal{N}'$ such that $\|u'\|^2 + \|v'\|^2 = 1$ and

$$B' \begin{bmatrix} u' \\ v' \end{bmatrix} = \begin{bmatrix} \mu' & \nu' \\ -\nu' & \mu' \end{bmatrix} \begin{bmatrix} u' \\ v' \end{bmatrix}, \quad \text{where} \;\; \mu', \nu' \in \mathsf{R}, \;\; \nu' \neq 0.$$

(The pair (u', v') correspond to nonreal eigenvalues $\mu' \pm i\nu'$.) Passing to a limit when $B' \longrightarrow \chi(A)$, in view of compactness of the unit sphere in $\mathsf{R}^{4n \times 1}$, we see that partial limits of the eigenvector w', or of the pairs of vectors (u', v') as the case may be, must exist. Using these partial limits it follows by Theorem 3.9.4 that $\widetilde{\mathcal{M}}$ contains eigenvectors of A corresponding to eigenvalues not in $T \cup \overline{T}$, or $\widetilde{\mathcal{M}}$ contains pairs of vectors (u, v), $u, v \in \mathsf{R}^{4n \times 1}$ such that $\|u\|^2 + \|v\|^2 = 1$ and

$$\chi(A) \begin{bmatrix} u \\ v \end{bmatrix} = \begin{bmatrix} \mu & \nu \\ -\nu & \mu \end{bmatrix} \begin{bmatrix} u \\ v \end{bmatrix},$$

where $\mu, \nu \in \mathsf{R}$, $\nu \neq 0$, and $\mu \pm i\nu \notin T \cup \overline{T}$. In both cases we obtain a contradiction to the statement (verified above) that $\widetilde{\mathcal{M}}$ is the root subspace of $\chi(A)$ corresponding to the eigenvalues in $T \cup \overline{T}$. Thus, we have proved that

$$\widetilde{\mathcal{M}}' \subseteq \mathcal{N}_{\mathcal{U} \cup \overline{\mathcal{U}}}'. \tag{5.2.4}$$

The converse inclusion follows from the equality of dimensions

$$\dim_{\mathsf{R}} \widetilde{\mathcal{M}}' = \dim_{\mathsf{R}} \mathcal{N}'_{\mathcal{U} \cup \overline{\mathcal{U}}}. \tag{5.2.5}$$

Indeed, by Theorem 5.15.1 (applied to the characteristic polynomial of $\chi(A)$), the sum of algebraic multiplicities of eigenvalues of $\chi(A)$ in $T \cup \overline{T}$ coincides with the sum of those of B' contained in $T' \cup \overline{T}'$ (if δ'_0 is sufficiently small). Thus,

$$\dim_{\mathsf{R}} \widetilde{\mathcal{M}} = \dim_{\mathsf{R}} \widetilde{\mathcal{M}}' \le \dim_{\mathsf{R}} \mathcal{N}'_{\mathcal{U} \cup \overline{\mathcal{U}}} = \dim_{\mathsf{R}} \widetilde{\mathcal{M}},$$

where the first equality follows by Theorem 3.9.3, the inequality follows by (5.2.4), and the latter equality follows from the remark immediately above. Thus, (5.2.5) holds true, and then equality in (5.2.4) holds true as well.

Let $\delta_0 = c_{n,n}\delta'_0$, and let $B \in \mathsf{H}^{n \times n}$ be such that $\|B - A\| \le \delta_0$. Then $\|\chi(B) - \chi(A)\| \le \delta'_0$, and applying (5.2.3) with $B' = \chi(B)$, we have

$$\|P_{\widetilde{\mathcal{M}}} - P_{\widetilde{\mathcal{M}}'}\| \le K'_0 (c_{n,n})^{-1} \|B - A\|. \tag{5.2.6}$$

Since $\widetilde{\mathcal{M}}'$ is a sum of root subspaces of $\chi(B)$, by choosing an orthonormal basis $u_1, \dots, u_r$ in the sum $\mathcal{M}'$ of the root subspaces of B corresponding to the eigenvalues in $\mathcal{U} \cup \overline{\mathcal{U}}$, we have that the columns of $[\chi(u_1) \ \dots \ \chi(u_r)]$ form an orthonormal basis in $\widetilde{\mathcal{M}}'$. Thus, using formula (3.9.1), we obtain:

$$
\begin{aligned}
P_{\widetilde{\mathcal{M}}} - P_{\widetilde{\mathcal{M}}'} &= [\chi(v_1)\ \chi(v_2)\ \dots\ \chi(v_r)][\chi(v_1)\ \chi(v_2)\ \dots\ \chi(v_r)]^T \\
&\quad - [\chi(u_1)\ \chi(u_2)\ \dots\ \chi(u_r)][\chi(u_1)\ \chi(u_2)\ \dots\ \chi(u_r)]^T \\
&= \chi([v_1\ v_2\ \dots\ v_r][v_1\ v_2\ \dots\ v_r]^* \\
&\quad - [u_1\ u_2\ \dots\ u_r][u_1\ u_2\ \dots\ u_r]^*) = \chi(P_{\mathcal{M}} - P_{\mathcal{M}'}).
\end{aligned}
$$

Therefore, $\|P_{\mathcal{M}} - P_{\mathcal{M}'}\| \le C_{n,n}\|P_{\widetilde{\mathcal{M}}} - P_{\widetilde{\mathcal{M}}'}\|$, and, combining this inequality with (5.2.6), the result follows.   □

Note that the degree of $p^{(A)}(t)$ need not be constant in a neighborhood of a given $A \in \mathsf{H}^{n \times n}$: if $a \in \mathsf{R}$, then $p^{(a)}(t) = t - a$, whereas $p^{(a+ib)}(t) = t^2 - 2at + a^2 + b^2$ for $b \in \mathsf{R} \setminus \{0\}$.

We mention that *sums of root subspaces of a given matrix $A \in \mathsf{H}^{n \times n}$ are isolated* (*in the sense of the gap metric*) *in the set of all invariant subspaces of A.* Indeed, suppose not. Then there is $\mathcal{M}$, the sum of root subspaces of A corresponding to a set $T \cup \overline{T}$ of roots of $p^{(A)}(t)$, such that some sequence $\{\mathcal{N}_m\}_{m=1}^\infty$ of distinct A-invariant subspaces converges to $\mathcal{M}$. Passing to a subsequence if necessary, we may assume that none of the $\mathcal{N}_m$'s is a sum of root subspaces for A and that in the decompositions (5.1.4) we have

$$\mathcal{N}_m = (\mathcal{N}_m \cap \mathcal{M}_1) \dotplus \cdots \dotplus (\mathcal{N}_m \cap \mathcal{M}_k), \qquad m = 1, 2, \dots,$$

with

$$\{0\} \ne \mathcal{N}_m \cap \mathcal{M}_{j_0} \ne \mathcal{M}_{j_0}, \quad \text{for} \quad m = 1, 2, \dots, \tag{5.2.7}$$

where the index j_0 is independent of m. Passing again to a subsequence of $\{\mathcal{N}_m\}_{m=1}^\infty$ if necessary, we may (and do) further assume that for a given index $j \in \{1, 2, \dots, k\}$, either $\mathcal{N}_m \cap \mathcal{M}_j = \{0\}$ for all m or $\mathcal{N}_m \cap \mathcal{M}_j \ne \{0\}$ for all m. If j is such that $\mathcal{N}_m \cap \mathcal{M}_j \ne \{0\}$, select $x_m \in \mathcal{N}_m \cap \mathcal{M}_j$, $\|x_m\| = 1$, and assume (without loss of generality) that the sequence $\{x_m\}_{m=1}^{\infty}$ converges to some $x \in \mathsf{H}^{n \times 1}$. By Theorem 3.9.4, $x \in \mathcal{M} \cap \mathcal{M}_j$, and since $\mathcal{M}$ is a sum of root subspaces, we must have $\mathcal{M}_j \subseteq \mathcal{M}$. Setting K to be the set of all indices j such that $\mathcal{N}_m \cap \mathcal{M}_j \neq \{0\}$, we now have

$$
\begin{aligned}
\dim \mathcal{N}_m &= \sum_{j \in K} \dim(\mathcal{N}_m \cap \mathcal{M}_j) \leq \sum_{j \in K} \dim \mathcal{M}_j \\
&= \dim \left(\sum_{j \in K} \mathcal{M}_j \right) \leq \dim \mathcal{M}.
\end{aligned}
\tag{5.2.8}
$$

But by Theorem 3.9.3, $\dim \mathcal{N}_m = \dim \mathcal{M}$, at least for m large enough. Thus, equality prevails in (5.2.8), a contradiction with (5.2.7).

5.3   EIGENVALUES AND EIGENVECTORS

Definition 5.3.1. Let $A \in \mathsf{H}^{n \times n}$. A vector $v \in \mathsf{H}^{n \times 1} \setminus \{0\}$ is said to be a *right eigenvector* of A corresponding to the *right eigenvalue* $\alpha \in \mathsf{H}$ if the equality

$$
Av = v\alpha
\tag{5.3.1}
$$

holds.

The set of all right eigenvalues of A is denoted $\sigma(A)$, the *spectrum* of A. Note that $\sigma(A)$ is closed under similarity of quaternions: if (5.3.1) holds, then

$$
A(v\lambda) = (v\lambda)(\lambda^{-1}\alpha\lambda), \quad \text{for all } \lambda \in \mathsf{H} \setminus \{0\},
\tag{5.3.2}
$$

so $v\lambda$ is a right eigenvector of A corresponding to the right eigenvalue $\lambda^{-1}\alpha\lambda$.

We introduce also the notion of left eigenvectors/eigenvalues.

Definition 5.3.2. A vector $u \in \mathsf{H}^{n \times 1} \setminus \{0\}$ is said to be a *left eigenvector* of A corresponding to the *left eigenvalue* $\alpha \in \mathsf{H}$ if

$$
Au = \alpha u.
\tag{5.3.3}
$$

There are no obvious general connections between left and right eigenvalues or left and right eigenvectors.

Example 5.3.3. Let

$$
A = \begin{bmatrix} 0 & \mathsf{i} \\ \mathsf{j} & 0 \end{bmatrix}.
$$

The equation for left eigenvalues λ of A boils down to the quadratic equation $\mathsf{i} = -\lambda\mathsf{j}\lambda$. A calculation shows that this equation has exactly two solutions $(\pm 1/\sqrt{2})(\mathsf{i} + \mathsf{j})$. On the other hand, $A^4 + I = 0$; therefore, the right eigenvalues μ of A must satisfy

$$
\mu^4 + 1 = 0.
\tag{5.3.4}
$$

In fact,

$$
\sigma(A) = \{\mu \in \mathsf{H} : \mu^4 + 1 = 0\}.
$$

Letting

$$
\mu = \mu_0 + \mu_1\mathsf{i} + \mu_2\mathsf{j} + \mu_3\mathsf{k}, \quad \text{with } \mu_0, \mu_1, \mu_2, \mu_3 \in \mathsf{R},
\tag{5.3.5}
$$

equation (5.3.4) is satisfied if and only if $\nu^2 + 1 = 0$, where $\nu = \mu^2$, i.e.,

$$\mathfrak{R}(\mu^2) = 0, \quad |\mathfrak{V}(\mu^2)| = 1. \tag{5.3.6}$$

Now (5.3.6) holds if and only if

$$\mu_0^2 = \mu_1^2 + \mu_2^2 + \mu_3^2 = \frac{1}{2}. \tag{5.3.7}$$

Thus the right eigenvalues of A are given by (5.3.5) and (5.3.7). In particular, $\sigma(A)$ does not contain any left eigenvalues of A. $\qquad\square$

However, *real* right eigenvalues are also left eigenvalues (with left and right eigenvectors being the same), and vice versa.

Proposition 5.3.4. *For $A \in \mathsf{H}^{n \times n}$ the following statements are equivalent:*

(1) *A is not invertible;*

(2) *zero is a left eigenvalue of A;*

(3) *zero is a right eigenvalue of A.*

Proof. Each of (2) and (3) amounts to the statement that $Av = 0$ for some $v \in \mathsf{H}^{n \times 1} \setminus \{0\}$. This is equivalent to noninvertibility of A, by Ex. 3.11.7. $\qquad\square$

In the sequel we will be interested in right eigenvalues and right eigenvectors, so we will use the terminology *eigenvalues* and *eigenvectors* having in mind right eigenvalues and right eigenvectors, respectively.

First, we establish the existence theorem.

Theorem 5.3.5. *For every $A \in \mathsf{H}^{n \times n}$, there exists an eigenvalue which is a complex number with nonnegative imaginary part.*

Proof. Write $A \in \mathsf{H}^{n \times n}$ and $u \in \mathsf{H}^{n \times 1}$ in the form

$$A = A_1 + A_2 \mathsf{j}, \quad u = u_1 + u_2 \mathsf{j}, \quad \text{where } A_1, A_2 \in \mathsf{C}^{n \times n}, \quad u_1, u_2 \in \mathsf{C}^{n \times 1}.$$

A straightforward calculation shows that the following three equations are equivalent, where λ is a complex number:

$$Au = u\lambda; \tag{5.3.8}$$

$$\begin{bmatrix} A_1 & A_2 \\ -\overline{A_2} & \overline{A_1} \end{bmatrix} \begin{bmatrix} u_1 \\ -\overline{u_2} \end{bmatrix} = \lambda \begin{bmatrix} u_1 \\ -\overline{u_2} \end{bmatrix}; \tag{5.3.9}$$

$$\begin{bmatrix} A_1 & A_2 \\ -\overline{A_2} & \overline{A_1} \end{bmatrix} \begin{bmatrix} u_2 \\ \overline{u_1} \end{bmatrix} = \overline{\lambda} \begin{bmatrix} u_2 \\ \overline{u_1} \end{bmatrix}. \tag{5.3.10}$$

But (5.3.9) has a solution for some λ with $\begin{bmatrix} u_1 \\ -\overline{u_2} \end{bmatrix} \neq 0$ because every complex matrix has a complex eigenvalue (as follows from the Fundamental Theorem of Algebra); hence, (5.3.10) and (5.3.8) hold as well for these values of λ, u_1, u_2. But if $\lambda \in \mathsf{C}$ is an eigenvalue of A, then so is $\overline{\lambda}$. Thus, λ and $\overline{\lambda}$ are eigenvalues of A for some $\lambda \in \mathsf{C}$. $\qquad\square$

We mention in passing that *every* $A \in \mathsf{H}^{n \times n}$ *has a left eigenvalue.* The proof of this fact is based on topological considerations and is beyond the scope of this book; see Wood [163] and Zhang [164] for more details.

For a matrix $A = [a_{ij}]_{i,j=1}^{n} \in \mathsf{H}^{n \times n}$, denote the *diagonal* of A by

$$D(A) = \{\lambda \in \mathsf{H} : \lambda = a_{ii} \text{ for some } i, \quad i = 1, 2, \ldots, n\}.$$

Note that in the case A is triangular, $D(A)$ does not generally coincide with $\sigma(A)$. For example, if $A = \mathsf{i}$, then $D(A) = \{\mathsf{i}\}$, but

$$\sigma(A) = \{b\mathsf{i} + c\mathsf{j} + d\mathsf{k} : b, c, d \in \mathsf{R} \text{ and } b^2 + c^2 + d^2 = 1\}.$$

Theorem 5.3.5 yields, in the usual manner (well known in complex linear algebra), results on triangulation and diagonalization by unitary similarity, as the next theorem shows. Observe that if $A \in \mathsf{H}^{n \times n}$ is hermitian, resp. skewhermitian, unitary, or normal, then for any unitary $U \in \mathsf{H}^{n \times n}$, the matrix U^*AU is hermitian, resp. skewhermitian, unitary, or normal, as the case may be.

Theorem 5.3.6. *Let* $A \in \mathsf{H}^{n \times n}$. *Then:*

(a) *(Schur's triangularization theorem) there exists a unitary* $U \in \mathsf{H}^{n \times n}$ *such that* U^*AU *is upper triangular and the diagonal* $D(A)$ *is complex;*

(b) *there exists a unitary* $U \in \mathsf{H}^{n \times n}$ *such that* U^*AU *is lower triangular and the diagonal* $D(A)$ *is complex;*

(c) *if* A *is hermitian, then there exists a unitary* $U \in \mathsf{H}^{n \times n}$ *such that* U^*AU *is diagonal and real;*

(d) *if* A *is skewhermitian, then there exists a unitary* $U \in \mathsf{H}^{n \times n}$ *such that* U^*AU *is diagonal and* $D(A) \subset \mathsf{iR}$;

(e) *if* A *is unitary, then there exists a unitary* $U \in \mathsf{H}^{n \times n}$ *such that* U^*AU *is diagonal and consists of unit complex numbers;*

(f) *if* A *is normal, then there exists a unitary* $U \in \mathsf{H}^{n \times n}$ *such that* U^*AU *is diagonal and complex.*

Proof. Part (a). Use induction on n, the case $n = 1$ being trivial. By Theorem 5.3.5, we find $u \in \mathsf{H}^{n \times 1}$, $\|u\| = 1$, such that $Au = u\lambda$ for some complex λ. Let $U \in \mathsf{H}^{n \times n}$ be a unitary matrix whose first column is u. (To construct such U, let $\{u_1, \ldots, u_n\}$ be a basis for $\mathsf{H}^{n \times 1}$ in which $u_1 = u$, and apply the Gram-Schmidt procedure to the basis $\{u_1, \ldots, u_n\}$.) We then have

$$AU = U \begin{bmatrix} \lambda & A_{12} \\ 0 & A_{22} \end{bmatrix}, \tag{5.3.11}$$

for some $A_{1,2} \in \mathsf{H}^{1 \times (n-1)}$, $A_{22} \in \mathsf{H}^{(n-1) \times (n-1)}$. Rewrite (5.3.11) in the form $U^*AU = \begin{bmatrix} \lambda & A_{12} \\ 0 & A_{22} \end{bmatrix}$, and use the induction hypothesis to complete the proof of Part (a).

For Part (b), apply the result of Part (a) to the adjoint matrix A^*, and take adjoints in the resulting equality $U^*A^*U = T$, where T is upper triangular.

For Parts (c), (d), (e), and (f), note that an upper (or lower) triangular matrix is normal only if it is diagonal. Indeed, suppose $A = [a_{i,j}]_{i,j=1}^{n} \in \mathsf{H}^{n \times n}$, where $a_{i,j} \in \mathsf{H}$ with $a_{i,j} = 0$ for $i > j$. Equating the (i,i) entry in A^*A with the (i,i) entry in AA^*, for $i = 1, 2, \ldots, n$, leads to the equalities (in writing down these equalities we took advantage of the property that $x^*x = xx^*$ for all $x \in \mathsf{H}$):

$$a_{1,1}^* a_{1,1} = a_{1,1}^* a_{1,1} + a_{1,2}^* a_{1,2} + \cdots + a_{1,n}^* a_{1,n}; \quad (5.3.12)$$

$$a_{1,2}^* a_{1,2} + a_{2,2}^* a_{2,2} = a_{2,2}^* a_{2,2} + a_{2,3}^* a_{2,3} + \cdots + a_{2,n}^* a_{2,n}; \quad (5.3.13)$$

etc.;

$$a_{1,n}^* a_{1,n} + a_{2,n}^* a_{2,n} + \cdots + a_{n,n}^* a_{n,n} = a_{n,n}^* a_{n,n}. \quad (5.3.14)$$

It is easy to see that we must have $a_{i,j} = 0$ for all $i < j$. Since hermitian, skewhermitian, and unitary matrices are, in particular, normal, the diagonal forms in (c), (d), (e), and (f) follow from the triangular form of (a) (or of (b)). Finally, observe that in the hermitian case the diagonal entries must be real, as dictated by the hermitian property of the matrix; similarly, the nature of diagonal entries is established in (c) and (d). $\qquad \square$

We have already seen above that, for triangular matrices, the spectrum of the matrix does not necessarily coincide with the diagonal. The exact relationship between the spectrum and the diagonal is described next.

Proposition 5.3.7. *If $A \in \mathsf{H}^{n \times n}$ is (upper or lower) triangular, then*

$$\sigma(A) = \{\alpha^{-1}\beta\alpha \, : \, \alpha \in \mathsf{H} \setminus \{0\}, \ \beta \in D(A)\}. \quad (5.3.15)$$

Proof. First, we prove the $\supseteq$ part of (5.3.15). We use induction on n, the case $n = 1$ being trivial. Suppose $A = [a_{ij}]_{i,j=1}^{n}$ is upper triangular, so $a_{ij} = 0$ if $i > j$. By the induction hypotheses (applied to the $(n-1) \times (n-1)$ matrix $A' := [a_{ij}]_{i,j=1}^{n-1}$, noting that the eigenvectors of A' can be trivially extended to eigenvectors of A, with the same eigenvalues), we have

$$\sigma(A) \supseteq \{\alpha^{-1} a_{jj}\alpha \, : \, \alpha \in \mathsf{H} \setminus \{0\}, \ \ j = 1, 2, \ldots, n - 1\}.$$

If $\mathfrak{R}(a_{nn}) = \mathfrak{R}(a_{j,j})$ and $|\mathfrak{V}(a_{nn})| = |\mathfrak{V}(a_{j,j})|$ for some $j = 1, 2, \ldots, n - 1$, then also $a_{nn} \in \sigma(A)$, and we are done. Otherwise, the equation

$$a_{n-1,n-1}x_1 + a_{n-1,n} = x_1 a_{nn}$$

has a (unique) solution $x_1 \in \mathsf{H}$, by Theorem 2.3.3(2). Again by Theorem 2.3.3(2), let x_2 be a (unique) solution of

$$a_{n-2,n-2}x_2 + a_{n-2,n-1}x_1 + a_{n-2,n} = x_2 a_{nn}, \quad x_2 \in \mathsf{H}.$$

Continuing this way, we obtain $x_1, \ldots x_{n-1} \in \mathsf{H}$. From the construction it follows that $\begin{bmatrix} x_{n-1} \\ \vdots \\ x_1 \\ 1 \end{bmatrix}$ is an eigenvector of A corresponding to the eigenvalue a_{nn}. This proves the $\supseteq$ part of (5.3.15).

Conversely, let v be an eigenvector of A corresponding to the eigenvalue γ. If $v = [v_j]_{j=1}^n$ and j_0 is the largest index such that $v_{j_0} \neq 0$, then $a_{j_0,j_0} v_{j_0} = v_{j_0} \gamma$, and $\gamma = v_{j_0}^{-1} a_{j_0,j_0} v_{j_0}$ belongs to the right-hand side of (5.3.15). $\square$

It follows that for a triangular matrix $A \in \mathsf{H}^{n \times n}$, the following three statements are equivalent: (1) $\sigma(A)$ is real; (2) $D(A)$ is real; (3) $\sigma(A) = D(A)$.

In particular (Theorem 5.3.6(c)), for every hermitian matrix H there is a unitary U such that $U^* H U$ is diagonal with the eigenvalues (perhaps repeated) of H on the main diagonal. Comparing with the canonical form of Theorem 4.1.6, the following characterization of classes of hermitian matrices in terms of eigenvalues is easily obtained.

Proposition 5.3.8. *A hermitian matrix H is positive definite, resp. positive semidefinite, negative definite, or negative semidefinite, if and only if all its eigenvalues are positive, resp. nonnegative, negative, or nonpositive.*

We conclude this section with a statement on linear independence of eigenvectors corresponding to nonsimilar eigenvalues.

Proposition 5.3.9. *Let $v_1, \ldots, v_p$ be eigenvectors of a matrix $A \in \mathsf{H}^{n \times n}$ that correspond to eigenvalues $\alpha_1, \ldots, \alpha_p$, respectively, and assume that $\alpha_1, \ldots, \alpha_p$ are pairwise nonsimilar (over H). Then $v_1, \ldots, v_p$ are linearly independent.*

Proof. Without loss of generality, we may assume that $\alpha_1, \ldots, \alpha_p$ are complex numbers with nonnegative imaginary parts (see (5.3.2)). In this case, the condition of pairwise nonsimilarity simply means that $\alpha_1, \ldots, \alpha_p$ are all distinct. Suppose $\sum_{j=1}^p v_j \lambda_j = 0$ for some $\lambda_1, \ldots, \lambda_p \in \mathsf{H}$. Let $q(t)$ be a polynomial with real coefficients such that $q(\alpha_1) \neq 0$ but $q(\alpha_j) = 0$ for $j = 2, 3, \ldots, p$. Then

$$0 = q(A) \sum_{j=1}^p v_j \lambda_j = \sum_{j=1}^p v_j q(\alpha_j) \lambda_j = v_1 q(\alpha_1) \lambda_1,$$

and in view of $v_1 \neq 0$, $q(\alpha_1) \neq 0$, we must have $\lambda_1 = 0$. Analogously, the equalities $\lambda_2 = \cdots = \lambda_p = 0$ are proved. $\square$

5.4   SOME PROPERTIES OF JORDAN BLOCKS

Jordan blocks are matrices of type

$$J_m(\lambda) = \begin{bmatrix} \lambda & 1 & 0 & \cdots & 0 \\ 0 & \lambda & 1 & \cdots & 0 \\ \vdots & \vdots & \ddots & \ddots & 0 \\ \vdots & \vdots & & \lambda & 1 \\ 0 & 0 & \cdots & 0 & \lambda \end{bmatrix} \in \mathsf{H}^{m \times m}, \quad \lambda \in \mathsf{H}.$$

Later in the text we will frequently work with Jordan blocks and matrices that are derived from Jordan blocks. The next two propositions are very useful tools in this work.

Proposition 5.4.1. *If $\alpha \in \mathsf{H}$, $\beta \in \mathsf{H}$ are not similar, then the equation*

$$J_m(\alpha)Y = Y J_p(\beta), \quad Y \in \mathsf{H}^{m \times p}, \tag{5.4.1}$$

has only the trivial solution $Y = 0$.

Proof. Write

$$Y = \begin{bmatrix} y_{1,1} & y_{1,2} & \cdots & y_{1,p} \\ y_{2,1} & y_{2,2} & \cdots & y_{2,p} \\ \vdots & \vdots & \cdots & \vdots \\ y_{m,1} & y_{m,2} & \cdots & y_{m,p} \end{bmatrix}, \quad y_{i,j} \in \mathsf{H}.$$

Equating the elements in the bottom row of the left-hand side and the right-hand side of equation (5.4.1), we obtain

$$\alpha y_{m,1} = y_{m,1}\beta, \quad \alpha y_{m,2} = y_{m,1} + y_{m,2}\beta, \quad \ldots, \quad \alpha y_{m,p} = y_{m,p-1} + y_{m,p}\beta.$$

Repeatedly using the fact (which follows from nonsimilarity of α and β) that $\alpha x = x\beta$, $x \in \mathsf{H}$, is possible only if $x = 0$, we see that $y_{m,1} = y_{m,2} = \cdots = y_{m,p} = 0$. Now equate the elements in the $(m-1)$th row of the left-hand side and the right-hand side of (5.4.1), which results in $y_{m-1,1} = y_{m-1,2} = \cdots = y_{m-1,p} = 0$. Continuing in this fashion, we eventually obtain $Y = 0$. $\qquad\square$

To formulate the second proposition, we introduce notation for upper triangular *Toeplitz matrices*: for $\alpha_1, \ldots, \alpha_q \in \mathsf{H}$, we let

$$\text{Toep}_q(\alpha_1, \ldots, \alpha_q) = \begin{bmatrix} \alpha_1 & \alpha_2 & \alpha_3 & \cdots & \alpha_q \\ 0 & \alpha_1 & \alpha_2 & \cdots & \alpha_{q-1} \\ 0 & 0 & \alpha_1 & \cdots & \alpha_{q-2} \\ \vdots & \ddots & \ddots & \ddots & \vdots \\ 0 & 0 & 0 & \cdots & \alpha_1 \end{bmatrix}$$

be the upper triangular Toeplitz matrix with $\alpha_1, \ldots, \alpha_q$ on the first, second, etc., superdiagonal, respectively.

Proposition 5.4.2.   (a) *If $\lambda \in \mathsf{R}$, then the general solution of the homogeneous matrix equation*

$$J_m(\lambda)X = X J_n(\lambda) \tag{5.4.2}$$

with unknown $X \in \mathsf{H}^{m \times n}$ is given by

$$X = \begin{cases} \begin{bmatrix} 0_{m \times (n-m)} & \text{Toep}_m(\alpha_1, \ldots, \alpha_m) \end{bmatrix}, & \begin{array}{l} \alpha_1, \ldots \alpha_m \in \mathsf{H} \\ \text{arbitrary} \\ \text{if } m \le n; \end{array} \\[2em] \begin{bmatrix} \text{Toep}_n(\alpha_1, \ldots, \alpha_n) \\ 0_{n \times (m-n)} \end{bmatrix}, & \begin{array}{l} \alpha_1, \ldots, \alpha_n \in \mathsf{H} \\ \text{arbitrary} \\ \text{if } m \ge n. \end{array} \end{cases} \tag{5.4.3}$$

(b) *If* $\lambda \in \mathsf{H} \setminus \mathsf{R}$, *then the general solution of* (5.4.2) *is given by*

$$X = \begin{cases} \begin{bmatrix} 0_{m \times (n-m)} & \mathrm{Toep}_m \left(\alpha_1, \ldots, \alpha_m \right) \end{bmatrix}, & \begin{aligned} & \alpha_1, \ldots \alpha_m \\ & \in \mathrm{Span}_\mathsf{R} \{1, \lambda\} \\ & \text{arbitrary if } m \leq n; \end{aligned} \\[4ex] \begin{bmatrix} \mathrm{Toep}_n \left(\alpha_1, \ldots, \alpha_n \right) \\ 0_{n \times (m-n)} \end{bmatrix}, & \begin{aligned} & \alpha_1, \ldots, \alpha_n \\ & \in \mathrm{Span}_\mathsf{R} \{1, \lambda\} \\ & \text{arbitrary if } m \geq n. \end{aligned} \end{cases} \qquad (5.4.4)$$

Proof. First consider the case $\lambda \in \mathsf{R}$. Then (5.4.2) is equivalent to

$$J_m(0)X = X J_n(0). \qquad (5.4.5)$$

Write $X = [x_{i,j}]_{i=1,j=1}^{m,n}$. Equating the entries of the first row and the first column in the left-hand side of (5.4.5) to the corresponding entries in the right-hand side yields the equalities

$$x_{2,1} = x_{3,1} = \cdots = x_{m,1} = 0 \quad \text{and} \quad x_{m,1} = x_{m,2} = \cdots = x_{m,n-1} = 0. \qquad (5.4.6)$$

Equating other entries on both sides of (5.4.5) also produces

$$x_{i+1,j} = x_{i,j+1}, \quad \text{for } i = 1, 2, \ldots m \quad \text{and } j = 1, 2, \ldots, n. \qquad (5.4.7)$$

Equalities (5.4.6) and (5.4.7) now yield the desired form, (5.4.3).

Now assume $\lambda \in \mathsf{H} \setminus \mathsf{R}$. Subtracting from λ its real part, we can (and do) assume without loss of generality, that $\mathfrak{R}(\lambda) = 0$. Let (q_1, q_2, q_3) be a units triple such that q_1 is a scalar multiple of λ: $\lambda = a q_1$, where $a \in \mathsf{R} \setminus \{0\}$. Write $X = X' + X''$, where $X' \in \mathrm{Span}_\mathsf{R} \{1, q_1\}$ and $X'' \in \mathrm{Span}_\mathsf{R} \{q_2, q_3\}$. Then the multiplication rules within the units triple force equation (5.4.2) to decompose into two equations:

$$J_m(\lambda)X' = X' J_n(\lambda), \qquad J_m(\lambda)X'' = X'' J_n(\lambda). \qquad (5.4.8)$$

Since $\mathrm{Span}_\mathsf{R} \{1, q_1\}$ is isomorphic to the field of complex numbers, the first equation in (5.4.8) is amenable to the standard analysis, as in the proof of Part (a), leading to the general form of solutions as in (5.4.4).

We will show that the second equation in (5.4.8) has only the trivial solution $X'' = 0$, thereby proving Proposition 5.4.2. Letting $X'' = X_2 q_2 + X_3 q_3$, where X_2, X_3 are real matrices, we rewrite the second equation in (5.4.8) in the following form:

$$J_m(a q_1) \left(X_2 q_2 + X_3 q_3 \right) = \left(X_2 q_2 + X_3 q_3 \right) J_n(a q_1). \qquad (5.4.9)$$

It will be convenient to denote the matrix $J_p(0)$ by K_p, so $J_m(a q_1) = a q_1 I_m + K_m$. We now compute the left-hand side of (5.4.9):

$$\begin{aligned} J_m(a q_1) \left(X_2 q_2 + X_3 q_3 \right) &= \left(a q_1 I_m + K_m \right) \left(X_2 q_2 + X_3 q_3 \right) \\ &= a q_1 X_2 q_2 + a q_1 X_3 q_3 + K_m X_2 q_2 + K_m X_3 q_3 \\ &= a X_2 q_3 - a X_3 q_2 + K_m X_2 q_2 + K_m X_3 q_3. \quad (5.4.10) \end{aligned}$$

Analogously,

$$\left(X_2 q_2 + X_3 q_3 \right) J_n(a q_1) = -a X_2 q_3 + a X_3 q_2 + X_2 q_2 K_n + X_3 q_3 K_n. \qquad (5.4.11)$$

Equating the right-hand sides of (5.4.10) and (5.4.11), we get

$$2aX_2q_3 - 2aX_3q_2 = -K_mX_2q_2 + X_2q_2K_n - K_mX_3q_3 + X_3q_3K_n,$$

or

$$-2aX_3 = -K_mX_2 + X_2K_n, \quad 2aX_2 = -K_mX_3 + X_3K_n. \qquad (5.4.12)$$

Substitute the value of X_3 from the first equation in (5.4.12) into the second:

$$X_2 = \frac{1}{(2a)^2}(-K_m^2X_2 + 2K_mX_2K_n - X_2K_n^2). \qquad (5.4.13)$$

We now repeatedly iterate equality (5.4.13), i.e., substitute for X_2 in the right-hand side of (5.4.13) its value given by (5.4.13). The result is equalities of the form

$$X_2 = \sum_{j=0}^{2p} a_{j,2p}K_m^jX_2K_n^{2p-j}, \quad p = 1, 2, \ldots,$$

for some real numbers $a_{j,2p}$. Taking p so large that $2p \geq m + n$, we see that for every $j = 0, \ldots, 2p$, at least one of the equalities $j \geq m$ or $2p - j \geq n$ holds, so $K_m^j = 0$ or $K_n^{2p-j} = 0$; hence, $X_2 = 0$. But then also $X_3 = 0$, and we are done. $\square$

We conclude this section with an easy observation concerning the minimal polynomials of Jordan blocks.

Proposition 5.4.3. *The minimal polynomial of $J_m(\lambda)$ is $(t - \lambda)^m$ if λ is real and $(t^2 - 2\Re(\lambda)t + |\lambda|^2)^m$ if λ is nonreal.*

Proof. In the case λ is real, we have

$$(J_m(\lambda)) - \lambda I)^k = \mathrm{Toep}_m\,\underbrace{(0,0,\ldots,0}_{k\ \mathrm{zeros}}, 1, 0, \ldots, 0), \quad k = 1, 2, \ldots, m, \qquad (5.4.14)$$

and the result follows. In the nonreal case, writing $\lambda = \Re(\lambda) + q\Im(\lambda)$, where $q^2 = -1$ and $\Im(\lambda) \in \mathsf{R} \setminus \{0\}$, a computation shows that

$$(J_m(\lambda))^2 - 2\Re(\lambda)J_m(\lambda) + |\lambda|^2I = 2q\Im(\lambda)J_m(0) + (J_m(0))^2,$$

and so

$$((J_m(\lambda))^2 - 2\Re(\lambda)J_m(\lambda) + |\lambda|^2I)^k$$
$$= (J_m(0))^k(2q\Im(\lambda)I_m + J_m(0))^k, \quad k = 1, 2, \ldots, m.$$

Since the matrix $2q\Im(\lambda)I_m + J_m(0)$ is invertible, the result now follows from (5.4.14) (where we set $\lambda = 0$). $\square$

5.5 JORDAN FORM

In this section we state a key result of quaternion linear algebra—the Jordan canonical form, Theorem 5.5.3 below.

We start with some preliminaries.

Definition 5.5.1. A matrix $A \in \mathsf{H}^{n \times n}$ is said to be *similar* to a matrix $B \in \mathsf{H}^{n \times n}$ if $B = S^{-1}AS$ for some invertible matrix $S \in \mathsf{H}^{n \times n}$.

We check easily that similarity is an equivalence relation, so we can use unambiguously the language of *similar matrices*.

Proposition 5.5.2. *If $A, B \in \mathsf{H}^{n \times n}$ are similar, i.e., $B = S^{-1}AS$ for some invertible S, then:*

(1) *A and B have the same minimal polynomial;*

(2) *A and B have the same eigenvalues;*

(3) *$x \in \mathsf{H}^{n \times 1}$ is an eigenvector of A corresponding to the eigenvalue λ if and only if $S^{-1}x$ is an eigenvector of B corresponding to the same eigenvalue λ;*

(4) *$\mathcal{M}$ is the root subspace of A corresponding to the eigenvalue λ if and only if $S^{-1}(\mathcal{M})$ is the root subspace of B corresponding to the same eigenvalue λ; in particular, the root subspaces of A and B corresponding to the same eigenvalue, have equal dimensions.*

Proof. (1) and (4) follow from the observation that $p(B) = S^{-1}p(A)S$ for every polynomial $p(t)$ with real coefficients. (2) and (3) follow from the observation that the eigenvector-eigenvalue equation $Ax = x\lambda$ for A can be rewritten in the form $(SBS^{-1})x = x\lambda$, or $BS^{-1}x = S^{-1}x\lambda$. $\qquad\square$

Note that the converse statements to Proposition 5.5.2 generally fail; cf. Ex. 5.16.15 and 5.16.16.

We now state our main result in this section.

Theorem 5.5.3. (a) *Let $A \in \mathsf{H}^{n \times n}$. Then there exists an invertible $S \in \mathsf{H}^{n \times n}$ such that $S^{-1}AS$ has the form*

$$S^{-1}AS = J_{m_1}(\lambda_1) \oplus \cdots \oplus J_{m_p}(\lambda_p), \quad \lambda_1, \ldots, \lambda_p \in \mathsf{H}, \qquad (5.5.1)$$

where $J_m(\lambda)$ is the $m \times m$ Jordan block having eigenvalue λ. The form (5.5.1) is uniquely determined by A up to an arbitrary permutation of blocks $\{J_{m_j}(\lambda_j)\}_{j=1}^{p}$ and up to a replacement of $\lambda_1, \ldots, \lambda_p$ with

$$\alpha_1^{-1}\lambda_1\alpha_1, \ldots, \alpha_p^{-1}\lambda_p\alpha_p$$

within the blocks $J_{m_1}(\lambda_1), \ldots, J_{m_p}(\lambda_p)$, respectively, where $\alpha_j \in \mathsf{H} \setminus \{0\}$, $j = 1, 2, \ldots, p$, are arbitrary.

(b) *For every $A \in \mathsf{H}^{n \times n}$ there is an invertible $S \in \mathsf{H}^{n \times n}$ such that $S^{-1}AS$ has the form (5.5.1) with $\lambda_1, \ldots, \lambda_p \in \mathsf{C}$ having nonnegative imaginary parts, and in such case the form (5.5.1) is unique up to an arbitrary permutation of blocks.*

Definition 5.5.4. The right-hand side of (5.5.1) is called the *Jordan form* of A.

Thus, the Jordan form represents a canonical form of a quaternion matrix under similarity.

Part (b) of Theorem 5.5.3 follows immediately from Part (a) upon the observation that every similarity class $\{\alpha^{-1}\lambda\alpha : \alpha \in \mathsf{H} \setminus \{0\}\}$, $\lambda \in \mathsf{H}$, contains unique complex number with nonnegative imaginary part (see Theorem 2.2.6).

We postpone the proof of Part (a) of Theorem 5.5.3 to the next section.

Next, we characterize various elements of the Jordan form of a matrix in terms of the matrix itself.

Theorem 5.5.5. *Let $A \in \mathsf{H}^{n \times n}$, and let $\lambda \in \sigma(A)$. Then:*

(a) *the number of Jordan blocks in the Jordan form of A corresponding to the eigenvalues similar to λ is equal to the maximal number of H-linearly independent eigenvectors corresponding to the eigenvalues similar to λ;*

(b) *the sum of the sizes of Jordan blocks in the Jordan form of A corresponding to the eigenvalues similar to λ is equal to the dimension of the root subspace of A corresponding to λ.*

Proof. It suffices to prove the theorem for A in the Jordan form (cf. Proposition 5.5.2). Thus, we set

$$A = J = J_{m_1}(\lambda_1) \oplus \cdots \oplus J_{m_p}(\lambda_p) \oplus \widetilde{J}, \qquad (5.5.2)$$

where $\lambda_1, \cdots, \lambda_p$ are similar to λ and $\widetilde{J}$ is the part of the Jordan form of A that contains Jordan blocks with eigenvalues not similar to λ. Clearly the p eigenvectors

$$e_1, e_{m_1+1}, e_{m_1+m_2+1}, \ldots, e_{m_1+m_2+\cdots+m_{p-1}+1} \qquad (5.5.3)$$

are H-linearly independent. Now let x be an eigenvector of J corresponding to an eigenvalue μ which is similar to λ, so $\mu = \alpha_q^{-1} \lambda_q \alpha_q$, $q = 1, 2, \ldots, p$ for some $\alpha_1, \ldots, \alpha_p \in \mathsf{H} \setminus \{0\}$. We then have

$$J_{m_q}(\lambda_q) x_q \alpha_q^{-1} = x_q \alpha_q^{-1} \lambda_q, \quad q = 1, 2, \ldots, p, \qquad \widetilde{J}\widetilde{x} = \widetilde{x}\mu,$$

where $x = \begin{bmatrix} x_1 \\ x_2 \\ \vdots \\ x_p \\ \widetilde{x} \end{bmatrix}$ is the partition of x consistent with (5.5.2). By Propositions

5.4.1 and 5.4.2, we have that $\widetilde{x} = 0$ and only the first component in each x_q may be nonzero. It follows that x is a linear combination of the vectors (5.5.3). Thus, there cannot be more than p linearly independent eigenvectors corresponding to eigenvalues similar to λ.

For Part (b) just observe that by Proposition 5.4.3, the Jordan blocks whose minimal polynomials are powers of $t - \lambda$ (if λ is real) or powers of $t^2 - 2\Re(\lambda)t + |\lambda|^2$ (if λ is nonreal) are exactly those that correspond to the eigenvalues similar to λ. $\square$

Definition 5.5.6. The *geometric multiplicity* of the root subspace $\mathcal{M}$ of $A \in \mathsf{H}^{n \times n}$ corresponding to a polynomial $p(t)^m$, where $p(t)$ is real irreducible, is defined as the maximal number of H-linearly independent eigenvectors of A in $\mathcal{M}$ or, equivalently, by Theorem 5.5.5, the number of Jordan blocks J in the Jordan form of A having the property that $p(D(J)) = 0$. We also say that the geometric multiplicity of an eigenvalue $\lambda \in \sigma(A)$ is, by definition, the geometric multiplicity of the root subspace corresponding to the polynomial (a factor of the minimal polynomial of A) of which λ is a root. The sizes of Jordan blocks that have eigenvalues similar to λ, each size repeated as many times as there are Jordan blocks of that size, in the Jordan form of A, are the *partial multiplicities* corresponding to λ. The *algebraic multiplicity* of a root subspace $\mathcal{R}$ of A and of any eigenvalue λ corresponding to $\mathcal{R}$ is defined as the (quaternion) dimension of $\mathcal{R}$.

By Theorem 5.5.5, the sum of the partial multiplicities corresponding to λ coincides with the (quaternion) dimension of the root subspace corresponding to λ. Clearly, for any given eigenvalue of A, the geometric multiplicity cannot exceed the algebraic multiplicity. It is easy to see that similar eigenvalues have the same geometric multiplicity and the same algebraic multiplicity.

Note that the set of eigenvectors of A in a root subspace corresponding to nonreal eigenvalues, together with the zero vector, is not necessarily a subspace of $\mathsf{H}^{n\times 1}$. For example, let $A = \begin{bmatrix} i & 0 \\ 0 & -i \end{bmatrix}$. Clearly, $\mathsf{H}^{2\times 1}$ is the (sole) root subspace of A, and $\begin{bmatrix} 1 \\ 0 \end{bmatrix}$ and $\begin{bmatrix} 0 \\ 1 \end{bmatrix}$ are eigenvectors. If the set of eigenvectors of A, together with zero, were a subspace of $\mathsf{H}^{2\times 1}$, then every nonzero vector in $\mathsf{H}^{2\times 1}$ would be an eigenvector. However, $\begin{bmatrix} a \\ b \end{bmatrix}$, $a \in \mathsf{H} \setminus \{0\}$, $b \in \mathsf{H}$, is an eigenvector of A if and only $-i(ba^{-1}) = (ba^{-1})i$. We have the following general statement.

Theorem 5.5.7. *Let $\mathcal{R}$ be a root subspace for $A \in \mathsf{H}^{n\times n}$ corresponding to a power of a real irreducible polynomial $p(x)$. Then the set of eigenvectors in $\mathcal{R}$, together with the zero vector, form a subspace of $\mathsf{H}^{n\times n}$ if and only if $p(x)$ is linear: $p(x) = x - a$ for some $a \in \mathsf{R}$, or $p(x)$ is quadratic and A has only one eigenvector (up to scaling) in $\mathcal{R}$.*

Proof. The "if" part is clear. For the "only if" part, let v_1 and v_2 be two linearly independent eigenvectors of A corresponding to eigenvalue λ, where λ is nonreal. Then $v_1 + v_2\alpha$, $\alpha \in \mathsf{H}$, is an eigenvector of A if and only if $\lambda\alpha = \alpha\lambda$, and since not all $\alpha \in \mathsf{H}$ satisfy this condition, the set of eigenvectors together with zero is not a subspace of $\mathsf{H}^{n\times 1}$. $\qquad\square$

As an easy consequence of Theorem 5.5.3, we identify the minimal polynomial of a quaternion matrix.

Theorem 5.5.8. *Let $A \in \mathsf{H}^{n\times n}$. Assume that $\beta_1, \ldots, \beta_u$ are the distinct real eigenvalues of A (if any) and $\beta_{u+1}, \ldots, \beta_{u+v} \in \mathsf{H}$ is a maximal set of nonreal pairwise nonsimilar eigenvalues of A (if any). Furthermore, in a Jordan form of A, let m_j be the largest size of Jordan blocks corresponding to the eigenvalue β_j, $j = 1, 2, \ldots, u + v$. Then the minimal polynomial of A is given by*

$$(t - \beta_1)^{m_1} \cdots (t - \beta_u)^{m_u} \prod_{j=1}^{v} (t^2 - 2\Re(\beta_{u+j})t + |\beta_{u+j}|^2)^{m_{u+j}}.$$

Indeed, A and its Jordan form have the same minimal polynomial. For a proof of Theorem 5.5.8 for the Jordan form, use Proposition 5.4.3.

Combining Theorem 5.5.8 with the result on continuous dependence of the roots of minimal polynomial (Theorem 5.2.5), it follows that the eigenvalues are also continuous functions of the matrix.

Theorem 5.5.9. *Let $A \in \mathsf{H}^{n\times n}$, and let $\lambda_1, \ldots, \lambda_s$ be all the distinct eigenvalues of A in the closed upper complex half-plane C_+. Then for every $\epsilon > 0$, there exists $\delta > 0$ such that if $B \in \mathsf{H}^{n\times n}$ satisfies $\|B - A\| < \delta$; then the eigenvalues of B are contained in the union*

$$\cup_{j=1}^{s} \{z \in \mathsf{C}_+ : |z - \lambda_j| < \epsilon\}.$$

Moreover, if ϵ is sufficiently small, then the sum of the algebraic multiplicities of B at eigenvalues in $\{z \in \mathsf{C}_+ : |z - \lambda_j| < \epsilon\}$ is equal to the algebraic multiplicity of A at λ_j, for $j = 1, 2, \ldots, s$.

Proof. To obtain the statement concerning algebraic multiplicities, we apply the map ω to the equality $A = S^{-1}KS$, where S is invertible and K is a Jordan form with complex numbers with nonnegative imaginary parts on the diagonal. Since ω is continuous, matrices that are close to A are transformed by ω to matrices as close to $(\omega(S))^{-1}\omega(K)\omega(S)$ as we wish. Now use Theorem 5.15.1 in the Appendix applied to $p(z) = \det(zI - \omega(K))$ (the characteristic polynomial of $\omega(K)$).   □

The Jordan form allows one to extend the definition of functions of quaternion matrices beyond polynomials with real coefficients by analogy with the standard concept of functions of complex matrices (see, e.g., Gantmacher [50], Lancaster and Tismenetsky [94], or Horn and Johnson [63]).

Theorem 5.5.10. (a) *Let*

$$f(t) = \sum_{j=0}^{\infty} f_j(t - t_0)^j$$

be a power series with real coefficients f_j centered at $t_0 \in \mathsf{R}$ and having nonzero radius of convergence d. Then the series

$$\sum_{j=0}^{\infty} f_j(A - t_0 I)^j \tag{5.5.4}$$

converges for any matrix $A \in \mathsf{H}^{n \times n}$ such that

$$(\mathfrak{R}(\lambda) - t_0)^2 + |\mathfrak{V}(\lambda)|^2 < d^2 \quad \text{for all eigenvalues } \lambda \text{ of } A.$$

(b) *Denoting the sum in (5.5.4) by $f(A)$ and letting (5.5.1) be the Jordan form of A, with $\lambda_1, \ldots, \lambda_p \in \mathsf{C}$ having nonnegative imaginary parts, we have*

$$f(A) = S\left(f(J_{m_1}(\lambda_1)) \oplus \cdots \oplus f(J_{m_p}(\lambda_p))\right)S^{-1},$$

where each $f(J_{m_j}(\lambda_j))$ is found by the formula

$$f(J_m(\lambda)) = \begin{bmatrix} f(\lambda) & f'(\lambda) & \dfrac{f''(\lambda)}{2!} & \cdots & \dfrac{f^{(m-1)}(\lambda)}{(m-1)!} \\ 0 & f(\lambda) & f'(\lambda) & \cdots & \dfrac{f^{(m-2)}(\lambda)}{(m-2)!} \\ \vdots & \vdots & \ddots & \ddots & \vdots \\ 0 & 0 & 0 & \cdots & f(\lambda) \end{bmatrix}. \tag{5.5.5}$$

Note that the algebraic properties expressed in (5.1.1) extend to nonpolynomial functions p and q given by power series, as in the right-hand side of (5.5.4), provided the spectrum of A is contained in the intersection of the disks of convergence of p and q. This can be easily verified using reduction to the Jordan form of A by similarity.

For example, the function $\log(I + A)$ is well defined as above for all $A \in \mathsf{H}^{n \times n}$ satisfying $|\lambda| < 1$ for every $\lambda \in \sigma(A)$.

Note that algebraic identities with real coefficients involving real power series carry over to the matrix case; this fact can be verified again using Theorem 5.5.3 and (5.1.1). So we have $(\sin A)^2 + (\cos A)^2 = I$ for all $A \in \mathsf{H}^{n \times n}$. In contrast, $e^{\mathrm{i}A}$ is generally not equal to $\cos A + \mathrm{i} \sin A$. For example,

$$\cos 1 + \mathsf{k} \sin 1 = e^{\mathsf{k}} = e^{\mathsf{ij}} \neq \cos \mathsf{j} + \mathrm{i} \sin \mathsf{j} = \frac{1}{2}(e + e^{-1}) + \mathsf{ij}\left(\frac{1}{2}(e - e^{-1})\right).$$

We conclude this section with a convenient characterization of uniqueness of invariant subspaces of fixed dimension.

Theorem 5.5.11. *For a fixed k, $1 \leq k \leq n - 1$, and a fixed matrix $A \in \mathsf{H}^{n \times n}$, the following statements are equivalent:*

(a) *A has a unique k-dimensional invariant subspace;*

(b) *A has only one eigenvector (up to scaling);*

(c) *the Jordan form of A has one Jordan block only.*

Proof. (b) $\implies$ (c) is obvious. If (c) holds, then setting up the eigenvalue/eigenvector equation $Ax = x\lambda$, $x \neq 0$, $\lambda \in \mathsf{H}$, and (without loss of generality) letting x be of the form $x = \mathrm{col}\,(x_{i-1}, x_1, \ldots, 1, 0, \ldots, 0)$, with 1 on the ith position, we obtain that $D(A) = \{\lambda\}$ on the main diagonal and that $\lambda x_1 + 1 = x_1 \lambda$. The latter equation is contradictory (Corollary 2.2.3) unless $i = 1$, i.e., $x = e_1$, and (b) follows.

Suppose (c) does not hold. We may (and do) assume that A is in the Jordan form and $D(A) \subset \mathsf{C}$. At this point we use known results on the structure of complex invariant subspaces of complex matrices; see, e.g., Gohberg et al. [54, Chapter 14] or Shayman [144] for more detailed analytic structure. It follows that A has either a continuum of complex k-dimensional invariant subspaces (in the case the geometric multiplicity of some eigenvalue of A is larger than one) or a finite number, but at least two, complex k-dimensional invariant subspaces (in the case the geometric multiplicity of every eigenvalue is equal to one). So (a) does not hold.

Finally, if (c) holds and if $\mathcal{M}$ is a k-dimensional A-invariant subspace, then (assuming $A = J_n(\lambda)$ for some $\lambda \in \mathsf{H}$) $\mathcal{M}$ cannot contain any vector $x = [x_i]_{i=1}^n \in \mathsf{H}^{n \times 1}$ having at least one nonzero component x_i with $i > k$; indeed, if $\mathcal{M}$ were to contain such a vector $x = [x_i]_{i=1}^n \in \mathsf{H}^{n \times 1}$, with $x_{i_0} \neq 0$, $x_i = 0$ for $i = i_0 + 1, \ldots, n$, and $i_0 > k$, then by scaling x and applying A repeatedly, one easily sees that $\mathcal{M}$ contains a linearly independent set of i_0 vectors of the form

$$\mathrm{col}\,(*, *, \ldots, 1, 0, \ldots, 0), \quad \mathrm{col}\,(*, \ldots, 1, 0, \ldots, 0), \ldots, \mathrm{col}\,(1, 0, \ldots, 0),$$

with the 1 appearing in the i_0th, $(i_0 - 1)$th, etc., 1st position, respectively, a contradiction to $\mathcal{M}$ having dimension k. It follows that $\mathcal{M} = \mathrm{Span}_{\mathsf{H}}\{e_1, \ldots, e_k\}$, hence, unique. $\qquad \square$

5.6   PROOF OF THEOREM 5.5.3

For the proof we need a slightly modified complex matrix representation of quaternions. Write a matrix $A \in \mathsf{H}^{n \times n}$ in the form

$$A = A_1 + \mathsf{j}A_2, \quad A_1, A_2 \in \mathsf{C}^{n \times n},$$

and define the map $\widetilde{\omega}_n : \mathsf{H}^{n \times n} \to \mathsf{C}^{2n \times 2n}$ by

$$\widetilde{\omega}_n(A) = \left[\begin{array}{cc} A_1 & \overline{A_2} \\ -A_2 & \overline{A_1} \end{array} \right]. \tag{5.6.1}$$

It is easy to see that the map $\widetilde{\omega}_n$ differs from the map $\omega_{n,n}$ introduced in Section 3.4 only by a simultaneous permutation of rows and columns:

$$\widetilde{\omega}_n(A) = P_n^{-1}(\omega_{n,n}(A))P_n, \tag{5.6.2}$$

where P_n is fixed, i.e., independent of $A \in \mathsf{H}^{n \times n}$ but dependent on n, permutation matrix. For example, if

$$A = \left[\begin{array}{cc} a_0 + ia_1 + ja_2 + ka_3 & b_0 + ib_1 + jb_2 + kb_3 \\ c_0 + ic_1 + jc_2 + kc_3 & d_0 + id_1 + jd_2 + kd_3 \end{array} \right] \in \mathsf{H}^{2 \times 2},$$

where $a_j, b_j, c_j, d_j \in \mathsf{R}$, then

$$\omega_{2,2}(A) = \left[\begin{array}{cccc} a_0 + ia_1 & a_2 + ia_3 & b_0 + ib_1 & b_2 + ib_3 \\ -a_2 + ia_3 & a_0 - ia_1 & -b_2 + ib_3 & b_0 - ib_1 \\ c_0 + ic_1 & c_2 + ic_3 & d_0 + id_1 & d_2 + id_3 \\ -c_2 + ic_3 & c_0 - ic_1 & -d_2 + id_3 & d_0 - id_1 \end{array} \right]$$

and

$$\widetilde{\omega}_2(A) = \left[\begin{array}{cccc} a_0 + ia_1 & b_0 + ib_1 & a_2 + ia_3 & b_2 + ib_3 \\ c_0 + ic_1 & d_0 + id_1 & c_2 + ic_3 & d_2 + id_3 \\ -a_2 + ia_3 & -b_2 + ib_3 & a_0 - ia_1 & b_0 - ib_1 \\ -c_2 + ic_3 & -d_2 + id_3 & c_0 - ic_1 & d_0 - id_1 \end{array} \right],$$

so

$$P_2 = \left[\begin{array}{cccc} 1 & 0 & 0 & 0 \\ 0 & 0 & 1 & 0 \\ 0 & 1 & 0 & 0 \\ 0 & 0 & 0 & 1 \end{array} \right].$$

In general, P_n is the $2n \times 2n$ matrix having 1s in the positions

$$(1,1), \ (2, n+1), \ (3,2), \ (4, n+2), \ \ldots, \ (2n-1, n), \ (2n, 2n),$$

and zeros elsewhere.

Because of (5.6.2), the algebraic properties of $\omega_{n,n}$ listed in Section 3.4 remain valid for $\widetilde{\omega}_n$; property (i) takes the following form.

(i') $\widetilde{\omega}_n$ is an isomorphism of the real algebra $\mathsf{H}^{n \times n}$ onto the real unital subalgebra

$$\widetilde{\Omega}_{2n} := \left\{ \left[\begin{array}{cc} A_1 & \overline{A_2} \\ -A_2 & \overline{A_1} \end{array} \right] : A_1, A_2 \in \mathsf{C}^{n \times n} \right\} \tag{5.6.3}$$

of $\mathsf{C}^{2n \times 2n}$, and $\widetilde{\omega}_n(I) = I$.

We can also check directly that $\widetilde{\Omega}_{2n}$ is a real unital subalgebra of $\mathsf{C}^{2n \times 2n}$ (Ex. 5.16.13).

We will show that for every matrix B in $\widetilde{\Omega}_{2n}$, there exists an invertible matrix T and a matrix in a (complex) Jordan form J in the same subalgebra such that $J = T^{-1}BT$; in other words, we will prove the following claim.

Claim 5.6.1. *For every* $B = \begin{bmatrix} B_1 & \overline{B_2} \\ -B_2 & \overline{B_1} \end{bmatrix}$, *where* $B_1, B_2 \in \mathsf{C}^{n \times n}$, *there exist*

an invertible matrix T *of the form* $T = \begin{bmatrix} T_1 & \overline{T_2} \\ -T_2 & \overline{T_1} \end{bmatrix}$, *where* $T_1, T_2 \in \mathsf{C}^{n \times n}$, *and a*

Jordan form $J_1 \in \mathsf{C}^{n \times n}$ *such that the equality*

$$\begin{bmatrix} J_1 & 0 \\ 0 & \overline{J_1} \end{bmatrix} = T^{-1} B T \tag{5.6.4}$$

holds.

The Jordan form J_1 can be chosen to have all eigenvalues with nonnegative imaginary parts.

Once Claim 5.6.1 is proved, the existence part of Theorem 5.5.3 follows easily. Indeed, observe first that if $T \in \widetilde{\Omega}_{2n}$ is invertible, then T^{-1} also belongs to $\widetilde{\Omega}_{2n}$. That is because $f(T) = 0$, where $f(t)$ is the minimal polynomial for T, and, therefore, T^{-1} can be expressed as a polynomial of T. Now, for a given $A \in \mathsf{H}^{n \times n}$, we let $B = \widetilde{\omega}_n(A)$ and apply the inverse map $(\widetilde{\omega}_n)^{-1}$ to equality (5.6.4). The following equation results:

$$J_1 = (\widetilde{\omega}_n)^{-1} (T^{-1}) A (T_1 + jT_2) = (T_1 + jT_2)^{-1} A (T_1 + jT_2),$$

and the existence of a Jordan form of any square-size quaternion matrix follows.

Proof of Claim 5.6.1. Denote

$$A := \begin{bmatrix} A_1 & \overline{A_2} \\ -A_2 & \overline{A_1} \end{bmatrix}, \quad A_1, A_2 \in \mathsf{C}^{n \times n}. \tag{5.6.5}$$

Since A is a complex matrix, there exists a nonsingular matrix P such that $P^{-1} A P = J$ is a Jordan form of A.

Step 1. Let $\alpha_1, \ldots, \alpha_m \in \mathsf{C}$ be the distinct eigenvalues of A. Then each column of P is a column vector v of $2n$ components satisfying one and only one of the following two relations:

$$\text{(i) } Av = v\alpha_j, \qquad \text{(ii) } Av = w + v\alpha_j,$$

where w is the column of P adjacent to v on the left. All $2n$ columns of P are linearly independent, and for each α_j there is at least one column of type (i).

Step 2. Define the column vector $v^\star$ relative to $v \in \mathsf{C}^{2n \times 1}$ as follows: if

$$v^T = [v_{1,1} \quad v_{2,1} \quad \cdots \quad v_{n,1} \quad w_{1,1} \quad w_{2,1} \quad \cdots \quad w_{n,1}],$$

where $v_{k,1}, w_{k,1} \in \mathsf{C}$ for $k = 1, 2, \ldots, n$, then

$$v^\star = [-\overline{w_{1,1}} \quad -\overline{w_{2,1}} \quad \cdots \quad -\overline{w_{n,1}} \quad \overline{v_{1,1}} \quad \overline{v_{2,1}} \quad \cdots \quad \overline{v_{n,1}}]^T.$$

If v is not the zero vector, then v and $v^\star$ are linearly independent, because if $c_1 v + c_2 v^\star = 0$, where $c_1, c_2 \in \mathsf{C} \setminus \{0\}$, then it follows that

$$(c_1 \overline{c_1} + c_2 \overline{c_2}) w_{k,1} = 0, \quad (c_1 \overline{c_1} + c_2 \overline{c_2}) v_{k,1} = 0, \quad \text{for } k = 1, 2, \ldots, n,$$

so $c_1 = c_2 = 0$, a contradiction. Notice the following properties of the * operation:

$$Av^\star = (Av)^\star; \qquad \text{if} \quad Av = v\alpha, \quad \text{then} \quad Av^\star = v^\star \overline{\alpha}. \tag{5.6.6}$$

Note also that $(v^\star)^\star = -v$ and

$$(c_1 w_1 + c_2 w_2 + \cdots + c_k w_k)^\star = \overline{c_1} w_1^\star + \overline{c_2} w_2^\star + \cdots + \overline{c_k} w_k^\star,$$

where $w_1, \ldots, w_k \in \mathsf{C}^{2n \times 1}$, $c_1, \ldots, c_k \in \mathsf{C}$.

Now let α be an eigenvalue of A, and suppose first that α is real. If v_1 is an eigenvector of A corresponding to α, then v_1 and $v_1^\star$ are linearly independent vectors of type (i) for α and either exhaust the number of such linearly independent column vectors or else there exists another, say, v_2, which is linearly independent of v_1 and $v_1^\star$ and satisfies $Av_2 = v_2\alpha$. Then $v_1, v_1^\star, v_2, v_2^\star$ are linearly independent. Indeed, suppose

$$c_1 v_1 + c_2 v_1^\star + c_3 v_2 + c_4 v_2^\star = 0, \quad c_1, c_2, c_3, c_4 \in \mathsf{C}. \tag{5.6.7}$$

We may assume c_3, c_4 are both nonzero (if, say, $c_3 = 0$, then (5.6.7) yields

$$\overline{c_1} v_1^\star - \overline{c_2} v_1 - c_4 v_2 = 0,$$

and we must have $c_1 = c_2 = c_4 = 0$). Together with (5.6.7) we also have

$$\overline{c_1} v_1^\star - \overline{c_2} v_1 + \overline{c_3} v_2^\star - \overline{c_4} v_2 = 0.$$

Solving this equation for $v_2^\star$ and substituting in (5.6.7) gives

$$(\overline{c_3}_1 + c_4 \overline{c_2}) v_1 + (\overline{c_3} c_2 - c_4 \overline{c_1}) v_1^\star + (c_3 \overline{c_3} + c_4 \overline{c_4}) v_2 = 0,$$

so $c_3 = c_4 = 0$, a contradiction. Either $v_1, v_1^\star, v_2, v_2^\star$ exhaust the number of linearly independent vectors of type (i) for α, or they do not. In the latter case, we continue this process and eventually obtain a set of linearly independent vectors of the form $v_1, v_1^\star, \ldots, v_k, v_k^\star$, which provide a basis for the vectors of type (i) corresponding to α. Note that $2k$ is also the number of columns in P that are eigenvectors of A corresponding to α; in particular, the number of such columns is even.

Suppose now α is complex nonreal. Then, if the matrix P contains a set of column vectors $v_1, v_2, \ldots, v_t$ such that $Av_j = v_j \alpha$ ($j = 1, 2, \ldots, t$), then $Av_j^\star = v_j^\star \overline{\alpha}$ ($j = 1, 2, \ldots, t$). So $\overline{\alpha}$ is also an eigenvalue of A, and a basis of the set of vectors of type (i) for $\overline{\alpha}$ can be taken in the form $v_1^\star, \ldots, v_t^\star$, where $v_1, \ldots, v_t$ is a basis of the set of vectors of type (i) for α.

Step 3. Consider vectors v of type (ii): $(A - \alpha_j I)v = w$, which implies $(A - \overline{\alpha_j} I)v^\star = w^\star$.

Writing α for α_j for brevity, assume first that α is real. By Step 2, there is a basis $\mathcal{B} := \{v_1, v_2, \ldots, v_{2k}\}$ for

$$\mathrm{Ker}_{\mathsf{C}} (A - \alpha I) := \{x \in \mathsf{C}^{2n \times 1} : (A - \alpha I)x = 0\}$$

such that $v \in \mathcal{B} \Rightarrow v^\star \in \mathcal{B}$, and for every pair of vectors $v, v^\star$ in $\mathcal{B}$ at least one of them is a column of P. Let $v_1^{(1)}, \ldots, v_p^{(1)}$ be the columns of P such that

$$(A - \alpha I)v_j^{(1)} \neq 0, \quad (A - \alpha I)^2 v_j^{(1)} = 0, \quad \text{for} \quad j = 1, 2, \ldots, p. \tag{5.6.8}$$

If (5.6.8) holds, then we say that the v'_j's are the (first) *generalized eigenvectors* associated with the eigenvectors $(A - \alpha I)v_j^{(1)}$. Clearly the vectors

$$v_1, \ldots, v_{2k}, v_1^{(1)}, \ldots, v_p^{(1)}$$

form a basis for $\text{Ker}_{\mathbb{C}}(A - \alpha I)^2$.

Consider $v_1^{(1)}$. Since $v_1, \ldots, v_{2k}, v_1^{(1)}$ are linearly independent, so are

$$v_1, \ldots, v_{2k}, v_1^{(1)}, (v_1^{(1)})^\star \tag{5.6.9}$$

(see the argument in Step 2, the case of real α). But $(v_1^{(1)})^\star$ is a generalized eigenvector associated with the eigenvector $(A - \alpha I)(v_1^{(1)})^\star$. We now replace a suitable pair of vectors $v_j, v_j^\star$ in $\mathcal{B}$ with

$$(A - \alpha I)v_1^{(1)}, \quad (A - \alpha I)(v_1^{(1)})^\star = ((A - \alpha I)v_1^{(1)})^\star,$$

so that the new set of vectors is again a basis for $\text{Ker}_{\mathbb{C}}(A - \alpha I)$. For simplicity of notation, we again denote the new set of vectors by $v_1, \ldots, v_{2k}$.

If (5.6.9) (with the new set $v_1, \ldots, v_{2k}$) is a basis for $\text{Ker}_{\mathbb{C}}(A - \alpha I)^2$, we stop. Otherwise, there is $v_j^{(1)}$, which is linearly independent of (5.6.9), and we repeat the procedure in the preceding paragraph. Continuing this way we eventually obtain a basis in $\text{Ker}_{\mathbb{C}}(A - \alpha I)^2$ of the following form (rearranging the vectors $v_1, \ldots, v_{2k}$ if necessary):

$$\begin{aligned} \mathcal{B}_1 \quad := \quad & \{v_1^{(1)}, (v_1^{(1)})^\star, \ldots, v_1^{(q)}, (v_1^{(q)})^\star, \\ & v_1, v_2 = v_1^\star, v_3, v_4 = v_3^\star, \ldots, v_{2q-1}, v_{2q} = v_{2q-1}^\star, \ldots, v_{2k}\}, \end{aligned}$$

for some integer $q \leq k$, where

$$(A - \alpha I)v_1^{(j)} = v_{2j-1}, \quad \text{for } j = 1, 2, \ldots, q.$$

In particular, the dimension of $\text{Ker}_{\mathbb{C}}(A - \alpha I)^2$ is even.

If

$$\text{Ker}_{\mathbb{C}}(A - \alpha I)^2 \neq \text{Ker}_{\mathbb{C}}(A - \alpha I)^3,$$

then we continue this process, working with $v_1^{(2)}, \ldots, v_r^{(2)}$, the columns of P such that

$$(A - \alpha I)^2 v_j^{(2)} \neq 0, \quad (A - \alpha I)^3 v_j^{(2)} = 0, \quad \text{for } j = 1, 2, \ldots, r. \tag{5.6.10}$$

The vectors $v_j^{(2)}$ satisfying (5.6.10) are called the (second) *generalized eigenvectors* associated with the eigenvector $(A - \alpha I)^2 v_j^{(2)}$. Thus, the set $\mathcal{B}_1 \cup \{v_1^{(2)}, (v_1^{(2)})^\star\}$ is linearly independent. Consider the four vectors

$$(A - \alpha I)v_1^{(2)}, \quad (A - \alpha I)(v_1^{(2)})^\star = ((A - \alpha I)v_1^{(2)})^\star, \tag{5.6.11}$$

$$(A - \alpha I)^2 v_1^{(2)}, \quad (A - \alpha I)^2(v_1^{(2)})^\star = ((A - \alpha I)^2 v_1^{(2)})^\star. \tag{5.6.12}$$

The equalities in (5.6.11) and (5.6.12) follow from (5.6.6) applied to $A - \alpha I$ and $(A - \alpha I)^2$ in place of A, bearing in mind that $A - \alpha I$ and $(A - \alpha I)^2$ also have the form as in (5.6.5). The vectors (5.6.11) and (5.6.12) are linearly independent. Indeed, if

$$c_1(A - \alpha I)v_1^{(2)} + c_2(A - \alpha I)(v_1^{(2)})^\star + c_3(A - \alpha I)^2 v_1^{(2)}$$

$$+ c_4(A - \alpha I)^2 (v_1^{(2)})^\star = 0, \tag{5.6.13}$$

where $c_1, c_2, c_3, c_4 \in \mathsf{C}$, then applying $A - \alpha I$ to this equality, we obtain

$$c_1(A - \alpha I)^2 v_1^{(2)} + c_2(A - \alpha I)^2 (v_1^{(2)})^\star = 0.$$

Hence, $c_1 = c_2 = 0$, but then (5.6.13) gives $c_3 = c_4 = 0$ as well. We now replace four suitable vectors in $\mathcal{B}_1$ with the vectors (5.6.11) and (5.6.12), so that the new set is again a basis for $\mathrm{Ker}_\mathsf{C}\,(A - \alpha I)^2$. We continue this process until we obtain a suitable basis in the root subspace $\mathrm{Ker}_\mathsf{C}\,(A - \alpha I)^{2n}$. Note that all subspaces $\mathrm{Ker}_\mathsf{C}\,(A - \alpha I)^i$, $i = 1, 2, \ldots, 2n$, are even dimensional.

Now let α be complex nonreal. If $u_1, \ldots, u_s$ are the columns of P that represent eigenvectors and associated generalized eigenvectors of A corresponding to α, then s is the (complex) dimension of the root subspace $\mathcal{R}(\alpha) := \mathrm{Ker}_\mathsf{C}\,(A - \alpha I)^{2n}$ of A corresponding to α. Also, the columns

$$u_1^\star, \ldots, u_s^\star \tag{5.6.14}$$

represent eigenvectors and associated generalized eigenvectors of A corresponding to $\overline{\alpha} \neq \alpha$. Clearly, the set (5.6.14) is linearly independent. We claim that (5.6.14) is a basis for the root subspace $\mathcal{R}(\overline{\alpha})$. Indeed, suppose not. Then the dimension of $\mathcal{R}(\overline{\alpha})$ exceeds s, and by taking a basis for $\mathcal{R}(\overline{\alpha})$ that consists of eigenvectors and associated generalized eigenvectors and by applying the map * to this basis, we obtain a linearly independent set in $\mathcal{R}(\alpha)$ of more than s vectors, which is impossible. Thus (5.6.14) is a basis for $\mathcal{R}(\overline{\alpha})$.

Step 4. Applying Steps 2 and 3 to all real eigenvalues and to all nonreal eigenvalues with positive imaginary parts of A, we eventually obtain a set of $2n$ linearly independent vectors arranged in pairs $w_1, w_1^*, \ldots, w_n, w_n^* \in \mathsf{C}^{2n \times 1}$, i.e., a basis for $\mathsf{C}^{2n \times 1}$, such that the set $\{w_1, \ldots, w_n\}$ consists of eigenvectors and associated generalized eigenvectors of A; then, necessarily, $\{w_1, \ldots, w_n\}$ consists of eigenvectors and associated generalized eigenvectors as well; if $w_{i_1}, \ldots, w_{i_k}$ is a chain consisting of an eigenvector w_{i_1} corresponding to an eigenvalue λ and associated generalized eigenvectors $w_{i_2}, \ldots, w_{i_k}$, then $w_{i_1}^\star$ is an eigenvector corresponding to the eigenvalue $\overline{\alpha}$ and $w_{i_2}^\star, \ldots, w_{i_k}^\star$ are associated generalized eigenvectors. Set

$$\widetilde{P} = [w_1 \ w_2 \ \ldots \ w_n \ w_1^* \ w_2^* \ \ldots \ w_n^*] \in \mathsf{C}^{n \times n},$$

where the vectors $w_1, \ldots, w_n$ are arranged so that the generalized eigenvectors are immediately to the right of the eigenvector they are associated with. Then clearly $\widetilde{P}$ is given by $\widetilde{P} = \begin{bmatrix} B & -\overline{C} \\ C & \overline{B} \end{bmatrix}$ for some $B, C \in \mathsf{C}^{n \times n}$, and the equality $A\widetilde{P} = \widetilde{P}\widetilde{J}$ holds with a matrix $\widetilde{J}$ in the Jordan form such that $\widetilde{J} = \begin{bmatrix} J_0 & 0 \\ 0 & \overline{J_0} \end{bmatrix}$ for some Jordan form matrix $J_0 \in \mathsf{C}^{n \times n}$.

Claim 5.6.1 is proved. □

Proof of uniqueness of a Jordan form. Uniqueness of a Jordan form for quaternion matrices can be proved by applying the map ω (or $\widetilde{\omega}$), thereby reducing the proof to the case of complex matrices. We will offer an independent proof, which will allow us to demonstrate techniques that may be useful on other occasions as well.

Let J and $\widetilde{J}$ be two similar $n \times n$ matrices in the Jordan form. We may assume without loss of generality, that the diagonals $D(J)$ and $D(\widetilde{J})$ consist of complex numbers with nonnegative imaginary parts. We have to prove that J and $\widetilde{J}$ are obtained from each other by a permutation of Jordan blocks. Observe that similar matrices have same eigenvalues, and for each eigenvalue λ_0 they have the same geometric multiplicities (maximal number of linearly independent eigenvectors, or what is the same, the number of Jordan blocks corresponding to eigenvalues similar to λ_0). Similar matrices also have the same algebraic multiplicities (dimension of the root subspace corresponding to λ_0 or, what is the same, the totality of the sizes of Jordan blocks that have eigenvalues similar to λ_0). In view of these observations, we may (and do) further assume that

$$J = J_1 \oplus \cdots \oplus J_k, \qquad \widetilde{J} = \widetilde{J}_1 \oplus \cdots \oplus \widetilde{J}_k, \tag{5.6.15}$$

where for each index ℓ, J_ℓ and $\widetilde{J}_\ell$ are matrices in the Jordan form of size $n_\ell \times n_\ell$ with λ_ℓ on the diagonal; here $\lambda_1, \ldots, \lambda_k$ are distinct complex numbers with nonnegative imaginary parts. Moreover,

$$J_\ell = J_{m_{\ell,1}}(\lambda_\ell) \oplus J_{m_{\ell,2}}(\lambda_\ell) \oplus \cdots \oplus J_{m_{\ell,p_\ell}}(\lambda_\ell), \quad \text{for } \ell = 1, 2, \ldots, k, \tag{5.6.16}$$

$$\widetilde{J}_\ell = J_{\widetilde{m}_{\ell,1}}(\lambda_\ell) \oplus J_{\widetilde{m}_{\ell,2}}(\lambda_\ell) \oplus \cdots \oplus J_{\widetilde{m}_{\ell,p_\ell}}(\lambda_\ell), \quad \text{for } \ell = 1, 2, \ldots, k, \tag{5.6.17}$$

where

$$m_{\ell,1} \geq m_{\ell,2} \geq \cdots \geq m_{\ell,p_\ell}, \quad \text{for } \ell = 1, 2, \ldots, k,$$

$$\widetilde{m}_{\ell,1} \geq \widetilde{m}_{\ell,2} \geq \cdots \geq \widetilde{m}_{\ell,p_\ell}, \quad \text{for } \ell = 1, 2, \ldots, k,$$

and the following equalities hold:

$$\sum_{j=1}^{p_\ell} m_{\ell,j} = \sum_{j=1}^{p_\ell} \widetilde{m}_{\ell,j}, \quad \text{for } \ell = 1, 2, \ldots, k.$$

Thus, the number of Jordan blocks corresponding to each eigenvalue λ_ℓ is the same in J and $\widetilde{J}$. Write down the similarity between J and $\widetilde{J}$:

$$JQ = Q\widetilde{J}, \quad Q \in \mathsf{H}^{n \times n} \text{ is invertible.} \tag{5.6.18}$$

Partition Q conformally with the partitions (5.6.15), (5.6.16), and (5.6.17) of J and $\widetilde{J}$:

$$Q = [Q_{i,j}]_{i,j=1}^k, \qquad Q_{\ell,\ell} = [Q_{j,q}^{(\ell)}]_{j,q=1}^{p_\ell}, \quad \text{for } \ell = 1, 2, \ldots, k,$$

where $Q_{j,q}^{(\ell)}$ is of size $m_{\ell,j} \times \widetilde{m}_{\ell,q}$. We have $J_i Q_{i,\ell} = Q_{i,\ell} \widetilde{J}_\ell$, so by Proposition 5.4.1 $Q_{i,\ell} = 0$ if $i \neq \ell$. Thus, in fact, Q is block diagonal, and (5.6.18) boils down to

$$J_\ell Q_{\ell,\ell} = Q_{\ell,\ell} \widetilde{J}_\ell, \qquad \ell = 1, 2, \ldots, k, \tag{5.6.19}$$

where $Q_{\ell,\ell}$ are invertible.

The uniqueness of Jordan form will be proved if we show that

$$m_{\ell,j} = \widetilde{m}_{\ell,j}, \quad \text{for } j = 1, 2, \ldots, p_\ell \text{ and } \ell = 1, 2, \ldots, k.$$

Suppose not, and for some ℓ_0 we have

$$m_{\ell_0,1} = \widetilde{m}_{\ell_0,1}, m_{\ell_0,} = \widetilde{m}_{\ell_0,2}, \ldots, m_{\ell_0,u-1} = \widetilde{m}_{\ell_0,u-1},$$

but $m_{\ell_0,u} \neq \widetilde{m}_{\ell_0,u}$. Say $m_{\ell_0,u} > \widetilde{m}_{\ell_0,u}$. (If the opposite inequality holds, then rewrite (5.6.19) in the form

$$\widetilde{J}_\ell Q_{\ell,\ell}^{-1} = Q_{\ell,\ell}^{-1} J_\ell, \qquad \ell = 1, 2, \ldots, k,$$

and argue analogously with the roles of $\widetilde{J}_\ell$ and J_ℓ interchanged.) Then it follows from Proposition 5.4.2 and from the equation

$$J_{\ell_0} Q_{\ell_0,\ell_0} = Q_{\ell_0,\ell_0} \widetilde{J}_{\ell_0},$$

or

$$J_{m_{\ell_0,j}}(\lambda_{\ell_0}) Q_{j,q}^{(\ell_0)} = Q_{j,q}^{(\ell_0)} J_{m_{\ell_0,q}}(\lambda_{\ell_0}), \quad j, q = 1, 2, \ldots, p_{\ell_0},$$

that the matrices $Q_{j,q}^{(\ell_0)}$ have the form of (5.4.3) or (5.4.4). We obtain, therefore, that each of the u rows

$$m_{\ell_0,1}\text{th}, \ (m_{\ell_0,1} + m_{\ell_0,2})\text{th}, \ \ldots, \ (m_{\ell_0,1} + m_{\ell_0,2} + \cdots + m_{\ell_0,u})\text{th} \qquad (5.6.20)$$

of Q_{ℓ_0,ℓ_0} contain at most $u-1$ nonzero elements, and those nonzero elements must be in the positions

$$m_{\ell_0,1}\text{th}, (m_{\ell_0,1} + m_{\ell_0,2})\text{th}, \ \ldots, \ (m_{\ell_0,1} + m_{\ell_0,2} + \cdots + m_{\ell_0,u-1})\text{th}.$$

(If it happens that $u = 1$, then the $m_{\ell_0,1}$th row of Q_{ℓ_0,ℓ_0} is zero.) Thus, the rows (5.6.20) (understood as left quaternion vectors) are linearly dependent, which contradicts invertibility of Q_{ℓ_0,ℓ_0}.

This completes the proof of Theorem 5.5.3. □

5.7 JORDAN FORMS OF MATRIX REPRESENTATIONS

By the *complex Jordan form*, resp. *real Jordan form*, of a complex, resp. real, matrix X, we mean the familiar Jordan form under similarity $X \mapsto S^{-1} X S$, where S is an invertible complex, resp. real, matrix. Thus, for a complex matrix the complex Jordan form is the direct sum of blocks of type $J_m(\lambda)$, where $\lambda \in \mathsf{C}$, and for a real matrix, the real Jordan form is the direct sum of blocks of type $J_m(\lambda)$, where $\lambda \in \mathsf{R}$, and of type $J_{2m}(a \pm ib)$, where $a, b \in \mathsf{R}$ and $b \neq 0$ (Theorem 15.1.1).

Theorem 5.7.1. (a) *If*

$$J_{m_1}(\alpha_1) \oplus \cdots \oplus J_{m_r}(\alpha_r), \qquad \alpha_1 = a_1 + ib_1, \ldots, \alpha_r = a_r + ib_r \in \mathsf{C} \qquad (5.7.1)$$

is a Jordan form of $A \in \mathsf{H}^{n \times n}$, *then*

$$\begin{bmatrix} J_{m_1}(\alpha_1) & 0 \\ 0 & J_{m_1}(\overline{\alpha_1}) \end{bmatrix} \oplus \cdots \oplus \begin{bmatrix} J_{m_r}(\alpha_r) & 0 \\ 0 & J_{m_r}(\overline{\alpha_r}) \end{bmatrix}$$

is the complex Jordan form of $\omega_n(A)$.

(b) *Let* (5.7.1) *be a Jordan form of* $A \in \mathsf{H}^{n \times n}$, *where* $\alpha_1, \ldots, \alpha_s$ *are real and* $\alpha_{s+1}, \ldots, \alpha_r$ *are nonreal. Then the real Jordan form of* $\chi_{n,n}(A)$ *is*

$$(J_{m_1}(\alpha_1) \oplus J_{m_1}(\alpha_1) \oplus J_{m_1}(\alpha_1) \oplus J_{m_1}(\alpha_1))$$

$$\oplus \quad \cdots \quad \oplus \left(J_{m_s}(\alpha_s) \oplus J_{m_s}(\alpha_s) \oplus J_{m_s}(\alpha_s) \oplus J_{m_s}(\alpha_s)\right)$$
$$\oplus \quad \left(J_{2m_{s+1}}(a_{s+1} \pm ib_{s+1}) \oplus J_{2m_{s+1}}(a_{s+1} \pm ib_{s+1})\right)$$
$$\oplus \quad \cdots \quad \oplus \left(J_{2m_r}(a_r \pm ib_r) \oplus J_{2m_r}(a_r \pm ib_r)\right). \tag{5.7.2}$$

Proof. Part (a) is a by-product of the proof of Theorem 5.5.3.

For part (b), observe that a permutation similarity in the matrix $\chi\left(J_{m_j}(\alpha_j)\right)$ ($\alpha_j \in \mathsf{R}$) yields

$$J_{m_j}(\alpha_j) \oplus J_{m_j}(\alpha_j) \oplus J_{m_j}(\alpha_j) \oplus J_{m_j}(\alpha_j).$$

Indeed, we have the equality

$$\chi\left(J_{m_j}(\alpha_j)\right)\left[e_1 \; e_5 \; \ldots e_{4m_j-3} \; e_2 \; e_6 \; \cdots \; e_{4m_j-2} \; \cdots \; e_4 \; e_8 \; \cdots \; e_{4m_j}\right]$$

$$= \left[e_1 \; e_5 \; \ldots e_{4m_j-3} \; e_2 \; e_6 \; \cdots \; e_{4m_j-2} \; \cdots \; e_4 \; e_8 \; \cdots \; e_{4m_j}\right]$$

$$\cdot \left(J_{m_j}(\alpha_j) \oplus J_{m_j}(\alpha_j) \oplus J_{m_j}(\alpha_j) \oplus J_{m_j}(\alpha_j)\right). \tag{5.7.3}$$

Also, for $j = s+1, \ldots, r$ we have

$$\chi\left(J_{m_j}(\alpha_j)\right)$$

$$= \begin{bmatrix} \begin{bmatrix} a_j & -b_j \\ b_j & a_j \end{bmatrix} & 0_2 & I_2 & 0_2 & \cdots & & 0_2 \\ 0_2 & \begin{bmatrix} a_j & -b_j \\ b_j & a_j \end{bmatrix} & 0_2 & I_2 & \cdots & & 0_2 \\ \vdots & \vdots & \ddots & \ddots & \ddots & & \vdots \\ 0_2 & 0_2 & 0_2 & 0_2 & \cdots & & \begin{bmatrix} a_j & -b_j \\ b_j & a_j \end{bmatrix} \end{bmatrix}. \tag{5.7.4}$$

A permutation similarity of the matrix in the right-hand side of (5.7.4) yields $J_{2m_j}(a_j \pm ib_j) \oplus J_{2m_j}(a_j \pm ib_j)$, and, thus, (Ex. 5.16.11):

$$\chi\left(J_{m_j}(\alpha_j)\right) = \widetilde{P}_{4m_j}^{-1}\left(J_{2m_j}(a_j \pm ib_j) \oplus J_{2m_j}(a_j \pm ib_j)\right)\widetilde{P}_{4m_j}, \tag{5.7.5}$$

where $\widetilde{P}_{4m_j}$ is a certain permutation matrix.

Part (b) is now obvious from (5.7.3) and (5.7.5). $\square$

Corollary 5.7.2. *The matrices $A \in \mathsf{H}^{n \times n}$, A^*, A_ϕ for every nonstandard involution ϕ, are all similar (over the quaternions).*

Proof. Let $A = SJS^{-1}$, where J is given by (5.7.1) with $\alpha_1, \ldots, \alpha_s$ real and $\alpha_{s+1}, \ldots, \alpha_r$ nonreal. Since $\chi_{n,n}(A^*) = (\chi_{n,n}(A))^T$, we see by Theorem 5.7.1 that the real Jordan form of $\chi_{n,n}(A^*)$ coincides with the real Jordan form of a matrix that is transpose to (5.7.2). But every real Jordan block is similar to its transpose:

$$J_m(\alpha)^T = F_m^{-1} J_m(\alpha) F_m, \quad J_{2m}(\alpha \pm i\beta)^T = F_{2m}^{-1} J_{2m}(\alpha \pm i\beta) F_{2m},$$

where $\alpha, \beta \in \mathsf{R}$, $\beta \neq 0$. (More generally, every real or complex matrix is similar to its transpose, a well-known fact that can be found, e.g., in Horn and Johnson [62].) Thus, the real Jordan form of $\chi_{n,n}(A^*)$ coincides with (5.7.2). By Theorem 5.7.1, A and A^* have the same Jordan form.

For a nonstandard involution ϕ, first observe that any Jordan matrix J (i.e., a direct sum of Jordan blocks with generally quaternion diagonal) is similar to J_ϕ. Indeed, if

$$J = J_{p_1}(\gamma_1) \oplus \cdots \oplus J_{p_s}(\gamma_s), \quad \gamma_1, \ldots, \gamma_s \in \mathsf{H},$$

then each γ_j is similar to $(\gamma_j)_\phi$: $(\gamma_j)_\phi = \lambda_j^{-1} \gamma_j \lambda_j$ for some $\lambda_j \in \mathsf{H} \setminus \{0\}$, and we have

$$J_{p_j}(\gamma_j)_\phi = (\lambda_j F_{p_j})^{-1} J_{p_1}(\gamma_1) (\lambda_j F_{p_j}), \quad j = 1, 2, \ldots, s.$$

Now, if $A = SJS^{-1}$ for some Jordan matrix J and invertible matrix S, then

$$A_\phi = S_\phi^{-1} J_\phi S_\phi = S_\phi^{-1} T J T^{-1} S_\phi = S_\phi^{-1} T S^{-1} A S T^{-1} S_\phi,$$

where the invertible matrix $T \in \mathsf{H}^{n \times n}$ is a similarity matrix between J_ϕ and J. $\quad \square$

5.8   COMPARISON WITH REAL AND COMPLEX SIMILARITY

It is instructive to compare Theorem 5.5.3 with results on similarity of real matrices (over the reals) and complex matrices (over the complexes).

Theorem 5.8.1.   (a) *Every square-size quaternion matrix is similar to a complex matrix.*

(b) *A square-size quaternion matrix A is similar to a real matrix if and only if for every m and every nonreal eigenvalue $\lambda \in \mathsf{H}$, the number of Jordan blocks $J_m(\alpha)$ with $\alpha \in \mathsf{H}$ similar to λ in the Jordan form of A is even.*

Proof.   Part (a) is obvious from Theorem 5.5.3(b): the Jordan form there is complex. For the proof of the Part (b) "if," we may assume that the Jordan form of A is given by the right-hand side of (5.5.1), where $\lambda_1, \ldots, \lambda_p \in \mathsf{C}$, with the imaginary parts of the λ_j's nonnegative. Using similarity (over quaternions), we replace one of each pair of identical Jordan blocks $J_m(\lambda)$, $J_m(\lambda)$ with nonreal λ by $J_m(\overline{\lambda})$. Then, observe that the Jordan block with a real eigenvalue is real, and $J_m(\lambda) \oplus J_m(\overline{\lambda})$ is similar to a real Jordan block:

$$\begin{aligned}
&[e_1 + ie_2 \quad e_3 + ie_4 \quad \ldots \quad e_{2m-1} + ie_{2m} \\
&\quad e_1 - ie_2 \quad e_3 - ie_4 \quad \ldots \quad e_{2m-1} - ie_{2m}] \\
&\cdot \left(J_m(\lambda) \oplus J_m(\overline{\lambda}) \right) = J_{2m}(\mathfrak{R}(\lambda) \pm i\mathfrak{I}(\lambda)) \\
&\cdot [e_1 + ie_2 \quad e_3 + ie_4 \quad \ldots \quad e_{2m-1} + ie_{2m} \\
&\quad e_1 - ie_2 \quad e_3 - ie_4 \quad \ldots \quad e_{2m-1} - ie_{2m}].
\end{aligned} \quad (5.8.1)$$

Now assume A is similar to a real matrix. Then A is similar to a real Jordan form. Formula (5.8.1) now shows that the Jordan blocks corresponding to nonreal eigenvalues come in pairs $J_m(\lambda)$, $J_m(\overline{\lambda})$, where $\lambda \in \mathsf{C}$ has positive imaginary part. Since $\overline{\lambda}$ is similar to λ (over H), the condition in (b) follows. $\quad \square$

Definition 5.8.2. For a matrix $A \in \mathsf{H}^{n \times n}$, its *quaternion similarity class* containing A consists of all matrices $B \in \mathsf{H}^{n \times n}$ for which $B = S^{-1}AS$ for some invertible $S \in \mathsf{H}^{n \times n}$. Analogously, for a real, resp. complex, square-size matrix A, the *real*, resp. *complex, similarity class* containing A consists of those real, resp. complex, matrices B for which $B = S^{-1}AS$ for some invertible real, resp. complex, matrix S.

Clearly, the real, resp. complex, similarity class containing a given real, resp. complex, matrix A is no larger than the set of real, resp. complex, matrices in the quaternion similarity class containing the same matrix A.

Theorem 5.8.3. (a) *For all real matrices $A \in \mathsf{R}^{n \times n}$, the real similarity class containing A coincides with the set*

$$\mathcal{Q}_\mathsf{R}(A) := \{B \in \mathsf{R}^{n \times n} : B = S^{-1}AS \text{ for some } S \in \mathsf{H}^{n \times n}\}. \tag{5.8.2}$$

(b) *For all complex matrices $A \in \mathsf{C}^{n \times n}$, the set*

$$\mathcal{Q}_\mathsf{C}(A) := \{B \in \mathsf{C}^{n \times n} : B = S^{-1}AS \text{ for some } S \in \mathsf{H}^{n \times n}\} \tag{5.8.3}$$

comprises a finite number p of complex similarity classes. The number p is found by the formula

$$p = \prod_{j=1}^{s} \left(\prod_{k=1}^{r_j} (p_{\lambda_j, k} + 1) \right), \tag{5.8.4}$$

where $\{\lambda_1, \ldots, \lambda_s\}$ is a set of distinct nonreal complex eigenvalues of A which is maximal (by inclusion) with respect to the property that it does not contain any pair of complex conjugate numbers, and $p_{\lambda_j, k}$ is the number of Jordan blocks in the complex Jordan form of A of size k with eigenvalue λ_j or $\overline{\lambda_j}$.

Thus, two real matrices which are similar over H are actually similar over R. However, two complex matrices which are similar over H are generally not similar over C; this happens if and only if all eigenvalues of the matrices are real.

For example, if

$$A = J_2(\mathrm{i}) \oplus J_2(\mathrm{i}) \oplus J_3(\mathrm{i}) \oplus J_4(-\mathrm{i}) \oplus J_1(2\mathrm{i}) \oplus J_3(-2\mathrm{i}),$$

then one can take $\{\lambda_1, \lambda_2\} = \{\mathrm{i}, -2\mathrm{i}\}$, and $p_{\lambda_1, k}$ is equal to 2 if $k = 2$, to 1 if $k = 3$ or $k = 4$, and to 0 for all other values of k, whereas $p_{\lambda_2, k}$ is equal to 1 if $k = 1$ or $k = 3$ and to 0 for all other values of k. The number p in this example is 48.

In Theorem 5.8.3(b), a typical complex similarity class contained in the set $\mathcal{Q}_\mathsf{C}(A)$ is constructed as follows. For every pair of values (j, k) as in (5.8.4) such that $p_{\lambda_j, k} \neq 0$, select an integer $\ell_{\lambda_j, k}$ such that $0 \leq \ell_{\lambda_j, k} \leq p_{\lambda_j, k}$. Then the complex similarity class corresponding to this selection of the $\ell_{\lambda_j, k}$'s consists of those complex matrices X whose complex Jordan form is

$$J_0 \oplus \oplus_{j=1}^{s} \oplus_{\{k \,:\, p_{\lambda_j, k} \neq 0\}} \left(\underbrace{J_k(\lambda_j) \oplus \cdots \oplus J_k(\lambda_j)}_{\ell_{\lambda_j, k} \text{ times}} \oplus \underbrace{J_k(\overline{\lambda_j}) \oplus \cdots \oplus J_k(\overline{\lambda_j})}_{p_{\lambda_j, k} - \ell_{\lambda_j, k} \text{ times}} \right), \tag{5.8.5}$$

where J_0 is the part of the complex Jordan form of A that consists of Jordan blocks with real eigenvalues (if any).

Proof. Part (a). Clearly, the real similarity class containing A is contained in $\mathcal{Q}_\mathsf{R}(A)$. Conversely, let $B \in \mathcal{Q}_\mathsf{R}(A)$, so $B = S^{-1}AS$, $S \in \mathsf{H}^{n \times n}$ invertible. Applying the real matrix representation map $\chi_{n,n}$, we obtain

$$\chi_{n,n}(B) = (\chi_{n,n}(S))^{-1} \chi_{n,n}(A) \chi_{n,n}(S). \tag{5.8.6}$$

It is easy to see that $\chi(X)$ is similar (over the reals) to $X^{\oplus 4}$ for any real square-size matrix X; in fact, the similarity is provided by a suitable permutation matrix. Thus $A^{\oplus 4}$ is similar to $B^{\oplus 4}$ in view of (5.8.6). By Corollary 15.1.4 A and B are similar (over the reals).

Part (b). Denote by $\mathcal{Q}_{\mathsf{C}}(\{\ell_{\lambda_j,k}\})$ the set of all complex matrices X having the complex Jordan form (5.8.5) for a fixed selection $\{\ell_{\lambda_j,k}\}$, as described above. Clearly, each $\mathcal{Q}_{\mathsf{C}}(\{\ell_{\lambda_j,k}\})$ is a complex similarity class, and one of those complex similarity classes contains A. On the other hand, by Theorem 5.5.3, all matrices $X \in \mathsf{C}^{n \times n}$ that belong to the union $\cup \mathcal{Q}_{\mathsf{C}}(\{\ell_{\lambda_j,k}\})$ (taken over all possible selections $\{\ell_{\lambda_j,k}\}$) are similar (over the quaternions). This shows the inclusion

$$\cup \mathcal{Q}_{\mathsf{C}}(\{\ell_{\lambda_j,k}\}) \subseteq \mathcal{Q}_{\mathsf{C}}(A).$$

The same Theorem 5.5.3 shows that the opposite inclusion holds as well. This proves Part (b). □

5.9   DETERMINANTS

The notion of determinant is familiar for matrices over a (commutative) field F:

$$\det(A) = \sum_{\tau} \left((-1)^{\text{sign}\,(\tau)}\, a_{1,\tau(1)} a_{2,\tau(2)} \cdots a_{n,\tau(n)} \right), \qquad (5.9.1)$$

where $A = [a_{i,j}]_{i,j=1}^n$ is a square-size matrix with entries in F, and the sum in (5.9.1) is taken over all $n!$ permutations τ of $\{1, 2, \ldots, n\}$. The determinant has many well-known remarkable properties, for example, multiplicativity: $\det(AB) = \det(A)\det(B)$ for all $A, B \in \mathsf{F}^{n \times n}$. However, the above definition of determinant is not useful in the context of matrices over noncommutative division rings, such as H. In this section we discuss a notion of determinants of square-size quaternion matrices.

Recall the complex matrix representations

$$\omega_{n,n} : \mathsf{H}^{n \times n} \to \mathsf{C}^{2n \times 2n}, \quad \widetilde{\omega}_n : \mathsf{H}^{n \times n} \to \mathsf{C}^{2n \times 2n},$$

defined in (3.4.1) and (5.6.1).

Definition 5.9.1. We define the *determinant* of $A \in \mathsf{H}^{n \times n}$ based on complex matrix representations as follows:

$$\det_{\mathsf{C}}(A) := \det(\omega_{n,n}(A)) = \det(\widetilde{\omega}_n(A)),$$

where in the right-hand side we use the standard determinant function (5.9.1) for the determinant of $2n \times 2n$ complex matrices, and the equality $\det(\omega_{n,n}(A)) = \det(\widetilde{\omega}_n(A))$ follows from (5.6.2).

For example, $\det_{\mathsf{C}}(\alpha) = |\alpha|^2$ for all $\alpha \in \mathsf{H}$.

This determinant shares many properties with the standard determinant (5.9.1) for matrices over a field. Some of these properties are listed below.

Theorem 5.9.2. *For all $A, B \in \mathsf{H}^{n \times n}$ we have:*

(1) $\det_{\mathsf{C}}(A)$ *is a homogeneous polynomial of degree $2n$ with real coefficients of the $4n^2$ real variables $a_{i,j}^{(k)}$, $k = 0, 1, 2, 3$; $i, j = 1, 2, \ldots, n$, where*

$$a_{i,j} = a_{i,j}^{(0)} + a_{i,j}^{(1)}\mathsf{i} + a_{i,j}^{(2)}\mathsf{j} + a_{i,j}^{(3)}\mathsf{k}$$

is the (i, j)th entry of A;

(2) $\det_C (A)$ *is real and nonnegative, and* $\det_C (A) = 0$ *if and only if A is not invertible;*

(3) $\det_C (AB) = \det_C (A) \cdot \det_C (B)$;

(4) *if A^{-1} exists, then* $\det_C (A^{-1}) = (\det_C (A))^{-1}$;

(5) *if A and B are similar, then* $\det_C (A) = \det_C (B)$;

(6) *if $\lambda_1, \ldots, \lambda_n$ are the eigenvalues of A which are complex numbers with non-negative imaginary parts, and which are counted according to their algebraic multiplicity, then*

$$\det_C (A) = |\lambda_1|^2 |\lambda_2|^2 \cdots |\lambda_n|^2; \qquad (5.9.2)$$

(7) *if A is upper triangular or lower triangular with $\lambda_1, \ldots, \lambda_n$ ($\lambda_j \in H$ for $j = 1, 2, \ldots, n$) on the main diagonal, then (5.9.2) holds.*

Proof. (3) follows from the definition of $\det_C$ and the multiplicative property of the standard determinant. (2) follows from Claim 5.6.1: in the notation of this claim, we have

$$\det_C (B_1 + jB_2) = \det(\widetilde{\omega}_n (B_1 + jB_2)) = \det(B) = \det (J_1 \overline{J_1}) = |\det (J_1)|^2.$$

Also, $\det (J_1) = 0$ if and only if $\det(B) = 0$, which is equivalent to noninvertibility of B or that of $B_1 + jB_2$. Now (2) follows from the definition of $\det_C$ and the inequality $\det_C (A) \geq 0$ for all $A \in H^{n \times n}$. Parts (4), (5), and (7) are an easy consequence of the definition of $\det_C$ and well-known properties of the standard determinant. Finally, for (6) write $A = T^{-1}JT$ for some invertible matrix T, where J is the Jordan form of A with $\lambda_1, \ldots, \lambda_n$ on the diagonal, perhaps in a different order. Then

$$\det_C (A) = \det_C (J) = |\det (J)|^2 = |\lambda_1|^2 |\lambda_2|^2 \cdots |\lambda_n|^2,$$

as required.   □

With respect to elementary row and column operations, the quaternion determinant behaves slightly differently from the standard determinant (5.9.1).

Theorem 5.9.3. *Let $A \in H^{n \times n}$. Then:*

(a) *if B is obtained from A by a permutation of its rows or columns, then*

$$\det_C (B) = \det_C (A); \qquad (5.9.3)$$

(b) *if B is obtained from A by multiplying one of its rows by $\lambda \in H$ on the left or by multiplying one of its columns by $\lambda \in H$ on the right, then*

$$\det_C (B) = |\lambda|^2 \det_C (A);$$

(c) *if B is obtained from A by adding one of its rows, multiplied by $\lambda \in H$ on the left, to another row, then (5.9.3) holds;*

(d) *if B is obtained from A by adding one of its columns multiplied by $\lambda \in H$ on the right to another column, then (5.9.3) holds.*

Proof. For (a), because every permutation is a product of transpositions, it suffices to consider the case when the permutation is actually a transposition. If B is obtained from A by interchange of its ith and jth rows (columns), then $\widetilde{\omega}(B)$ is obtained from $\widetilde{\omega}(A)$ by interchange of its ith and jth rows and interchange of its $(n+i)$th and $(n+j)$th rows (columns). Thus, $\det(\widetilde{\omega}(B)) = \det\widetilde{\omega}(A)$, and (a) follows.

Part (b). We have $B = QA$ or $B = AQ$, where

$$Q = \operatorname{diag}(1, 1, \ldots, \lambda, 1, \ldots, 1) \in \mathsf{H}^{n \times n},$$

with λ on the ith place (if it is the ith row or column that gets multiplied by λ). Now by Theorem 5.9.2(3), (7):

$$\det_{\mathsf{C}}(B) = \det_{\mathsf{C}}(Q)\det_{\mathsf{C}}(A) = |\lambda|^2 \det_{\mathsf{C}}(A).$$

For Part (c) note that $B = QA$, where now Q has 1s on the diagonal and λ in exactly one off-diagonal position. By Theorem 5.9.2(7) $\det_{\mathsf{C}}(Q) = 1$, and (5.9.3) follows. Part (d) is proved analogously. $\qquad\square$

5.10   DETERMINANTS BASED ON REAL MATRIX REPRESENTATIONS

By analogy with the determinant $\det_{\mathsf{C}}(\cdot)$ of the preceding section, we introduce in this section the notion of determinant based on the real matrix representations. Namely, for $A \in \mathsf{H}^{n \times n}$ we define

$$\det_{\mathsf{R}}(A) := \det(\chi_{n,n}(A)),$$

where $\det(\chi_{n,n}(A))$ is the standard determinant of the $4n \times 4n$ real matrix $\chi_{n,n}(A)$.

For example, a straightforward calculation shows that $\det_{\mathsf{R}}(\alpha) = |\alpha|^4$ for every $\alpha \in \mathsf{H}$.

Before we proceed with the study of $\det_{\mathsf{R}}(\cdot)$, we point out some properties of invertible quaternion matrices, which are of interest in their own right.

Proposition 5.10.1. (a) *A matrix $A \in \mathsf{H}^{n \times n}$ is invertible if and only if it is an exponential, i.e., $A = e^B$ for some $B \in \mathsf{H}^{n \times n}$.*

(b) *The set of invertible matrices of size $n \times n$ is (pathwise) connected, open, and dense in $\mathsf{H}^{n \times n}$.*

Proof. If $A = e^B$, then A is invertible with the inverse e^{-B}. Conversely, assume A is invertible, and let J be its Jordan form. Since $e^{S^{-1}XS} = S^{-1}e^X S$ for any $X \in \mathsf{H}^{n \times n}$ and any invertible $S \in \mathsf{H}^{n \times n}$, it suffices to prove that J is an exponential, which, in turn, reduces to the proof that every invertible Jordan block $J_k(\alpha)$ with $\alpha \in \mathsf{C}$ is an exponential. This is easily verified using formula (5.5.5): $J_k(\alpha) = e^B$, where B is the upper triangular Toeplitz $k \times k$ matrix having the first row

$$\left[\log\alpha, \quad \frac{1}{\alpha}, \quad \frac{-1}{2\alpha^2}, \quad \cdots, \quad \frac{(-1)^k}{(k-1)\alpha^{k-1}}\right].$$

For Part (b), to prove that invertible matrices are dense, in view of the Jordan form (5.5.1), it is enough to show that any Jordan block can be approximated (in the norm $\| \cdot \|$) as close as we wish by invertible matrices; indeed, $J_k(0)$ can be approximated by $J_k(\varepsilon)$, where ε is close to zero. Connectivity follows from Part (a): if $A = e^B$, then A is connected to I within the set of invertible matrices by the path $e^{(1-t)B}$, $0 \leq t \leq 1$. Finally, if $A \in \mathsf{H}^{n \times n}$ is invertible and $C \in \mathsf{H}^{n \times n}$ is such that $\|C - A\| < \|A^{-1}\|^{-1}$, then C is invertible as well:

$$
\begin{aligned}
C^{-1} &= [(C - A) + A]^{-1} = A^{-1}[A^{-1}(C - A) + I]^{-1} \\
&= [I - (A^{-1}(C - A)) + (A^{-1}(C - A))^2 - \ldots]A^{-1},
\end{aligned}
$$

where the series converges because $\|A^{-1}(C - A)\| \leq \|A^{-1}\| \, \|C - A\| < 1$. $\qquad \square$

We now return to $\det_{\mathsf{R}}(\cdot)$. It turns out that $\det_{\mathsf{R}}(\cdot)$ takes only nonnegative values; this and other properties of the determinant are summarized in the next proposition.

Proposition 5.10.2. (1) $\det_{\mathsf{R}}(\cdot)$ *is multiplicative*:

$$
\det_{\mathsf{R}}(AB) = \det_{\mathsf{R}}(A) \det_{\mathsf{R}}(B), \qquad \text{for all} \quad A, B \in \mathsf{H}^{n \times n};
$$

(2) $\det_{\mathsf{R}}(A) \neq 0$ *if and only if A is invertible*;
(3) *for every $A \in \mathsf{H}^{n \times n}$, we have $\det_{\mathsf{R}}(A) \geq 0$.*

Proof. (1) follows from the multiplicativity of the map $\chi_{n,n}$ and the multiplicative property of the determinant of $\chi(\cdot)$. If $\det_{\mathsf{R}}(A) \neq 0$, then $\chi_{n,n}(A)$ is invertible in $\mathsf{R}^{4n \times 4n}$, and it is easy to see that its inverse also belongs to the subalgebra $\{\chi_{n,n}(X) : X \in \mathsf{H}^{n \times n}\}$. Thus, A is invertible in $\mathsf{H}^{n \times n}$ (use the multiplicative property of $\chi_{n,n}(\cdot)$). If $\det_{\mathsf{R}}(A) = 0$, then $\chi_{n,n}(A)$ is not invertible, and so A is not invertible as well. For the proof of (3), note that the set

$$
\{\det_{\mathsf{R}}(A) : A \in \mathsf{H}^{n \times n} \quad \text{is} \quad \text{invertible}\} \tag{5.10.1}
$$

is contained in $\mathsf{R} \setminus \{0\}$, and, being connected (by Proposition 5.10.1(b) and continuity of the determinant function), we must have that the set (5.10.1) is contained in the set of positive real numbers. Now the denseness property of Proposition 5.10.1(b) implies that the range of $\det_{\mathsf{R}}(\cdot)$ is contained in the nonnegative real numbers. $\qquad \square$

At this point we make use of the following result concerning the axiomatic description of quaternion determinants (proved by Cohen and De Leo [28] based on results of Dieudonné [32]):

Theorem 5.10.3. *There is a unique functional $\mathcal{F}$ that maps $\mathsf{H}^{n \times n}$ into nonnegative real numbers, which is multiplicative and satisfies the scaling condition $\mathcal{F}(qI_n) = |q|^n$ for every $q \in \mathsf{H}$.*

Combining Theorem 5.10.3 with Theorem 5.9.2 and Proposition 5.10.2, we obtain that $\sqrt{\det_{\mathsf{C}}(A)} = \sqrt[4]{\det_{\mathsf{R}}(A)}$ for every $A \in \mathsf{H}^{n \times n}$; in other words, $(\det_{\mathsf{C}}(\cdot))^2 = \det_{\mathsf{R}}(\cdot)$.

5.11   LINEAR MATRIX EQUATIONS

As an application of the Jordan form, in this section we study certain linear matrix equations with matrix quaternion coefficients and unknowns.

Theorem 5.11.1. *Let $A \in \mathsf{H}^{m \times m}$, $B \in \mathsf{H}^{n \times n}$, $C \in \mathsf{H}^{m \times n}$. Then the equation*

$$AX - XB = C, \qquad X \in \mathsf{H}^{m \times n} \tag{5.11.1}$$

has unique solution if and only if

$$\sigma(A) \cap \sigma(B) = \emptyset. \tag{5.11.2}$$

Otherwise, (5.11.1) has either infinitely many solutions or no solutions.

Proof. Equation (5.11.1) is a system of real linear equations in $4mn$ real unknowns, the real constituents of the entries of X. Thus, by the standard theory of systems of linear equations over a field, we need to show that the homogeneous equation

$$AX - XB = 0 \tag{5.11.3}$$

has only the trivial solution $X = 0$ if and only if (5.11.2) holds.

Suppose (5.11.2) holds. We prove that (5.11.3) has only the trivial solution. Applying a transformation

$$A \to T^{-1}AT, \quad B \to S^{-1}BS, \quad X \to T^{-1}XS, \tag{5.11.4}$$

we can assume that both A and B are in the Jordan forms:

$$A = \oplus_{j=1}^{k} J_{m_j}(\alpha_j); \qquad B = \oplus_{j=1}^{\ell} J_{p_j}(\beta_j), \tag{5.11.5}$$

where $\alpha_1, \ldots \alpha_k, \beta_1, \ldots, \beta_\ell \in \mathsf{H}$. Partition X conformally with (5.11.5): $X = [X_{i,j}]_{i,j=1}^{k,\ell}$, where $X_{i,j} \in \mathsf{H}^{m_i \times p_j}$. Then equation (5.11.3) boils down to the system

$$J_{m_i}(\alpha_i)X_{i,j} - X_{i,j}J_{p_j}(\beta_j) = 0, \quad i = 1, 2, \ldots, k; \quad j = 1, 2, \ldots, \ell.$$

Thus the proof that (5.11.3) has only the trivial solution reduces to Proposition 5.4.1.

Conversely, if $\alpha \in \sigma(A) \cap \sigma(B)$, then by Corollary 5.7.2 there exist nonzero $x, y \in \mathsf{H}^{n \times 1}$ such that

$$Ax = x\alpha, \qquad B^*y = y\alpha^*.$$

Then $y^*B = \alpha y^*$, and $X := xy^* \neq 0$ satisfies the equation (5.11.3). □

Equation (5.11.1) is, generally speaking, not linear in the sense of quaternion vector spaces, because the left-hand side is not quaternion linear as a function of X, regardless of whether $\mathsf{H}^{m \times n}$ is given the structure of left or right quaternion vector space. It is, however, linear over R, when $\mathsf{H}^{m \times n}$ is given the standard structure of $4mn$-dimensional real vector space, in terms of the four real components of each entry in an $m \times n$ quaternion matrix.

Theorem 5.11.2. *Let $A \in \mathsf{H}^{m \times m}$, $B \in \mathsf{H}^{n \times n}$, and assume that $\lambda_1, \ldots, \lambda_k$ are all distinct common eigenvalues of A and B, where the λ_j's are taken to be complex numbers with nonnegative imaginary parts, with $\lambda_1, \ldots, \lambda_\ell$ real and $\lambda_{\ell+1}, \ldots, \lambda_k$ nonreal (the cases when $\ell = 0$ or $\ell = k$ are not excluded). Let*

$$m_{1,j}, m_{2,j}, \ldots, m_{p_j,j} \quad \text{and} \quad n_{1,j}, n_{2,j}, \ldots, n_{q_j,j}$$

be the partial multiplicities of λ_j as an eigenvalue of A and B, respectively. Then the real dimension of the solution set of $AX - XB = 0$ as a real vector space is equal to

$$4\sum_{j=1}^{\ell}\sum_{i=1}^{p_j}\sum_{r=1}^{q_j}\min\{m_{i,j},n_{r,j}\} + 2\sum_{j=\ell+1}^{k}\sum_{i=1}^{p_j}\sum_{r=1}^{q_j}\min\{m_{i,j},n_{r,j}\}.$$

Proof. Applying a transformation of the type (5.11.4), we may suppose that A and B are in their Jordan forms:

$$A = J_{m_1}(\lambda_1) \oplus \cdots \oplus J_{m_r}(\lambda_r), \quad B = J_{n_1}(\mu_1) \oplus \cdots \oplus J_{n_s}(\mu_s).$$

Partition X conformally: $X = [X_{i,j}]_{i=1,j=1}^{r,s}$, where $X_{i,j} \in \mathsf{H}^{m_i \times n_j}$. Then the equation $AX = XB$ decomposes into rs independent equations

$$J_{m_i}(\lambda_i)X_{i,j} = X_{i,j}J_{n_j}(\mu_j), \quad \text{for } i = 1, 2, \ldots, r, \quad \text{and } j = 1, 2, \ldots. \quad (5.11.6)$$

Now use Propositions 5.4.1 and 5.4.2 for each equation, as appropriate, in (5.11.6). $\qquad \square$

The equation $AXB - X = C$ can be analyzed along similar lines.

Theorem 5.11.3. *Let $A \in \mathsf{H}^{m \times m}$, $B \in \mathsf{H}^{n \times n}$, $C \in \mathsf{H}^{m \times n}$. Then the equation*

$$AXB - X = C, \qquad X \in \mathsf{H}^{m \times n} \quad (5.11.7)$$

has unique solution if and only if

$$\lambda\mu \neq 1 \quad \text{for all } \lambda \in \sigma(A), \ \mu \in \sigma(B). \quad (5.11.8)$$

Otherwise, (5.11.7) has either infinitely many solutions or no solutions.

Proof. Analogous to the proof of Theorem 5.11.1, we need to verify the following statement: *The homogeneous equation*

$$AXB = X, \qquad X \in \mathsf{H}^{m \times n} \quad (5.11.9)$$

has only the trivial solution $X = 0$ if and only if (5.11.8) holds.

Suppose (5.11.8) holds. We will show that (5.11.9) has only the trivial solution. As in the proof of Theorem 5.11.1, the proof reduces to the case when $A = J_m(\alpha)$, $B = J_p(\beta)$:

$$J_m(\alpha)\, X\, J_p(\beta) = X, \quad X \in \mathsf{H}^{m \times p}, \quad (5.11.10)$$

with $\alpha, \beta \in \mathsf{H}$. If $\beta \neq 0$, then (5.11.10) is equivalent to $J_m(\alpha)\, X = X\, (J_p(\beta))^{-1}$, and we are done by Theorem 5.11.1; note that the eigenvalues of $(J_p(\beta))^{-1}$ are precisely the inverses of the eigenvalues of $J_p(\beta)$; cf. Ex. 5.16.8. Analogously, we are done if $\alpha \neq 0$. In the remaining case $\alpha = \beta = 0$, iterate equation (5.11.10):

$$\begin{aligned} X &= J_m(0)\, X\, J_p(0) = (J_m(0))^2\, X\, (J_p(0))^2 \\ &= \cdots = (J_m(0))^q\, X\, (J_p(0))^q, \end{aligned} \quad (5.11.11)$$

which is equal to zero for $q \geq \min\{m, p\}$.

Conversely, suppose $\lambda_0\,\mu_0 = 1$ for some eigenvalues λ_0 of A and μ_0 of B. We have to show that (5.11.9) has nontrivial solutions. Applying a transformation

(5.11.4), we may assume that both A and B are in the Jordan forms (5.11.5), where without loss of generality, we further assume that $\alpha_1 = \lambda_0$, $\beta_1 = \mu_0$. Partitioning $X = [X_{i,j}]_{i,j=1}^{k,\ell}$ conformally with (5.11.5), we see that (5.11.9) amounts to a system of homogeneous equations

$$J_{m_i}(\alpha_i)\, X_{i,j}\, J_{p_j}(\beta_j) = X_{i,j}, \qquad i = 1, 2, \ldots, k; \quad j = 1, 2, \ldots, \ell.$$

It remains to observe that the equation

$$J_{m_1}(\alpha_1)\, X_{1,1}\, J_{p_1}(\beta_1) = X_{1,1},$$

or, equivalently,

$$J_{m_1}(\alpha_1)\, X_{1,1} = X_{1,1}\, (J_{p_1}(\beta_1))^{-1},$$

has nontrivial solutions $X_{1,1}$ by Theorem 5.11.1. $\square$

The equation $AXB - X = C$ is also linear in the sense of vector spaces over R, but not linear in the sense of quaternions. The analogue of Theorem 5.11.2 for the equation $AXB = X$ is as follows.

Theorem 5.11.4. *Let $A \in \mathsf{H}^{m \times m}$, $B \in \mathsf{H}^{n \times n}$, and assume that $\lambda_1, \ldots, \lambda_k$ are all distinct nonzero eigenvalues of A, where the λ_j's are taken to be complex numbers with nonnegative imaginary parts, such that $\lambda_1^{-1}, \ldots, \lambda_k^{-1}$ are eigenvalues of B. Assume furthermore that $\lambda_1, \ldots, \lambda_\ell$ are real and $\lambda_{\ell+1}, \ldots, \lambda_k$ are nonreal (the cases when $\ell = 0$ or $\ell = k$ are not excluded). Let*

$$m_{1,j}, m_{2,j}, \ldots, m_{p_j,j} \quad \text{and} \quad n_{1,j}, n_{2,j}, \ldots, n_{q_j,j}$$

be the partial multiplicities of λ_j and λ_j^{-1} as eigenvalues of A and B, respectively, for $j = 1, 2, \ldots, k$. Then the real dimension of the solution set of $AX - XB = 0$ as a real vector space is equal to

$$4 \sum_{j=1}^{\ell} \sum_{i=1}^{p_j} \sum_{r=1}^{q_j} \min\{m_{i,j}, n_{r,j}\} + 2 \sum_{j=\ell+1}^{k} \sum_{i=1}^{p_j} \sum_{r=1}^{q_j} \min\{m_{i,j}, n_{r,j}\}.$$

The proof of Theorem 5.11.4 is similar to that of Theorem 5.11.2 and is omitted; see Ex. 5.16.14.

5.12   COMPANION MATRICES AND POLYNOMIAL EQUATIONS

Definition 5.12.1. An $n \times n$ matrix of the form

$$C = \begin{bmatrix} 0 & 1 & 0 & 0 & \cdots & 0 \\ 0 & 0 & 1 & 0 & \cdots & 0 \\ \vdots & \vdots & \vdots & \ddots & \cdots & \vdots \\ 0 & 0 & 0 & 0 & \cdots & 1 \\ -\alpha_0 & -\alpha_1 & -\alpha_2 & -\alpha_3 & \cdots & -\alpha_{n-1} \end{bmatrix}, \qquad \alpha_0, \ldots, \alpha_{n-1} \in \mathsf{H},$$

is called a *companion matrix*.

Its eigenvectors have a specific form, as follows.

Proposition 5.12.2. *If* $x = \begin{bmatrix} x_1 \\ x_2 \\ \vdots \\ x_n \end{bmatrix} \in \mathsf{H}^{n \times 1}$ *is an eigenvector of* C, *then*

$x_1 \neq 0$, *and there exists* $z_0 \in \mathsf{H}$ *such that*

$$x_j = x_1 z_0^{j-1} \quad \text{for} \quad j = 2, 3, \ldots, n. \tag{5.12.1}$$

In fact, z_0 *is the corresponding eigenvalue.*

Proof. The eigenvalue-eigenvector equation reads

$$\begin{bmatrix} 0 & 1 & 0 & 0 & \cdots & 0 \\ 0 & 0 & 1 & 0 & \cdots & 0 \\ \vdots & \vdots & \vdots & \ddots & \cdots & \vdots \\ 0 & 0 & 0 & 0 & \cdots & 1 \\ -\alpha_0 & -\alpha_1 & -\alpha_2 & -\alpha_3 & \cdots & -\alpha_{n-1} \end{bmatrix} \begin{bmatrix} x_1 \\ x_2 \\ \vdots \\ x_n \end{bmatrix} = \begin{bmatrix} x_1 \\ x_2 \\ \vdots \\ x_n \end{bmatrix} z_0, \tag{5.12.2}$$

and we have $x_j = x_{j-1} z_0$, for $j = 2, 3, \ldots, n$. Thus, equalities (5.12.1) follow. If x_1 were zero, then (5.12.1) would imply $x = 0$, an impossibility. $\qquad\square$

Equating the bottom components in the left- and right-hand sides of (5.12.2) gives

$$-\alpha_0 x_1 - \alpha_1 x_1 z_0 - \cdots - \alpha_{n-1} x_1 z_0^{n-1} = x_1 z_0^n;$$

hence, scaling x so that $x_1 = 1$, we obtain that z_0 is a solution of the polynomial equation

$$z^n + \sum_{j=0}^{n-1} \alpha_j z^j. \tag{5.12.3}$$

Theorem 5.12.3. *If* $z_0 \in \mathsf{H}$ *is a solution of equation* (5.12.3), *then* $z_0 \in \sigma(C)$. *Conversely, if* $z_0 \in \sigma(C)$, *then there exists* $\alpha \in \mathsf{H} \setminus \{0\}$ *such that* $\alpha^{-1} z_0 \alpha$ *is a solution of* (5.12.3).

Proof. The converse part we have seen above (α appears because of the need to scale x which results in replacing the eigenvalue by a similar one). The direct part is proved by reversing the argument. $\qquad\square$

Using Theorem 5.12.3 and the Jordan form of C, the following information about solutions of polynomial equations is obtained.

Theorem 5.12.4. *Let* $\alpha_0, \ldots, \alpha_n \in \mathsf{H}$ *with* $\alpha_n \neq 0$. *Then the equation*

$$\alpha_n z^n + \alpha_{n-1} z^{n-1} + \cdots + \alpha_1 z + \alpha_0 = 0, \quad z \in \mathsf{H}, \tag{5.12.4}$$

has at least one solution z, *and the number of quaternion similarity equivalence classes that contain a solution of* (5.12.4) *does not exceed* n.

Recall that similarity equivalence classes are sets of quaternions of the form $\{\beta^{-1} \lambda \beta : \beta \in \mathsf{H} \setminus \{0\}\}$, $\lambda \in \mathsf{H}$.

Proof. The proof quickly reduces to the case $\alpha_n = 1$ by replacing z with $\alpha^{-1} z \alpha$. The statement concerning existence of solution follows from existence of eigenvalues of the companion matrix. The statement concerning similarity equivalence classes containing solutions follows from the fact that the companion matrix cannot have eigenvalues in more than n similarity equivalence classes, as is evident from the uniqueness of its Jordan form. $\qquad\square$

As it turns out, for any given similarity equivalence class, one of the three possibilities occurs:

(a) the entire class consists of zeros (perhaps not all zeros) of (5.12.4);

(b) the class contains exactly one zero of (5.12.4);

(c) the class does not contain any zeros of (5.12.4).

We will not prove this statement here and refer the reader to the original paper by Janovska and Opfer [72]. By Theorem 5.12.4, the total number of classes for which (a) or (b) holds cannot exceed n. A more refined result is proved by Pogorui and Shapiro [119]: *the total quantity of the classes of type* (a) *and of the double number of the classes of type* (a) *does not exceed* n, *the degree of the polynomial.*

The following example is given in Janovska and Opfer [72].

Example 5.12.5. Let

$$p(z) := z^6 + \mathsf{j}z^5 + \mathsf{i}z^4 - z^2 - \mathsf{j}z - i.$$

Then the polynomial $p(z)$ has the zeros

$$1, \ -1, \ \frac{1}{2}(1 - \mathsf{i} - \mathsf{j} - \mathsf{k}), \ \frac{1}{2}(-1 + \mathsf{i} - \mathsf{j} - \mathsf{k}),$$

$$\{q \in \mathsf{H} : \mathfrak{R}(q) = 0, \ |\mathfrak{V}(q)| = 1\}.$$

Thus, (a) holds true for the similarity equivalence classes $\{1\}$, $\{-1\}$, $\{q \in \mathsf{H} : \mathfrak{R}(q) = 0, \ |\mathfrak{V}(q)| = 1\}$, (b) holds true for the two similarity equivalence classes

$$\left\{q \in \mathsf{H} : \mathfrak{R}(q) = \frac{1}{2}, \ |\mathfrak{V}(q)| = \frac{\sqrt{3}}{2}\right\}$$

and

$$\left\{q \in \mathsf{H} : \mathfrak{R}(q) = -\frac{1}{2}, \ |\mathfrak{V}(q)| = \frac{\sqrt{3}}{2}\right\},$$

and (c) holds true for all other similarity equivalence classes. $\qquad\square$

The next proposition is instructive.

Proposition 5.12.6. *Let* $a, b \in \mathsf{H}$ *be nonzero with zero real parts. Then the equation*

$$z^2 - (a + b)z + ab = 0, \qquad z \in \mathsf{H}, \tag{5.12.5}$$

has precisely two distinct solutions if $|a| \neq |b|$, *has the unique solution* $z' = b$ *if* $|a| = |b|$ *and* $a + b \neq 0$, *and has the solutions* $\{z' \in \mathsf{H} : \mathfrak{R}(z') = 0, \ |z'| = |b|\}$ *if* $a + b = 0$.

Proof. The case when $a+b=0$ is clear: the equation (5.12.5) reads $z^2+|b|^2 = 0$. Assume now $a+b \neq 0$. First note that b is indeed a solution of (5.12.5). By scaling a and b, i.e., replacing $a \mapsto ra$, $b \mapsto rb$, where $r > 0$, we may assume $|b| = 1$. By simultaneously applying an automorphism of H to a and b (Theorem 2.4.4), we may further assume that $b = \mathsf{i}$, $a = a_1\mathsf{i}+a_2\mathsf{j}$, where a_1, a_2 are real and either $a_2 \neq 0$ or $a_2 = 0$, $a_1 \neq -1$. Let $z' = z_0 + z_1\mathsf{i} + z_2\mathsf{j} + z_3\mathsf{k}$, $z_0, z_1, z_2, z_3 \in \mathsf{R}$, be a solution of (5.12.5). Then we have the equality

$$(z')^2 = ((a_1 + 1)\mathsf{i} + a_2\mathsf{j})z' + (a_1 + a_2\mathsf{k}). \tag{5.12.6}$$

Writing out the real coefficients of $1, \mathsf{i}, \mathsf{j}, \mathsf{k}$ in both sides of (5.12.6), we see that (5.12.6) boils down to the following four equations:

$$\begin{aligned}
z_0^2 - z_1^2 - z_2^2 - z_3^2 &= -(a_1 + 1)z_1 - a_2z_2 + a_1, & (5.12.7) \\
2z_0z_1 &= (a_1 + 1)z_0 + a_2z_3, & (5.12.8) \\
2z_0z_2 &= a_2z_0 - (a_1 + 1)z_3, & (5.12.9) \\
2z_0z_3 &= -a_2z_1 + (a_1 + 1)z_2 + a_2. & (5.12.10)
\end{aligned}$$

Suppose first $z_0 \neq 0$. Then (5.12.8) and (5.12.9) give

$$z_2 = \frac{1}{2}\left(a_2 - \frac{(a_1 + 1)z_3}{z_0}\right), \quad z_1 = \frac{1}{2}\left((a_1 + 1) + \frac{a_2z_3}{z_0}\right), \tag{5.12.11}$$

and substituting in (5.12.10) yields, after some simple algebra, the equality

$$z_3 = \frac{a_2z_0}{2z_0^2 + \frac{1}{2}a_2^2 + \frac{1}{2}(a_1 + 1)^2}. \tag{5.12.12}$$

Substituting (5.12.11) and (5.12.12) in (5.12.7), we obtain, after clearing denominators, a cubic equation in the variable $y := z_0^2$ of the form

$$4y^3 + c_2y^2 + c_1y + c_0 = 0, \qquad c_0, c_1, c_2 \in \mathsf{R}.$$

This equation cannot have positive solutions y_0: if it did, then we would obtain at least three pairwise nonsimilar solutions of (5.12.5), namely, b and the two solutions obtained by setting $z_0 = \pm\sqrt{y_0}$ in (5.12.11) and (5.12.12). However, this contradicts Theorem 5.12.4. Thus, the assumption $z_0 \neq 0$ leads to a contradiction.

Now suppose $z_0 = 0$. Then we have, from (5.12.7)–(5.12.10): $z_3 = 0$ and, consequently,

$$-z_1^2 - z_2^2 = -(a_1 + 1)z_1 - a_2z_2 + a_1, \qquad 0 = -a_2z_1 + (a_1 + 1)z_2 + a_2. \tag{5.12.13}$$

If $a_2 = 0$, then we conclude that $z_2 = 0$, and the first equation in (5.12.13) gives two values for z_1: $z_1' = 1$ (thus, $z' = b$) and $z_1' = a_1$. So we obtain exactly two zeros of (5.12.5) if $|a| \neq |b|$ and exactly one zero of (5.12.5) if $|a| = |b|$. Consider now the case when $a_2 \neq 0$. Then we solve the second equation in (5.12.13) for z_1:

$$z_1 = \frac{(a_1 + 1)z_2 + a_2}{a_2}.$$

Substituting in the first equation in (5.12.13) gives rise to a quadratic equation for z_2 whose solutions are $z_2' = 0$ and

$$z_2'' = \frac{a_2(-1 + a_1^2 + a_2^2)}{(a_1 + 1)^2 + a_2^2}.$$

Thus, if $|a| = 1$, then also $z_2'' = 0$, and it is easy to see that we obtain $z' = b$, the unique solution of (5.12.5). If $|a| \neq 1$, then we have two distinct values for z_2' and z_2'' and, consequently, exactly two solutions of (5.12.5). $\qquad\square$

Let $\mathcal{K}(\mathsf{H})$ be the set of all nonempty compact subsets of H, considered as a metric space with the Hausdorff distance as a metric. The Hausdorff distance between two compact subsets $\mathcal{L}_1, \mathcal{L}_2 \subset \mathsf{H}$ is defined by

$$d(\mathcal{L}_1, \mathcal{L}_2) = \max \left\{ \max_{x \in \mathcal{L}_1} \left(\min_{y \in \mathcal{L}_2} |x - y| \right), \quad \max_{x \in \mathcal{L}_2} \left(\min_{y \in \mathcal{L}_1} |y - x| \right) \right\}.$$

See Ex. 5.16.29. Define the map $\mathcal{Z} : \mathsf{H}^n \to \mathcal{K}(\mathsf{H})$ by

$$\mathcal{Z}(\alpha_0, \cdots, \alpha_{n-1}) = \{\text{the zero set of } z^n + \alpha_{n-1}z^{n-1} + \cdots + \alpha_0\}.$$

Proposition 5.12.6 shows that the map $\mathcal{Z}$ is generally discontinuous, in contrast to the continuity of the complex roots of real or complex polynomials (Theorem 5.15.1).

Open Problem 5.12.7. *Study the (dis)continuity properties of the map $\mathcal{Z}$. For example: What are the points of continuity of $\mathcal{Z}$? Cf. Theorem 5.5.9.*

5.13 EIGENVALUES OF HERMITIAN MATRICES

We record here some properties of eigenvalues of hermitian quaternion matrices that will be used later in the text. First note that the eigenvalues of hermitian matrices are real (Theorem 5.3.6(c)). Indeed, if $A = A^* \in \mathsf{H}^{n \times n}$ and $Ax = x\lambda$ for some nonzero $x \in \mathsf{H}^{n \times 1}$ and $\lambda \in \mathsf{H}$, then $x^*Ax = x^*x\lambda$. But x^*Ax is real and x^*x is positive, so λ must be real as well.

Theorem 5.13.1. *Let $A, B \in \mathsf{H}^{n \times n}$ be hermitian matrices such that $A - B$ is positive semidefinite. If*

$$\lambda_1 \geq \lambda_2 \geq \cdots \geq \lambda_n, \qquad \mu_1 \geq \mu_2 \geq \cdots \geq \mu_n$$

are the eigenvalues of A and B, respectively, arranged in nonincreasing order, then

$$\lambda_j \geq \mu_j, \qquad \text{for } j = 1, 2, \ldots, n.$$

Theorem 5.13.2. *If $A \in \mathsf{H}^{n \times}$ is hermitian and $S \in \mathsf{H}^{n \times n}$, then the number of positive, resp. negative, eigenvalues of SAS^*, counted with algebraic multiplicities, does not exceed the number of positive, resp. negative, eigenvalues of A, also counted with algebraic multiplicities.*

The proofs of Theorems 5.13.1 and 5.13.2 are essentially the same as for the complex hermitian matrices (see Sections 4.3, 7.7, and 4.5 in Horn and Johnson [62], for example), and will not be repeated here.

5.14 DIFFERENTIAL AND DIFFERENCE EQUATIONS

Consider a system of linear differential equations with constant coefficients

$$A_\ell x^{(\ell)}(t) + A_{\ell-1}x^{(\ell-1)}(t) + \cdots + A_1 x'(t) + A_0 x(t) = 0, \qquad t \in \mathsf{R}, \qquad (5.14.1)$$

where $A_\ell, \ldots, A_1, A_0 \in \mathsf{H}^{n \times n}$, and $x(t)$ is an unknown ℓ times continuously differentiable $\mathsf{H}^{n \times 1}$-valued function of the real independent variable t.

Definition 5.14.1. We say that the system (5.14.1) is *forward stable* if all solutions $x(t)$ tend to zero as $t \rightarrow +\infty$ and *forward bounded* if every solution is a bounded function as $t \rightarrow +\infty$.

Criteria will be given for stability and for boundedness in terms of location of the spectrum of the companion matrix, in compete analogy with the well-known case of equations (5.14.1) with complex coefficients.

Assuming A_ℓ is invertible, let

$$
C = \begin{bmatrix}
0 & I_n & 0 & \cdots & 0 \\
0 & 0 & I_n & \cdots & 0 \\
\vdots & \vdots & \vdots & \ddots & \vdots \\
0 & 0 & 0 & \cdots & I_n \\
-A_\ell^{-1}A_0 & -A_\ell^{-1}A_1 & -A_\ell^{-1}A_2 & \cdots & -A_\ell^{-1}A_{\ell-1}
\end{bmatrix} \in \mathsf{H}^{\ell n \times \ell n}
$$

be the *companion matrix* of system (5.14.1).

Theorem 5.14.2.   (a) *The system* (5.14.1) *is forward stable if and only if all eigenvalues of C have negative real parts.*

(b) *The system* (5.14.1) *is forward bounded if and only if all eigenvalues of C have nonpositive real parts, and for those eigenvalues with zero real parts, the geometric multiplicity coincides with the algebraic multiplicity.*

The proof is based on the following lemma.

Lemma 5.14.3. *Let $X \in \mathsf{H}^{m \times m}$. Then:*

(a) $\lim_{t \rightarrow +\infty} e^{tX} = 0$ *if and only if all eigenvalues of X have negative real parts;*

(b) $\|e^{tX}\|$ *is bounded as $t \rightarrow +\infty$ if and only if all eigenvalues of X have nonpositive real parts, and for those eigenvalues with zero real parts the geometric multiplicity coincides with the algebraic multiplicity.*

Proof. Part (a). If all eigenvalues of X have negative real parts, then

$$
\lim_{t \rightarrow \infty} \|e^{tX}\| = 0
$$

(as one can readily see by reducing X to its Jordan form), and forward stability of (5.14.1) follows. Conversely, let λ_0 be an eigenvalue of X with nonnegative real part and corresponding eigenvector v_0. We may (and do) choose λ_0 to be a complex number with nonnegative imaginary part. If $S^{-1}XS = J$ is a Jordan form with complex eigenvalues having nonnegative imaginary parts, then $S^{-1}v_0$ is an eigenvector of J corresponding to the eigenvalue λ_0. Now

$$
e^{tX}v_0 = S(e^{tJ}(S^{-1}v_0)) = S(S^{-1}v_0 e^{t\lambda_0}) = v_0 e^{t\lambda_0} \not\rightarrow \infty, \qquad (5.14.2)
$$

as $t \rightarrow \infty$.

Part (b). The "if" statement follows from formula (5.5.5), as the condition stated in (b) means that the Jordan blocks in the Jordan form of X that correspond to eigenvalues λ with zero real parts (if any) are in fact 1×1, and so, on these Jordan blocks the function e^{tX} is simply $e^{t\lambda}$, which is bounded as $t \rightarrow \infty$.

Part (b). The "only if" statement: Formula (5.5.5) (applied to $f(s) = e^{ts}$, where for a moment t is considered fixed, and $s \in \mathsf{R}$ is the independent variable) shows that

$$e^{t(J_m(\lambda))} = \begin{bmatrix} e^{t\lambda} & te^{t\lambda} & \dfrac{t^2 e^{t\lambda}}{2!} & \cdots & \dfrac{t^{m-1}e^{t\lambda}}{(m-1)!} \\[2mm] 0 & e^{t\lambda} & te^{t\lambda} & \cdots & \dfrac{t^{m-2}e^{t\lambda}}{(m-2)!} \\[2mm] \vdots & \vdots & \ddots & \ddots & \vdots \\[2mm] 0 & 0 & 0 & \cdots & e^{t\lambda} \end{bmatrix}.$$

So, if the real part of λ is positive, or if $m \geq 2$, we have that $e^{t(J_m(\lambda))}$ is unbounded as $t \to \infty$. $\qquad\square$

Proof of Theorem 5.14.2. Let

$$y(t) = \begin{bmatrix} x(t) \\ x'(t) \\ \vdots \\ x^{(\ell-1)}(t) \end{bmatrix} \in \mathsf{H}^{\ell n \times 1}.$$

It is easily verified that $x(t)$ is a solution of (5.14.1) if and only if $y(t)$ is a solution of

$$y'(t) = Cy(t). \tag{5.14.3}$$

The general solution of (5.14.3) is

$$y(t) = e^{tC}y(0). \tag{5.14.4}$$

If C has an eigenvalue λ_0 with nonnegative real part and corresponding eigenvector v_0, then a solution $y(t) = e^{tC}v_0$ does not tend to zero as $t \to \infty$ (cf. (5.14.2)). Notice that because of the structure of C, the vector v_0 necessarily has the form

$$v_0 = \begin{bmatrix} v_1 \\ v_1\lambda_0 \\ \vdots \\ v_1\lambda_0^{\ell-1} \end{bmatrix}, \text{ where } v_1 \text{ is the first component of } v_0. \text{ Therefore, it follows that}$$

the first component of $y(t)$, i.e., $x(t)$, does not tend to zero as $t \to \infty$. Thus, (5.14.1) is not forward stable. This proves (a).

The proof of (b) is analogous to that of part (a), using Lemma 5.14.3(b). $\qquad\square$

A parallel development can be given to systems of difference equations. Consider a system of linear difference equations with constant coefficients

$$A_\ell x_j + A_{\ell-1}x_{j-1} + \cdots + A_0 x_{j-\ell} = 0, \quad j = \ell, \ell+1, \ldots. \tag{5.14.5}$$

Here $A_0, \ldots, A_\ell \in \mathsf{H}^{n \times n}$ are given matrices, and $\{x_j\}_{j=0}^\infty$ is a sequence of vectors in $\mathsf{H}^{n \times 1}$ to be found. Assuming A_ℓ is invertible, we define the companion matrix $C \in \mathsf{H}^{\ell n \times \ell n}$ as before. Letting

$$z_j = \begin{bmatrix} x_j \\ x_{j-1} \\ \vdots \\ x_{j-\ell} \end{bmatrix} \in \mathsf{H}^{\ell n \times 1}, \quad j = \ell, \ell+1, \ldots,$$

we see that (5.14.5) is equivalent to the system

$$z_{j+1} = Cz_j, \quad j = \ell, \ell+1, \ldots. \tag{5.14.6}$$

The solutions of (5.14.6) are

$$z_j = C^{j-\ell} z_\ell, \quad j = \ell+1, \ell+2, \ldots.$$

Theorem 5.14.4. (a) *All solutions of system* (5.14.5) *tend to zero as* $j \to \infty$ *if and only if all eigenvalues of* C *have norm less than one.*

(b) *All solutions of system* (5.14.5) *are bounded if and only if all eigenvalues of* C *have norm less than or equal to one, and for those eigenvalues with norm one, the geometric multiplicity coincides with the algebraic multiplicity.*

The following lemma is used in the proof.

Lemma 5.14.5. *Let* $X \in \mathsf{H}^{m \times m}$. *Then:*

(a) $\lim_{p \to \infty} X^p = 0$ *if and only if all eigenvalues of* X *have norm less than one;*

(b) *the sequence* $\|X^p\|$, $p = 1, 2, \ldots$, *is bounded if and only if all eigenvalues of* X *have norm less than or equal to one, and for those eigenvalues with norm one, the geometric multiplicity coincides with the algebraic multiplicity.*

The proof of Lemma 5.14.5 is analogous to that of Lemma 5.14.3, using the following formula for powers of Jordan blocks (a particular case of (5.5.5)):

$$J_m(\lambda)^p = \begin{bmatrix} \lambda^p & p\lambda^{p-1} & \binom{p}{2}\lambda^{p-2} & \cdots & \binom{p}{m-1}\lambda^{p-m+1} \\ 0 & \lambda^p & p\lambda^{p-1} & \cdots & \binom{p}{m-2}\lambda^{p-m+2} \\ \vdots & \vdots & \ddots & \ddots & \vdots \\ 0 & 0 & 0 & \cdots & \lambda^p \end{bmatrix}.$$

5.15   APPENDIX: CONTINUOUS ROOTS OF POLYNOMIALS

Continuity of complex roots of polynomial equations with real or complex coefficients is expressed in the following well-known result.

Theorem 5.15.1. *Let* $p(z) := \sum_{j=0}^{n} a_j z^j$ *be a polynomial, where* $a_0, \ldots, a_n \in \mathsf{F}$, $\mathsf{F} = \mathsf{R}$ *or* $\mathsf{F} = \mathsf{C}$ *and* $a_n \neq 0$. *Let* $\lambda_1, \ldots, \lambda_k$ *be the distinct roots in the complex plane of the polynomial equation*

$$p(z) = 0. \tag{5.15.1}$$

Then for every sufficiently small $\epsilon > 0$, *there is* $\delta > 0$ *such that if* $|a'_j - a_j| < \delta$, $a_j \in \mathsf{F}$, *for* $j = 0, 1, \ldots, n$, *then all the roots of the polynomial equation*

$$\sum_{j=0}^{n} a'_j z^j = 0 \tag{5.15.2}$$

are contained in the union of open disks

$$\cup_{j=1}^{k} \{\lambda \in \mathsf{C} : |\lambda - \lambda_j| < \epsilon\}.$$

Moreover, the number of roots of (5.15.2) *in the disk* $\{\lambda \in \mathsf{C} : |\lambda - \lambda_i| < \epsilon\}$ *counted with their multiplicities is equal to the multipicity of* λ_i *as a root of* (5.15.1).

Proof. Clearly the real case $\mathsf{F} = \mathsf{R}$ is just a particular situation of the more general complex case, so we prove Theorem 5.15.1 for $\mathsf{F} = \mathsf{C}$ only.

Let $\epsilon > 0$ be so small that the disks

$$E_{\epsilon/2}(\lambda_1) := \left\{ \lambda \in \mathsf{C} : |\lambda - \lambda_1| \leq \frac{\epsilon}{2} \right\}, \ldots, E_{\epsilon/2}(\lambda_k) := \left\{ \lambda \in \mathsf{C} : |\lambda - \lambda_k| \leq \frac{\epsilon}{2} \right\}$$

do not pairwise intersect. Then, in particular, the polynomial $p(z) := \sum_{j=0}^{n} a_j z^j$ has no roots on the circles

$$\mathcal{E}_{\epsilon/2} := \cup_{i=1}^{k} \left\{ \lambda \in \mathsf{C} : |\lambda - \lambda_1| = \frac{\epsilon}{2} \right\}.$$

Using continuity of $p(z)$ (as a function of z and of the coefficients a_j) and compactness of $\mathcal{E}_{\epsilon/2}$, we see that there is δ (which depends on $p(z)$ and ϵ) such that for every $z_0 \in \mathcal{E}_{\epsilon/2}$ the following implication holds:

$$|a'_j - a_j| < \delta \implies |\sum_{j=0}^{n}(a'_j - a_j)z_0^j| < \min\{|p(z)| : z \in \mathcal{E}_{\epsilon/2}\}.$$

We now use Rouché's theorem, which is given in most textbooks on complex analysis (see, e.g., Churchill and Brown [27]), to obtain that the number of roots, counted with multiplicities, of $p(z)$ and of $\sum_{j=0}^{n} a'_j z^j$ in $E_{\epsilon/2}(\lambda_i)$ is the same for every fixed i, $i = 1, 2, \ldots, k$. This proves the theorem. $\qquad\square$

5.16 EXERCISES

Ex. 5.16.1. Verify formula (5.1.6).

Ex. 5.16.2. Verify that $\sigma(A) = \{\mu \in \mathsf{H} : m^4 = -1\}$ for $A = \begin{bmatrix} 0 & \mathsf{i} \\ \mathsf{j} & 0 \end{bmatrix}$.

Ex. 5.16.3. For a fixed $A \in \mathsf{H}^{n \times n}$, show that the set of A- invariant subspaces forms a lattice: If subspaces $\mathcal{M}_1, \mathcal{M}_2 \in \mathsf{H}^{n \times 1}$ are A-invariant, then so are $\mathcal{M}_1 + \mathcal{M}_2$ and $\mathcal{M}_1 \cap \mathcal{M}_2$.

Ex. 5.16.4. Prove that the following statements are equivalent for $A \in \mathsf{H}^{n \times n}$ and a subspace $\mathcal{M} \in \mathsf{H}^{n \times 1}$:

1. $\mathcal{M}$ is A-invariant;

2. $\mathcal{M}^{\perp}$ is A^*-invariant;

3. for any nonstandard involution ϕ the ϕ-orthogonal companion $\mathcal{M}^{\perp\phi}$ is A_ϕ-invariant.

Ex. 5.16.5. Characterize all matrices $A \in \mathsf{H}^{n \times n}$ for which $e^{\mathsf{i}A} = \cos A + \mathsf{i}\sin A$. Hint: Complex matrices have this property.

Ex. 5.16.6. Show that if $\lambda \in \mathsf{H}$ is an eigenvalue or a left eigenvalue of $A \in \mathsf{H}^{n \times n}$, then $|\lambda| \leq \|A\|$.

Ex. 5.16.7. Provide details for the proofs of Theorem 5.14.4 and Lemma 5.14.5.

Ex. 5.16.8. Prove that if $A \in \mathsf{H}^{n \times n}$ is invertible, then

$$\sigma(A^{-1}) = \{\lambda^{-1} : \lambda \in \sigma(A)\}.$$

Ex. 5.16.9. Show explicitly that the Jordan blocks $J_m(\lambda)$ and $J_m(\lambda^*)$ are similar (over the quaternions); here $\lambda \in \mathsf{H}$.

Ex. 5.16.10. Show that the matrix $\begin{bmatrix} 1 & i \\ j & k \end{bmatrix}$ is not similar to its transpose $\begin{bmatrix} 1 & j \\ i & k \end{bmatrix}$. Hint: One of these two matrices is invertible, the other is not.

Ex. 5.16.11. Find explicitly the permutation matrix $\widetilde{P}_{4m_j}$ of (5.7.5).

Ex. 5.16.12. Prove that if $A \in \mathsf{H}^{n \times n}$ is invertible, then the inverse A^{-1} is given by $A^{-1} = p(A)$, where $p(t)$ is a certain polynomial with real coefficients. Hint: Consider the minimal polynomial of A.

Ex. 5.16.13. Verify directly that $\widetilde{\Omega}_{2n}$ is a real unital subalgebra of $\mathsf{C}^{2n \times 2n}$.

Ex. 5.16.14. Provide details for the proof of Theorem 5.11.4. Hint: Use the fact that $(J_m(\lambda))^{-1}$, $\lambda \in \mathsf{H} \setminus \{0\}$, is similar to $J_m(\lambda^{-1})$.

Ex. 5.16.15. (a) Show that if two matrices $A, B \in \mathsf{H}^{n \times n}$, $n \leq 3$, have the same minimal polynomial, then they are similar.
(b) Show by example that the statement in (a) fails if $n \geq 4$.

Ex. 5.16.16. (a) Show that if two matrices $A, B \in \mathsf{H}^{n \times n}$, $n \leq 6$, have the same minimal polynomial and the same geometric multiplicity for every eigenvalue, then they are similar.
(b) Show by example that the statement in (a) fails if $n \geq 7$.

Ex. 5.16.17. The system (5.14.1) is said to be *backward bounded* if all solutions are bounded as $t \longrightarrow -\infty$. Prove the following analogue of Theorem 5.14.2(b): the system (5.14.1) is backward bounded if and only if all eigenvalues of C have nonnegative real parts, and for those eigenvalues with zero real parts, the geometric multiplicity coincides with the algebraic multiplicity.

Ex. 5.16.18. Consider polynomial equation with *real* coefficients

$$z^n + \sum_{j=0}^{n-1} a_j z^j = 0, \qquad z \in \mathsf{H}. \tag{5.16.1}$$

Let $a_1, \ldots, a_k$ be all the distinct solutions of (5.16.1) in the closed upper complex half-plane, with $a_1, \ldots a_p$ real and $a_{p+1}, \ldots, a_k$ complex nonreal. Show that all solutions of (5.16.1) are given by $a_1, \ldots, a_p, \alpha_{p+1}, \ldots, \alpha_k$, where $\mathfrak{R}(\alpha_j) = \mathfrak{R}(a_j)$, $|\mathfrak{V}(\alpha_j)| = \mathfrak{I}(a_j)$, for $j = p+1, p+2, \ldots, k$.

Ex. 5.16.19. Find the Jordan forms of the following two matrices:

$$A_1 = \begin{bmatrix} i+j & 1 \\ 0 & k+1 \end{bmatrix}; \qquad A_2 = \begin{bmatrix} i & 1 & 0 \\ 0 & j & 1 \\ 0 & 0 & k \end{bmatrix}.$$

Ex. 5.16.20. Find the minimal polynomials of the matrices in Ex. 5.16.19.

Ex. 5.16.21. Find all invariant subspaces of the matrices

$$\text{(a)} \quad A = \begin{bmatrix} i & 1 & 0 \\ 0 & i & 0 \\ 0 & 0 & j \end{bmatrix}; \quad \text{(b)} \quad B = \begin{bmatrix} i & 1 & 0 \\ 0 & i & 0 \\ 0 & 1 & j \end{bmatrix}.$$

Ex. 5.16.22. Find all eigenvectors for each of the following matrices:

$$(a)\ \begin{bmatrix} j & 0 \\ 0 & j \end{bmatrix};\quad (b)\ \begin{bmatrix} j & 0 & 0 \\ 0 & j & 0 \\ 0 & 0 & j \end{bmatrix};\quad (c)\ \begin{bmatrix} i & 0 & 0 \\ 0 & j & 0 \\ 0 & 0 & k \end{bmatrix}.$$

Ex. 5.16.23. (a) Show that the degree of a minimal polynomial of an $n \times n$ quaternion matrix does no exceed $2n$. Hint: Use the Jordan form and Proposition 5.4.3.

(b) For every positive integer n, give an example of an $n \times n$ matrix whose minimal polynomial has degree $2n$.

Ex. 5.16.24. Find all solutions of the following matrix equations:

$$(a)\ \begin{bmatrix} i & 1 \\ 0 & i \end{bmatrix} X - X \begin{bmatrix} j & 0 \\ 1 & j \end{bmatrix} = 0;$$

$$(b)\ \begin{bmatrix} i & 1 \\ 0 & i \end{bmatrix} X - X \begin{bmatrix} 1 & 0 \\ 1 & 1 \end{bmatrix} = \begin{bmatrix} j & 0 \\ 0 & j \end{bmatrix};$$

$$(c)\ \begin{bmatrix} i & 1 \\ 0 & i \end{bmatrix} X \begin{bmatrix} k & 0 \\ 1 & k \end{bmatrix} - X = 0;$$

$$(d)\ \begin{bmatrix} i & 1 \\ 0 & i \end{bmatrix} X \begin{bmatrix} 2 & 0 \\ 1 & 2 \end{bmatrix} - X = \begin{bmatrix} k & 0 \\ 0 & k \end{bmatrix}.$$

Ex. 5.16.25. Show that, for a given n, the number of complex similarity classes in a fixed quaternion similarity class of $n \times n$ complex matrices does not exceed 2^n. Hint: The expression (5.8.4) achieves maximum when $s = n$, $r_j = 1$ for $j = 1, 2, \ldots, n$, and $p_{\lambda_j, k} = 1$.

Ex. 5.16.26. Prove that if system (5.14.1) is forward stable, then every nearby system is also forward stable, in other words, there exists $\varepsilon > 0$ such that every system

$$A'_\ell x^{(\ell)}(t) + A'_{\ell-1} x^{(\ell-1)}(t) + \cdots + A'_1 x'(t) + A'_0 x(t) = 0, \quad A'_\ell, \ldots, A'_1, A'_0 \in \mathsf{H}^{n \times n}$$

is forward stable, provided

$$\|A'_j - A_j\| < \varepsilon, \qquad \text{for } j = 0, 1, \ldots, \ell.$$

Ex. 5.16.27. Show by example that a property analogous to Ex. 5.16.26 generally fails for forward bounded systems (5.14.1).

Ex. 5.16.28. State and prove analogues of Ex. 5.16.26 and 5.16.27 for systems (5.14.5).

Ex. 5.16.29. Show that the quaternion solution set of every equation of the form (5.12.3) is compact.

Ex. 5.16.30. Let $A = \begin{bmatrix} 0 & x \\ y & 0 \end{bmatrix}$, $x, y \in \mathsf{H}$. Show that the eigenvalues of A are as follows:

$$\sigma(A) = \begin{cases} \{0\} & \text{if } xy = 0; \\ \{\pm\sqrt{xy}\} & \text{if } xy \text{ is real and positive;} \\ \sqrt{-xy} \cdot u & \text{if } xy < 0; \\ u^{-1}(\pm\sqrt{xy})u & \text{if } xy \notin \mathsf{R}, \end{cases}$$

where $u \in \mathsf{H}$ with $\mathfrak{R}(u) = 0$ and $|u| = 1$, but otherwise arbitrary. Hint: Use the complex matrix representation of A.

5.17 NOTES

The proof of Proposition 5.1.4 is standard for complex matrices.

Propositions 5.2.2 and 5.2.4 are taken from Rodman [137].

It turns out that sums of root subspaces (as well as the zero subspace) are the only invariant subspaces of matrices that enjoy the Lipschitz property as described in Theorem 5.2.5. This result was proved in Kaashoek et al. [75] for complex matrices and in Rodman [137] for quaternion matrices. In contrast, generally speaking, there may exist isolated invariant subspaces of a quaternion matrix other than sums of root subspaces or zero. Invariant subspaces of quaternion matrices that are well behaved (in various senses) under small perturbations of the matrices have been characterized in Rodman [137, 138].

The proof of Theorem 5.3.8 is taken from Zhang [164]. The result on existence of complex eigenvalues with nonnegative imaginary parts was proved in Brenner [18] and Lee [98].

Part of Proposition 5.3.7 (namely, that $D(A) \subseteq \sigma(A)$) is proved in Zhang [164].

The Jordan form for quaternion matrices is well known and can be found in many sources; see, e.g., Jacobson [71], Zhang [164], Farenick and Pidkowich [38], or Karow [78].

The proof in Section 5.6 is adapted from Wiegmann [160].

Theorems 5.5.8 and 5.7.1 and Theorem 5.11.1 and its proof are taken from Rodman [132].

The proof of Theorem 5.15.1 is standard in the literature; see, e.g., Theorem IV.1.1 in Stewart and Sun [149].

The part of Theorem 5.12.4 about the number of quaternion similarity equivalence classes that contain a solution of (5.12.4) is proved in Gordon and Motzkin [57].

Example 5.3.3 is taken from Zhang [164].

The analysis of Proposition 5.4.2, Part (a), is standard for complex matrices or, more generally, matrices over a field; see, e.g., Lancaster and Tismenetsky [94, Section 12.4] or Gantmacher [50, Chapter VIII].

Various notions of determinants for quaternion matrices have been studied in the literature, besides the determinant introduced in Section 5.9: Study determinant [150], Dieudonné determinant [32], also [5], double determinant (see Chen [24, 25] and Zhang [164]). See also Cohen and De Leo [28] and references there for an axiomatic approach to quaternion determinants.

Exposition in Section 5.14 follows a standard approach to stability and boundedness of linear differential and difference systems of equations with constant coefficients.

Exercise 5.16.30 is taken from [146].

Proposition 5.12.6 was suggested by Bolotnikov [15].

Chapter Six

Invariant neutral and semidefinite subspaces

In this chapter we study subspaces that are simultaneously neutral or semidefinite for one matrix and invariant for another. The presentation here does not use canonical forms for pairs of matrices, which is developed in later chapters.

In this and subsequent chapters we will often work with matrices that enjoy certain symmetry properties. For the reader's convenience, we collect the nomenclature associated with classes of such matrices $A \in \mathsf{H}^{n \times n}$, where $\mathsf{F} = \mathsf{R}$, $\mathsf{F} = \mathsf{C}$, or $\mathsf{F} = \mathsf{H}$, in three tables.

The notation $B \geq C$ or $C \leq B$, where $B, C \in \mathsf{F}^{n \times n}$ are hermitian matrices, indicates that the difference $B - C$ is positive semidefinite. For ϕ-skewhermitian matrices $D \in \mathsf{H}^{n \times n}$, where ϕ is a nonstandard involution, we use the notation $\beta(\phi)^{-1} D \geq 0$ to indicate that $\beta(\phi)^{-1} x_\phi D x \geq 0$ for all $x \in \mathsf{H}^{n \times 1}$.

	$H_\phi = H$ ϕ-hermitian	$H_\phi = -H$ ϕ-skewhermitian
	$\mathsf{F} = \mathsf{H}$	$\mathsf{F} = \mathsf{H}$
$A_\phi H = HA$	(H, ϕ)-symmetric	(H, ϕ)-skew-Hamiltonian
$A_\phi H = -HA$	(H, ϕ)-skewsymmetric	(H, ϕ)-Hamiltonian
$A_\phi HA = H$	(H, ϕ)-orthogonal	(H, ϕ)-symplectic
$\beta^{-1} A_\phi HA \geq \beta^{-1} H$		(H, ϕ)-expansive
$\beta^{-1} x_\phi A_\phi HAx \geq 0$ if $\beta^{-1} x_\phi Hx \geq 0$		(H, ϕ)-plus-matrix

	$H^* = H$ hermitian	$H^* = -H$ skewhermitian
	$\mathsf{F} \in \{\mathsf{R}, \mathsf{C}, \mathsf{H}\}$	$\mathsf{F} = \mathsf{H}$
$A^* H = HA$	$(H,^*)$-hermitian	$(H,^*)$-skew-Hamiltonian
$A^* H = -HA$	$(H,^*)$-skewhermitian	$(H,^*)$-Hamiltonian
$A^* HA = H$	$(H,^*)$-unitary	$(H,^*)$-symplectic
$A^* HA \geq H$	$(H,^*)$-expansive	
$x^* A^* HAx \geq 0$ if $x^* Hx \geq 0$	$(H,^*)$-plus-matrix	

	$H^T = H$ symmetric	$H^T = -H$ skewsymmetric
	$\mathsf{F} \in \{\mathsf{C}, \mathsf{R}\}$	$\mathsf{F} \in \{\mathsf{C}, \mathsf{R}\}$
$A^T H = HA$	$(H,^T)$-symmetric	$(H,^T)$-skew-Hamiltonian
$A^T H = -HA$	$(H,^T)$-skewsymmetric	$(H,^T)$-Hamiltonian
$A^T HA = H$	$(H,^T)$-orthogonal	$(H,^T)$-symplectic

Several remarks are in order here:

1. In the above tables, ϕ stands for a nonstandard involution, and $\beta = \beta(\phi)$.

2. In the real case ^{T} is the same as *.

3. The matrix H is generally not assumed to be invertible.

4. In the sequel, the * and ^{T} will be sometimes omitted in the above notation if it is clear from the context; e.g., H-Hamiltonian will be used in place of $(H,^*)$-Hamiltonian.

6.1 STRUCTURED MATRICES AND INVARIANT NEUTRAL SUBSPACES

Let ϕ be an involution; thus, either ϕ is the conjugation or ϕ is a nonstandard involution.

Definition 6.1.1. Given a ϕ-skewhermitian $H \in \mathsf{H}^{n \times n}$, a matrix $A \in \mathsf{H}^{n \times n}$ is said to be (H, ϕ)-*Hamiltonian* if $(HA)_\phi = HA$, i.e., HA is ϕ-hermitian. A matrix $W \in \mathsf{H}^{2n \times 2n}$ is said to be (H, ϕ)-*skew-Hamiltonian* if the equality $(HW)_\phi = -HW$ holds. For a ϕ-hermitian H, a matrix $A \in \mathsf{H}^{n \times n}$ is said to be (H, ϕ)-*symmetric* (for ϕ nonstandard) or $(H,^*)$-*hermitian* (for ϕ the conjugation) if $(HA)_\phi = HA$; the matrix A is said to be (H, ϕ)-*skewsymmetric* (for ϕ nonstandard) or $(H,^*)$-*skewhermitian* (for ϕ the conjugation) if $(HA)_\phi = -HA$.

We do not generally assume in this chapter that H is invertible; such assumption will be made as necessary for certain particular results.

Definition 6.1.2. If H is ϕ-skewhermitian, then a matrix $U \in \mathsf{H}^{n \times n}$ is said to be (H, ϕ)-*symplectic* if

$$U_\phi H U = H. \tag{6.1.1}$$

Analogously, if H is ϕ-hermitian, then a matrix $U \in \mathsf{H}^{n \times n}$ is said to be $(H,^*)$-*unitary* (for ϕ the conjugation) or (H, ϕ)-*orthogonal* (for ϕ nonstandard) if the equality

$$U_\phi H U = H \tag{6.1.2}$$

holds true.

It is easy to verify the following proposition.

Proposition 6.1.3. *If U is (H, ϕ)-symplectic, resp. $(H,^*)$-unitary or (H, ϕ)-orthogonal, and invertible, then so is U^{-1}; also, if U, V are (H, ϕ)-symplectic, resp. $(H,^*)$-unitary or (H, ϕ)-orthogonal, then so is UV.*

Assuming in addition that H is invertible and $H^{-1} = \pm H$, then if U is (H, ϕ)-symplectic, resp. $(H,^*)$-unitary or (H, ϕ)-orthogonal, then so is U_ϕ.

Proof. We provide details only for the second part. Indeed, taking inverses in the equality $U_\phi H U = H$, we get

$$\pm H = H^{-1} = U^{-1} H^{-1} (U_\phi)^{-1} = \pm U^{-1} H (U_\phi)^{-1};$$

hence, $U H U_\phi = H$, which proves that U_ϕ is (H, ϕ)-symplectic. □

Note that if H is ϕ-skewhermitian, if A is (H, ϕ)-Hamiltonian, resp. (H, ϕ)-skew-Hamiltonian, and U is (H, ϕ)-symplectic and invertible, then $U^{-1}AU$ is also

(H, ϕ)-Hamiltonian or (H, ϕ)-skew-Hamiltonian, as the case may be. An analogous statement holds for (H, ϕ)-symmetric and (H, ϕ)-skewsymmetric matrices, where H is ϕ-hermitian, with respect to (H, ϕ)-unitary matrices.

Fix a ϕ-hermitian or ϕ-skewhermitian matrix H. Recall that a subspace $\mathcal{M} \subseteq \mathsf{F}^{n \times 1}$ is called (H, ϕ)-*neutral* if $y_\phi H x = 0$ for all $x, y \in \mathcal{M}$ or, equivalently, if $x_\phi H x = 0$ for all $x \in \mathcal{M}$. The equivalence of these two definitions follows from the polarization identities (3.5.1) and (3.7.1).

The classes of matrices identified in the above tables are appropriately transformed under simultaneous conjugation of H and similarity of A.

Proposition 6.1.4. *Let $H \in \mathsf{F}^{n \times n}$ be as in the above tables, and assume that A belongs to one of the classes identified in the tables, e.g., A is (H, ϕ)-orthogonal. Then for every invertible $S \in \mathsf{F}^{n \times n}$, the matrix $S^{-1} A S$ belongs to the same class associated with $S_\phi H S$ or $S^* H S$, as the case may be.*

The proof is straightforward and is left to the reader.

Definition 6.1.5. Given an (H, ϕ)-Hamiltonian or (H, ϕ)-skew-Hamiltonian matrix $A \in \mathsf{H}^{n \times n}$, a subspace $\mathcal{M} \in \mathsf{H}^{n \times 1}$ is called *maximal A-invariant (H, ϕ)-neutral* if $\mathcal{M}$ is A-invariant and (H, ϕ)-neutral but no subspace strictly larger than $\mathcal{M}$ is A-invariant and H-neutral.

Theorem 6.1.6. *Let $H = \delta H_\phi \in \mathsf{H}^{n \times n}$, where $\delta = \pm 1$, and let ϕ be an involution. Let $A \in \mathsf{H}^{n \times n}$ be (H, ϕ)-Hamiltonian or (H, ϕ)-skew-Hamiltonian (if $\delta = -1$), or (H, ϕ)-symmetric or (H, ϕ)-skewsymmetric (if ϕ is nonstandard and $\delta = 1$), or $(H, *)$-hermitian or $(H, *)$-skewhermitian (if $\phi = *$ is the conjugation and $\delta = 1$). Then all maximal A-invariant (H, ϕ)-neutral subspaces have the same dimension.*

Proof. In the proof, we write H-neutral, short for (H, ϕ)-neutral. Denote by $[\cdot, \cdot]$ the H-inner product with respect to ϕ:

$$[x, y] := y_\phi H x = \delta \phi([y, x]), \qquad x, y \in \mathsf{H}^{n \times 1}.$$

Definition 6.1.7. If $\mathcal{L}$ is a subset of $\mathsf{H}^{n \times 1}$, we define

$$\mathcal{L}^{[\perp]} := \{x \in \mathsf{H}^{n \times 1} : [x, y] = 0 \quad \text{for all} \ \ y \in \mathcal{L}\},$$

the H-*orthogonal companion* of $\mathcal{L}$.

Clearly, $\mathcal{L}^{[\perp]}$ is a subspace of $\mathsf{H}^{n \times 1}$. We have

$$[Ax, y] = \tau \delta [x, Ay], \qquad \text{for all} \ \ x, y \in \mathsf{H}^{n \times 1}, \tag{6.1.3}$$

where $\tau = -1$ if A is (H, ϕ)-Hamiltonian and $\tau = 1$ if A is (H, ϕ)-skew-Hamiltonian. Note also that if $\mathcal{L} \subseteq \mathsf{H}^{n \times 1}$ is an A-invariant subspace, then $\mathcal{L}^{[\perp]}$ is A-invariant as well.

Let $\mathcal{L}$ and $\mathcal{M}$ be maximal A-invariant H-neutral subspaces of $\mathsf{H}^{n \times 1}$. We will prove they have the same dimension.

Step 1. Assume $\mathcal{L} \cap \mathcal{M} = \{0\}$.

Note that

$$\dim (\mathcal{L} \cap \mathcal{M}^{[\perp]}) \geq \dim \mathcal{L} - \dim \mathcal{M}. \tag{6.1.4}$$

Indeed, let $f_1, \ldots, f_\ell$ and $g_1, \ldots, g_m$ be bases for $\mathcal{L}$ and $\mathcal{M}$, respectively. Then $\mathcal{L} \cap \mathcal{M}^{[\perp]}$ consists of all vectors $\sum_{j=1}^{\ell} f_j \alpha_j$, where $\alpha_1, \ldots, \alpha_\ell \in \mathsf{H}$, such that

$$\sum_{j=1}^{\ell} [f_j, g_k] \alpha_j = 0, \quad k = 1, 2, \ldots, m.$$

The solution space of this system clearly has dimension at least $\ell - m$. Therefore, if

$$\mathcal{L} \cap \mathcal{M}^{[\perp]} = \mathcal{L}^{[\perp]} \cap \mathcal{M} = \{0\}, \tag{6.1.5}$$

then by (6.1.4) $\dim \mathcal{L} = \dim \mathcal{M}$, and we are done. Thus, assume (6.1.5) does not hold, say, $\mathcal{D} := \mathcal{L} \cap \mathcal{M}^{[\perp]} \neq \{0\}$. Clearly, $\mathcal{D}$ is A-invariant and H-neutral. Since $\mathcal{L} \cap \mathcal{M} = \{0\}$, the subspace $\mathcal{D} \dotplus \mathcal{M}$ is A-invariant and H-neutral and contains $\mathcal{M}$ properly, a contradiction with maximality of $\mathcal{M}$.

Step 2. Now assume $\mathcal{R} := \mathcal{L} \cap \mathcal{M} \neq \{0\}$.

Let $\mathcal{C} := \mathcal{L} + \mathcal{M}$, an A-invariant subspace, and let $\mathcal{V}$ be a direct complement of $\mathcal{R}$ in $\mathcal{C}$: $\mathcal{C} = \mathcal{R} \dotplus \mathcal{V}$. Define the projection P of $\mathcal{C}$ onto $\mathcal{V}$ along $\mathcal{R}$, and let $A_0 : \mathcal{V} \to \mathcal{V}$ be an H-linear transformation on $\mathcal{V}$ defined by $A_0 x = PAx$, $x \in \mathcal{V}$. We claim that

$$[A_0 x, y] = \tau \delta [x, A_0 y], \quad \text{for all } x, y \in \mathcal{V}. \tag{6.1.6}$$

To see this, observe first that

$$[Px, y] = [x - (I - P)x, y] = [x, y] = [x, Py + (I - P)y] = [x, Py], \tag{6.1.7}$$

for all $x, y \in \mathcal{C}$, in view of the H-neutrality of $\mathcal{L}$ and $\mathcal{M}$. Now (6.1.6) follows easily from (6.1.3), (6.1.7), and the definition of A_0. Indeed, for any $x, y \in \mathcal{V}$:

$$\begin{aligned}
[A_0 x, y] &= [PAx, y] = [Ax, Py] = [Ax, y] \\
&= \tau \delta [x, Ay] = \tau \delta [Px, Ay] = \tau \delta [x, PAy] = \tau \delta [x, A_0 y].
\end{aligned}$$

Define $\mathcal{L}_0 := \mathcal{L} \cap \mathcal{V}$, $\mathcal{M}_0 := \mathcal{M} \cap \mathcal{V}$. Then $\mathcal{L}_0 \cap \mathcal{M}_0 = \{0\}$. Clearly, $\mathcal{L}_0$ and $\mathcal{M}_0$ are H-neutral. These subspaces are also A_0-invariant. Indeed, let $v \in \mathcal{L}_0$. Then $Av \in \mathcal{C}$; hence, $Av = \hat{v} + r$, $r \in \mathcal{R}$, $\hat{v} \in \mathcal{V}$. Furthermore, $A_0 v = Av - r$, and as $\mathcal{R} \subseteq \mathcal{L}$, $A_0 v \in \mathcal{L}$. Obviously, $Av = PAv \in \mathcal{V}$, so $A v_0 \in \mathcal{L} \cap \mathcal{V}$ as claimed. A similar argument proves that $\mathcal{M}_0$ is A-invariant.

Fix a basis $\mathcal{B} := \{b_1, \ldots, b_k\}$ for $\mathcal{V}$, and represent A_0 as a $k \times k$ matrix with respect to this basis. Then A_0 is H_0-Hamiltonian (or H_0-skew-Hamiltonian, as the case may be), where the $k \times k$ matrix $H_0 = [(H_0)_{i,j}]_{i,j=1}^{k}$ is defined by $(H_0)_{i,j} = [b_i, b_j]$. Note that $(H_0)_\phi = \delta H_0$. Using the coordinate mapping with respect to the basis $\mathcal{B}$—which is an isomorphism—we transform the subspaces $\mathcal{L}_0, \mathcal{M}_0$ into subspaces $\widetilde{\mathcal{L}}_0, \widetilde{\mathcal{M}}_0 \subseteq \mathsf{H}^{k \times 1}$, respectively. Clearly, $\widetilde{\mathcal{L}}_0, \widetilde{\mathcal{M}}_0$ are A_0-invariant H_0-neutral, and

$$\dim \widetilde{\mathcal{L}}_0 = \dim \mathcal{L}_0, \quad \dim \widetilde{\mathcal{M}}_0 = \dim \mathcal{M}_0, \quad \widetilde{\mathcal{L}}_0 \cap \widetilde{\mathcal{M}}_0 = \{0\}. \tag{6.1.8}$$

(The latter equality follows from $\mathcal{L}_0 \cap \mathcal{M}_0 = \{0\}$.) Also, since $\mathcal{R} \subseteq \mathcal{L}$, we have

$$\mathcal{L} = (\mathcal{L} \cap \mathcal{R}) \dotplus (\mathcal{L} \cap \mathcal{V}) = \mathcal{R} \dotplus \mathcal{L}_0;$$

hence, $\dim \mathcal{L}_0 = \dim \mathcal{L} - \dim \mathcal{R}$ and, similarly, $\dim \mathcal{M}_0 = \dim \mathcal{M} - \dim \mathcal{R}$. So, it remains to show that $\mathcal{L}_0$, $\mathcal{M}_0$ are maximal A_0-invariant H-neutral subspaces in $\mathcal{V}$. Indeed, then we will have that $\widetilde{\mathcal{L}}_0, \widetilde{\mathcal{M}}_0$ are maximal A_0-invariant H_0-neutral subspaces in $\mathsf{H}^{k \times 1}$, and the equality $\dim \mathcal{L} = \dim \mathcal{M}$ will follow from Step 1 applied to the subspaces $\widetilde{\mathcal{L}}_0$ and $\widetilde{\mathcal{M}}_0$, in view of (6.1.8).

To prove this maximality property for $\mathcal{L}_0$ (analogously, it can be proved for $\mathcal{M}_0$), let $\mathcal{Q}_0$ be any A_0-invariant H-neutral subspace such that $\mathcal{L}_0 \subseteq \mathcal{Q}_0 \subseteq \mathcal{V}$, and define $\mathcal{Q} := \mathcal{R} \dotplus \mathcal{Q}_0$. Then $\mathcal{Q} \supseteq \mathcal{R} \dotplus \mathcal{L}_0 = \mathcal{L}$ and $\mathcal{Q}$ is H-neutral (this is because $\mathcal{R} = \mathcal{L} \cap \mathcal{M}$, $\mathcal{Q}_0 \subseteq \mathcal{C} = \mathcal{L} + \mathcal{M}$, $\mathcal{Q}_0$ is H-neutral, and both $\mathcal{M}$, $\mathcal{L}$ are H-neutral). We claim that $\mathcal{Q}$ is also A-invariant. To confirm this, let $q \in \mathcal{Q}$ and write $q = q_0 + r$, $q_0 \in \mathcal{Q}_0$, $r \in \mathcal{R}$. Then

$$Aq = Aq_0 + Ar = A_0 q_0 + (I - P)Aq_0 + Ar.$$

The last two terms are in $\mathcal{R}$ and the first term is in $\mathcal{Q}_0$. By the maximality property of $\mathcal{L}$ we must have $\mathcal{Q} = \mathcal{L}$; hence, $\mathcal{Q}_0 = \mathcal{L}_0$, and we are done. $\square$

Definition 6.1.8. The dimension identified in Theorem 6.1.6 is called the *order of neutrality* of the pair (H, A) and denoted $\gamma(H, A)$.

(This terminology was introduced in Lancaster et al. [85] for complex matrices that are selfadjoint with respect to an indefinite inner product.)

Open Problem 6.1.9. (a) *Express the order of neutrality in terms of parameters of the canonical forms for the pair* (H, A), *which are given in Chapters* 10 *and* 13 (*in the case of invertible* H).

(b) *For a given pair* (H, A) *of matrices as in Theorem* 6.1.6, *describe the set of all maximal* A-*invariant* (H, ϕ)-*neutral subspaces in terms of analytic manifolds.*

For Hamiltonian matrices, the order of neutrality has been identified in Rodman [140]. For Part (b), Open Problem 5.1.5 may be relevant.

We remark that a result analogous to Theorem 6.1.6 but with respect to a pair of ϕ-hermitian matrices (A, B) (where we take ϕ to be the conjugation) generally does not hold, in other words, the subspaces that are simultaneously A-neutral and B-neutral and are maximal with respect to this property need not have the same dimension. The following example illustrates that.

Example 6.1.10. Let

$$A = A^* = \begin{bmatrix} 0 & 0 & 1 & -1 \\ 0 & 1 & 0 & 0 \\ 1 & 0 & 0 & 0 \\ -1 & 0 & 0 & 0 \end{bmatrix}, \quad B = B^* = \begin{bmatrix} 0 & 0 & 0 & 1 \\ 0 & 0 & 1 & 0 \\ 0 & \pm 1 & 0 & 0 \\ \pm 1 & 0 & 0 & 0 \end{bmatrix} \in \mathsf{H}^{4 \times 4}.$$

Obviously, there exists a two-dimensional (A, B)- neutral subspace: $\mathrm{Span}_\mathsf{H} \{e_3, e_4\}$. The subspace $\mathrm{Span}_\mathsf{H} \{e_1\}$ is also (A, B)-neutral. But it is not contained in any two-dimensional (A, B)-neutral subspace. Indeed, if

$$\mathcal{M} := \mathrm{Span}_\mathsf{H} \{e_1, e_2 x_2 + e_3 x_3 + e_4 x_4\}, \qquad x_1, x_2, x_3 \in \mathsf{H},$$

were such a subspace, then we would have

$$0 = \begin{bmatrix} 1 & 0 & 0 & 0 \end{bmatrix} A \begin{bmatrix} 0 \\ x_1 \\ x_2 \\ x_3 \end{bmatrix} = x_2 - x_3,$$

$$0 = \begin{bmatrix} 1 & 0 & 0 & 0 \end{bmatrix} B \begin{bmatrix} 0 \\ x_1 \\ x_2 \\ x_3 \end{bmatrix} = x_3,$$

$$0 = \begin{bmatrix} 0 & x_1 & x_2 & x_3 \end{bmatrix}^* A \begin{bmatrix} 0 \\ x_1 \\ x_2 \\ x_3 \end{bmatrix} = |x_1|^2;$$

thus, $x_1 = x_2 = x_3 = 0$, a contradiction with $\mathcal{M}$ being two-dimensional. So, $\mathcal{M}$ is a maximal (A, B)-neutral subspace.

Note that in this example $B^{-1}A$ is $(B,{}^*)$-hermitian. However, the subspace $\mathrm{Span}_{\mathsf{H}} \{e_1\}$ is not $B^{-1}A$-invariant, whereas $\mathrm{Span}_{\mathsf{H}} \{e_3, e_4\}$ is, as confirmed by Theorem 6.1.6. $\qquad\square$

Open Problem 6.1.11. *Identify those pairs of hermitian matrices for which the maximal neutral subspaces have the same dimension. An analogous question occurs for pair of matrices that are ϕ-hermitian or ϕ-skewhermitian with respect to a nonstandard involution ϕ.*

6.2 INVARIANT SEMIDEFINITE SUBSPACES RESPECTING CONJUGATION

In this and the next three sections we study subspaces of $\mathsf{H}^{n \times 1}$ which are invariant for a matrix in certain class and simultaneously have definiteness properties with respect to the indefinite inner product induced by a hermitian matrix.

Let $H = H^* \in \mathsf{H}^{n \times n}$. The matrix H will be fixed throughout this and the next three sections. We denote by $[\,\cdot\,,\,\cdot\,]$ the indefinite scalar product in $\mathsf{H}^{n \times 1}$ induced by H:

$$[x, y] := \langle x, Hy \rangle = \langle Hx, y \rangle = y^* H x, \quad x, y \in \mathsf{H}^{n \times 1}.$$

Several classes of subspaces of $\mathsf{H}^{n \times 1}$ will be of main interest in this and the next sections. These were introduced in Section 4.2; we recall the definitions. The classes of *nonnegative, nonpositive, and neutral subspaces* defined by

$$\mathcal{S}_\nu(H) = \{\mathcal{V} : \mathcal{V} \text{ is a subspace of } \mathsf{H}^{n \times 1} \text{ such that } [x, x]\nu 0 \text{ for every } x \in \mathcal{V}\},$$

where the symbol ν is $\geq, \leq, =$, respectively, and the classes of *positive and negative subspaces* are defined by

$$\mathcal{S}_{\nu 0}(H) = \{\mathcal{V} : \mathcal{V} \text{ is a subspace of } \mathsf{H}^{n \times 1} \text{ such that } [x, x]\nu 0 \text{ for every } x \in \mathcal{V} \setminus \{0\}\},$$

where $\nu = >$ and $\nu = <$, respectively.

Definition 6.2.1. A matrix $A \in \mathsf{H}^{n \times n}$ is called an *H-plus-matrix* if $[Ax, Ax] \geq 0$ for every $x \in \mathsf{H}^{n \times 1}$ such that $[x, x] \geq 0$.

A useful characterization of H-plus-matrices is given in the next proposition.

Proposition 6.2.2. $A \in \mathsf{H}^{n \times n}$ *is an H-plus-matrix if and only if there exists a nonnegative μ such that*

$$[Ax, Ax] \geq \mu[x, x] \qquad \text{for all} \quad x \in \mathsf{H}^{n \times 1}. \qquad (6.2.1)$$

Proof. The 'if' part is clear from the definition of H-plus-matrices. Suppose A is an H-plus-matrix. Then the joint numerical range $WJ_*^{\mathsf{H}}(A^*HA, H) \subseteq \mathsf{R}^2$ does not intersect the set

$$\{(a, b) \in \mathsf{R}^2 : b \geq 0 \text{ and } a < 0\}. \qquad (6.2.2)$$

Since the set $WJ_*^{\mathsf{H}}(A^*HA, H)$ is compact and convex (Theorem 3.5.7), there is a line that separates $WJ_*^{\mathsf{H}}(A^*HA, H)$ from the convex set (6.2.2); see Theorem 6.6.2 in the Appendix. It is easy to see that the line can be taken to pass through the origin and not to be horizontal. Then the inverse of the slope of this line is a suitable value of μ. $\qquad \square$

Note that μ need not be unique (for a given A).

Example 6.2.3. Let

$$H = \begin{bmatrix} 1 & 0 \\ 0 & -1 \end{bmatrix}, \quad A = \begin{bmatrix} 1 & 0 \\ 0 & 0 \end{bmatrix}.$$

Then for $x = \begin{bmatrix} x_1 \\ x_2 \end{bmatrix}$, where $x_1, x_2 \in \mathsf{H}$, we have $[x, x] = |x_1|^2 - |x_2|^2$, $[Ax, Ax] = |x_1|^2$, so (6.2.1) holds true for any $\mu \in [0, 1]$. $\qquad \square$

Definition 6.2.4. An H-plus-matrix A is said to be *strict H-plus-matrix* if there exists $\mu > 0$ satisfying Proposition 6.2.2.

We will write $\mu(A)$ in place of μ if the dependence on A is to be emphasized.

Observe that these classes of matrices and their invariant subspaces transform with respect to simultaneous congruence of H and similarity of A.

Proposition 6.2.5. *Let $H = H^* \in \mathsf{H}^{n \times n}$. Then:*

(1) *if A is H-plus, resp. strict H-plus, and $S \in \mathsf{H}^{n \times n}$ is invertible, then $S^{-1}AS$ is S^*HS-plus, resp. strict S^*HS-plus;*

(2) *a subspace $\mathcal{M} \subseteq \mathsf{H}^{n \times 1}$ belongs to the class $\mathcal{S}_\nu(H)$ if and only if $S^{-1}(\mathcal{M})$ belongs to $\mathcal{S}_\mu(S^*HS)$, where ν is one of the symbols $\{\geq 0, > 0, \leq 0, < 0, 0\}$;*

(3) *a subspace $\mathcal{M}$ is A-invariant if and only if $S^{-1}(\mathcal{M})$ is $S^{-1}AS$-invariant.*

The proof is by straightforward verification.

The key result in this section is the following theorem on extension of H-nonnegative subspaces that are invariant for H-plus-matrices.

Theorem 6.2.6. *Assume H is invertible, and let A be an invertible H-plus-matrix. Then:*

(a) *for every A-invariant H-nonnegative subspace $\mathcal{M}_0 \subseteq H^{n \times 1}$ there exists an H-nonnegative A-invariant subspace $\widetilde{\mathcal{M}} \supseteq \mathcal{M}_0$ such that*

$$\dim \widetilde{\mathcal{M}} = \text{In}_+(H); \tag{6.2.3}$$

(b) *for every A-invariant H-nonpositive subspace $\mathcal{M}_0' \subseteq H^{n \times 1}$ there exists an H-nonpositive A-invariant subspace $\widetilde{\mathcal{M}}' \supseteq \mathcal{M}_0'$ such that*

$$\dim \widetilde{\mathcal{M}}' = \text{In}_-(H). \tag{6.2.4}$$

Note that by Theorem 4.2.6(a) the right-hand side of (6.2.3), resp. (6.2.4), represents the maximal dimension of an H-nonnegative, resp. H-nonpositive, subspace.

The hypothesis that A is invertible is essential in Theorem 6.2.6. The following example illustrates this.

Example 6.2.7. Let

$$A = \begin{bmatrix} 0 & 1 & 0 & p \\ -1 & 0 & 1 & q \\ 0 & 0 & 0 & 0 \\ 0 & 0 & 0 & 0 \end{bmatrix}, \qquad H = \begin{bmatrix} 0 & 0 & 1 & 0 \\ 0 & 0 & 0 & 1 \\ 1 & 0 & 0 & 0 \\ 0 & 1 & 0 & 0 \end{bmatrix},$$

where p and q are real numbers such that $4p + q^2 > 0$. Clearly, A is an H-plus-matrix; in fact, $[Ax, Ax] = 0$ for every $x \in H^{4 \times 1}$. Let

$$\mathcal{M}_0 = \text{Span}_H \begin{bmatrix} 1 & 0 & 1 & 0 \end{bmatrix}^T.$$

Then $A\mathcal{M}_0 = \{0\}$, in particular, $\mathcal{M}_0$ is A-invariant. Since $\begin{bmatrix} 1 & 0 & 1 & 0 \end{bmatrix} H(e_1 + e_3) = 2$, the subspace $\mathcal{M}_0$ is H-positive. On the other hand,

$$\text{Ker } A = \text{Span}_H \left\{ \begin{bmatrix} 1 \\ 0 \\ 1 \\ 0 \end{bmatrix}, \begin{bmatrix} 0 \\ -p \\ -q \\ 1 \end{bmatrix} \right\}$$

is not H-nonnegative (this is where the condition $4p + q^2 > 0$ is needed). Besides $\text{Ker } A$, there is a continuum of 2-dimensional A-invariant subspaces that contain $\mathcal{M}_0$, namely,

$$\text{Span}_H \left\{ \begin{bmatrix} 1 \\ 0 \\ 1 \\ 0 \end{bmatrix}, \begin{bmatrix} 1 \\ \alpha \\ 0 \\ 0 \end{bmatrix} \right\}, \qquad \alpha \in H, \quad \alpha^2 = -1.$$

But

$$\begin{bmatrix} 1 & \alpha^* & 0 & 0 \\ 1 & 0 & 1 & 0 \end{bmatrix} H \begin{bmatrix} 1 & 1 \\ \alpha & 0 \\ 0 & 1 \\ 0 & 0 \end{bmatrix} = \begin{bmatrix} 0 & 1 \\ 1 & 2 \end{bmatrix},$$

so none of these subspaces is H-nonnegative. $\qquad \square$

We shall see later that the condition of invertibility of A can be relaxed (for Part (a) of Theorem 6.2.6). Namely, the hypothesis that $\operatorname{Ran} A$ is not H-nonnegative, or that A is a strict H-plus-matrix, will suffice (Corollary 6.4.3 below).

The proof of Theorem 6.2.6 is rather involved. It will be given in a separate section.

Open Problem 6.2.8. *Extend if possible the result of Theorem* 6.2.6 *to cases when H and/or A are not necessarily invertible.*

Some results in this direction will be given later on in Section 8.6.

6.3   PROOF OF THEOREM 6.2.6

First observe that it will suffice to prove Part (a). Indeed, we note that the inverse matrix A^{-1} is $(-H)$-plus-matrix, as easily follows by post- and premultiplying the positive semidefinite matrix $A^*HA - H$ by $(A^*)^{-1}$ and A^{-1}, respectively, and observing that multiplying by -1 turns a positive semidefinite matrix into a negative semidefinite one. Then applying part (a) to the $(-H)$-plus-matrix A^{-1} yields part (b) for the H-plus-matrix A (and note that if A is invertible, then A and A^{-1} have exactly the same set of invariant subspaces).

Therefore, we will prove part (a) only. We need some preliminaries for the proof. First of all, we can assume that H has the form

$$H = \begin{bmatrix} I_p & 0 \\ 0 & -I_q \end{bmatrix}. \tag{6.3.1}$$

Indeed, by applying a congruence $H \to S^*HS$ for a suitable invertible matrix S, and simultaneously transforming $A \to S^{-1}AS$ and $\mathcal{M}_0 \to S^{-1}\mathcal{M}_0$, the form (6.3.1) can always be achieved, by Theorem 4.1.6.

With respect to the form (6.3.1) the H-nonnegative subspaces can be conveniently described, as in the next lemma.

Lemma 6.3.1. *Let H be given by* (6.3.1). *Then a subspace $\mathcal{M} \subseteq \mathsf{H}^{n \times 1}$ of dimension d is H-nonnegative if and only if $\mathcal{M}$ has the form*

$$\mathcal{M} = \operatorname{Ran} \begin{bmatrix} P \\ K \end{bmatrix}, \tag{6.3.2}$$

where $P^*P = I$ and $K \in \mathsf{H}^{q \times d}$ satisfies $\|K\| \leq 1$.

The case when $d = 0$ is not excluded; in this case $\mathcal{M}$ is interpreted as the zero subspace.

Proof. It is easy to see that every subspace of the form (6.3.2) is H-nonnegative. Indeed, let

$$\begin{bmatrix} x_1 \\ x_2 \end{bmatrix} = \begin{bmatrix} P \\ K \end{bmatrix} y$$

for some $y \in \mathsf{H}^{d \times 1}$. Then

$$
\left\langle H \begin{bmatrix} x_1 \\ x_2 \end{bmatrix}, \begin{bmatrix} x_1 \\ x_2 \end{bmatrix} \right\rangle = \left\langle \begin{bmatrix} I_p & 0 \\ 0 & -I_q \end{bmatrix} \begin{bmatrix} P \\ K \end{bmatrix} y, \begin{bmatrix} P \\ K \end{bmatrix} y \right\rangle
$$

$$
= \langle Py, Py \rangle - \langle Ky, Ky \rangle
$$

$$
= \langle P^*Py, y \rangle - \langle Ky, Ky \rangle
$$

$$
= \|y\|^2 - \|Ky\|^2 \geq 0, \tag{6.3.3}
$$

where the last inequality follows in view of $\|K\| \leq 1$.

Conversely, let $\mathcal{M} \in \mathcal{S}_{\geq 0}(H)$, dim $\mathcal{M} = d$. Let $f_1, \ldots, f_d$ be a basis of $\mathcal{M}$ (we ignore the trivial case $\mathcal{M} = \{0\}$) and partition:

$$
f_j = \begin{bmatrix} f_{1j} \\ f_{2j} \end{bmatrix}, \quad f_{1j} \in \mathsf{H}^{p \times 1}, \ f_{2j} \in \mathsf{H}^{q \times 1}.
$$

The vectors $f_{11}, \ldots, f_{1d}$ are linearly independent; otherwise, $\mathcal{M}$ would contain a nonzero vector of the form $\begin{bmatrix} 0_p \\ f \end{bmatrix}$, $f \in \mathsf{H}^{q \times 1} \setminus \{0\}$, a contradiction with $\mathcal{M}$ being H-nonnegative. Now let

$$
P = [f_{11} \ f_{12} \ \cdots \ f_{1d}] \, T \in \mathsf{H}^{p \times d},
$$

where the invertible matrix T is chosen so that $P^*P = I_d$. This choice is always possible, because in view of the linear independence of $f_{11}, \ldots, f_{1d}$, the matrix

$$
[f_{11} \ f_{12} \ \cdots \ f_{1d}]^* [f_{11} \ f_{12} \ \cdots \ f_{1d}]
$$

is invertible and, hence, positive definite. Finally, let

$$
K = [f_{21} \ f_{22} \ \cdots \ f_{2d}] T \in \mathsf{H}^{q \times d}.
$$

With these definitions of P and K, the verification of the formula (6.3.2) is immediate, because

$$
\mathcal{M} = \operatorname{Ran}[f_1 \ \cdots \ f_d] = \operatorname{Ran} \begin{bmatrix} f_{11} & \cdots & f_{1d} \\ f_{21} & \cdots & f_{2d} \end{bmatrix}
$$

$$
= \operatorname{Ran} \begin{bmatrix} f_{11} & \cdots & f_{1d} \\ f_{21} & \cdots & f_{2d} \end{bmatrix} T = \operatorname{Ran} \begin{bmatrix} P \\ K \end{bmatrix}.
$$

Using the equality $P^*P = I$, one verifies as in (6.3.3) that

$$
\left\langle H \begin{bmatrix} P \\ K \end{bmatrix} y, \begin{bmatrix} P \\ K \end{bmatrix} y \right\rangle = \|y\|^2 - \|Ky\|^2 \geq 0,
$$

for every $y \in \mathsf{H}^{d \times 1}$, and hence, $\|K\| \leq 1$. $\qquad \square$

Note that the representation (6.3.2) for a given $\mathcal{M}$ is not unique. To study the problem of describing this nonuniqueness, first of all observe that if

$$\mathrm{Ran}\begin{bmatrix} P \\ K_1 \end{bmatrix} = \mathrm{Ran}\begin{bmatrix} P \\ K_2 \end{bmatrix}, \qquad (6.3.4)$$

where $P^*P = I_d$, then $K_1 = K_2$; in other words, the matrix K in (6.3.2) is uniquely determined by $\mathcal{M}$ and P. The verification is easy: if (6.3.4) holds and $P^*P = I_d$, then, in particular, the columns of $\begin{bmatrix} P \\ K_1 \end{bmatrix}$ and of $\begin{bmatrix} P \\ K_2 \end{bmatrix}$ are linearly independent, and we have

$$\begin{bmatrix} P \\ K_1 \end{bmatrix} = \begin{bmatrix} P \\ K_2 \end{bmatrix} W$$

for some invertible matrix W. Consequently, $P = PW$. Multiplying this equality on the left by P^*, it follows that $W = I$. But then, also, $K_1 = K_2$. As for the nonuniqueness of P in (6.3.2), assume that

$$\mathrm{Ran}\begin{bmatrix} P_1 \\ K_1 \end{bmatrix} = \mathrm{Ran}\begin{bmatrix} P_2 \\ K_2 \end{bmatrix},$$

where $P_1^*P_1 = P_2^*P_2 = I_d$. Then

$$\begin{bmatrix} P_1 \\ K_1 \end{bmatrix} = \mathrm{Ran}\begin{bmatrix} P_2 \\ K_2 \end{bmatrix} W \qquad (6.3.5)$$

for a unique invertible matrix W. In fact, $W = P_2^*P_1$, as can be verified by multiplying the equality $P_1 = P_2W$ on the left by P_2^*. Moreover, (6.3.5) implies $P_1W^{-1} = P_2$, and, therefore, $W^{-1} = P_1^*P_2$. We see that $W^{-1} = W^*$; in other words, W is unitary. The following result is obtained.

Lemma 6.3.2. *If*

$$\mathrm{Ran}\begin{bmatrix} P_1 \\ K_1 \end{bmatrix} = \mathrm{Ran}\begin{bmatrix} P_2 \\ K_2 \end{bmatrix}, \qquad (6.3.6)$$

where P_1 and P_2 are $p \times d$ matrices such that $P_1^*P_1 = P_2^*P_2 = I$, then $P_1 = P_2W$, $K_1 = K_2W$ for some unitary $d \times d$ matrix W. Such unitary matrix W is unique.

In particular, Lemma 6.3.2 applies to an H-nonnegative subspace of the form (6.3.6). However, for the validity of the lemma, it is not necessary that $\|K_1\| \le 1$, $\|K_2\| \le 1$.

Corollary 6.3.3. *Every maximal H-nonnegative subspace $\mathcal{M}$ can be written uniquely in the form*

$$\mathcal{M} = \mathrm{Ran}\begin{bmatrix} I_p \\ K \end{bmatrix}, \qquad (6.3.7)$$

where $K \in \mathsf{H}^{q \times p}$ satisfies $\|K\| \le 1$. Conversely, every subspace of the form (6.3.7) is maximal H-nonnegative.

Indeed, the uniqueness of (6.3.7) follows from Lemma 6.3.2. Existence of (6.3.7) follows from Lemma 6.3.1 in which P is unitary, because in view of Theorem 4.2.6(a) the dimension of $\mathcal{M}$ is equal to p.

Next, we express containment of an H-nonnegative subspace in a maximal such subspace, in terms of the representation (6.3.2).

Lemma 6.3.4. *Let*

$$\mathcal{M}_0 = \mathrm{Ran} \begin{bmatrix} P_0 \\ K_0 \end{bmatrix} \in \mathcal{S}_{\geq 0}(H), \quad \mathcal{M} = \mathrm{Ran} \begin{bmatrix} I_p \\ K \end{bmatrix} \in \mathcal{S}_{\geq 0}(H),$$

where $P_0 \in \mathsf{H}^{p \times d}$, $K_0 \in \mathsf{H}^{q \times d}$, $K \in \mathsf{H}^{q \times p}$ are such that $P_0^* P_0 = I$ and $\|K_0\| \leq 1$, $\|K\| \leq 1$. (In particular, $\mathcal{M}$ is maximal H-nonnegative.) Then $\mathcal{M}_0 \subseteq \mathcal{M}$ if and only if $K_0 = K P_0$.

Proof. If $K_0 = K P_0$, then obviously

$$\begin{bmatrix} P_0 \\ K_0 \end{bmatrix} = \begin{bmatrix} I \\ K \end{bmatrix} P_0,$$

and, therefore, $\mathcal{M}_0 \subseteq \mathcal{M}$. Conversely, if $\mathcal{M}_0 \subseteq \mathcal{M}$, then there exists a matrix B such that

$$\begin{bmatrix} P_0 \\ K_0 \end{bmatrix} = \begin{bmatrix} I \\ K \end{bmatrix} B.$$

It is immediate from this equality that in fact $B = P_0$. $\qquad\square$

Proof of Theorem 6.2.6. Let $\mathcal{M}_0$ be as in Theorem 6.2.6. Represent $\mathcal{M}_0$ according to Lemma 6.3.1:

$$\mathcal{M}_0 = \mathrm{Ran} \begin{bmatrix} P_0 \\ K_0 \end{bmatrix},$$

where $P_0^* P_0 = I$, $\|K_0\| \leq 1$. On the other hand, consider the set S of all maximal H-nonnegative subspace $\mathcal{M}$ that contain $\mathcal{M}_0$. Writing

$$\mathcal{M} = \mathrm{Ran} \begin{bmatrix} I \\ K \end{bmatrix} \in S,$$

in view of Lemma 6.3.4 we can identify S with the set of all $q \times p$ matrices K such that $\|K\| \leq 1$ and $K_0 = K P_0$. Note that S is closed and bounded convex set of H^{pq} (if we identify the vector space of $q \times p$ quaternion matrices with H^{pq}). Next, observe that

$$A(\mathcal{M}) := \{Ax : x \in \mathcal{M}\} \in S \quad \text{for all} \quad \mathcal{M} \in S.$$

Indeed, the hypothesis that A is an H-plus-matrix implies that $A(\mathcal{M}) \in S_{\geq 0}(H)$ for every $\mathcal{M} \in \mathcal{S}_{\geq 0}(H)$. Since A is invertible, $\dim A(\mathcal{M}) = \dim \mathcal{M} = \mathrm{In}_+(H)$ for every maximal H-nonnegative subspace $\mathcal{M}$, which implies by Theorem 4.2.6 that $A(\mathcal{M})$ is maximal H-nonnegative as well. Finally, since $A(\mathcal{M}_0) = \mathcal{M}_0$, we obviously have $A(\mathcal{M}) \supseteq \mathcal{M}_0$ for every subspace $\mathcal{M} \supseteq \mathcal{M}_0$.

In other words, A maps S into itself. Identifying S with the set

$$\mathcal{K} := \{K \in \mathsf{H}^{q \times p} \,|\, \|K\| \leq 1 \text{ and } K_0 = K P_0\},$$

the map $\widetilde{A}$ induced by A is a continuous function of $K \in \mathcal{K}$. To verify this, write

$$A = \begin{bmatrix} A_{11} & A_{12} \\ A_{21} & A_{22} \end{bmatrix},$$

where A_{11} is $p \times p$ and A_{22} is $q \times q$, and note that the equality

$$A\left(\operatorname{Ran} \begin{bmatrix} I \\ K \end{bmatrix}\right) = \operatorname{Ran} \begin{bmatrix} I \\ \widetilde{A}(K) \end{bmatrix},$$

where $\|K\| \leq 1$, $K_0 = KP_0$, can be rewritten in the form

$$\begin{bmatrix} A_{11} & A_{12} \\ A_{21} & A_{22} \end{bmatrix} \begin{bmatrix} I \\ K \end{bmatrix} X = \begin{bmatrix} I \\ \widetilde{A}(K) \end{bmatrix},$$

for some invertible matrix X. It follows that $X = (A_{11} + A_{12}K)^{-1}$ and

$$\widetilde{A}(K) = (A_{21} + A_{22}K)(A_{11} + A_{12}K)^{-1}.$$

(Note that the invertibility of $A_{11} + A_{12}K$ is guaranteed for every $K \in \mathcal{K}$.) This is obviously a continuous function of K in $\mathcal{K}$. Now the fixed point theorem (Theorem 6.6.3 in the Appendix) guarantees existence of $K' \in \mathsf{H}^{q \times p}$ such that $\|K'\| \leq 1$, $K_0 = K'P_0$ and $\widetilde{A}(K') = K'$. Then

$$\widetilde{\mathcal{M}} = \operatorname{Ran} \begin{bmatrix} I \\ K' \end{bmatrix}$$

satisfies all the requirements of Theorem 6.2.6. $\qquad \square$

6.4 UNITARY, DISSIPATIVE, AND EXPANSIVE MATRICES

Theorem 6.2.6 and its proof have many important corollaries. As in the preceding two sections, we fix in this section an invertible hermitian matrix $H \in \mathsf{H}^{n \times n}$, and let $[x, y] = y^* H x$, where $x, y \in \mathsf{H}^{n \times 1}$.

To start with, every H-unitary matrix is obviously invertible and an H-plus-matrix.

Theorem 6.4.1. *Assume $H = H^* \in \mathsf{H}^{n \times n}$ is invertible. Let A be H-unitary, and let $\mathcal{M}_0 \subseteq \mathsf{H}^{n \times 1}$ be an A-invariant H-nonnegative (resp. H-nonpositive) subspace. Then there exists an A-invariant H-nonnegative (resp. H-nonpositive) subspace $\mathcal{M}$ such that $\mathcal{M} \supseteq \mathcal{M}_0$ and $\dim \mathcal{M} = \operatorname{In}_+(H)$ (resp. $\dim \mathcal{M} = \operatorname{In}_-(H)$).*

The part of Theorem 6.4.1 concerning H-nonpositive subspaces follows by applying Theorem 6.2.6, Part (b).

We remark next that the proof of Theorem 6.2.6 shows the following result to be true (relaxation of the invertibility condition in Theorem 6.2.6, Part (a)).

Theorem 6.4.2. *Assume H is invertible. Let A be an H-plus-matrix having the property that $Ax \neq 0$ for all $x \in \mathsf{H}^{n \times 1} \setminus \{0\}$ such that $[x, x] \geq 0$. Let $\mathcal{M}_0 \subseteq \mathsf{H}^{n \times 1}$ be an A-invariant subspace that is H-nonnegative. Then there exists H-nonnegative A-invariant subspace $\widetilde{\mathcal{M}} \supseteq \mathcal{M}_0$ such that $\dim \widetilde{\mathcal{M}} = \operatorname{In}_+(H)$.*

Corollary 6.4.3. *Assume H is invertible. Let A be a strict H-plus-matrix. Then for every A-invariant H-nonnegative subspace $\mathcal{M}_0 \subseteq \mathsf{H}^{n \times 1}$, there exists an H-nonnegative A-invariant subspace $\widetilde{\mathcal{M}} \supseteq \mathcal{M}_0$ such that $\dim \widetilde{\mathcal{M}} = \mathrm{In}_+(H)$.*

Proof. We verify that the conditions of Theorem 6.4.2 are satisfied. Assume $Ax = 0$ for some $x \in \mathsf{H}^{n \times 1} \setminus \{0\}$ such that $[x, x] \geq 0$. Since A is strict H-plus, the matrix $A^* H A - \mu H$ is positive semidefinite for some $\mu > 0$. It follows that $-\mu[x, x] \geq 0$, and so $[x, x] = 0$. Now

$$x^*(A^* H A - \mu H)x = 0,$$

and the positive semidefiniteness of $A^* H A - \mu H$ implies $(A^* H A - \mu H)x = 0$ (cf. Ex. 4.6.10), which, in turn, yields $Hx = 0$. Thus, $x = 0$ in view of invertibility of H, a contradiction. $\qquad\square$

Next, we use linear fractional transformations.

Definition 6.4.4. A matrix $A \in \mathsf{H}^{n \times n}$ is said to be H-*expansive* if $[Ax, Ax] \geq [x, x]$ for every $x \in \mathsf{H}^{n \times 1}$ or, equivalently, if $A^* H A - H$ is positive semidefinite. A matrix $B \in \mathsf{H}^{n \times n}$ is called H-*dissipative* if $\mathfrak{R}([Bx, x]) \leq 0$ for every $x \in \mathsf{H}^{n \times 1}$.

The dissipativity condition can be easily interpreted in terms of negative semidefiniteness: A matrix B is H-dissipative if and only if

$$HB + B^* H \leq 0 \quad \text{(negative semidefinite)}. \tag{6.4.1}$$

Note that A is H-expansive, resp. H-dissipative, if and only if $S^{-1} A S$ is $S^* H S$-expansive, resp. $S^* H S$-dissipative, for any invertible matrix $S \in \mathsf{H}^{n \times n}$.

Lemma 6.4.5.   (a) *Let $A \in \mathsf{H}^{n \times n}$ be H-expansive, and assume that $\eta \notin \sigma(A)$, where $\eta = \pm 1$. Then*

$$B := -(A + \eta I)(A - \eta I)^{-1} \tag{6.4.2}$$

is H-dissipative and $-1 \notin \sigma(B)$.

(b) *Let $B \in \mathsf{H}^{n \times n}$ be H-dissipative, and let ρ be a negative number not an eigenvalue of B. Then*

$$A := (B - \rho I)^{-1}(\rho I + B)\eta, \qquad \eta = \pm 1, \tag{6.4.3}$$

is H-expansive and $\eta \notin \sigma(A)$.

Proof. The proof of (a) is by straightforward verification: if A is H-expansive with $\pm 1 = \eta \notin \sigma(A)$ and (6.4.2) holds, then

$$
\begin{aligned}
HB + B^* H &= -H(A + \eta I)(A - \eta I)^{-1} - (A^* - \eta I)^{-1}(A^* + \eta I)H \\
&= -(A^* - \eta I)^{-1}\big((A^* - \eta I)H(A + \eta I) \\
&\quad + (A^* + \eta I)H(A - \eta I)\big)(A - \eta I)^{-1} \\
&= (A^* - \eta I)^{-1}(-2A^* H A + 2H)(A - \eta I)^{-1},
\end{aligned}
$$

and since $A^* H A - H$ is positive semidefinite, the matrix

$$(A^* - \eta I)^{-1}(-2A^* H A + 2H)(A - \eta I)^{-1}$$

is negative semidefinite. If we had $Bx = -x$ for some $x \in \mathsf{H}^{n \times 1}$, then

$$(A + \eta I)x = (A - \eta I)x,$$

which is possible only if $x = 0$. So $-1 \notin \sigma(B)$.

For Part (b), if B is H-dissipative with $\rho \notin \sigma(B)$ and A is given by (6.4.3), then

$$
\begin{aligned}
A^* H A - H \; &= \; (\rho I + B^*)(B^* - \rho I)^{-1} H (B - \rho I)^{-1}(\rho I + B) - H \\
&= \; (B^* - \rho I)^{-1} \left((\rho I + B^*) H (\rho I + B) \right. \\
&\quad \left. - (B^* - \rho I) H (B - \rho I) \right) (B - \rho I)^{-1} \\
&= \; (B^* - \rho I)^{-1} (2\rho B^* H + 2\rho H B)(B - \rho I)^{-1},
\end{aligned}
$$

and the positive semidefiniteness of $A^* H A - H$ follows. One checks easily that η cannot be an eigenvalue of A. $\square$

The transformations in (6.4.2) and (6.4.3) are, in fact, the inverses of each other if $\rho = -1$. Thus, denote

$$g(t) = -(t + \eta)(t - \eta)^{-1}, \quad t \in \mathsf{R},$$

and

$$h(t) = \eta(t - \rho)^{-1}(t + \rho), \quad t \in \mathsf{R},$$

where $\eta = \pm 1$. Then for every matrix $A \in \mathsf{H}^{n \times n}$ not having eigenvalue η, the number -1 is not an eigenvalue of the matrix $B = g(A)$, and

$$A = h(B), \quad \text{for} \quad \rho = -1. \tag{6.4.4}$$

Here the matrices $g(A)$ and $h(B)$ are understood in the sense of functions of matrices; they are also given by formulas (6.4.2) and (6.4.3), respectively. The verification of (6.4.4) is easy:

$$h(g(t)) = \left(-\frac{t + \eta}{t - \eta} - \rho \right)^{-1} \left(-\frac{t + \eta}{t - \eta} + \rho \right) \eta, \tag{6.4.5}$$

which, for $\rho = -1$, after some simple algebra is seen to be equal to t. Therefore, by the properties of functions of matrices, $h(g(A)) = A$.

Theorem 6.4.6. *Assume $H = H^* \in \mathsf{H}^{n \times n}$ is invertible. Let $B \in \mathsf{H}^{n \times n}$ be H-dissipative.*

(a) *Let $\mathcal{M}_0 \subseteq \mathsf{H}^{n \times 1}$ be a B-invariant H-nonnegative subspace. Then there exists a B-invariant maximal H-nonnegative subspace $\mathcal{M}$ such that $\mathcal{M} \supseteq \mathcal{M}_0$.*

(b) *Let $\mathcal{M}_0' \subseteq \mathsf{H}^{n \times 1}$ be a B-invariant H-nonpositive subspace. Then there exists a B-invariant maximal H-nonpositive subspace $\mathcal{M}'$ such that $\mathcal{M}' \supseteq \mathcal{M}_0'$.*

Proof. Assume $\mathcal{M}_0$ is H-nonnegative. Let A be given by (6.4.3). Then A is H-expansive and, in particular, A is a strict H-plus-matrix. Note also that $\mathcal{M}_0$ is A-invariant, because A is a function of B. By Corollary 6.4.3 there exists an A-invariant subspace $\mathcal{M}$ which is maximal H-nonnegative and contains $\mathcal{M}_0$. Since B is a function of A (given by formula (6.4.2)), $\mathcal{M}$ is also B-invariant. This proves part (a) of Theorem 6.4.6. If $\mathcal{M}_0'$ is H-nonpositive, apply the already proved part (a) of Theorem 6.4.6 to the $(-H)$-dissipative matrix $-B$. $\qquad\square$

We conclude this section with an open problem concerning spectral properties of restrictions of H-expansive matrices to invariant maximal semidefinite subspaces.

The following result is known for complex matrices (proved in Iohvidov et al. [70] in the context of linear operators on infinite dimensional spaces).

Theorem 6.4.7. *Assume $H = H^* \in \mathsf{C}^{n \times n}$ is invertible, and let $A \in \mathsf{C}^{n \times n}$ be H-expansive. Then:*

(1) *there exist A-invariant H-nonnegative subspaces $\mathcal{M} \subseteq \mathsf{C}^{n \times 1}$ of dimension $\mathrm{In}_+(H)$ and such that $|\lambda| \geq 1$ for every eigenvalue λ of $A|_{\mathcal{M}}$;*

(2) *there exist A-invariant H-nonpositive subspaces $\mathcal{M} \subseteq \mathsf{C}^{n \times 1}$ of dimension $\mathrm{In}_-(H)$ and such that $|\lambda| \leq 1$ for every eigenvalue λ of $A|_{\mathcal{M}}$;*

(3) *every subspace $\mathcal{M}$ with the properties as in (1), resp. (2), contains all root subspaces of A corresponding to its eigenvalues λ with $|\lambda| > 1$, resp. $|\lambda| < 1$.*

The hypothesis on invertibility of H is essential in Theorem 6.4.7 (take $H = 0$ to see that).

Open Problem 6.4.8. *Extend if possible the result of Theorem 6.4.7 to quaternion matrices.*

6.5   INVARIANT SEMIDEFINITE SUBSPACES: NONSTANDARD INVOLUTION

In this section we fix a nonstandard involution ϕ. We study invariant semidefinite subspaces with respect to ϕ.

Let $H \in \mathsf{H}^{n \times n}$ is a ϕ-skewhermitian matrix, i.e., $H_\phi = -H$. Recall the definitions of (H, ϕ)-*nonnegative*, (H, ϕ)-*nonpositive*, (H, ϕ)-*positive*, and (H, ϕ)-*negative* matrices given in Definition 4.2.2. A subspace $\mathcal{M} \subseteq \mathsf{H}^{n \times}$ is said to be (H, ϕ)-*neutral* if $\beta(\phi)^{-1} y_\phi H x = 0$ for all $x, y \in \mathcal{M}$.

The next observation follows from Proposition 4.1.7.

Proposition 6.5.1. *Let $H_0 = H_0^* \in \mathsf{H}^{n \times n}$. Then a subspace $\mathcal{M} \subseteq \mathsf{H}^{n \times 1}$ is $(\beta(\phi)H_0, \phi)$-nonnegative, resp. $(\beta(\phi)H_0, \phi)$-nonpositive, $(\beta(\phi)H_0, \phi)$-positive, $(\beta(\phi)H_0, \phi)$-negative, or $(\beta(\phi)H_0, \phi)$-neutral, if and only if $\mathcal{M}$ is $(H_0, {}^*)$-nonnegative, resp. $(H_0, {}^*)$-nonpositive, $(H_0, {}^*)$-positive, $(H_0, {}^*)$-negative, or $(H_0, {}^*)$-neutral.*

Proposition 6.5.1 allows us to obtain results concerning semidefinite or neutral subspaces with respect to a nonstandard involution from the corresponding results with respect to the conjugation. As an example of using this approach, see Theorem 6.5.5 below.

Let $H \in \mathsf{H}^{n \times n}$ be ϕ-skewhermitian. It will be convenient to introduce the (H, ϕ)-inner product by the formula

$$[x, y]_{H, \phi} = \beta(\phi)^{-1} y_\phi H x, \quad \text{for all} \ \ x, y \in \mathsf{H}^{n \times 1}.$$

Note that $[x, x]_{H, \phi}$ is real-valued.

Definition 6.5.2. A matrix $A \in \mathsf{H}^{n \times n}$ is said to be (H, ϕ)-*plus-matrix* if the inequality $[Ax, Ax]_{H, \phi} \geq 0$ holds true for every $x \in \mathsf{H}^{n \times 1}$ such that $[x, x]_{H, \phi} \geq 0$.

By analogy with Proposition 6.2.2, we obtain that *A is an* (H, ϕ)-*plus-matrix if and only if there exists a nonnegative* μ *such that*

$$[Ax, Ax]_{H, \phi} \geq \mu[x, x]_{H, \phi}, \quad \text{for all} \quad x \in \mathsf{H}^{n \times 1}.$$

The proof of this statement can be obtained analogously to that of Proposition 6.2.2 (using Theorem 3.7.13 instead of Theorem 3.5.7).

Definition 6.5.3. A matrix $A \in \mathsf{H}^{n \times n}$ is called a *strict* (H, ϕ)-*plus-matrix* if μ can be chosen to be positive (note that in general μ is not unique).

We write $\mu = \mu(A)$ if we wish to emphasize the particular (H, ϕ)-plus-matrix A to which the number μ refers.

Proposition 6.5.4. *Let* $H = -H_\phi \in \mathsf{H}^{n \times n}$. *Then:*

(1) *if A is (H, ϕ)-plus, resp. strict (H, ϕ)-plus, and $S \in \mathsf{H}^{n \times n}$ is invertible, then $S^{-1}AS$ is $(S_\phi HS, \phi)$-plus, resp. strict $(S_\phi HS, \phi)$-plus; moreover, $\mu(A) = \mu(S^{-1}AS)$;*

(2) *a subspace $\mathcal{M} \subseteq \mathsf{H}^{n \times 1}$ is (H, ϕ)-nonnegative, resp. (H, ϕ)-negative, (H, ϕ)-nonpositive, (H, ϕ)-positive, or (H, ϕ)-neutral, if and only if the subspace $S^{-1}(\mathcal{M})$ is $(S_\phi HS, \phi)$-nonnegative, resp. $(S_\phi HS, \phi)$-negative, $(S_\phi HS, \phi)$-nonpositive, $(S_\phi HS, \phi)$-positive, or $(S_\phi HS, \phi)$-neutral;*

(3) *a subspace $\mathcal{M}$ is A-invariant if and only if $S^{-1}(\mathcal{M})$ is $S^{-1}AS$-invariant.*

Proof. We prove only (1), leaving proof of the rest for the reader.

Assume A is (strictly) (H, ϕ)-plus. Set $B = S^{-1}AS$. Then the following string of equalities and an inequality proves (1):

$$
\begin{aligned}
[Bx, Bx]_{S_\phi HS, \phi} &= \beta(\phi)^{-1} x_\phi B_\phi (S_\phi HS) Bx \\
&= \beta(\phi)^{-1} x_\phi (S_\phi A_\phi S_\phi^{-1})(S_\phi HS) S^{-1} ASx \\
&= \beta(\phi)^{-1} (Sx)_\phi A_\phi HA(Sx) \\
&\geq \mu(A)\, \beta(\phi)^{-1} (Sx)_\phi HSx \; = \; \mu(A)\, [x, x]_{S_\phi HS, \phi},
\end{aligned}
$$

for all $\in \mathsf{H}^{n \times 1}$. $\qquad\square$

An analogue of Theorem 6.2.6 on extension of invariant semidefinite subspaces holds for (H, ϕ)-plus-matrices.

Theorem 6.5.5. *Assume H is invertible, and let A be an invertible (H, ϕ)-plus-matrix. Denote by $(\mathrm{In}_+ (H), \mathrm{In}_- (H), 0)$ the $\beta(\phi)$-signature of the ϕ-skewhermitian matrix H. Then:*

(a) *for every A-invariant (H, ϕ)-nonnegative subspace $\mathcal{M}_0 \subseteq \mathsf{H}^{n \times 1}$ there exists an (H, ϕ)-nonnegative A-invariant subspace $\widetilde{\mathcal{M}} \supseteq \mathcal{M}_0$ such that*

$$\dim \widetilde{\mathcal{M}} = \mathrm{In}_+ (H); \tag{6.5.1}$$

(b) *for every A-invariant (H, ϕ)-nonpositive subspace $\mathcal{M}'_0 \subseteq \mathsf{H}^{n \times 1}$, there exists an (H, ϕ)-nonpositive A-invariant subspace $\widetilde{\mathcal{M}}' \supseteq \mathcal{M}'_0$ such that*

$$\dim \widetilde{\mathcal{M}}' = \mathrm{In}_-(H). \tag{6.5.2}$$

Proof. In view of Proposition 6.5.4 and Theorem 4.1.2(b,) we may (and do) assume that $H = \mathrm{diag}\,(\beta(\phi)I_p, \beta(\phi)(-I_q), 0_{n-p-q})$, for some integers $p, q, 0 \leq p, q \leq p + q \leq n$. It remains to apply Proposition 6.5.1 with $H_0 = \mathrm{diag}\,(I_p, -I_q, 0_{n-p-q})$ and refer to Theorem 6.2.6. $\qquad\square$

6.6   APPENDIX: CONVEX SETS

Definition 6.6.1. A set $X \subseteq \mathsf{R}^d$ is said to be *convex* it contains the line segment between any two points in the set: $x, y \in X$ implies $tx + (1 - t)y \in X$ for every $t \in [0, 1]$.

We present here two standard and important results concerning convex sets.

The first one has to do with separation between convex sets. We formulate the result in the form that is used in the text, without attempting to put it in a more general formulation.

Theorem 6.6.2. *Let $X \subseteq \mathsf{R}^2$ be a convex compact set. Then:*

(a) *if $Y \subseteq \mathsf{R}^2$ is a convex set, not necessarily compact, such that $Y \cap X = \emptyset$, then there exists a line separating X and Y: there exist real numbers a, b, c with a, b not both zero such that*

$$
\begin{aligned}
X &\subseteq \{(x_1, x_2) \in \mathsf{R}^2 : ax_1 + bx_2 \geq c\}, \\
Y &\subseteq \{(x_1, x_2) \in \mathsf{R}^2 : ax_1 + bx_2 \leq c\};
\end{aligned}
$$

(b) *if, in addition to the hypotheses in part* (a), *Y is compact, then there exists a line strictly separating X and Y: there exist real numbers a, b, c_1, c_2 with a, b not both zero and $c_1 < c_2$ such that*

$$
\begin{aligned}
X &\subseteq \{(x_1, x_2) \in \mathsf{R}^2 : ax_1 + bx_2 \geq c_2\}, \\
Y &\subseteq \{(x_1, x_2) \in \mathsf{R}^2 : ax_1 + bx_2 \leq c_1\}.
\end{aligned}
$$

A proof can be found in many books on convex analysis; see, e.g., Rockafellar [128, Theorems 11.3, 11.4] and Lay [96, Section 4].

The second result is the Brouwer fixed point theorem. Again, it is presented in the form to be used in the book (not in full generality).

Theorem 6.6.3. *Let $X \subseteq \mathsf{R}^d$ be a nonempty compact convex set, and let $\mathcal{F} : X \longrightarrow X$ be a continuous map. Then $\mathcal{F}$ has a fixed point, i.e., there exists $x \in X$ such that $\mathcal{F}(x) = x$.*

See, e.g., Border [17, Corollary 6.6] or Webster [158, Section 7.5], for a proof.

6.7 EXERCISES

Ex. 6.7.1. If A and B are two commuting (H, ϕ)-Hamiltonian matrices, then AB is (H, ϕ)-skew-Hamiltonian.

Ex. 6.7.2. Let ϕ be the conjugation or a nonstandard involution, and let $H \in \mathsf{H}^{n \times n}$ be ϕ-skewhermitian. Show that if $f(t)$ is a polynomial of $t \in \mathsf{R}$ with real coefficients such that $f(-t) = -f(t)$ and if A is (H, ϕ)-Hamiltonian, then so is $f(A)$.

Ex. 6.7.3. With ϕ and H as in Ex 6.7.2, show that if $f(t)$ is a polynomial of $t \in \mathsf{R}$ with real coefficients such that $f(-t) = f(t)$, and if A is (H, ϕ)-Hamiltonian, then $f(A)$ is (H, ϕ)-skew-Hamiltonian.

Ex. 6.7.4. Let $A \in \mathsf{H}^{n \times n}$ be an H-plus-matrix, where $H = H^*$ is not negative definite. Show that the set of all nonnegative μ such that the equality $[Ax, Ax] \geq [x, x]$ holds for all $x \in \mathsf{H}^{n \times 1}$ forms a closed bounded interval in the half-line $[0, \infty)$. Hint: μ belongs to the set precisely when the hermitian matrix $A^* H A - \mu H$ is positive semidefinite. Now consider μ as a parameter and use continuity of the smallest eigenvalue of $A^* H A - \mu H$ as a function of μ.

Ex. 6.7.5. Show that the classes of (quaternion) H-plus-matrices, strictly H-plus-matrices, and H-expansive matrices are closed under multiplication (for a fixed hermitian $H \in \mathsf{H}^{n \times n}$). In other words, if $A, B \in \mathsf{H}^{n \times n}$ belong to one of those classes, so does AB.

Ex. 6.7.6. Show that if X is (H, ϕ)-skewsymmetric, (H, ϕ)-Hamiltonian, $(H, {}^*)$-skewhermitian, $(H, {}^*)$-Hamiltonian, $(H, {}^T)$-skewsymmetric, or $(H, {}^T)$-Hamiltonian, then e^X is (H, ϕ)-orthogonal, (H, ϕ)-symplectic, $(H, {}^*)$-unitary, $(H, {}^*)$-symplectic, $(H, {}^T)$-orthogonal, or $(H, {}^T)$-symplectic, respectively.

Ex. 6.7.7. Assume $n \geq 2$, and let $\mathcal{U}$ be one of the following four sets of matrices, for a fixed $H \in \mathsf{H}^{n \times n}$:

(a) the $(H, {}^*)$-unitary matrices, where $H = H^*$;

(b) the $(H, {}^*)$-symplectic matrices, where $H = -H^*$;

(c) the (H, ϕ)-orthogonal matrices, where ϕ is a nonstandard involution and $H = H_\phi$;

(d) the (H, ϕ)-symplectic matrices, where ϕ is a nonstandard involution and $H = -H_\phi$.

Prove that $\mathcal{U}$ is unbounded, except in cases when $H = H^*$ is (positive or negative) definite or $H = -H_\phi$ is $\beta(\phi)$-(positive or negative) definite. Show that in the exceptional cases the set $\mathcal{U}$ is bounded.

Ex. 6.7.8. Provide details for the proof of the first part of Proposition 6.1.3.

Ex. 6.7.9. Find all 2×2 quaternion H-plus matrices for $H = \begin{bmatrix} 0 & 1 \\ 1 & 0 \end{bmatrix}$.

Ex. 6.7.10. Find all A_j-invariant $(H_j,{}^*)$-neutral subspaces for each of the following pairs of matrices (A_j, H_j):

(1) $\quad H_1 = \begin{bmatrix} 0 & 0 & 1 \\ 0 & 1 & 0 \\ 1 & 0 & 0 \end{bmatrix}, \quad A_1 = \begin{bmatrix} 0 & 0 & 0 \\ 0 & 0 & 0 \\ 1 & 0 & 0 \end{bmatrix},$

(2) $\quad H_2 = A_2 = \begin{bmatrix} 0 & 0 & 1 \\ 0 & 1 & 0 \\ 1 & 0 & 0 \end{bmatrix},$

(3) $\quad H_3 = \begin{bmatrix} 0 & 0 & -\mathrm{i} \\ 0 & 1 & 0 \\ \mathrm{i} & 0 & 0 \end{bmatrix}, \quad A_3 = \begin{bmatrix} 0 & 0 & \mathrm{j} \\ 0 & 0 & 0 \\ 0 & 0 & 0 \end{bmatrix},$

(4) $\quad H_4 = H_3, \quad A_4 = \begin{bmatrix} 0 & 0 & \mathrm{j} \\ 0 & \mathrm{j} & 0 \\ \mathrm{j} & 0 & 0 \end{bmatrix},$

(5) $\quad H_5 = \begin{bmatrix} 0 & 0 & \mathrm{i} \\ 0 & \mathrm{i} & 0 \\ \mathrm{i} & 0 & 0 \end{bmatrix}, \quad A_5 = \begin{bmatrix} 0 & 0 & 0 \\ 0 & \mathrm{j} & 0 \\ 0 & 0 & 0 \end{bmatrix},$

(6) $\quad H_6 = H_5, \quad A_6 = \begin{bmatrix} \mathrm{i} & 0 & 0 \\ 0 & \mathrm{i} & 0 \\ 0 & 0 & \mathrm{i} \end{bmatrix}.$

Ex. 6.7.11. Find all A_j-invariant (H_j, ϕ)-neutral subspaces for each of the following pairs of matrices (A_j, H_j), where ϕ is a nonstandard involution, $\beta = \beta(\phi)$, and the nonzero quaternion x is such that $\beta x = -x\beta$:

(1) $\quad H_1 = \begin{bmatrix} 0 & 0 & \beta \\ 0 & \beta & 0 \\ \beta & 0 & 0 \end{bmatrix}, \quad A_1 = \begin{bmatrix} 0 & 0 & 0 \\ 0 & 0 & 0 \\ x & 0 & 0 \end{bmatrix},$

(2) $\quad H_2 = H_1, \quad A_2 = \begin{bmatrix} 0 & 0 & x \\ 0 & x & 0 \\ x & 0 & 0 \end{bmatrix};$

(3) $\quad H_3 = \begin{bmatrix} 0 & 1 & 0 \\ -1 & 0 & 0 \\ 0 & 0 & \beta \end{bmatrix}, \quad A_3 = \begin{bmatrix} 0 & 0 & 0 \\ \beta & 0 & 0 \\ 0 & 0 & 0 \end{bmatrix},$

(4) $\quad H_4 = H_3, \quad A_4 = \begin{bmatrix} 0 & 0 & -\beta \\ \beta & 0 & 0 \\ 0 & 1 & 0 \end{bmatrix};$

(5) $\quad H_5 = \begin{bmatrix} 0 & 0 & x \\ 0 & x & 0 \\ x & 0 & 0 \end{bmatrix}, \quad A_5 = \begin{bmatrix} 0 & 0 & 1 \\ 0 & 1 & 0 \\ 1 & 0 & 0 \end{bmatrix},$

(6) $\quad H_6 = H_5, \quad A_6 = \begin{bmatrix} \beta & x^{-1}\beta & 0 \\ 0 & 0 & x^{-1}\beta \\ 0 & 0 & -\beta \end{bmatrix}.$

Ex. 6.7.12. Let $H \in \mathsf{H}^{n\times n}$ be hermitian. A matrix $A \in \mathsf{H}^{n\times n}$ is said to be H-*contractive*, resp. H-*strictly contractive*, if $[Ax, Ax] \leq [x, x]$ for all $x \in \mathsf{H}^{n\times n}$, resp. $[Ax, Ax] < [x, x]$ for all $x \in \mathsf{H}^{n \times n} \setminus \{0\}$; here $[x, y] = y^* H x$, $x, y \in \mathsf{H}^{n \times n}$ is the indefinite inner product induced by H. Matrices that are H-*strictly expansive* are defined similarly, replacing $<$ with $>$.

Show that if A is H-expansive, resp. H-strictly expansive, and invertible, then A^{-1} is H-contractive, resp. H-strictly contractive; conversely, if A is H-contractive, resp. H-strictly contractive, and invertible, then A^{-1} is H-expansive, resp. H-strictly expansive.

Ex. 6.7.13. Consider the following four sets of pairs of matrices:

$$v_n(\delta, \xi) := \{(H, A) \in \mathsf{H}^{n \times n} : H = \delta H^*, \ HA = \xi A^* H\},$$

where $\delta = \pm 1$, $\xi = \pm 1$.

(a) Show that the sets $v_n(\delta, \xi)$ are closed.

(b) Prove that the order of neutrality $\gamma(H, A)$ is upper semicontinuous on each of the four sets $v_n(\delta, \xi)$: if $\{(H_m, A_m)\}_{m=1}^{\infty}$ is a sequence of pair of matrices such that $(H_m, A_m) \in v_n(\delta, \xi)$ and $\gamma(H_m, A_m) \geq k$ for $m = 1, 2, \ldots$ and if there exist the limits

$$H_0 = \lim_{m \to \infty} H_m, \quad A_0 = \lim_{m \to \infty} H_m,$$

then also $\gamma(H, A) \geq k$.

Ex. 6.7.14. Repeat Ex. 6.7.13 for the four sets

$$\{(H, A) \in \mathsf{H}^{n \times n} : H = \delta H_\phi, \ HA = \xi A_\phi H\},$$

where $\delta = \pm 1$, $\xi = \pm 1$, and ϕ is a nonstandard involution.

Ex. 6.7.15. (a) Let A and H be as in one of the tables at the beginning of this chapter (except expansive and plus-matrix). Prove that $\operatorname{Ker} H$ is A-invariant. Hint: Use the reduction $H \mapsto S_\phi HS$ (or $H \mapsto S_\phi HS$), $A \mapsto S^{-1}AS$ to put H in one of the canonical forms of Theorems 4.1.6 and 4.1.2.

(b) Deduce that if A and H are as in Theorem 6.1.6, then the order of neutrality of (H, A) is equal to the dimension of $\operatorname{Ker} H$ plus the order of neutrality of (H_1, A_1), where H_1 is the invertible part of H, and A_1 is the corresponding part of A.

6.8 NOTES

Real $(H,^T)$-Hamiltonian matrices and complex $(H,^*)$-skewhermitian and $(H,^T)$-skew-Hamiltonian matrices have been extensively studied in the literature (e.g., Faßbender et al. [41], Faßbender and Ikramov [39], Mehl et al. [108], and Rodman [131]). In particular, it was proved in Faßbender et al. [41] that every real $(H,^T)$-skew-Hamiltonian matrix has a real $(H,^T)$-Hamiltonian square root. For quaternion matrices, the problem of existence of such square roots in various classes was treated in Rodman [139]. In contrast with the real case, Hamiltonian square roots of skew-Hamiltonian quaternion matrices do not always exist.

Quadratic forms on finite dimensional quaternion vector spaces have been studied in depth in Brieskorn [20].

The proof of Theorem 6.1.6 follows closely the arguments given in Lancaster et al. [87] in the context of complex matrices.

For complex matrices with ϕ the conjugation, the notion of order of neutrality was studied in Lancaster and Rodman [89] and (suitably modified for infinite dimensional spaces) in Lancaster et al. [85] and Zizler [165] in connection with definitizable operators on Krein spaces.

Theorem 6.2.6 and its proof is adapted from Gohberg et al. [53].

The classes of plus-matrices and plus-operators (the analogue of plus-matrices acting in infinite dimensional Krein spaces) and expansive and dissipative operators have been studied in depth in the context of indefinite inner products in complex vector spaces; see Iohvidov et al. [70] and Azizov and Iohvidov [13]. Note that in the literature the generally adopted definition of dissipative operators is different from the one given in this chapter; namely, an operator is said to be dissipative if its imaginary part is positive semidefinite. In the context of operators on complex vector spaces, the two definitions are essentially equivalent, but in the context of quaternion vector spaces we have to adopt the definition given in the text.

In the complex case, Theorem 6.4.7 is proved in Iohvidov et al. [70] in the more general setting of operators on infinite dimensional spaces; see Theorem 11.2 there.

Invariant nonnegative (nonpositive, neutral) subspaces play an important role in applications, and for real and complex matrices have been extensively studied in the literature; see, e.g., Gohberg et al. [52], [53], Lancaster and Rodman [89], Rodman [130], Mehl et al. [110], Freiling et al. [46], and Ran and Rodman [123]. In particular, in the context of real and complex matrices, they appear in linear control systems (see, e.g., Lancaster and Rodman [91], and in studies of symmetric factorizations of matrix valued functions with symmetries: Gohberg et al. [55], Ran and Rodman [124], [126], and Ran [122].

Ex. 6.7.9 for complex matrices is given in Gohberg et al. [53, Example 10.2.3].

Chapter Seven

Smith form and Kronecker canonical form

In this chapter we study polynomials with quaternion matrix coefficients. The exposition is focused on two major results. One is the Smith form, which asserts that every quaternion matrix polynomial can be brought to a diagonal form under pre- and postmultiplication by unimodular matrix polynomials, with the appropriate divisibility relations among the diagonal entries. The other is the Kronecker canonical form for quaternion matrix polynomials of first degree under pre- and postmultiplication by invertible constant matrices. The Kronecker form generalizes the Jordan canonical for matrices. Complete and detailed proofs are given for both the Smith form and the Kronecker form.

7.1 MATRIX POLYNOMIALS WITH QUATERNION COEFFICIENTS

Let $\mathsf{H}(t)$ be the noncommutative ring of polynomials with quaternion coefficients, with the real independent variable t. Note that t commutes with the quaternions. Therefore, for every fixed $t_0 \in \mathsf{R}$, the evaluation map

$$f \mapsto f(t_0), \quad f(t) \in \mathsf{H}(t)$$

is well defined as a unital homomorphism of real algebras $\mathsf{H}(t) \longrightarrow \mathsf{H}$.

Let $p(t), q(t) \in \mathsf{H}(t)$.

Definition 7.1.1. A polynomial $q(t)$ is called a *divisor* of $p(t)$ if $p(t) = q(t)s(t)$ and $p(t) = r(t)q(t)$ for some $s(t), r(t) \in \mathsf{H}(t)$. A polynomial q is said to be a *total divisor* of p if $\alpha q(t)\alpha^{-1}$ is a divisor of $p(t)$ for every $\alpha \in \mathsf{H} \setminus \{0\}$ or, equivalently, if $q(t)$ is a divisor of $\beta p(t)\beta^{-1}$ for all $\beta \in \mathsf{H} \setminus \{0\}$.

This definition is equivalent to the following: $q(t)$ is a total divisor of $p(t)$ if and only if the containment

$$q(t)\mathsf{H}(t) \cap \mathsf{H}(t)q(t) \supseteq \mathsf{H}(t)p(t)\mathsf{H}(t) \tag{7.1.1}$$

holds true (Ex. 7.7.2).

Let $\mathsf{H}(t)^{m \times n}$ be the set of all $m \times n$ matrices with entries in $\mathsf{H}(t)^{m \times n}$, which will be called *matrix polynomials* with the standard operations of addition, right and left multiplication by quaternions, and matrix multiplication: If $A(t) \in \mathsf{H}(t)^{m \times n}$ and $B(t) \in \mathsf{H}(t)^{n \times p}$, then $A(t)B(t) \in \mathsf{H}(t)^{m \times p}$.

In this section we derive the Smith form (Theorem 7.1.4 below) for matrix polynomials with quaternion entries. Our exposition follows the general outline for the standard proof of the Smith form for matrix polynomials over fields.

Definition 7.1.2. A matrix polynomial $A(t) \in \mathsf{H}(t)^{n \times n}$ is said to be *elementary* if it can be represented as a product (in any order) of diagonal $n \times n$ polynomials with constant nonzero quaternions on the diagonal and of $n \times n$ polynomials with 1s on the diagonal and a sole nonzero off diagonal entry.

Constant permutation matrices are elementary, as follows from the equality (given in Jacobson [71])

$$\begin{bmatrix} 0 & 1 \\ 1 & 0 \end{bmatrix} = \begin{bmatrix} 1 & 1 \\ 0 & 1 \end{bmatrix} \begin{bmatrix} 1 & 0 \\ -1 & 1 \end{bmatrix} \begin{bmatrix} 1 & 1 \\ 0 & 1 \end{bmatrix} \begin{bmatrix} -1 & 0 \\ 0 & 1 \end{bmatrix}.$$

Definition 7.1.3. A matrix polynomial $A(t) \in \mathsf{H}(t)^{m \times n}$ is said to be *unimodular* if

$$A(t)B(t) = B(t)A(t) \equiv I \qquad (7.1.2)$$

for some matrix polynomial $B(t) \in \mathsf{H}(t)^{n \times m}$. (In this case, clearly $m = n$.)

Obviously, every elementary matrix polynomial is unimodular. We shall see later that the converse also holds.

Theorem 7.1.4. *Let $A(t) \in \mathsf{H}(t)^{m \times n}$. Then there exist elementary matrix polynomials $D(t) \in \mathsf{H}(t)^{m \times m}$, $E(t) \in \mathsf{H}(t)^{n \times n}$, and monic (i.e., with leading coefficient equal to 1) scalar polynomials $a_1(t), a_2(t), \ldots, a_r(t) \in \mathsf{H}(t)$, $0 \leq r \leq \min\{m, n\}$, such that*

$$D(t)A(t)E(t) = \mathrm{diag}\,(a_1(t), a_2(t), \ldots, a_r(t), 0, \ldots, 0), \qquad (7.1.3)$$

where $a_j(t)$ is a total divisor of $a_{j+1}(t)$, for $j = 1, 2, \ldots, r - 1$.

Definition 7.1.5. The right-hand side of (7.1.3) will be called the *Smith form* of $A(t)$.

Proof. Step 1. We prove the existence of the diagonal form (7.1.3) with monic polynomials $a_1(t), \ldots a_r(t)$, but not necessarily with the property that $a_j(t)$ is a total divisor of $a_{j+1}(t)$, for $j = 1, 2, \ldots, r - 1$.

If $A(t) \equiv 0$ we are done. Otherwise, let α be the minimal degree of nonzero entries of all matrix polynomials of the form $D_1(t)A(t)E_1(t)$, where $D_1(t)$ and $E_1(t)$ are elementary matrix polynomials. Thus,

$$D_2(t)A(t)E_2(t) = B(t)$$

for some elementary matrix polynomials $D_2(t)$ and $E_2(t)$, where

$$B(t) = [b_{p,q}(t)]_{p=1, q=1}^{m,n}$$

is such that the degree of $b_{1,1}(t)$ is equal to α. Consider $b_{1,q}(t)$ with $q = 2, \ldots, n$. Division with remainder

$$b_{1,q}(t) = b_{1,1}(t)c_{1,q}(t) + d_{1,q}(t), \quad c_{1,q}(t), d_{1,q}(t) \in \mathsf{H}(t),$$

where the degree of $d_{1,q}(t)$ is smaller than α, shows that we must have $d_{1,q}(t) \equiv 0$, for otherwise the product $B(t)F(t)$, where the elementary matrix polynomial $F(t)$ has 1 on the main diagonal, $-c_{1,q}(t)$ in the $(1, q)$th position, and zero everywhere else, will have a nonzero entry of degree less than α, a contradiction with the definition of α. Thus,

$$b_{1,q}(t) = b_{1,1}(t)c_{1,q}(t), \quad q = 2, 3, \ldots, n,$$

for some $c_{1,q}(t) \in \mathsf{H}(t)$. Now it is easy to see that there exists an elementary matrix polynomial $E_3(t)$ of the form $E_3(t) = I + \widehat{E}(t)$, where $\widehat{E}(t)$ may have nonzero entries only in the positions $(1,2), \ldots, (1,n)$, such that $B(t)E_3(t)$ has zero entries in the positions $(1,2), \ldots, (1,n)$. Similarly, we prove that there exists an elementary matrix polynomial $D_3(t)$ such that $D_3(t)B(t)E_3(t)$ has zero entries in all positions $(1,q)$, $q \neq 1$ and $(p,1)$, $p \neq 1$. Now induction on $m+n$ completes the proof of existence of (7.1.3). The base of induction, namely, the cases when $m = 1$ or $n = 1$, can be dealt with analogously, and the case $m = n = 1$ is trivial.

Step 2. Now, clearly, r is an invariant of the diagonal form (7.1.3) with monic polynomials $a_j(t)$. Indeed, r coincides with the maximal rank of the matrix $A(t)$, over all real t. In verifying this fact, use the property that each $a_j(t)$, being a monic polynomial of real variable with quaternion coefficients, may have only a finite number of real roots (Theorem 5.12.4).

Step 3. Consider a diagonal form (7.1.3) with monic $a_1(t), \ldots, a_r(t)$ that have the following additional property: The degrees $\delta_1, \ldots, \delta_r$ of $a_1(t), \ldots, a_r(t)$, respectively, are such that for any other diagonal form

$$\widetilde{D}(t)A(t)\widetilde{E}(t) = \operatorname{diag}(b_1(t), b_2(t), \ldots, b_r(t), 0, \ldots, 0),$$

where $\widetilde{D}(t)$ and $\widetilde{E}(t)$ are elementary matrix polynomials, and $b_1(t), b_2(t), \ldots, b_r(t)$ are monic scalar polynomials, the degrees $\delta'_1, \ldots, \delta'_r$ of $b_1(t), \ldots, b_r(t)$, respectively, satisfy the inequalities:

$$\delta_1 \leq \delta'_1; \quad \text{if } \delta_1 = \delta'_1, \text{ then } \delta_2 \leq \delta'_2;$$
$$\text{if } \delta_j = \delta'_j \text{ for } j = 1, 2, \text{ then } \delta_3 \leq \delta'_3; \quad \ldots \quad ;$$
$$\text{if } \delta_j = \delta'_j \text{ for } j = 1, 2, \ldots, r-1, \text{ then } \delta_r \leq \delta'_r.$$

Obviously, a diagonal form $\operatorname{diag}(a_1(t), a_2(t), \ldots, a_r(t), 0, \ldots, 0)$ of $A(t)$ with these properties does exist. We claim that then $a_j(t)$ is a total divisor of $a_{j+1}(t)$, for $j = 1, 2, \ldots, r-1$. Suppose not, and assume that $a_j(t)$ is not a total divisor of $a_{j+1}(t)$, for some j. Then there exists $\alpha \in \mathsf{H} \setminus \{0\}$ such that

$$\alpha a_{j+1}(t)\alpha^{-1} = d(t)a_j(t) + s(t),$$

where $d(t), s(t) \in \mathsf{H}(t)$, $s(t) \neq 0$, is such that the degree of s is smaller than δ_j, the degree of a_j. (If it happens that

$$\alpha a_{j+1}(t)\alpha^{-1} = a_j(t)d(t) + s(t),$$

the subsequent argument is completely analogous.) We have

$$\begin{bmatrix} 1 & 0 \\ d & \alpha \end{bmatrix} \begin{bmatrix} a_j & 0 \\ 0 & a_{j+1} \end{bmatrix} \begin{bmatrix} 1 & 0 \\ -\alpha^{-1} & \alpha^{-1} \end{bmatrix} = \begin{bmatrix} a_j & 0 \\ -s & \alpha a_{j+1}\alpha^{-1} \end{bmatrix}.$$

Since the degree of s is smaller than δ_j, Step 1 of the proof shows that for some elementary 2×2 matrix polynomials $D'(t)$ and $E'(t)$, we have

$$D'(t) \begin{bmatrix} a_j(t) & 0 \\ 0 & a_{j+1}(t) \end{bmatrix} E'(t) = \begin{bmatrix} a'_j(t) & 0 \\ 0 & a'_{j+1}(t) \end{bmatrix},$$

where $a'_j(t)$ and $a'_{j+1}(t)$ are monic polynomials and the degree of $a'_j(t)$ is smaller than δ_j. Using

$$\begin{bmatrix} a'_j(t) & 0 \\ 0 & a'_{j+1}(t) \end{bmatrix} \qquad \text{in place of} \qquad \begin{bmatrix} a_j(t) & 0 \\ 0 & a_{j+1}(t) \end{bmatrix}$$

in

$$\operatorname{diag}\left(a_1(t), a_2(t), \ldots, a_r(t), 0, \ldots, 0\right),$$

we obtain a contradiction with the choice of $a_1(t), \ldots, a_r(t)$. $\qquad\square$

The well-known Smith form for matrix polynomials with real or complex coefficients (or, more generally, with coefficients in a field) is defined analogously to (7.1.3); see, e.g., Gantmacher [50] or Gohberg et al. [55, Chapter S1]. Thus, for $\mathsf{F} = \mathsf{R}$ or $\mathsf{F} = \mathsf{C}$, consider $A(t) \in \mathsf{F}^{m \times n}$, a matrix with entries in $\mathsf{F}(t)$, the commutative ring of polynomials with coefficients in F. Then there exist elementary (or, what is the same, unimodular) matrix polynomials $D(t) \in \mathsf{F}(t)^{m \times m}$, $E(t) \in \mathsf{H}(t)^{n \times n}$, and monic scalar polynomials $a_1(t), a_2(t), \ldots, a_r(t) \in \mathsf{F}(t)$, $0 \le r \le \min\{m, n\}$, such that

$$D(t)A(t)E(t) = \operatorname{diag}\left(a_1(t), a_2(t), \ldots, a_r(t), 0, \ldots, 0\right), \qquad (7.1.4)$$

where $a_j(t)$ is a divisor of $a_{j+1}(t)$, for $j = 1, 2, \ldots, r-1$. (In the context of real and complex fields, the notion of total divisibility coincides with the standard notion of divisibility of polynomials.)

Definition 7.1.6. The right-hand side of (7.1.4) will be called the F-*Smith form* of $A(t)$.

The following example (taken from Pereira [116, Example 3.4.4]; see also Pereira et al. [117]) is also instructive.

Example 7.1.7. Consider

$$A(t) = \begin{bmatrix} t + \mathsf{i} & 0 \\ 0 & t + \mathsf{i} \end{bmatrix}. \qquad (7.1.5)$$

This is not a Smith form, because $t + \mathsf{i}$ is not a total divisor of $t + \mathsf{i}$. On the other hand, calculations show that

$$\begin{bmatrix} \dfrac{\mathsf{k}}{2} & \dfrac{\mathsf{i}}{2} \\ t\mathsf{j} + \mathsf{k} & t - \mathsf{i} \end{bmatrix} \cdot A(t) \cdot \begin{bmatrix} -\mathsf{j} & -\dfrac{\mathsf{k}}{2}(t + \mathsf{i}) \\ -1 & \dfrac{\mathsf{i}}{2}(t - \mathsf{i}) \end{bmatrix} = \begin{bmatrix} 1 & 0 \\ 0 & t^2 + 1 \end{bmatrix}, \qquad (7.1.6)$$

and

$$\begin{bmatrix} \dfrac{\mathsf{k}}{2} & \dfrac{\mathsf{i}}{2} \\ t\mathsf{j} + \mathsf{k} & t - \mathsf{i} \end{bmatrix} \cdot \begin{bmatrix} -\mathsf{j} & -\dfrac{\mathsf{k}}{2}(t + \mathsf{i}) \\ -1 & \dfrac{\mathsf{i}}{2}(t - \mathsf{i}) \end{bmatrix} = \begin{bmatrix} 0 & \dfrac{1}{2}\mathsf{i} \\ 2\mathsf{i} & t \end{bmatrix}$$

$$= \begin{bmatrix} t & -\dfrac{1}{2}\mathsf{i} \\ -2\mathsf{i} & 0 \end{bmatrix}^{-1}. \qquad (7.1.7)$$

It follows from (7.1.7) and Ex. 7.7.7 that the matrix polynomials

$$
\begin{bmatrix} \dfrac{k}{2} & \dfrac{i}{2} \\[2mm] tj + k & t - i \end{bmatrix}
\quad \text{and} \quad
\begin{bmatrix} -j & -\dfrac{k}{2}(t + i) \\[2mm] -1 & -\dfrac{i}{2}(t - i) \end{bmatrix}
$$

are unimodular and that $\operatorname{diag}(1, t^2 + 1)$ is the Smith form for $A(t)$. Note that (7.1.5) is the C-Smith form of $A(t)$, but $\operatorname{diag}(1, t^2 + 1)$ is not. $\qquad \square$

In contrast, the R-Smith form of a *real* matrix polynomial is automatically its (quaternion) Smith form.

We tackle the problem of uniqueness (or, more precisely, nonuniqueness) of the Smith form in the next section. Here, we continue with corollaries from the Smith form.

Corollary 7.1.8. *A matrix polynomial $A(t)$ is elementary if and only if $A(t)$ is unimodular.*

Proof. The "only if" part was already observed. Assume that $A(t)$ is such that (7.1.2) holds for some matrix polynomial $B(t)$. Clearly, $A(t) \in \mathsf{H}(t)^{n \times n}$ for some n. Without loss of generality, we may assume that $A(t)$ is in the form (7.1.3). Now equation (7.1.2) implies that $r = n$, and for every $j = 1, 2, \dots, n$ we have $a_j(t) b_j(t) \equiv 1$ for some $b_j(t) \in \mathsf{H}(t)$. Hence, $a_j(t)$ is a nonzero constant and we are done. $\qquad \square$

Definition 7.1.9. For $\mathsf{F} \in \{\mathsf{R}, \mathsf{C}, \mathsf{H}\}$, we say that matrix polynomials $A(t), B(t) \in \mathsf{F}(t)^{m \times n}$ are F-*equivalent* or simply *equivalent* if $\mathsf{F} = \mathsf{H}$, if $A(t) = D(t) B(t) E(t)$ for some elementary (or unimodular) matrix polynomials $D(t) \in \mathsf{F}^{m \times m}$, $E(t) \in \mathsf{F}^{n \times n}$.

Clearly, this is an equivalence relation. The Smith form is a simple form for a matrix polynomial under F-equivalence.

For matrix polynomials of degree one, the notion of equivalence simplifies to strict equivalence (to be studied in Section 7.3).

Theorem 7.1.10. *Let $A_1 t + A_0, B_1 t + B_0 \in \mathsf{H}(t)^{n \times n}$ and assume that A_1 and B_1 are invertible. Then $A_1 t + A_0$ and $B_1 t + B_0$ are equivalent if and only if*

$$
P(A_1 t + A_0)Q = B_1 t + B_0
$$

for some constant invertible matrices $P, Q \in \mathsf{H}^{n \times n}$.

The proof is almost verbatim the same as for matrix polynomials with coefficients in a field (a well-known result—see, for instance, Gantmacher [50, Chapter VI] or Gohberg et al. [54, Appendix]) and will not be reproduced here.

Definition 7.1.11. The *rank* $r(A(t))$ of a matrix polynomial $A(t) \in \mathsf{H}(t)^{m \times n}$ is defined as the maximal rank of quaternion matrices $A(t_0)$, over all $t_0 \in \mathsf{R}$.

It is easy to see that the rank of $A(t)$ coincides with integer r of Theorem 7.1.4. Also, if the rank of $A(t_0)$ is smaller than a certain integer q for infinitely many real values of t_0, then $r(A(t)) < q$. This follows from the fact that a nonidentically zero quaternion polynomial $s(t) \in \mathsf{H}(t)$ can have only finitely many real zeros, by Theorem 5.12.4.

Corollary 7.1.12. *Let* $A(t) \in \mathsf{H}(t)^{m \times n}$.

(a) *Assume that* $r(A(t)) < n$. *(This condition is automatically satisfied if* $m < n$.) *Then there exists* $Y(t) \in \mathsf{H}(t)^{n \times (n - r(A(t)))}$ *such that*

$$A(t)Y(t) \equiv 0 \qquad (7.1.8)$$

and $Y(t)$ *is left-invertible:* $\widetilde{Y}(t)Y(t) \equiv I$ *for some* $\widetilde{Y}(t) \in \mathsf{H}(t)^{(n - r(A(t))) \times n}$; *in particular,* $r(Y(t)) = n - r(A(t))$.

(b) *Assume that* $r(A(t)) < m$. *(This condition is automatically satisfied if* $m > n$.) *Then there exists* $Y(t) \in \mathsf{H}(t)^{(m - r(A(t))) \times m}$ *such that*

$$Y(t)A(t) \equiv 0$$

and $Y(t)$ *is right-invertible; in particular,* $r(Y(t)) = m - r(A(t))$.

Proof. Part (a). Without loss of generality, we may assume that $A(t)$ is in the form of the right-hand side of (7.1.3). (Indeed, if $D(t)A(t)E(t) = B(t)$ for some elementary matrix polynomials $D(t)$ and $E(t)$, and if (7.1.8) holds true, then $B(t) \cdot (E(t)^{-1}Y(t)) \equiv 0$.) Now the hypothesis of the corollary implies that the $n - r(A(t))$ right-most columns of $A(t)$ are zero, and we may take $Y(t) = \begin{bmatrix} 0 \\ I \end{bmatrix}$. The proof of (b) is analogous. $\qquad \square$

7.2   NONUNIQUENESS OF THE SMITH FORM

Definition 7.2.1. We say that two scalar polynomials $a(t), b(t) \in \mathsf{H}(t)^{1 \times 1}$ are H-*similar* if there exists $\alpha \in \mathsf{H} \setminus \{0\}$ such that $\alpha^{-1}a(t)\alpha = b(t)$ for all real t.

Clearly, H-similarity is an equivalence relation, and H-similar polynomials must have the same degree.

Also, if

$$\mathrm{diag}\,(a_1(t), a_2(t), \ldots, a_r(t), 0, \ldots, 0) \qquad (7.2.1)$$

is a Smith form of $A(t) \in \mathsf{H}^{m \times n}$, then so is

$$\mathrm{diag}\,(b_1(t), b_2(t), \ldots, b_r(t), 0, \ldots, 0), \qquad (7.2.2)$$

for every choice of scalar polynomials $b_1(t), \ldots, b_r(t)$ such that $b_j(t)$ is H-similar to $a_j(t)$, for $j = 1, 2, \ldots, r$. As it turns out, the converse is generally not true; in other words, all Smith forms of a given matrix polynomial need not be related by H-similarity, as (7.2.1) and (7.2.2) are. The following example (taken from Pereira [116]) illustrates this phenomenon.

In contrast, it is well known that the Smith forms of matrix polynomials over a *field* are unique; see, e.g., Gantmacher [50].

Example 7.2.2. Let

$$A(t) = \begin{bmatrix} t - \mathrm{i} & 0 \\ 0 & t - 2\mathrm{j} \end{bmatrix} \in \mathsf{H}(t)^{2 \times 2}.$$

Then

$$\Gamma(t) := \begin{bmatrix} 1 & 0 \\ 0 & \left(t + \dfrac{3\mathrm{i} + 4\mathrm{j}}{5} \right)(t - 2\mathrm{j}) \end{bmatrix}$$

is a Smith form for $A(t)$. Indeed, consider

$$U(t) \ := \ \frac{1}{5} \begin{bmatrix} i - 2j & 2j - i \\ -5t - 8i + 6j & 5t + 3i + 4j \end{bmatrix} \in \mathsf{H}(t)^{2\times 2},$$

$$V(t) \ := \ \frac{1}{5} \begin{bmatrix} 5 & (i-2j)t - 4 - 2k \\ 5 & (i-2j)t + 1 - 2k \end{bmatrix} \in \mathsf{H}(t)^{2\times 2},$$

and

$$W(t) := \begin{bmatrix} (i-2j)t + 1 - 2k & (-i+2j)t + 4 - 2k \\ -5 & 5 \end{bmatrix} \in \mathsf{H}(t)^{2\times 2}.$$

Then

$$U(t)^{-1} = \begin{bmatrix} t - i & \frac{1}{5}i - \frac{2}{5}j \\ t - 2j & \frac{1}{5}i - \frac{2}{5}j \end{bmatrix} \quad \text{and} \quad V(t)W(t) = \begin{bmatrix} 5 & 0 \\ 0 & 5 - 4k \end{bmatrix},$$

so $U(t)$ and $V(t)$ are unimodular (Ex. 7.7.7). Also, the equality $\Gamma(t) = U(t)A(t)V(t)$ is verified; thus, $\Gamma(t)$ is indeed a Smith form for $A(t)$ (Corollary 7.1.8).

On the other hand,

$$\widehat{\Gamma}(t) := \begin{bmatrix} 1 & 0 \\ 0 & (t-i)(t-2j) \end{bmatrix}$$

is also a Smith form for $A(t)$. To verify this, let

$$\widehat{U}(t) := \begin{bmatrix} \dfrac{5}{9} & \dfrac{1}{3}(i+2j) \\ -\dfrac{1}{3}((i+2j)t + 4 - 2k) & -i + t \end{bmatrix}$$

and

$$\widehat{V}(t) := \begin{bmatrix} -\dfrac{1}{5}(3i+6j) & -t + 2j \\ 1 & \dfrac{1}{3}((-i-2j)t - 1 + 2k) \end{bmatrix}.$$

Then

$$\widehat{U}(t)^{-1} = \begin{bmatrix} \dfrac{1}{5}(-(3i+6j)t - 3 + 6k) & -1 \\ t - 2j & -\dfrac{1}{3}(i+2j) \end{bmatrix},$$

$$\widehat{V}(t)^{-1} = \begin{bmatrix} \dfrac{5}{9}(t-i) & \dfrac{1}{3}((i+2j)t + 4 - 2k) \\ -\dfrac{1}{3}(i+2j) & 1 \end{bmatrix},$$

so, $\widehat{U}(t)$ and $\widehat{V}(t)$ are unimodular. The equality $\widehat{\Gamma}(t) = \widehat{U}(t)A(t)\widehat{V}(t)$ shows that $\widehat{\Gamma}(t)$ is indeed a Smith form for $A(t)$.

However,

$$\left(t + \frac{3i + 4j}{5} \right)(t - 2j) \quad \text{and} \quad (t-i)(t-2j)$$

are not H-similar, because the quaternions

$$\frac{1}{5}(3i + 4j)(-2j) \quad \text{and} \quad (-i)(-2j)$$

have different real parts and, therefore, cannot be similar.   □

We point out a particular situation when uniqueness of the Smith form is guaranteed.

Theorem 7.2.3. *If* $A(t) \in H^{m \times n}$ *has Smith forms*

$$\text{diag}\,(a_1(t), a_2(t), \ldots, a_r(t), 0, \ldots, 0) \tag{7.2.3}$$

$$\text{and} \quad \text{diag}\,(b_1(t), b_2(t), \ldots, b_r(t), 0, \ldots, 0), \tag{7.2.4}$$

and the $a_j(t)$*'s and* $b_j(t)$*'s are polynomials with real coefficients, then* $a_j(t) = b_j(t)$
for $j = 1, 2, \ldots r$.

Proof. Apply the map χ to the equality

$$\text{diag}\,(a_1(t), a_2(t), \ldots, a_r(t), 0, \ldots, 0)$$

$$= U(t)\,\text{diag}\,(b_1(t), b_2(t), \ldots, b_r(t), 0, \ldots, 0)\,V(t),$$

where $U(t) \in H(t)^{m \times m}$ and $V(t) \in H^{n \times n}$ are unimodular. We obtain that the matrix polynomials

$$\chi(\text{diag}\,(a_1(t), a_2(t), \ldots, a_r(t), 0, \ldots, 0))$$

$$\text{and} \quad \chi(\text{diag}\,(b_1(t), b_2(t), \ldots, b_r(t), 0, \ldots, 0))$$

are R-equivalent. But their R-Smith forms are easily seen to be

$$\text{diag}\,(a_1(t)I_4, \ldots, a_r(t)I_4, 0, \ldots, 0)), \quad \text{resp.} \quad \text{diag}\,(b_1(t)I_4, \ldots, b_r(t)I_4, 0, \ldots, 0)).$$

The result now follows from the uniqueness of the R-Smith form.   □

We conclude this section with an open problem.

Open Problem 7.2.4. *Study connections between the* C-*Smith form of a matrix polynomial* $A(t) \in C(t)^{m \times n}$ *and its Smith forms* (*over the quaternions*).
For example, given a complex $m \times n$ *matrix polynomial* $A(t)$, *determine* (*in terms of the* C-*Smith form of* $A(t)$) *the number of* C-*equivalence classes in the set*

$$\{B(t) \in C^{m \times n} : B(t) \text{ is } H-\text{equivalent to } A(t)\}.$$

The results of Pereira et al. [117] may be relevant here.

The corresponding problem for real matrix polynomials admits a simple solution.

Theorem 7.2.5. *Two real matrix polynomials are* R-*equivalent if and only if they are* H-*equivalent.*

Proof. We need only to show that if two Smith forms (7.2.3) and (7.2.4) with real polynomials $a_1(t), \ldots, a_r(t)$, $b_1(t), \ldots, b_r(t)$ are H-equivalent, then they are R-equivalent. But this is immediate from Theorem 7.2.3.   □

7.3   STATEMENT OF THE KRONECKER FORM

In this section we present the Kronecker form of a pair of quaternion matrices, with a complete proof (to follow) modeled after the standard proof for complex (or real) matrices; see, e.g., Gantmacher [50, 49], Thompson [151], and also Gohberg et al. [54]. Although an alternative proof can be given using the map ω (see the proof of the Jordan form in Section 5.6), we prefer a direct proof, as it can be applicable to matrices over more general division rings.

Consider two matrix polynomials of degree at most one, often called *matrix pencils*: $A_1 + tB_1$ and $A_2 + tB_2$, where $A_1, B_1, A_2, B_2 \in \mathsf{H}^{m \times n}$.

Definition 7.3.1. The matrix pencils $A_j + tB_j$, $j = 1, 2$, are called *strictly equivalent* if

$$A_1 = PA_2Q, \quad B_1 = PB_2Q$$

for some invertible matrices $P \in \mathsf{H}^{m \times m}$ and $Q \in \mathsf{H}^{n \times n}$.

We develop here canonical form of matrix pencils under strict equivalence.

The Kronecker form of matrix polynomials $A + tB$ is formulated in Theorem 7.3.2 below.

Introduce special matrix polynomials:

$$L_{\varepsilon \times (\varepsilon+1)}(t) = [0_{\varepsilon \times 1} \ \ I_\varepsilon] + t[I_\varepsilon \ \ 0_{\varepsilon \times 1}] \in \mathsf{H}^{\varepsilon \times (\varepsilon+1)}.$$

Here ε is a positive integer. Also, standard real symmetric matrices F_m, G_m, and $\widetilde{G}_m$ defined by (1.2.3), (1.2.4), and (1.2.5), respectively, will be used.

Theorem 7.3.2. *Every pencil $A + tB \in \mathsf{H}(t)^{m \times n}$ is strictly equivalent to a matrix pencil with the block diagonal form:*

$$
\begin{aligned}
0_{u \times v} \quad & \oplus \quad L_{\varepsilon_1 \times (\varepsilon_1+1)} \oplus \cdots \oplus L_{\varepsilon_p \times (\varepsilon_p+1)} \oplus L^T_{\eta_1 \times (\eta_1+1)} \oplus L^T_{\eta_q \times (\eta_q+1)} \\
& \oplus \quad (I_{k_1} + tJ_{k_1}(0)) \oplus \cdots \oplus (I_{k_r} + tJ_{k_r}(0)) \\
& \oplus \quad (tI_{\ell_1} + J_{\ell_1}(\alpha_1)) \oplus \cdots \oplus (tI_{\ell_s} + J_{\ell_s}(\alpha_s)),
\end{aligned}
\tag{7.3.1}
$$

where $\varepsilon_1 \leq \cdots \leq \varepsilon_p$; $\eta_1 \leq \cdots \leq \eta_q$; $k_1 \leq \cdots \leq k_r$, are positive integers, and $\alpha_1, \ldots, \alpha_s \in \mathsf{H}$.

Moreover, the integers u, v, and ε_i, η_j, k_w are uniquely determined by the pair A, B, and the part

$$(tI_{\ell_1} + J_{\ell_1}(\alpha_1)) \oplus \cdots \oplus (tI_{\ell_s} + J_{\ell_s}(\alpha_s))$$

is uniquely determined by A and B up to an arbitrary permutation of the diagonal blocks and up to replacing α_j with any quaternion similar to α_j in each $J_{\ell_j}(\alpha_j)$.

In particular, the α_j's can be chosen to be complex numbers with nonnegative imaginary parts.

The following terminology is used in connection with the Kronecker form (7.3.1) of the matrix pencil $A + tB$.

Definition 7.3.3. The integers $\varepsilon_1 \leq \cdots \leq \varepsilon_p$ and $\eta_1 \leq \cdots \leq \eta_q$ are called the *left indices* and the *right indices*, respectively, of $A + tB$. The integers $k_1 \leq \cdots \leq k_r$ are called the *indices*, or *partial multiplicities*, at infinity of $A + tB$. The quaternions $-\alpha_1, \ldots, -\alpha_s$ are called the *eigenvalues* of $A + tB$.

The eigenvalues of $A + tB$ are uniquely determined up to permutation and similarity. Thus, the set of eigenvalues of $tI + J_m(\alpha)$ consists of all quaternions similar to $-\alpha$.

Definition 7.3.4. The part

$$0_{u \times v} \oplus L_{\varepsilon_1 \times (\varepsilon_1 + 1)} \oplus \cdots \oplus L_{\varepsilon_p \times (\varepsilon_p + 1)} \oplus L^T_{\eta_1 \times (\eta_1 + 1)} \oplus L^T_{\eta_q \times (\eta_q + 1)}$$

is termed the *singular part* of the form (7.3.1), and

$$(I_{k_1} + tJ_{k_1}(0)) \oplus \cdots \oplus (I_{k_r} + tJ_{k_r}(0)) \oplus (tI_{\ell_1} + J_{\ell_1}(\alpha_1)) \oplus \cdots \oplus (tI_{\ell_s} + J_{\ell_s}(\alpha_s))$$

is the *regular part*.

If we define the *left kernel* of a matrix $Z \in \mathsf{H}^{m \times n}$ by

$$\operatorname{Kel} Z = \{x \in \mathsf{H}^{1 \times m} : xZ = 0\},$$

then the parameters u and v in (7.3.1) can be identified as:

$$u = \dim_{\mathsf{H}} \bigcap_{t \in \mathsf{R}} \operatorname{Kel} (A + tB), \tag{7.3.2}$$

$$v = \dim_{\mathsf{H}} \bigcap_{t \in \mathsf{R}} \operatorname{Ker} (A + tB). \tag{7.3.3}$$

Note that if B is invertible, then the parts $0_{u \times v}$, $L_{\epsilon_j \times (\epsilon_j + 1)}$, $L^T_{\eta_j \times (\eta_j + 1)}$, and $I + tJ_{\ell_j}(0)$ are absent in the Kronecker form; moreover, the eigenvalues of $A + tB$ are exactly the negatives of the eigenvalues of $B^{-1}A$ (or the eigenvalues of AB^{-1}). Indeed, we have

$$P(A + tB)Q = tI + J, \qquad J := J_{\ell_1}(\alpha_1)) \oplus \cdots \oplus J_{\ell_s}(\alpha_s)$$

for some invertible matrices P and Q, so

$$PBQ = I, \qquad PAQ = J,$$

and

$$P(AB^{-1})P^{-1} = PA \cdot QP \cdot P^{-1} = PAQ = J.$$

Thus, J is the Jordan form of AB^{-1}. Analogously, one shows that J is the Jordan form of $B^{-1}A$.

Definition 7.3.5. For a fixed eigenvalue α of $A + tB$, let $i_1 < \cdots < i_w$ be all the subscripts in (7.3.1) such that $\alpha_{i_1}, \cdots, \alpha_{i_w}$ are similar to $-\alpha$; then the integers $\ell_{i_1}, \ldots, \ell_{i_w}$ are called the *indices*, or *partial multiplicities*, of the eigenvalue α of $A + tB$. Note that there may be several indices at infinity that are equal to a fixed positive integer; the same remark applies to the indices of a fixed eigenvalue of $A + tB$, to the left indices of $A + tB$, and to the right indices of $A + tB$.

Looking ahead, we indicate the following corollary, to be used in the sequel.

Corollary 7.3.6. *Let ϕ be an involution, either the conjugation or a nonstandard one. Then a matrix pencil $A + tB$ is strictly equivalent to $A_\phi + tB_\phi$ if and only if*

$$\dim_{\mathsf{H}} \bigcap_{t \in \mathsf{R}} \operatorname{Kel} (A + tB) = \dim_{\mathsf{H}} \bigcap_{t \in \mathsf{R}} \operatorname{Ker} (A + tB),$$

and the right indices of $A + tB$, arranged in the nondecreasing order, coincide with its left indices, also arranged in the nondecreasing order. In particular, $A + tB$ is strictly equivalent to $A_\phi + tB_\phi$ if the singular part in the Kronecker form of $A + tB$ is absent.

Proof. Let $A_0 + tB_0$ be the Kronecker form of $A + tB$ given by (7.3.1). Then $(A_0)_\phi + t(B_0)_\phi$ is strictly equivalent to $A_\phi + tB_\phi$. Since by Corollary 5.7.2 $(J_m(\alpha))_\phi$ is similar to $J_m(\alpha)$, we obtain that the Kronecker form of $(A_0)_\phi + t(B_0)_\phi$ is

$$
\begin{aligned}
0_{v \times u} \quad &\oplus \quad L^T_{\varepsilon_1 \times (\varepsilon_1 + 1)} \oplus \cdots \oplus L^T_{\varepsilon_p \times (\varepsilon_p + 1)} \oplus L_{\eta_1 \times (\eta_1 + 1)} \oplus L_{\eta_q \times (\eta_q + 1)} \\
&\oplus \quad (I_{k_1} + tJ_{k_1}(0)) \oplus \cdots \oplus (I_{k_r} + tJ_{k_r}(0)) \\
&\oplus \quad (tI_{\ell_1} + J_{\ell_1}(\alpha_1)) \oplus \cdots \oplus (tI_{\ell_s} + J_{\ell_s}(\alpha_s)).
\end{aligned} \tag{7.3.4}
$$

In view of the uniqueness of the Kronecker form, it follows that the Kronecker form of $(A_0)_\phi + t(B_0)_\phi$ coincides with that of $A + tB$ if and only if $u = v$, $p = q$, and $\varepsilon_j = \eta_j$ for $j = 1, 2, \ldots, p$. $\qquad \square$

7.4   PROOF OF THEOREM 7.3.2: EXISTENCE

We start with a reduction theorem, which is the key to the proof of existence in Theorem 7.3.2.

Theorem 7.4.1. *Let $A + tB \in \mathsf{H}(t)^{m \times n}$ be a matrix pencil such that the rank of $A + tB$ is smaller than n. Assume that*

$$
Ax = Bx = 0, \quad x \in \mathsf{H}^{n \times 1} \quad \Longrightarrow \quad x = 0. \tag{7.4.1}
$$

Then $A + tB$ is strictly equivalent to a direct sum

$$
L_{\varepsilon \times (\varepsilon + 1)}(t) \oplus (A' + tB') \tag{7.4.2}
$$

for some integer $\varepsilon > 0$.

Proof. With some changes, the proof follows the proof of the reduction theorem in Gantmacher [49, Chapter XII] and [48, Chapter 2]; see also Gohberg et al. [54, Appendix] or Rodman [132].

By Corollary 7.1.12, there exists a nonzero polynomial $y(t) \in \mathsf{H}(t)^{n \times 1}$ such that

$$
(A + tB)y(t) \equiv 0. \tag{7.4.3}
$$

Let $\varepsilon - 1 \geq 1$ be the smallest degree of a nonzero $y(t)$ for which (7.4.3) holds. (We cannot have $\varepsilon - 1 = 0$ in view of hypothesis (7.4.1)). Write

$$
y(t) = \sum_{j=0}^{\varepsilon - 1} t^j (-1)^j y_j, \quad y_0, \ldots, y_{\varepsilon-1} \in \mathsf{H}^{n \times 1}, \quad y_{\varepsilon-1} \neq 0.
$$

Then equation (7.4.3) reads

$$
\begin{bmatrix}
A & 0_{m \times n} & \cdots & 0_{m \times n} \\
B & A & \cdots & 0_{m \times n} \\
0_{m \times n} & B & \cdots & \vdots \\
\vdots & \vdots & \ddots & A \\
0_{m \times n} & 0_{m \times n} & \cdots & B
\end{bmatrix}
\begin{bmatrix}
y_0 \\
-y_1 \\
\vdots \\
(-1)^{\varepsilon-1} y_{\varepsilon-1}
\end{bmatrix} = 0. \tag{7.4.4}
$$

Denote by $Z_{(\varepsilon+1)m\times\varepsilon n}(A+tB)$ the $(\varepsilon+1)m \times \varepsilon n$ matrix in the left-hand side of (7.4.4). In view of (7.4.4) we have

$$\text{rank } Z_{(\varepsilon+1)m\times\varepsilon n}(A+tB) < \varepsilon n, \tag{7.4.5}$$

and since $\varepsilon - 1$ is the smallest possible degree of a nonzero vector $y(x)$ for which (7.4.3) holds, we have

$$\text{rank } Z_{(q+1)m\times qn}(A+tB) = qn, \qquad q = 1, 2, \ldots, \varepsilon - 1. \tag{7.4.6}$$

Next, we prove that the vectors

$$Ay_1, \ldots, Ay_{\varepsilon-1} \in \mathsf{H}^{m\times 1} \tag{7.4.7}$$

are linearly independent. Assume the contrary, and let

$$Ay_h = \sum_{j=1}^{h-1} Ay_j\alpha_{h-j}, \qquad \alpha_1, \ldots, \alpha_{h-1} \in \mathsf{H},$$

for some h, $2 \le h \le \varepsilon - 1$. (If Ay_1 were equal to 0, then in view of (7.4.4) we would have $Ay_0 = By_0 = 0$; hence, $y_0 = 0$ by (7.4.1), and so $y(t)/t$ would be a polynomial of degree smaller than $\varepsilon - 1$ still satisfying (7.4.3), a contradiction with the choice of $\varepsilon - 1$.) Equation (7.4.4) gives

$$By_{h-1} = \sum_{j=1}^{h-1} By_{j-1}\alpha_{h-j},$$

in other words,

$$B\widetilde{y}_{h-1} = 0, \qquad \widetilde{y}_{h-1} := y_{h-1} - \sum_{j=0}^{h-2} y_j\alpha_{h-1-j}.$$

Introducing the vectors

$$\widetilde{y}_{h-2} := y_{h-2} - \sum_{j=0}^{h-3} y_j\alpha_{h-2-j}, \quad \widetilde{y}_{h-3} := y_{h-3} - \sum_{j=0}^{h-4} y_j\alpha_{h-3-j}, \quad \ldots,$$

$$\widetilde{y}_1 := y_1 - y_0\alpha_1, \quad \widetilde{y}_0 := y_0,$$

we obtain, using (7.4.4), the equalities

$$A\widetilde{y}_{h-1} = B\widetilde{y}_{h-2}, \quad \ldots, \quad A\widetilde{y}_1 = B\widetilde{y}_0, \quad A\widetilde{y}_0 = 0.$$

Thus,

$$\widetilde{y}(t) = \sum_{j=0}^{h-1} t^j(-1)^j\widetilde{y}_j$$

is a nonzero (because $\widetilde{y}_0 \ne 0$) polynomial of degree smaller than $h \le \varepsilon - 1$ satisfying (7.4.3), a contradiction with the choice of $\varepsilon - 1$.

Now it is easy to see that the vectors $y_0, \ldots, y_{\varepsilon-1}$ are linearly independent. Indeed, if

$$y_0\alpha_0 + \cdots + y_{\varepsilon-1}\alpha_{\varepsilon-1} = 0, \qquad \alpha_j \in \mathsf{H},$$

then applying A and using the linear independence of (7.4.7), we obtain $\alpha_1 = \cdots = \alpha_{\varepsilon-1} = 0$, and since $y_0 \neq 0$, the equality $\alpha_0 = 0$ follows as well.

Now let $Q \in \mathsf{H}^{n \times n}$ be an invertible matrix whose first ε columns are $y_0, \ldots, y_{\varepsilon-1}$ (in that order), and let $P \in \mathsf{H}^{m \times m}$ be an invertible matrix whose first $\varepsilon - 1$ columns are $Ay_1, \ldots, Ay_{\varepsilon-1}$. The existence of such P and Q follows from the replacement Theorem 3.1.1, where we take $v_1 = e_1, \ldots, v_n = e_n$ (for P), and $v_1 = e_1, \ldots, v_n = e_m$ (for Q), the standard bases in $\mathsf{H}^{n \times 1}$ and in $\mathsf{H}^{m \times 1}$, respectively.

Define the matrices A' and B' by the equality

$$(A + tB)Q = P(A' + tB').$$

Using (7.4.4), it is easily seen that

$$A' + tB' = \left[\begin{array}{cc} L_{(\varepsilon-1)\times\varepsilon}(t) & D + tF \\ 0_{(m-(\varepsilon-1))\times\varepsilon} & A'' + tB'' \end{array} \right], \tag{7.4.8}$$

for some matrices

$$A'', B'' \in \mathsf{H}^{(m-(\varepsilon-1))\times(n-\varepsilon)} \quad \text{and} \quad D, F \in \mathsf{H}^{(\varepsilon-1)\times(n-\varepsilon)}.$$

We obviously have

$$Z_{\varepsilon m \times (\varepsilon-1)n}(A + tB)$$

$$= P' \cdot \left[\begin{array}{c} Z_{\varepsilon(\varepsilon-1)\times(\varepsilon-1)\varepsilon}(L_{(\varepsilon-1)\times\varepsilon}(t)) \\ 0 \\ * \\ Z_{\varepsilon(m-(\varepsilon-1))\times(\varepsilon-1)(n-\varepsilon)}(A'' + tB'') \end{array} \right] \cdot Q' \tag{7.4.9}$$

for some invertible matrices P' and Q'. One easily checks that the matrix

$$Z_{\varepsilon(\varepsilon-1)\times(\varepsilon-1)\varepsilon}(L_{(\varepsilon-1)\times\varepsilon}(t))$$

is invertible. Since by (7.4.6)

$$\text{rank } Z_{\varepsilon m \times (\varepsilon-1)n}(A + tB) = (\varepsilon - 1)n,$$

in view of (7.4.9) we have

$$\text{rank } Z_{\varepsilon(m-(\varepsilon-1))\times(\varepsilon-1)(n-\varepsilon)}(A'' + tB'') = (\varepsilon - 1)(n - \varepsilon). \tag{7.4.10}$$

Finally, we shall prove that by applying a suitable strict equivalence, one can reduce the matrix polynomial in the right-hand side of (7.4.8) to the form (7.4.2). More precisely, we shall show that for some quaternion matrices X and Y of suitable sizes, the equality

$$L_{(\varepsilon-1)\times\varepsilon}(t) \oplus (A'' + tB'') = \left[\begin{array}{cc} I_{\varepsilon-1} & Y \\ 0 & I_{m-(\varepsilon-1)} \end{array} \right] \left[\begin{array}{cc} L_{(\varepsilon-1)\times\varepsilon}(t) & D + tF \\ 0_{(m-(\varepsilon-1))\times\varepsilon} & A'' + tB'' \end{array} \right]$$

$$\cdot \left[\begin{array}{cc} I_\varepsilon & -X \\ 0 & I_{n-\varepsilon} \end{array} \right] \tag{7.4.11}$$

holds. Equality (7.4.11) can be rewritten in the form

$$L_{(\varepsilon-1)\times\varepsilon}(t)X = D + tF + Y(A'' + tB''). \tag{7.4.12}$$

Introduce notation for the entries of D, F, X, for the rows of Y, and for the columns of A'' and B'':

$$D = [d_{i,k}], \quad F = [f_{i,k}], \quad X = [x_{j,k}],$$

where

$$i = 1, 2, \ldots, \varepsilon - 1; \quad k = 1, 2, \ldots, n - \varepsilon; \quad j = 1, 2, \ldots, \varepsilon;$$

and

$$Y = \begin{bmatrix} y_1 \\ y_2 \\ \vdots \\ y_{\varepsilon-1} \end{bmatrix}, \quad A'' = \begin{bmatrix} a_1 & a_2 & \cdots & a_{n-\varepsilon} \end{bmatrix},$$

$$B'' = \begin{bmatrix} b_1 & b_2 & \cdots & b_{n-\varepsilon} \end{bmatrix}.$$

Then, equating the entries in the kth column of both sides of (7.4.12), we see that (7.4.12) is equivalent to the following system of scalar equations, with unknowns $x_{j,k}$ and y_j:

$$\begin{aligned} x_{2,k} + tx_{1,k} &= d_{1,k} + tf_{1,k} + y_1 a_k + ty_1 b_k, & (7.4.13) \\ & k = 1, 2, \ldots, n - \varepsilon, \\ x_{3,k} + tx_{2,k} &= d_{2,k} + tf_{2,k} + y_2 a_k + ty_2 b_k, \quad k = 1, 2, \ldots, n - \varepsilon, \end{aligned}$$

$$\vdots$$

$$\begin{aligned} x_{\varepsilon,k} + tx_{\varepsilon-1,k} &= d_{\varepsilon-1,k} + tf_{\varepsilon-1,k} + y_{\varepsilon-1} a_k + ty_{\varepsilon-1} b_k, & (7.4.14) \\ & k = 1, 2, \ldots, n - \varepsilon. \end{aligned}$$

We first solve the system of equations

$$\begin{aligned} y_1 a_k - y_2 b_k &= f_{2,k} - d_{1,k}, & k = 1, 2, \ldots, n - \varepsilon, \\ y_2 a_k - y_3 b_k &= f_{3,k} - d_{2,k}, & k = 1, 2, \ldots, n - \varepsilon, \end{aligned}$$

$$\vdots$$

$$y_{\varepsilon-2} a_k - y_{\varepsilon-1} b_k = f_{\varepsilon-1,k} - d_{\varepsilon-2,k}, \quad k = 1, 2, \ldots, n - \varepsilon. \quad (7.4.15)$$

Indeed, (7.4.15) can be rewritten in the form

$$\begin{bmatrix} y_1 & -y_2 & \cdots & (-1)^{\varepsilon-1} y_{\varepsilon-2} & (-1)^\varepsilon y_{\varepsilon-1} \end{bmatrix}$$

$$\cdot Z_{(\varepsilon-1)(m-\varepsilon+1) \times (\varepsilon-2)(n-\varepsilon)} (A'' + tB'')$$

$$= \begin{bmatrix} [f_{2,k} - d_{1,k}]_{k=1}^{n-\varepsilon} & [f_{3,k} - d_{2,k}]_{k=1}^{n-\varepsilon} \end{bmatrix}$$

$$\cdots [f_{\varepsilon-2,k} - d_{\varepsilon-3,k}]_{k=1}^{n-\varepsilon} \quad [f_{\varepsilon-1,k} - d_{\varepsilon-2,k}]_{k=1}^{n-\varepsilon}].$$

This equation can be always solved for $y_1, \ldots, y_{\varepsilon-1}$, because by (7.4.10) and Proposition 3.2.8 the matrix

$$Z_{(\varepsilon-1)(m-\varepsilon+1) \times (\varepsilon-2)(n-\varepsilon)} (A'' + tB'')$$

is left-invertible. Once (7.4.15) is satisfied, a solution of (7.4.14) is easily obtained by setting

$$x_{1,k} = f_{1,k} + y_1 b_k, \quad x_{2,k} = f_{2,k} + y_2 b_k, \quad \ldots, \quad x_{\varepsilon-1,k} = f_{\varepsilon-1,k} + y_{\varepsilon-1} b_k.$$

(If $\varepsilon = 2$, then the system (7.4.14) has only equations (7.4.13), which can be easily solved for $x_{1,k}$, $x_{2,k}$.) This concludes the proof of Theorem 7.4.1. □

The dual statement of Theorem 7.4.1 reads as follows.

Theorem 7.4.2. *Let $A + tB \in H(x)^{m \times n}$ be a matrix pencil such that the rank of $A + tB$ is smaller than m. Assume that*

$$yA = yB = 0, \quad y \in H^{1 \times m} \implies y = 0. \tag{7.4.16}$$

Then $A + tB$ is strictly equivalent to a direct sum

$$(L_{\varepsilon \times (\varepsilon+1)}(t))^T \oplus (A' + tB') \tag{7.4.17}$$

for some integer $\varepsilon > 0$.

The proof is reduced to Theorem 7.4.1 upon considering the matrix pencil $A^* + tB^*$.

We are now ready to prove Theorem 7.3.2. First, we prove the existence of the form (7.3.1). The proof proceeds by induction on the sizes m and n. If $\text{rank}\,(A + tB) < n$ and (7.4.1) holds true, or if $\text{rank}\,(A+tB) < m$ and (7.4.16) holds true, then we can use Theorem 7.4.1 or 7.4.2, as appropriate, to reduce the existence proof to matrices of smaller sizes, and we are done by induction. If neither of (7.4.1) and (7.4.16) holds true, then by selecting a basis $\{x_1, \ldots, x_v\}$ for the solution set of $Ax = Bx = 0$ and a basis $\{y_1, \ldots, y_u\}$ for the solution set of $yA = yB = 0$, and appropriately changing bases in $H^{n \times 1}$ and in $H^{1 \times m}$, we see that $A + tB$ has the form

$$A + tB = \begin{bmatrix} 0_{u,v} & 0 \\ 0 & A' + tB' \end{bmatrix},$$

and we are done, again by the induction hypothesis. It remains to consider the case when $m = n$ and $\text{rank}\,(A + t_0 B) = n$ for some real t_0. In other words, the matrix $A + t_0 B$ is invertible. Now, an application of the Jordan form of $(A + t_0 B)^{-1} B$ (Theorem 5.5.3) easily completes the proof of existence of the form (7.3.1). Indeed, with $A_0 := A + t_0 B$ and $t' = t - t_0$ we have

$$
\begin{aligned}
A + tB &= A_0 + t'B = A_0(I + t'A_0^{-1}B) \\
&= A_0 S^{-1}(I + t'J)S = A_0 S^{-1}(I - t_0 J + tJ)S
\end{aligned}
$$

for some invertible $S \in H^{n \times n}$ and some matrix J in the Jordan form. Using the property that $(J_k(\alpha))^{-1}$, $\alpha \in H \setminus \{0\}$, is similar to $J_k(\alpha^{-1})$, one easily transforms $I - t_0 J + tJ$ to the form as in (7.3.1), with the parts $0_{u \times v}$, $L_{\epsilon_j \times (\epsilon_j+1)}$, and $L^T_{\eta_j \times (\eta_j+1)}$ absent.

7.5 PROOF OF THEOREM 7.3.2: UNIQUENESS

An independent proof of uniqueness can be given along the lines as in Gantmacher [50, 49] or Gohberg et al. [54, Appendix]. We give a proof based on reduction to the case of complex (or real) matrices.

We start with the Kronecker forms of matrix representations. This information is of independent interest. It will be convenient to use the following notation: if $X \in H^{\delta_1 \times \delta_2}$, then $X^{\oplus m}$ stands for the $m\delta_1 \times m\delta_2$ matrix $X \oplus \cdots \oplus X$, where X is repeated m times.

Theorem 7.5.1. *Let*

$$0_{u \times v} \quad \oplus \quad L_{\varepsilon_1 \times (\varepsilon_1 + 1)} \oplus \cdots \oplus L_{\varepsilon_p \times (\varepsilon_p + 1)} \oplus L^T_{\eta_1 \times (\eta_1 + 1)} \oplus L^T_{\eta_q \times (\eta_q + 1)}$$
$$\oplus \quad (I_{k_1} + t J_{k_1}(0)) \oplus \cdots \oplus (I_{k_r} + t J_{k_r}(0))$$
$$\oplus \quad (t I_{\ell_1} + J_{\ell_1}(\alpha_1)) \oplus \cdots \oplus (t I_{\ell_s} + J_{\ell_s}(\alpha_s)), \tag{7.5.1}$$

be the Kronecker form, as in Theorem 7.3.2, of a matrix pencil $A + tB$, $A, B \in \mathsf{H}^{m \times n}$, where $\alpha_1, \ldots, \alpha_s$ are taken to be complex numbers. Then:

(1) *the pencil of real matrices $\chi_{m,n} A + t \chi_{m,n}(B)$ has the real Kronecker form*

$$0_{4u \times 4v} \quad \oplus \quad L^{\oplus 4}_{\varepsilon_1 \times (\varepsilon_1 + 1)} \oplus \cdots \oplus L^{\oplus 4}_{\varepsilon_p \times (\varepsilon_p + 1)}$$
$$\oplus \quad (L^T_{\eta_1 \times (\eta_1 + 1)})^{\oplus 4} \oplus \cdots \oplus (L^T_{\eta_q \times (\eta_q + 1)})^{\oplus 4}$$
$$\oplus \quad (I_{k_1} + t J_{k_1}(0))^{\oplus 4} \oplus \cdots \oplus (I_{k_r} + t J_{k_r}(0))^{\oplus 4}$$
$$\oplus \quad (t I_{\ell_1} + J_{\ell_1}(\alpha_1))^{\oplus 4} \oplus \cdots \oplus (t I_{\ell_w} + J_{\ell_w}(\alpha_w))^{\oplus 4}$$
$$\oplus \quad \left(t I_{2\ell_{w+1}} + J_{2\ell_{w+1}}(\Re(\alpha_{w+1}) \pm i \Im(\alpha_{w+1})) \right)$$
$$\oplus \quad \left(t I_{2\ell_{w+1}} + J_{2\ell_{w+1}}(\Re(\alpha_{w+1}) \pm i \Im(\alpha_{w+1})) \right)$$
$$\oplus \quad \cdots \oplus (t I_{2\ell_s} + J_{2\ell_s}(\Re(\alpha_s) \pm i \Im(\alpha_s)))$$
$$\oplus \quad (t I_{2\ell_s} + J_{2\ell_s}(\Re(\alpha_s) \pm i \Im(\alpha_s))),$$

where $\alpha_1, \ldots, \alpha_w$ are real and $\alpha_{w+1}, \ldots, \alpha_s$ are nonreal;

(2) *the pencil of complex matrices $\omega_{m,n}(A) + t \omega_{m,n}(B)$ has the complex Kronecker form*

$$0_{2u \times 2v} \quad \oplus \quad L^{\oplus 2}_{\varepsilon_1 \times (\varepsilon_1 + 1)} \oplus \cdots \oplus L^{\oplus 2}_{\varepsilon_p \times (\varepsilon_p + 1)}$$
$$\oplus \quad (L^T_{\eta_1 \times (\eta_1 + 1)})^{\oplus 2} \oplus \cdots \oplus (L^T_{\eta_q \times (\eta_q + 1)})^{\oplus 2}$$
$$\oplus \quad (I_{k_1} + t J_{k_1}(0))^{\oplus 2} \oplus \cdots \oplus (I_{k_r} + t J_{k_r}(0))^{\oplus 2}$$
$$\oplus \quad (t I_{\ell_1} + J_{\ell_1}(\alpha_1))^{\oplus 2} \oplus \cdots \oplus (t I_{\ell_w} + J_{\ell_w}(\alpha_w))^{\oplus 2}$$
$$\oplus \quad \left(t I_{\ell_{w+1}} + J_{\ell_{w+1}}(\alpha_{w+1}) \right) \oplus \left(t I_{\ell_{w+1}} + J_{\ell_{w+1}}(\overline{\alpha_{w+1}}) \right)$$
$$\oplus \quad \cdots \oplus (t I_{\ell_s} + J_{\ell_s}(\alpha_s)) \oplus (t I_{\ell_s} + J_{\ell_s}(\overline{\alpha_s})). \tag{7.5.2}$$

Proof. The blocks $t I_{\ell_u} + J_{\ell_u}(\alpha_u)$ are taken care of in Theorem 5.7.1. Note that the other blocks in 7.5.1 are real. Now it is easy to see from the definitions of the maps χ and ω that the pencils $\chi_{m,n}(A) + t \chi(B)$ and $\omega_{m,n}(A) + t \omega_{m,n}(B)$ have the Kronecker form as claimed (a permutation of columns (resp. rows) is required to put $L_{\varepsilon_v \times (\varepsilon_v + 1)}$ (resp. $L^T_{\eta \times \eta + 1}$) in a canonical form). $\square$

Now the proof of uniqueness in Theorem 7.5.1 is not difficult. Suppose that a matrix pencil $A + tB$, where $A, B \in \mathsf{H}^{m \times n}$ is strictly equivalent to (7.3.1) (with the α_j's taken to be complex numbers) as well as strictly equivalent to

$$0_{u' \times v'} \quad \oplus \quad L_{\varepsilon'_1 \times (\varepsilon'_1 + 1)} \oplus \cdots \oplus L_{\varepsilon'_{p'} \times (\varepsilon'_{p'} + 1)} \oplus L^T_{\eta'_1 \times (\eta'_1 + 1)} \oplus L^T_{\eta'_{q'} \times (\eta'_{q'} + 1)}$$
$$\oplus \quad (I_{k'_1} + t J_{k'_1}(0)) \oplus \cdots \oplus (I_{k'_{r'}} + t J_{k'_{r'}}(0))$$
$$\oplus \quad (t I_{\ell'_1} + J_{\ell'_1}(\alpha'_1)) \oplus \cdots \oplus (t I_{\ell'_{s'}} + J_{\ell'_{s'}}(\alpha'_{s'})), \tag{7.5.3}$$

where

$$\varepsilon_1' \leq \cdots \leq \varepsilon_{p'}'; \quad \eta_1' \leq \cdots \leq \eta_{q'}'; \quad k_1' \leq \cdots \leq k_{r'}'; \quad \ell_1' \leq \cdots \leq \ell_{s'}'$$

are positive integers and $\alpha_1', \ldots, \alpha_{s'}' \in \mathsf{C}$. Then by (7.3.2) and (7.3.3) we have $u' = u$ and $v' = v$. Moreover, denoting the pencils (7.3.1) and (7.5.3) by $E + tG$ and $E' + tG'$, respectively, we have

$$P(E + tG)Q = E' + tG'$$

for some invertible matrices P and Q. Applying the map $\omega_{m,n}$ to this equality, we see that the complex matrix pencils $\omega_{m,n}E + t\omega_{m,n}G$ and $\omega_{m,n}E' + t\omega_{m,n}G'$ are strictly equivalent. Now formula (7.5.2) (applied to $\omega_{m,n}E + t\omega_{m,n}G$ and to $\omega_{m,n}E' + t\omega_{m,n}G'$) together with the uniqueness statement for Kronecker form of complex matrix pencils (Theorem 15.1.3) imply the desired uniqueness in Theorem 7.3.2. $\qquad\square$

7.6 COMPARISON WITH REAL AND COMPLEX STRICT EQUIVALENCE

Theorems 5.8.1 and 5.8.3 have straightforward analogues for matrix pencils and strict equivalence.

Theorem 7.6.1. *A quaternion matrix pencil $A + tB$ is strictly equivalent to a real matrix pencil if and only if for every positive integer m and every nonreal $\lambda \in \mathsf{H}$, the number of blocks $tI_m + J_m(\alpha)$ with α similar to λ in the Kronecker form of $A + tB$ is even.*

The proof (which is omitted) is similar to that of Theorem 5.8.1, using the Kronecker form of the pencil and equality (5.8.1).

Definition 7.6.2. We say that two real matrix pencils $A + tB, A' + tB' \in \mathsf{R}^{m \times n}$ are *strictly equivalent over the reals*, or R-*strictly equivalent*, if $P(A + tB)Q = A' + tB'$ for some real invertible matrices P and Q. Analogously, the notion of C-strictly equivalent complex matrix pencils is defined. Finally, two quaternion matrix pencils $A + tB, A' + tB' \in \mathsf{H}^{m \times n}$ are said to be *strictly equivalent over quaternions*, or H-*strictly equivalent*, if $P(A + tB)Q = A' + tB'$ for some invertible quaternion matrices P and Q.

In particular, we will apply the notion of H-strict equivalence to real and complex matrix pencils.

Theorem 7.6.3. (a) *Two real matrix pencils are R-strictly equivalent if and only if they are H-strictly equivalent.*

(b) *Let $A + tB$ be a complex matrix pencil, and let*

$$
\begin{aligned}
0_{u \times v} \quad &\oplus \quad L_{\varepsilon_1 \times (\varepsilon_1 + 1)} \oplus \cdots \oplus L_{\varepsilon_p \times (\varepsilon_p + 1)} \oplus L_{\eta_1 \times (\eta_1 + 1)}^T \oplus L_{\eta_q \times (\eta_q + 1)}^T \\
&\oplus \quad (I_{k_1} + tJ_{k_1}(0)) \oplus \cdots \oplus (I_{k_r} + tJ_{k_r}(0)) \\
&\oplus \quad (tI_{\ell_1} + J_{\ell_1}(\alpha_1)) \oplus \cdots \oplus (tI_{\ell_s} + J_{\ell_s}(\alpha_s)) \quad\quad (7.6.1)
\end{aligned}
$$

be the complex Kronecker form of $A + tB$. Then the set of all complex matrix pencils that are H-strictly equivalent to $A + tB$ consists of exactly p classes

$\mathcal{E}_1, \ldots, \mathcal{E}_p$ *of mutually* C*-strictly equivalent complex matrix pencils so that complex matrix pencils from different* $\mathcal{E}_j$*'s are not* C*-strictly equivalent. The integer* p *is given by*

$$p = \prod_{j=1}^{u} \left(\prod_{k=1}^{r_j} (p_{\lambda_j, k} + 1) \right), \tag{7.6.2}$$

where $\{\lambda_1, \ldots, \lambda_u\}$ *is a set of distinct nonreal complex numbers among* $\{\alpha_1, \ldots, \alpha_s\}$ *that is maximal (by inclusion) with respect to the property that it does not contain any pair of complex conjugate numbers, and* $p_{\lambda_j, k}$ *is the total number of Jordan blocks in* (7.6.1) *of size* k *with eigenvalue* λ_j *or* $\overline{\lambda_j}$*;* r_j *being the largest size of a Jordan block with eigenvalue* λ_j *or* $\overline{\lambda_j}$*.*

In Theorem 7.6.3(b), the classes $\mathcal{E}_1, \ldots, \mathcal{E}_p$ are constructed as in Theorem 5.8.3 (b). Namely, for every pair of values (j, k) as in (7.6.2) such that $p_{\lambda_j, k} \neq 0$, select an integer $\ell_{\lambda_j, k}$ satisfying the inequalities $0 \leq \ell_{\lambda_j, k} \leq p_{\lambda_j, k}$. Then the class $\mathcal{E}_i$ corresponding to this selection of the $\ell_{\lambda_j, k}$'s consists of those complex matrix pencils whose complex Kronecker form is

$$K_0 \quad \oplus \quad \oplus_{j=1}^{u} \oplus_{\{k \,:\, p_{\lambda_j,k} \neq 0\}} \left(\underbrace{(tI + J_k(\lambda_j)) \oplus \cdots \oplus (tI + J_k(\lambda_j))}_{\ell_{\lambda_j,k} \text{ times}} \right.$$

$$\left. \oplus \underbrace{(tI + J_k(\overline{\lambda_j})) \oplus \cdots \oplus (tI + J_k(\overline{\lambda_j}))}_{p_{\lambda_j,k} - \ell_{\lambda_j,k} \text{ times}} \right), \tag{7.6.3}$$

where K_0 is the part of the complex Kronecker form of $A + tB$ that consists of blocks other than $tI_{\ell_j} + J_{\ell_j}(\alpha_j)$ with nonreal α_j.

The proof is analogous to that of Theorem 5.8.3, using Theorem 7.5.1. We omit the proof.

7.7   EXERCISES

Ex. 7.7.1. Provide details for the proof of Theorem 7.6.1.

Ex. 7.7.2. Show that $q(t) \in \mathsf{H}(t)$ is a total divisor of $p(t)\mathsf{H}(t)$ if and only if (7.1.1) holds.

Ex. 7.7.3. Prove equalities (7.3.2) and (7.3.3).

Ex. 7.7.4. Find the Smith form of the following upper triangular $n \times n$ matrix polynomial:

$$\begin{bmatrix} p_1(t) & p_2(t) & 0 & \ldots & 0 \\ 0 & \alpha_1 & p_3(t) & \ldots & 0 \\ \vdots & \vdots & \ddots & \ddots & \vdots \\ 0 & 0 & \ldots & \alpha_{n-2} & p_n(t) \\ 0 & 0 & \ldots & 0 & \alpha_{n-1} \end{bmatrix},$$

where $\alpha_1, \ldots, \alpha_{n-1} \in \mathsf{H}$ and $p_1(t), \ldots, p_n(t)$ are scalar polynomials with quaternion coefficients.

Ex. 7.7.5. Show that a complex matrix pencil $A + tB$ has the property that a complex matrix pencil is H-strictly equivalent to $A + tB$ precisely when it is C-strictly equivalent to $A + tB$, if and only if $A + tB$ has no nonreal eigenvalues.

Ex. 7.7.6. If $\mathrm{diag}\,(a_1(t), \ldots, a_r(t), 0, \ldots, 0)$ is the Smith form of $A(t) \in \mathsf{H}^{n \times n}$. Find the Smith form of the following quaternion matrix polynomials:

(1) $A(p(t))$, where $p(t)$ is a scalar polynomial with real coefficients;

(2) $\underbrace{A(t) \oplus \cdots \oplus A(t)}_{p \ \text{times}}$.

Ex. 7.7.7. (a) Show that if $A(t) \in \mathsf{H}(t)^{n \times n}$ is a square-size matrix polynomial such that $A(t)B(t) \equiv I$ for some $B(t) \in \mathsf{H}^{n \times n}$, then $A(t)$ is unimodular and, necessarily, $A(t)^{-1} = B(t)$.

(b) Repeat Part (a) but replace $A(t)B(t) \equiv I$ with $B(t)A(t) \equiv I$. Hint: Use the Smith form for $B(t)$.

Ex. 7.7.8. Find all Smith forms of the diagonal matrix polynomial $\mathrm{diag}\,(t - \mathsf{i}, t - \mathsf{i}, \ldots, t - \mathsf{i}) \in \mathsf{H}^{n \times n}$.

7.8   NOTES

The Smith form (Theorem 7.1.4) in the more general context of matrix polynomials with coefficients in a division ring was proved in Nakayama [115]; see also Jacobson [71]. For matrix polynomials over a (commutative) field, and more generally over commutative principal ideal domains, the Smith form is widely known; see, e.g., Gantmacher [49], Gohberg et al. [54, 55], and Lancaster and Tismenetsky [94].

The Smith form is treated in Jacobson [71, Chapter 3]; see also Guralnick et al. [59], Guralnick and Levy [58], Levy and Robson [100] in a more general context of noncommutative principal ideal domains, and Pereira et al. [117] (in a different setup of Laurent polynomials).

A complete set of invariants for equivalence (in the more general context of matrices over noncommutative principal ideal domains Λ) does not seem to be known; see Guralnick et al. [59], where it is proved that if the rank of $A \in \Lambda^{m \times n}$ is at least two, then the isomorphism class of the left module $\Lambda^n / (\Lambda^m A)$ forms a complete set of invariants for the equivalence relation. Examples given in Guralnick and Levy [58] and Levy and Robson [100] show that the result fails for matrices of rank one.

The Kronecker canonical form is well known for real and complex matrices and, more generally, for matrices over fields (if the field is not algebraically closed, a rational normal from may be used instead of the Jordan form) and can be found in many sources in literature. For quaternion matrices, the Kronecker canonical form is also known; see, e.g., Sergeichuk [141] and Djoković [33].

Ex. 7.7.6 (1) is found in Gohberg et al. [53], in the context of matrix polynomials with complex coefficients.

The exposition in this chapter follows Rodman [132].

Chapter Eight

Pencils of hermitian matrices

The main theme in this and the next chapter is the canonical forms of pairs of quaternion matrices, or matrix pencils, under simultaneous congruence, where the matrices are either hermitian or skewhermitian. For convenience of exposition, the material is divided between two chapters: In the current chapter the case of two hermitian matrices is studied, whereas the next chapter is devoted to pairs of skewhermitian matrices and to mixed pairs, when one of the matrices if hermitian and the other is skewhermitian. We also present here applications of the canonical form to the problems of existence of positive definite or semidefinite (nontrivial) linear combinations of two hermitian matrices and to the related problem of simultaneous diagonalizability. Other applications include treatment of H-expansive and H-plus-matrices, in particular for the cases when the hermitian matrix H is singular.

In this and subsequent chapters we will be often concerned with congruence of matrix pencils or, what is the same, simultaneous congruence of two matrices. Thus, it will be useful to introduce the concept of matrix pencils congruence in a general context.

Definition 8.0.1. Let F be one of R, C, or H. Two matrix pencils $A + tB$ and $A' + tB'$, where $A, B, A', B' \in \mathsf{F}^{m \times m}$, are said to be F-*congruent*, or simply *congruent* in the case when $\mathsf{F} = \mathsf{H}$, if $A + tB = S^*(A' + tB')S$ for some invertible matrix $S \in \mathsf{F}^{m \times m}$. In this case we also say that the pairs (A', B') and (A, B) are F-*simultaneously congruent,* or simply *simultaneously congruent* if $\mathsf{F} = \mathsf{H}$.

Clearly, F-congruence of matrix pencils is an equivalence relation.

8.1 CANONICAL FORMS

Definition 8.1.1. A matrix pencil $A + tB$, $A, B \in \mathsf{H}^{n \times n}$, is said to be *hermitian* if $A = A^*$ and $B = B^*$ or, equivalently, if $A + tB$ is hermitian for every real t.

Obviously, any matrix pencil which is congruent to a hermitian matrix pencil is itself hermitian.

Canonical forms for hermitian matrix pencils under strict equivalence and congruence are given by the following theorem. We use the standard real symmetric $m \times m$ matrices

$$F_m := \begin{bmatrix} 0 & \cdots & \cdots & 0 & 1 \\ \vdots & & & 1 & 0 \\ \vdots & & & & \vdots \\ 0 & 1 & & & \vdots \\ 1 & 0 & \cdots & \cdots & 0 \end{bmatrix}, \tag{8.1.1}$$

$$G_m := \begin{bmatrix} 0 & \cdots & \cdots & 1 & 0 \\ \vdots & & & 0 & 0 \\ \vdots & & & & \vdots \\ 1 & 0 & & & \vdots \\ 0 & 0 & \cdots & \cdots & 0 \end{bmatrix} = \begin{bmatrix} F_{m-1} & 0 \\ 0 & 0 \end{bmatrix}. \tag{8.1.2}$$

Theorem 8.1.2. (a) *Every hermitian matrix pencil $A+tB$, where $A, B \in \mathsf{H}^{n\times n}$, is strictly equivalent to a hermitian matrix pencil of the following form:*

$$0_u \quad \oplus \quad \left(t \begin{bmatrix} 0 & 0 & F_{\varepsilon_1} \\ 0 & 0 & 0 \\ F_{\varepsilon_1} & 0 & 0 \end{bmatrix} + G_{2\varepsilon_1+1} \right)$$

$$\oplus \quad \cdots \quad \oplus \left(t \begin{bmatrix} 0 & 0 & F_{\varepsilon_p} \\ 0 & 0 & 0 \\ F_{\varepsilon_p} & 0 & 0 \end{bmatrix} + G_{2\varepsilon_p+1} \right)$$

$$\oplus \quad (F_{k_1} + tG_{k_1}) \oplus \cdots \oplus (F_{k_r} + tG_{k_r})$$

$$\oplus \quad ((t + \alpha_1) F_{\ell_1} + G_{\ell_1}) \oplus \cdots \oplus ((t + \alpha_q) F_{\ell_q} + G_{\ell_q})$$

$$\oplus \quad \left(\begin{bmatrix} 0 & (t+\beta_1)F_{m_1} \\ (t+\beta_1^*)F_{m_1} & 0 \end{bmatrix} + \begin{bmatrix} 0 & G_{m_1} \\ G_{m_1} & 0 \end{bmatrix} \right)$$

$$\oplus \quad \cdots \quad \oplus \left(\begin{bmatrix} 0 & (t+\beta_s)F_{m_s} \\ (t+\beta_s^*)F_{m_s} & 0 \end{bmatrix} + \begin{bmatrix} 0 & G_{m_s} \\ G_{m_s} & 0 \end{bmatrix} \right). \tag{8.1.3}$$

Here, the numbers $\varepsilon_1 \le \cdots \le \varepsilon_p$ and $k_1 \le \cdots \le k_r$ are positive integers, α_j are real numbers, β_j are nonreal quaternions, and F_m, G_m are the $m \times m$ matrices given by (1.2.3) and (1.2.4).

The form (8.1.3) is uniquely determined by $A+tB$ up to an arbitrary permutation of the blocks in each of the parts

$$\oplus_{j=1}^q \left((t+\alpha_j)F_{\ell_j} + G_{\ell_j} \right)$$

and

$$\oplus_{j=1}^s \left(\begin{bmatrix} 0 & (t+\beta_j)F_{m_j} \\ (t+\beta_j^*)F_{m_j} & 0 \end{bmatrix} + \begin{bmatrix} 0 & G_{m_j} \\ G_{m_j} & 0 \end{bmatrix} \right), \tag{8.1.4}$$

and up to replacement of β_j in every block

$$\begin{bmatrix} 0 & (t+\beta_j)F_{m_j} \\ (t+\beta_j^*)F_{m_j} & 0 \end{bmatrix} + \begin{bmatrix} 0 & G_{m_j} \\ G_{m_j} & 0 \end{bmatrix} \tag{8.1.5}$$

by a similar quaternion.

(b) *Every hermitian matrix pencil $A + tB$, where $A, B \in \mathsf{H}^{n \times n}$, is congruent to* *a hermitian matrix pencil of the form*

$$0_u \quad \oplus \quad \left(t \begin{bmatrix} 0 & 0 & F_{\varepsilon_1} \\ 0 & 0 & 0 \\ F_{\varepsilon_1} & 0 & 0 \end{bmatrix} + G_{2\varepsilon_1+1} \right)$$

$$\oplus \quad \cdots \quad \oplus \quad \left(t \begin{bmatrix} 0 & 0 & F_{\varepsilon_p} \\ 0 & 0 & 0 \\ F_{\varepsilon_p} & 0 & 0 \end{bmatrix} + G_{2\varepsilon_p+1} \right)$$

$$\oplus \quad \delta_1 \left(F_{k_1} + t G_{k_1} \right) \oplus \cdots \oplus \delta_r \left(F_{k_r} + t G_{k_r} \right)$$

$$\oplus \quad \eta_1 \left((t+\alpha_1) F_{\ell_1} + G_{\ell_1} \right) \oplus \cdots \oplus \eta_q \left((t+\alpha_q) F_{\ell_q} + G_{\ell_q} \right)$$

$$\oplus \quad \left(\begin{bmatrix} 0 & (t+\beta_1)F_{m_1} \\ (t+\beta_1^*)F_{m_1} & 0 \end{bmatrix} + \begin{bmatrix} 0 & G_{m_1} \\ G_{m_1} & 0 \end{bmatrix} \right)$$

$$\oplus \quad \cdots \quad \oplus \quad \left(\begin{bmatrix} 0 & (t+\beta_s)F_{m_s} \\ (t+\beta_s^*)F_{m_s} & 0 \end{bmatrix} + \begin{bmatrix} 0 & G_{m_s} \\ G_{m_s} & 0 \end{bmatrix} \right) . \quad (8.1.6)$$

Here, $\varepsilon_1 \leq \cdots \leq \varepsilon_p$, the numbers $k_1 \leq \cdots \leq k_r$ and $\ell_1, \ldots, \ell_q$ are positive integers, α_j are real numbers, β_j are nonreal quaternions, and $\delta_1, \ldots, \delta_r, \eta_1, \ldots, \eta_q$ are signs, each equal to $+1$ or -1.

The form (8.1.6) is uniquely determined by $A+tB$ up to an arbitrary permutation of the blocks in each of the parts

$$\oplus_{j=1}^r (\delta_j (F_{k_j} + t G_{k_j})), \qquad \oplus_{j=1}^q \left(\eta_j \left((t+\alpha_j) F_{\ell_j} + G_{\ell_j} \right) \right),$$

and (8.1.4) and up to replacement of β_j in every block (8.1.5) by a similar quaternion.

The next section is dedicated to the proof of Theorem 8.1.2.

The signs $\delta_1, \ldots, \delta_r, \eta_1, \ldots, \eta_q$ form the *sign characteristic* of the pencil $A+tB$. It associates a sign ± 1 to every partial multiplicity of $A+tB$ corresponding to real eigenvalues and infinity.

Remark 8.1.3. One can take the β_j's in Theorem 8.1.2 to be complex numbers with positive imaginary parts.

Remark 8.1.4. An alternative version of Theorem 8.1.2 is obtained by using

$$\widetilde{G}_\alpha := F_\alpha G_\alpha F_\alpha = \begin{bmatrix} 0 & 0 \\ 0 & F_{\alpha-1} \end{bmatrix} \tag{8.1.7}$$

in place of G_α and

$$\begin{bmatrix} 0 & \widetilde{G}_\alpha \\ \widetilde{G}_\alpha & 0 \end{bmatrix} \quad \text{in place of} \quad \begin{bmatrix} 0 & G_\alpha \\ G_\alpha & 0 \end{bmatrix} \quad (\beta \neq \beta^*).$$

The verification of the alternative version is left as Ex. 8.7.1, and we indicate only the following relevant, easily verifiable equalities:

$$F_{2\varepsilon+1} \left(t \begin{bmatrix} 0 & 0 & F_\varepsilon \\ 0 & 0 & 0 \\ F_\varepsilon & 0 & 0 \end{bmatrix} + G_{2\varepsilon+1} \right) = \left(t \begin{bmatrix} 0 & 0 & F_\varepsilon \\ 0 & 0 & 0 \\ F_\varepsilon & 0 & 0 \end{bmatrix} + \widetilde{G}_{2\varepsilon+1} \right) F_{2\varepsilon+1};$$

$$F_k \left(F_k + t G_k \right) = \left(F_k + t \widetilde{G}_k \right) F_k;$$

$$F_{2m}\left(\begin{bmatrix} 0 & (t+\beta)F_m \\ (t+\beta^*)F_m & 0 \end{bmatrix} + \begin{bmatrix} 0 & G_m \\ G_m & 0 \end{bmatrix}\right)$$

$$= \left(\begin{bmatrix} 0 & (t+\beta^*)F_m \\ (t+\beta)F_m & 0 \end{bmatrix} + \begin{bmatrix} 0 & \widetilde{G}_m \\ \widetilde{G}_m & 0 \end{bmatrix}\right)F_{2m}, \quad \beta \neq \beta^*.$$

We also indicate an explicit congruence of the block

$$\begin{bmatrix} 0 & (t+\beta_j)F_{m_j} \\ (t+\beta_j^*)F_{m_j} & 0 \end{bmatrix} + \begin{bmatrix} 0 & G_{m_j} \\ G_{m_j} & 0 \end{bmatrix} \qquad (j=1,\dots,s) \qquad (8.1.8)$$

to the analogous block in which β_j is replaced by β_j^*. Denoting the block (8.1.8) by $K_{2m_j}(\beta_j)$, we have:

$$\begin{bmatrix} 0 & I_{m_j} \\ I_{m_j} & 0 \end{bmatrix} K_{2m_j}(\beta_j) \begin{bmatrix} 0 & I_{m_j} \\ I_{m_j} & 0 \end{bmatrix} = K_{2m_j}\left(\beta_j^*\right).$$

Remark 8.1.5. Another alternative version of Theorem 8.1.2 is obtained by using $-\widetilde{G}_m$ and/or $-G_m$ in place of $\widetilde{G}_m$ and/or G_m. This observation is based on equalities

$$\Xi_\ell(1)((t+\alpha)F_\ell - G_\ell)\Xi_\ell(1) = (t+\alpha)F_\ell + \widetilde{G}_\ell,$$
$$\Xi_\ell(1)(F_\ell - tG_\ell)\Xi_\ell(1) = F_\ell + \widetilde{G},$$

where $\Xi_\ell(1)$ is given by (1.2.6). Noting that $\Xi_\ell(1)^* = (-1)^{\ell-1}\Xi_\ell(1)$, we obtain: *Upon replacement of G_{ℓ_j}, resp. $\widetilde{G}_{\ell_j}$, by $-G_{\ell_j}$, resp. $-\widetilde{G}_{\ell_j}$, in the block $(t+\alpha_j)F_{\ell_j} + G_{\ell_j}$, resp. $(t+\alpha_j)F_{\ell_j} + \widetilde{G}_{\ell_j}$, in (8.1.6), the sign η_j remains the same if ℓ_j is odd and reverses to its negative if ℓ_j is even. This rule applies to the blocks $F_{k_i} + tG_{k_i}$ as well.*

Remark 8.1.6. The blocks $\eta(F_k + tG_k)$ and $\eta((t+\alpha)F_k + G_k)$ ($\eta = \pm 1$, α real), which appear in (8.1.6), can be conveniently represented in a unified form

$$(\sin\theta)F_k - (\cos\theta)G_k + t((\cos\theta)F_k + (\sin\theta)G_k), \qquad (8.1.9)$$

where $0 \leq \theta < 2\pi$.

The parameters θ, α, and η are related by $\alpha = \tan\theta$ and by the real number $(-1)^{k-1}\cos\theta$ having the sign of η (assuming $\cos\theta \neq 0$; if $\cos\theta = 0$, then (8.1.9) coincides with $\eta(F_k + tG_k)$). To verify this statement, we present the following proposition (in a more general form than is actually needed here).

Proposition 8.1.7. *Let $H_1 = [h_{ij}^{(1)}]_{i,j=1}^k$ and $H_2 = [h_{ij}^{(2)}]_{i,j=1}^k$ be real triangular Hankel matrices:*

$$h_{ij}^{(\ell)} = \begin{cases} p_{i+j-1}^{(\ell)} & \text{if } i+j \leq k+1, \\ \\ 0 & \text{otherwise}, \end{cases}$$

where $p_1^{(\ell)},\dots,p_k^{(\ell)}$ are real numbers, $\ell = 1,2$. Assume that $p_k^{(1)} \neq 0$ and

$$x := p_{k-1}^{(2)}p_k^{(1)} - p_{k-1}^{(1)}p_k^{(2)} \neq 0. \qquad (8.1.10)$$

Then the canonical form of the real symmetric matrix pencil $tH_1 + H_2$, under $\mathbb{R}$-congruence, is given by

$$\eta((t+\alpha)F_k + G_k), \qquad (8.1.11)$$

where $\alpha := p_k^{(2)}/p_k^{(1)}$ *and*

$$
\eta = \begin{cases}
\operatorname{sign} p_k^{(1)} & \text{if } x > 0, \\
\operatorname{sign} p_k^{(1)} & \text{if } x < 0 \text{ and } k \text{ odd}, \\
-\operatorname{sign} p_k^{(1)} & \text{if } x < 0 \text{ and } k \text{ even}.
\end{cases} \tag{8.1.12}
$$

The proof will be given in terms of H-hermitian matrices and their canonical forms, as described later in Chapter 10. For this reason the proof of Proposition 8.1.7 is postponed until Chapter 10.

The alternative form (8.1.9) is just a particular case of Proposition 8.1.7 (assuming $\cos\theta \neq 0$), with

$$
p_k^{(1)} = \cos\theta, \quad p_{k-1}^{(1)} = \sin\theta, \quad p_k^{(2)} = \sin\theta, \quad p_{k-1}^{(2)} = -\cos\theta,
$$

and

$$
p_j^{(1)} = p_j^{(2)} = 0 \qquad \text{for } j = 1, 2, \ldots, k-2.
$$

Remark 8.1.8. Note that the canonical forms (8.1.3) and (8.1.6) can be chosen to involve only real and complex numbers. It turns out that a version of these canonical forms exists that involves real numbers only. We will provide explicit formulas. Let $\beta \in \mathsf{H} \setminus \mathsf{R}$, and write $\beta = \mu - q\nu$, where $\mu, \nu \in \mathsf{R}$, $\nu \neq 0$, and $q^2 = -1$. We will work with the following matrices:

$$
C_{2m} := \frac{1}{\sqrt{2}} \left(\begin{bmatrix} 1 & 1 \\ -q & q \end{bmatrix} \oplus \begin{bmatrix} 1 & 1 \\ -q & q \end{bmatrix} \oplus \cdots \oplus \begin{bmatrix} 1 & 1 \\ -q & q \end{bmatrix} \right)
$$

of size $2m \times 2m$, where m is a positive integer,

$$
D_{2m} := [\, e_1 \quad e_3 \quad \cdots \quad e_{2m-1} \quad e_2 \quad e_4 \quad \cdots \quad e_{2m} \,] \in \mathsf{R}^{2m \times 2m},
$$

and $T_{2m} = C_{2m}D_{2m}$. Let $Z(t, \mu, \nu)$ be the real matrix pencil given by (1.2.9). A computation shows that

$$
T_{2m}^* Z_{2m}(t, \mu, \nu) T_{2m} = \begin{bmatrix} 0_m & K_m \\ K_m^* & 0_m \end{bmatrix},
$$

where $K_m := qG_m + (qt + q\mu + \nu)F_m$. Therefore,

$$
\begin{bmatrix} I_m & 0 \\ 0 & -q^{-1}I_m \end{bmatrix} T_{2m}^* Z_{2m}(t, \mu, \nu) T_{2m} \begin{bmatrix} I_m & 0 \\ 0 & q^{-1}I_m \end{bmatrix}
$$

$$
= \begin{bmatrix} 0_m & G_m + (t + \beta)F_m \\ G_m + (t + \beta^*)F_m & 0_m \end{bmatrix},
$$

and so the block (8.1.8) in (8.1.6) can be replaced by a real symmetric block

$$
Z_{2m_j}(t, \mathfrak{R}(\beta), -q^{-1}\mathfrak{V}(\beta)).
$$

8.2   PROOF OF THEOREM 8.1.2

We start with the proof of Part (a).

Thus, let $A + tB$ a hermitian matrix pencil, and let $A_0 + tB_0$ be its Kronecker form of (7.3.1). Then

$$A_0 + tB_0 = P(A + tB)Q \qquad (8.2.1)$$

for some invertible $P, Q \in \mathsf{H}^{n \times n}$. Taking conjugate transposes in (8.2.1), we have

$$A_0^* + tB_0^* = Q^*(A + tB)P^*.$$

Clearly, $A_0^* + tB_0^*$ is also strictly equivalent to $A + \lambda B$. But

$$
\begin{aligned}
A_0^* + tB_0^* \;=\; & 0_{v \times u} \oplus L_{\varepsilon_1 \times (\varepsilon_1 + 1)}^T \oplus \cdots \oplus L_{\varepsilon_p \times (\varepsilon_p + 1)}^T \\
\oplus \;& L_{\eta_1 \times (\eta_1 + 1)} \oplus \cdots \oplus L_{\eta_q \times (\eta_q + 1)} \\
\oplus \;& (I_{k_1} + tJ_{k_1}(0)^T) \oplus \cdots \oplus (I_{k_r} + tJ_{k_r}(0)^T) \\
\oplus \;& (tI_{\ell_1} + J_{\ell_1}(\alpha_1)^*) \oplus \cdots \oplus (tI_{\ell_s} + J_{\ell_s}(\alpha_s)^*),
\end{aligned}
$$

which, in view of the similarity between $J_m(\alpha)^*$ and $J_m(\alpha^*)$, as established by the equality

$$F_m J_m(\alpha^*) F_m = (J_m(\alpha))^*,$$

is strictly equivalent to

$$
\begin{aligned}
0_{r \times u} \quad \oplus \;& L_{\varepsilon_1 \times (\varepsilon_1 + 1)}^T \oplus \cdots \oplus L_{\varepsilon_p \times (\varepsilon_p + 1)}^T \\
\oplus \;& L_{\eta_1 \times (\eta_1 + 1)} \oplus \cdots \oplus L_{\eta_q \times (\eta_q + 1)} \\
\oplus \;& (I_{k_1} + tJ_{k_1}(0)) \oplus \cdots \oplus (I_{k_r} + tJ_{k_r}(0)) \\
\oplus \;& (tI_{\ell_1} + J_{\ell_1}(\alpha_1^*)) \oplus \cdots \oplus (tI_{\ell_s} + J_{\ell_s}(\alpha_s^*)). \qquad (8.2.2)
\end{aligned}
$$

Now the uniqueness of the Kronecker form implies that $r = u$; $p = q$; $\varepsilon_j = \eta_j$ for $j = 1, \ldots, q$ and, also, that the number of blocks $tI_{\ell_u} + J_{\ell_u}(\alpha_t)$ with α_u nonreal is even and these blocks appear in pairs: $tI_{\ell_u} + J_{\ell_u}(\alpha_u)$, $tI_{\ell_u} + J_{\ell_u}(\alpha_u^*)$.

The goal of this analysis is to obtain an *hermitian* matrix pencil which is strictly equivalent to (8.2.2). To this end, symmetric expressions can be obtained from the blocks of (8.2.2) by applying suitable strict equivalence transformations. Thus, for the singular part of the form (8.2.2), it is easily verified that

$$(L_{\varepsilon \times (\varepsilon + 1)} \oplus (F_{\varepsilon + 1} L_{\varepsilon \times (\varepsilon + 1)}^T F_\varepsilon)) F_{2\varepsilon + 1} = t \begin{bmatrix} 0 & 0 & F_\varepsilon \\ 0 & 0 & 0 \\ F_\varepsilon & 0 & 0 \end{bmatrix} + G_{2\varepsilon + 1}.$$

Similarly, for the regular part:

$$(I_{k_u} + tJ_{k_u}(0)) F_{k_u} = tG_{k_u} + F_{k_u}; \quad (tI_{\ell_u} + J_{\ell_u}(\alpha)) F_{\ell_u} = (t + \alpha) F_{\ell_u} + G_{\ell_u},$$

where α is real; and

$$((tI_{\ell_u} + J_{\ell_u}(\alpha)) \oplus (tI_{\ell_u} + J_{\ell_u}(\alpha^*))) F_{2\ell_u}$$

$$= \begin{bmatrix} 0 & (t + \alpha) F_{\ell_u} \\ (t + \alpha^*) F_{\ell_u} & 0 \end{bmatrix} + \begin{bmatrix} 0 & G_{\ell_u} \\ G_{\ell_u} & 0 \end{bmatrix} \qquad (8.2.3)$$

for nonreal α.

Applying these transformations to the blocks in (8.2.2), the form (8.1.3) is obtained.

The uniqueness statement in Part (a) is immediate from the uniqueness statement in Theorem 7.3.2 and the above analysis.

Next, we prove existence in Part (b). For this a technical lemma will be needed.

Lemma 8.2.1. *Let $p(t)$ be a real scalar polynomial having one of the forms $p(t) = t - a$ with $a > 0$, or $p(t) = (t - a)(t - \bar{a})$ with $a \in \mathsf{C} \setminus \mathsf{R}$. Then for every positive integer m, there exists a scalar polynomial $f_m(t)$ with real coefficients such that $t \equiv (f_m(t))^2$ modulo a real polynomial multiple of $(p(t))^m$.*

Proof. For $m = 1$, we let $f_1(t) = \sqrt{a}$ if $p(t) = t - a$, and

$$f_1(t) = \{|a| + t\} \{2(|a| + \Re(a))\}^{-1/2}$$

if $p(t) = (t - a)(t - \bar{a})$. Continue by induction on m. If $f_m(t)$ has already been found, then

$$t = f_m(t)^2 + u(t)(p(t))^m \tag{8.2.4}$$

for some real polynomial $u(t)$. In particular, in view of the form of $p(t)$, (8.2.4) implies that $f_m(t)$ and $p(t)$ are relatively prime. Therefore, there exist polynomials $g(t)$ and $h(t)$ with real coefficients such that

$$2g(t)f_m(t) + h(t)p(t) = u(t).$$

Now set

$$f_{m+1}(t) = f_m(t) + g(t)(p(t))^m.$$

$\square$

We now return to the proof of existence in Part (b).

Let $A + tB$ be a hermitian matrix pencil, with matrices of size $n \times n$. By Theorem 8.1.2, there exist invertible matrices $P, Q \in \mathsf{H}^{n \times n}$ such that

$$A + tB = P(A_0 + tB_0)Q, \tag{8.2.5}$$

where $A_0 + tB_0$ is the canonical form of (8.1.3). Replacing $A + tB$ by the congruent pencil $(Q^*)^{-1}(A + \lambda B)Q^{-1}$, it may (and will) be assumed without loss of generality that $Q = I$. Then the equalities $A = A^*$, $B = B^*$ imply

$$P(A_0 + tB_0) = A^* + tB^* = (A_0^* + tB_0^*)P^* = (A_0 + tB_0)P^*,$$

i.e.,

$$PA_0 = A_0 P^*, \quad PB_0 = B_0 P^*. \tag{8.2.6}$$

Therefore, for every polynomial $f(t)$ with real coefficients we also have

$$f(P)A_0 = A_0(f(P))^*, \quad f(P)B_0 = B_0(f(P))^*. \tag{8.2.7}$$

The argument now proceeds in several cases, depending on the nature of eigenvalues of P.

Case 8.2.2. *P has only one distinct complex eigenvalue with nonnegative real part, and this eigenvalue is real.*

Assume first that the eigenvalue of P is positive. Let $(t - \gamma)^m$, $\gamma > 0$, be the minimal polynomial of P^{-1}. By Lemma 8.2.1 there exists a polynomial $f_m(t)$ with real coefficients such that $t = (f_m(t))^2$ modulo a real polynomial multiple of $(t - \gamma)^m$. Then $P^{-1} = (f_m(P^{-1}))^2$, and since P^{-1} itself is a polynomial of P with real coefficients, we have $P^{-1} = (f(P))^2$ for some polynomial $f(t)$ with real coefficients. Let $R = f(P)$, and use (8.2.7) to obtain

$$
\begin{aligned}
R(A + tB)R^* &= RP(A_0 + tB_0)R^* = RP(A_0 + tB_0)(f(P))^* \\
&= f(P)Pf(P)(A_0 + tB_0) = A_0 + tB_0, \quad (8.2.8)
\end{aligned}
$$

where the second equality is valid by virtue of (8.2.7), and the existence of (8.1.6) follows.

Now assume that the eigenvalue of P is negative. Then, arguing as in the preceding paragraph, it is found that $(-P)^{-1} = (f(-P))^2$ for some polynomial $f(t)$ with real coefficients. As in (8.2.8) it follows that

$$
R(A + tB)R^* = -(A_0 + tB_0),
$$

where $R = f(-P)$. Thus, to complete the proof of existence of the form (8.1.6) in Case 8.2.2, it remains to show that each of the blocks

$$
Z_1 := t \begin{bmatrix} 0 & 0 & F_\varepsilon \\ 0 & 0 & 0 \\ F_\varepsilon & 0 & 0 \end{bmatrix} + G_{2\varepsilon+1} \quad (8.2.9)
$$

and

$$
Z_2 := \begin{bmatrix} 0 & (t + \beta)F_m \\ (t + \beta^*)F_m & 0 \end{bmatrix} + \begin{bmatrix} 0 & G_m \\ G_m & 0 \end{bmatrix}, \quad \beta \in \mathsf{H} \setminus \mathsf{R}, \quad (8.2.10)
$$

is congruent to its negative. This is easily seen because for the block (8.2.10) we have

$$
\begin{bmatrix} I_m & 0 \\ 0 & -I_m \end{bmatrix} Z_2 \begin{bmatrix} I_m & 0 \\ 0 & -I_m \end{bmatrix} = -Z_2,
$$

and for the block (8.2.9) we have

$$
\begin{bmatrix} I_\varepsilon & 0 \\ 0 & -I_{\varepsilon+1} \end{bmatrix} Z_1 \begin{bmatrix} I_\varepsilon & 0 \\ 0 & -I_{\varepsilon+1} \end{bmatrix} = -Z_1. \quad (8.2.11)
$$

Case 8.2.3. *P has exactly one distinct complex eigenvalue with nonnegative imaginary part, and this eigenvalue is nonreal.*

Let $\alpha \neq \overline{\alpha}$ be the complex eigenvalue of P^{-1} with nonnegative imaginary part. If $g(t) = ((t - \alpha)(t - \overline{\alpha}))^w$ is the minimal polynomial of P^{-1}, then P^{-1} satisfies $g(P^{-1}) = 0$. Using Lemma 8.2.1, we find a polynomial $g_w(t)$ with real coefficients such that $t \equiv (g_w(t))^2$ modulo a real polynomial multiple of $g(t)$. Then $P^{-1} = (g_w(P^{-1}))^2$. Since P is itself a matrix root of a polynomial with real coefficients, namely, $\widehat{g}(P) = 0$, where

$$
\widehat{g}(t) = ((t - \alpha^{-1})(t - (\overline{\alpha})^{-1}))^w,
$$

it follows that P^{-1} is also a polynomial of P with real coefficients. Now argue as in Case 8.2.2 (when the eigenvalue of P is positive).

Case 8.2.4. *All other possibilities* (*not covered in Cases 8.2.2 and 8.2.3*).

In this case, using the quaternion Jordan form of Theorem 5.5.3, P can be written in the form

$$P = S \operatorname{diag}(P_1, P_2, \ldots, P_r) S^{-1}, \qquad (8.2.12)$$

where $S \in \mathsf{H}^{n \times n}$ is invertible and $P_1, \ldots, P_r$ are matrices of sizes $n_1 \times n_1, \ldots, n_r \times n_r$, respectively, such that

$$\lambda \in \sigma(P_i) \implies \mu \lambda \mu^{-1} \notin \sigma(P_j) \quad \text{for} \quad j \neq i \text{ and for any } \mu \in \mathsf{H} \setminus \{0\} \qquad (8.2.13)$$

(i.e., any similarity class of quaternion eigenvalues of P is confined to just one block, P_j). We also have $r \geq 2$ (the situations when $r = 1$ are covered in Cases 8.2.2 and 8.2.3). Substituting (8.2.12) in the equality $A + tB = P(A_0 + tB_0)$, we obtain

$$\left(P_1^{-1} \oplus \cdots \oplus P_r^{-1}\right)\left(\widetilde{A} + t\widetilde{B}\right) = \widetilde{A}_0 + t\widetilde{B}_0, \qquad (8.2.14)$$

where

$$\widetilde{A} = S^{-1}A(S^*)^{-1}, \qquad \widetilde{B} = S^{-1}B(S^*)^{-1},$$

$$\widetilde{A}_0 = S^{-1}A_0(S^*)^{-1}, \qquad \widetilde{B}_0 = S^{-1}B_0(S^*)^{-1}.$$

Partition the matrix $\widetilde{A}$:

$$\widetilde{A} = [M_{ij}]_{i,j=1}^r,$$

where M_{ij} is of the size $n_i \times n_j$. Since $\widetilde{A}$ and $\widetilde{A}_0$ are hermitian, (8.2.14) implies

$$P_i^{-1}M_{ij} = M_{ij}(P_j^*)^{-1}.$$

In view of (8.2.13),

$$\sigma(P_i^{-1}) \cap \sigma((P_j^*)^{-1}) = \emptyset \quad \text{for} \quad i \neq j.$$

Now by Theorem 5.11.1 we have $M_{ij} = 0$ $(i \neq j)$. In other words,

$$\widetilde{A} = M_{11} \oplus \cdots \oplus M_{rr}.$$

Similarly, $\widetilde{B} = N_1 \oplus \cdots \oplus N_{rr}$, where N_{ii} is of size $n_i \times n_i$.

Now, induction is used on n to complete the proof that every hermitian matrix pencil is congruent to a pencil of the form (8.1.6); the basis of induction, when A and B are scalars, is trivially verified. Indeed, by the induction hypothesis, each pencil $M_{ii} + tN_{ii}$ is congruent to a pencil of the form (8.1.6); therefore, the same is true for

$$\widetilde{A} + t\widetilde{B} = (M_{11} + tN_{11}) \oplus \cdots \oplus (M_{rr} + tN_{rr}).$$

This completes the proof of existence of the form (8.1.6).

It remains to prove the uniqueness in Part (b). It will be proved using the real matrix representation given by the map $\chi_{n,n}$. Let $A_0 + tB_0$ be the hermitian matrix pencil given by (8.1.6), let $A_0' + tB_0'$ be a hermitian matrix pencil given by the same formula (8.1.6) but with possibly different signs $\delta_1', \ldots, \delta_r', \eta_1', \ldots, \eta_q'$ in place of $\delta_1, \ldots, \delta_r, \eta_1, \ldots, \eta_q$, respectively, and assume that $A_0 + tB_0$ and $A_0' + tB_0'$ are congruent:

$$A_0' + tB_0' = S^*(A_0 + tB_0)S, \quad S \in \mathsf{H}^{n \times n} \text{ invertible.} \qquad (8.2.15)$$

We have to prove that for every positive integer k the equality

$$\sum_{j=1,2,\ldots,k_r \text{ such that } k_j=k} \delta'_j = \sum_{j=1,2,\ldots,k_r \text{ such that } k_j=k} \delta_j \qquad (8.2.16)$$

holds true, and for every real number α and every positive integer k the equality

$$\sum_{j=1,2,\ldots,q \text{ such that } \alpha_j=\alpha \text{ and } \ell_j=k} \eta_j = \sum_{j=1,2,\ldots,q \text{ such that } \alpha_j=\alpha \text{ and } \ell_j=k} \eta'_j \qquad (8.2.17)$$

holds true. We focus on (8.2.16); the consideration of (8.2.17) is completely analogous. To this end apply the map $\chi_{n,n}$ to equality (8.2.15). We then obtain that the real symmetric matrix pencils $\chi_{n,n}(A_0)+t\chi_{n,n}(B_0)$ and $\chi_{n,n}(A'_0)+t\chi_{n,n}(B'_0)$ are R-congruent. It is easy to see that the part of $\chi_{n,n}(A_0)+t\chi_{n,n}(B_0)$ that corresponds to the eigenvalue at infinity is R-congruent to

$$\sum_{j=1}^{r} \delta_j \left((F_{k_j} + tG_{k_j})^{\oplus 4} \right). \qquad (8.2.18)$$

Analogously, the part of $\chi_{n,n}(A'_0)+t\chi_{n,n}(B'_0)$ corresponding to the eigenvalue at infinity is R-congruent to

$$\sum_{j=1}^{r} \delta'_j \left((F_{k_j} + tG_{k_j})^{\oplus 4} \right). \qquad (8.2.19)$$

It follows that (8.2.18) and (8.2.19) are R-congruent, and the uniqueness of the canonical form of a pair of real symmetric matrices under R-congruence (see Theorem 15.2.1) yields the desired equality (8.2.16). □

We remark that a direct, independent proof of uniqueness can be obtained by following the arguments in the the proof of the corresponding result in the complex case in Lancaster and Rodman [92].

8.3   POSITIVE SEMIDEFINITE LINEAR COMBINATIONS

Definition 8.3.1. If $A, B \in \mathsf{H}^{n \times n}$ are hermitian and $aA+bB$ is positive semidefinite for some real numbers a, b not both zero, then we say that the pair (A, B) admits a *positive semidefinite linear combination*. Similarly, we say that the pair (A, B) admits a *positive definite linear combination* if $aA + bB$ is positive definite for some real numbers a, b.

As we shall see later in this section, these properties play a role in the important problem of simultaneous diagonalization.

We start with an elementary proposition.

Proposition 8.3.2. *Let $A, B \in \mathsf{H}^{n \times n}$ be hermitian. Then:*

(a) *if (A, B) admits a positive semidefinite, resp. definite, linear combination, then so do all pairs simultaneously congruent to (A, B), i.e., of the form (S^*AS, S^*BS) for some invertible $S \in \mathsf{H}^{n \times n}$;*

(b) *if (A, B) admits a positive semidefinite, resp. definite, linear combination, then so do all pairs of the form $(\alpha A + \beta B, \gamma A + \delta B)$, where $\alpha, \beta, \gamma, \delta$ are real numbers such that $\alpha\delta - \beta\gamma \neq 0$.*

Proof. Part (a) follows from a particular case of the inertia theorem: a hermitian matrix is positive semidefinite, resp. definite, if and only if any congruent matrix is as well. For Part (b), note that

$$a'(\alpha A + \beta B) + b'(\gamma A + \delta B) = (a'\alpha + b'\gamma)A + (a'\beta + b'\delta)B, \qquad a', b', a, b \in \mathsf{R},$$

and

$$aA + bB = a'(\alpha A + \beta B) + b'(\gamma A + \delta)B, \qquad a, b \in \mathsf{R}, \tag{8.3.1}$$

where in (8.3.1) a', b' are found from the equation

$$\begin{bmatrix} \alpha & \gamma \\ \beta & \delta \end{bmatrix} \begin{bmatrix} a' \\ b' \end{bmatrix} = \begin{bmatrix} a \\ b \end{bmatrix}.$$

Thus, every linear combination with real coefficients of A and B is that of $\alpha A + \beta B$ and $\gamma A + \delta B$, and vice versa, and part (b) follows. $\qquad\square$

As an application of Theorem 8.1.2, a criterion for having a positive semidefinite linear combination can now be given.

Theorem 8.3.3. *Let $A, B \in \mathsf{H}^{n \times n}$ be hermitian. Then the pair (A, B) admits a positive semidefinite linear combination if and only if the following property holds true.*

(α) For any $x \in \mathsf{H}^{n \times 1}$ such that $\langle Ax, x \rangle = \langle Bx, x \rangle = 0$, the two vectors Ax and Bx are linearly dependent over the reals.

We relegate the proof of Theorem 8.3.3 to the next section.

The following result provides a criterion for existence of a positive (or negative) definite linear combination.

Theorem 8.3.4. (a) *Let $A, B \in \mathsf{H}^{n \times n}$ be hermitian. Then the pair (A, B) admits a positive definite linear combination if and only if the following property holds true:*

$$\langle Ax, x \rangle = \langle Bx, x \rangle = 0, \quad x \in \mathsf{H}^{n \times 1} \qquad \Longrightarrow \qquad x = 0. \tag{8.3.2}$$

Proof. If $C := aA + bB$ is positive definite for some real a and b, then (8.3.2) obviously holds true, because $\langle Cx, x \rangle = 0$ is possible only if $x = 0$. Conversely, if (8.3.2) holds true, then $0 \notin WJ_*^H(A, B)$, the joint numerical range of A and B. Then the convexity (Theorem 3.5.7) and compactness of $WJ_*^H(A, B)$ imply by virtue of the convex separation theorem, 6.6.2, that

$$WJ_*^H(A, B) \subseteq \{(x_1, x_2) : ax_1 + bx_2 \geq c\},$$

where a, b, c are real numbers such that a and b are not both zero and $c > 0$. Then $aA + bB$ is positive definite. $\qquad\square$

It is a well-known fact that if A and B are two complex hermitian (or real symmetric) matrices and if the pair (A, B) admits a positive definite linear combination, then A and B are *simultaneously diagonalizable* by congruence, i.e., there exist real diagonal matrices D_1 and D_2 such that the hermitian pencil $A + tB$ is congruent to $D_1 + tD_2$ (with the real congruence matrix if A and B are real). The result is valid also for quaternion hermitian matrices, and can be easily obtained by inspection of (8.1.3). It generally fails if the positive definiteness hypothesis is omitted.

Example 8.3.5. Let

$$A = \begin{bmatrix} 0 & 1 \\ 1 & 0 \end{bmatrix}, \qquad B = \begin{bmatrix} 1 & 0 \\ 0 & 0 \end{bmatrix}.$$

Then there does not exist an invertible $S \in \mathsf{H}^{n \times n}$ such that both S^*AS and S^*BS are diagonal. Indeed, this follows from the uniqueness of the canonical form of the matrix pencil $A + tB$ in Theorem 8.1.2, since $\begin{bmatrix} t & 1 \\ 1 & 0 \end{bmatrix}$ represents one of the primitive canonical blocks in that theorem and, therefore, cannot be simultaneously congruent to a direct sum of primitive blocks.   □

Theorem 8.1.2(b) leads to results concerning simultaneous diagonalizability with the positive definiteness hypothesis relaxed.

Theorem 8.3.6. *Let* $A, B \in \mathsf{H}^{n \times n}$ *be hermitian matrices. Assume that there exists a linear combination*

$$C = \alpha A + \beta B, \qquad \alpha, \beta \in \mathsf{R},$$

such that C *is positive semidefinite, and*

$$\operatorname{Ker}(C) \subseteq \operatorname{Ker}(A) \cap \operatorname{Ker}(B). \tag{8.3.3}$$

Then A *and* B *are simultaneously diagonalizable by congruence, with a quaternion congruence matrix.*

In particular, the hypotheses of Theorem 8.3.6 are satisfied if C is positive definite.

Proof. We may clearly assume that at least one of α and β is nonzero (otherwise the theorem is trivial). Applying the transformation of Proposition 8.3.2, we may further assume that $C = B$. Finally, take $A + tB$ in the form (8.1.6). The condition that B is positive semidefinite easily implies that

$$\begin{aligned} A + tB &= 0_{u \times u} &\oplus& \quad \delta_1\left(F_{k_1} + tG_{k_1}\right) \oplus \cdots \oplus \delta_r\left(F_{k_r} + tG_{k_r}\right) \\ &&\oplus& \quad \operatorname{diag}\left(t + \alpha_1, \ldots, t + \alpha_q\right), \end{aligned} \tag{8.3.4}$$

where $k_j \leq 2$. If there is a term in (8.3.4) with $k_j = 2$, then $\operatorname{Ker}(B)$ is not contained in $\operatorname{Ker}(A)$, a contradiction to (8.3.3). Thus, we must have all $k_j = 1$, and the right-hand side of (8.3.4) is diagonal.   □

Example 8.3.5 shows that condition (8.3.3) is essential in Theorem 8.3.6.

8.4   PROOF OF THEOREM 8.3.3

It will be convenient to dispose of a particular case first.

Lemma 8.4.1. *Let $A, B \in \mathsf{H}^{n \times n}$ be hermitian with block forms*

$$A = \begin{bmatrix} D_1 & 0 \\ 0 & D_2 \end{bmatrix}, \quad B = \begin{bmatrix} I_p & 0 \\ 0 & -I_q \end{bmatrix},$$

where $D_1 \in \mathsf{H}^{p \times p}$ and $D_2 \in \mathsf{H}^{q \times q}$ are hermitian. If property (α) is satisfied, then the pair (A, B) admits a positive semidefinite linear combination.

Proof. Applying unitary similarity to each of D_1 and D_2, we may assume that D_1 and D_2 are diagonal. The proof proceeds by induction on p and q. The basis of induction, i.e., $p = 0$ or $q = 0$, is trivial (then $\pm B$ is positive definite). The case $p = q = 1$ is handled next; thus, let

$$A = \begin{bmatrix} d_1 & 0 \\ 0 & d_2 \end{bmatrix}, \quad B = \begin{bmatrix} 1 & 0 \\ 0 & -1 \end{bmatrix}, \quad d_1, d_2 \in \mathsf{R}.$$

If A is positive or negative semidefinite, we are done; otherwise, replacing A with $-A$ and multiplying A by a positive number, we can assume without loss of generality, that $A = \mathrm{diag}\,(1, d)$, where $d < 0$. If $d \geq -1$, then $A - B$ is positive semidefinite; if $d < -1$, then $-A + B$ is positive semidefinite.

Suppose at least one of p and q is larger than 1; say $p > 1$. We may assume that the largest eigenvalue of D_1, call it γ, is in the top left corner of D_1. Replace A with $A - \gamma B$. This transformation does not change the property (α) and the property of having a (nontrivial) positive semidefinite linear combination. Then

$$A = \begin{bmatrix} 0_{1 \times 1} & 0 & 0 \\ 0 & \widetilde{D}_1 & 0 \\ 0 & 0 & \widetilde{D}_2 \end{bmatrix},$$

where $\widetilde{D}_1$ is negative semidefinite. Applying the induction hypothesis to

$$\widetilde{A} := \begin{bmatrix} \widetilde{D}_1 & 0 \\ 0 & \widetilde{D}_2 \end{bmatrix}, \quad \widetilde{B} := \begin{bmatrix} I_{p-1} & 0 \\ 0 & -I_q \end{bmatrix},$$

we see that there exist $\alpha, \beta \in \mathsf{R}$ not both zeros such that $\alpha \widetilde{A} + \beta \widetilde{B}$ is positive semidefinite. Since $\widetilde{D}_1$ is negative semidefinite, it follows that $\beta \geq 0$. But then $\alpha A + \beta B$ is positive semidefinite as well. $\square$

Proof of Theorem 8.3.3. Assume $\alpha A + \beta B$ is positive semidefinite for some $\alpha, \beta \in \mathsf{R}$, not both zero. If $\langle Ax, x \rangle = \langle Bx, x \rangle = 0$, then $\langle (\alpha A + \beta B)x, x \rangle = 0$, and in view of positive semidefiniteness of $\alpha A + \beta B$ we have $(\alpha A + \beta B)x = 0$. So the R-linear dependence of Ax and Bx follows.

Conversely, assume that (α) holds true. Without loss of generality, it may be assumed that the hermitian pencil $A + tB$ has the form (8.1.6). Clearly, condition (α) holds true for every constituent diagonal block in the direct sum (8.1.6).

Now consider whether condition (α) holds true for each of the possible diagonal blocks in (8.1.6). Clearly, (α) holds true for $0_{u \times u}$.

Let

$$A' + tB' = \lambda \begin{bmatrix} 0 & 0 & F_\varepsilon \\ 0 & 0 & 0 \\ F_\varepsilon & 0 & 0 \end{bmatrix} + G_{2\varepsilon+1},$$

and let $x = e_1 \in \mathsf{H}^{(2\varepsilon+1)\times 1}$, the first unit coordinate vector. Then $\langle A'x, x \rangle = \langle B'x, x \rangle = 0$, but $A'x$ and $B'x$ are linearly independent. Thus, (α) does not hold true for the pair (A', B').

Let

$$A' + tB' = F_k + tG_k \quad \text{or} \quad A' + tB' = (t + \alpha)F_k + G_k, \quad (\alpha \in \mathsf{R}).$$

Then, using $x = e_1$ again, it follows that (α) does not hold for the pair (A', B'), unless $k \leq 2$.

Let

$$A' + tB' = \left[\begin{array}{cc} 0 & (t + \beta)F_m + G_m \\ (t + \overline{\beta})F_m + G_m & 0 \end{array} \right], \quad \beta \in \mathsf{C} \setminus \mathsf{R}.$$

Then, using $x = e_1$ we see that (α) does not hold for (A', B').

Next, consider the four possibilities (8.4.1)–(8.4.4) for the pair (A', B'). It will be proved that, in each of the four cases, the pair (A', B') does not have property (α):

$$
\begin{aligned}
A' &= \left[\begin{array}{cc} 0 & 1 \\ 1 & 0 \end{array} \right] \oplus \pm \left[\begin{array}{cc} 1 & \alpha \\ \alpha & 0 \end{array} \right], \\
B' &= \left[\begin{array}{cc} 1 & 0 \\ 0 & 0 \end{array} \right] \oplus \pm \left[\begin{array}{cc} 0 & 1 \\ 1 & 0 \end{array} \right], \quad \alpha \in \mathsf{R}. \tag{8.4.1}
\end{aligned}
$$

$$
\begin{aligned}
A' &= \left[\begin{array}{cc} 0 & 1 \\ 1 & 0 \end{array} \right] \oplus \left[\begin{array}{cc} 0 & -1 \\ -1 & 0 \end{array} \right], \\
B' &= \left[\begin{array}{cc} 1 & 0 \\ 0 & 0 \end{array} \right] \oplus \left[\begin{array}{cc} -1 & 0 \\ 0 & 0 \end{array} \right], \quad \alpha \in \mathsf{R}. \tag{8.4.2}
\end{aligned}
$$

$$
\begin{aligned}
A' &= \left[\begin{array}{cc} 1 & \alpha \\ \alpha & 0 \end{array} \right] \oplus \left[\begin{array}{cc} -1 & -\alpha \\ -\alpha & 0 \end{array} \right], \\
B' &= \left[\begin{array}{cc} 0 & 1 \\ 1 & 0 \end{array} \right] \oplus \left[\begin{array}{cc} 0 & -1 \\ -1 & 0 \end{array} \right], \quad \alpha \in \mathsf{R}. \tag{8.4.3}
\end{aligned}
$$

$$
\begin{aligned}
A' &= \left[\begin{array}{cc} 1 & \alpha \\ \alpha & 0 \end{array} \right] \oplus \pm \left[\begin{array}{cc} 1 & \beta \\ \beta & 0 \end{array} \right], \\
B' &= \left[\begin{array}{cc} 0 & 1 \\ 1 & 0 \end{array} \right] \oplus \pm \left[\begin{array}{cc} 0 & 1 \\ 1 & 0 \end{array} \right], \quad \alpha, \beta \in \mathsf{R}, \quad \alpha \neq \beta. \tag{8.4.4}
\end{aligned}
$$

Now, in each of the four cases, a vector $x \in \mathsf{R}^4$ is formulated for which $\langle Ax, x \rangle = \langle Bx, x \rangle = 0$, and Ax, Bx are R-linearly independent:

$$x = e_2 + e_4 \quad \text{if } (A', B') \text{ are given by (8.4.1) or (8.4.4)}$$

(we denote by e_p the unit coordinate vector with 1 in the pth position and zeros elsewhere), and

$$x = e_1 + e_3 \quad \text{if } (A', B') \text{ are given by (8.4.2) or (8.4.3)}.$$

Using the above analysis and, if necessary, replacing (A, B) by $(-A, -B)$, we see that the pair (A, B) can be taken in one of the forms

$$
\begin{aligned}
A + tB \;=\; & 0_{u \times u} \oplus \delta_1 \oplus \cdots \oplus \delta_r \oplus \eta_1 \left(t + \alpha_1 \right) \oplus \cdots \oplus \eta_q \left(t + \alpha_q \right) \\
& \oplus \left(\begin{bmatrix} 0 & 1 \\ 1 & 0 \end{bmatrix} + t \begin{bmatrix} 1 & 0 \\ 0 & 0 \end{bmatrix} \right) \\
& \oplus \cdots \oplus \left(\begin{bmatrix} 0 & 1 \\ 1 & 0 \end{bmatrix} + t \begin{bmatrix} 1 & 0 \\ 0 & 0 \end{bmatrix} \right),
\end{aligned}
\tag{8.4.5}
$$

where δ_j and η_j are signs ± 1, and $\alpha_j \in \mathsf{R}$, or

$$
\begin{aligned}
A + tB \;=\; & 0_{u \times u} \oplus \delta_1 \oplus \cdots \oplus \delta_r \oplus \eta_1 \left(t + \alpha_1 \right) \oplus \cdots \oplus \eta_q \left(t + \alpha_q \right) \\
& \oplus \left(\begin{bmatrix} 1 & \gamma \\ \gamma & 0 \end{bmatrix} + t \begin{bmatrix} 0 & 1 \\ 1 & 0 \end{bmatrix} \right) \\
& \oplus \cdots \oplus \left(\begin{bmatrix} 1 & \gamma \\ \gamma & 0 \end{bmatrix} + t \begin{bmatrix} 0 & 1 \\ 1 & 0 \end{bmatrix} \right),
\end{aligned}
\tag{8.4.6}
$$

where δ_j and η_j are signs ± 1 and $\alpha_j, \gamma \in \mathsf{R}$. Without loss of generality, it may be assumed that $u = 0$, i.e., the term $0_{u \times u}$ does not appear in (8.4.5) and (8.4.6).

Suppose first that (8.4.5) holds, and assume that at least one of the blocks $\eta_j \left(t + \alpha_j \right)$ and at least one of the blocks $\begin{bmatrix} 0 & 1 \\ 1 & 0 \end{bmatrix} + t \begin{bmatrix} 1 & 0 \\ 0 & 0 \end{bmatrix}$ is present. If one of the signs η_j is -1, say, $\eta_1 = -1$, then we obtain a contradiction with the hypothesis that (A, B) satisfies property (α), because the pair

$$
(A', B') = \left(\begin{bmatrix} q & 0 & 0 \\ 0 & 0 & 1 \\ 0 & 1 & 0 \end{bmatrix}, \begin{bmatrix} -1 & 0 & 0 \\ 0 & 1 & 0 \\ 0 & 0 & 0 \end{bmatrix} \right), \quad q \in \mathsf{R},
\tag{8.4.7}
$$

does not satisfy property (α), as can be easily seen by considering the vector $x = \begin{bmatrix} 1 & 1 & -q/2 \end{bmatrix}^T$. Indeed, $\langle A'x, x \rangle = \langle B'x, x \rangle = 0$, but $A'x = \begin{bmatrix} q & -q/2 & 1 \end{bmatrix}^T$ and $B'x = \begin{bmatrix} -1 & 1 & 0 \end{bmatrix}^T$ are linearly independent. Thus, all $\eta_j = 1$. But then B is positive semidefinite, and we are done in this case.

If no blocks $\eta_j \left(t + \alpha_j \right)$ are present, then B is positive semidefinite, and it may be assumed that no blocks $\begin{bmatrix} 0 & 1 \\ 1 & 0 \end{bmatrix} + t \begin{bmatrix} 1 & 0 \\ 0 & 0 \end{bmatrix}$ are present. We may further assume (adding if necessary a suitable scalar multiple of B to A), that all the α_j's are different from zero. Now the result follows from Lemma 8.4.1 (with the roles of A and B interchanged), upon applying a suitable simultaneous congruence to A and B.

Next, suppose that (8.4.6) holds. Then

$$
\begin{aligned}
B + t(A - \gamma B) \;=\; & t\delta_1 \oplus \cdots \oplus t\delta_r \\
& \oplus \left(\eta_1 + t(\eta_1 \alpha_1 - \eta_1 \gamma) \right) \oplus \cdots \oplus \left(\eta_q + t(\eta_q \alpha_q - \eta_q \gamma) \right) \\
& \oplus \left(\begin{bmatrix} 0 & 1 \\ 1 & 0 \end{bmatrix} + t \begin{bmatrix} 1 & 0 \\ 0 & 0 \end{bmatrix} \right) \\
& \oplus \cdots \oplus \left(\begin{bmatrix} 0 & 1 \\ 1 & 0 \end{bmatrix} + t \begin{bmatrix} 1 & 0 \\ 0 & 0 \end{bmatrix} \right).
\end{aligned}
\tag{8.4.8}
$$

This form can be easily reduced, after a suitable congruence with a diagonal congruence matrix to a form of type (8.4.5), to complete the proof. $\qquad \square$

8.5   COMPARISON WITH REAL AND COMPLEX CONGRUENCE

Theorem 8.5.1. *Let $A, B, A', B' \in \mathsf{R}^{m \times m}$ be real symmetric matrices. Then, the real matrix pencils $A + tB$ and $A' + tB'$ are H-congruent if and only if $A + tB$ and $A' + tB'$ are R-congruent.*

Proof. The "if" part being trivial, we prove the "only if" part. Let an invertible matrix $S \in \mathsf{H}^{m \times m}$ be such that equality

$$S^*(A + tB)S = A' + tB' \tag{8.5.1}$$

holds true, and write $S = S_0 + \mathsf{i}S_1 + \mathsf{j}S_2 + \mathsf{k}S_3$, where S_0, S_1, S_2, S_3 are real matrices. Then (8.5.1) takes the form

$$
\begin{bmatrix}
S_0^T & S_1^T & S_2^T & S_3^T \\
S_1^T & -S_0^T & -S_3^T & S_2^T \\
S_2^T & S_3^T & -S_0^T & -S_1^T \\
S_3^T & -S_2^T & S_1^T & -S_0^T
\end{bmatrix}
\begin{bmatrix}
A + tB & 0 & 0 & 0 \\
0 & A + tB & 0 & 0 \\
0 & 0 & A + tB & 0 \\
0 & 0 & 0 & A + tB
\end{bmatrix}
$$

$$
\cdot
\begin{bmatrix}
S_0 & S_1 & S_2 & S_3 \\
S_1 & -S_0 & S_3 & -S_2 \\
S_2 & -S_3 & -S_0 & S_1 \\
S_3 & S_2 & -S_1 & -S_0
\end{bmatrix}
$$

$$
=
\begin{bmatrix}
A' + tB' & 0 & 0 & 0 \\
0 & A' + tB' & 0 & 0 \\
0 & 0 & A' + tB' & 0 \\
0 & 0 & 0 & A' + tB'
\end{bmatrix}. \tag{8.5.2}
$$

The matrix

$$
U :=
\begin{bmatrix}
S_0 & S_1 & S_2 & S_3 \\
S_1 & -S_0 & S_3 & -S_2 \\
S_2 & -S_3 & -S_0 & S_1 \\
S_3 & S_2 & -S_1 & -S_0
\end{bmatrix}
$$

is invertible. Indeed, the equality $Ux = 0$, $x = [x_0 \ x_1 \ x_2 \ x_3]^T \in \mathsf{R}^4$, is equivalent to $S(x_0 - \mathsf{i}x_1 - \mathsf{j}x_2 - \mathsf{k}x_3) = 0$, and, therefore, in view of the invertibility of S, x must be the zero vector. Now apply Corollary 15.3.7 to conclude that $A + tB$ and $A' + tB'$ are R-congruent. $\qquad\square$

An analogue of Theorem 8.5.1 holds in the complex case, with the transposes replaced by conjugate transposes.

Theorem 8.5.2. *Let $A, B, A', B' \in \mathsf{C}^{m \times m}$ be hermitian matrices. Then, the complex matrix pencils $A + tB$ and $A' + tB'$ are H-congruent if and only if $A + tB$ and $A' + tB'$ are C-congruent.*

It will be convenient to prove a lemma first.

Lemma 8.5.3. *Every complex hermitian matrix pencil $A_0 + tB_0$ is C-congruent to its conjugate $\overline{A_0} + t\overline{B_0}$.*

Proof. It is easy to see that we need to consider only the case when $A_0 + tB_0$ is in the canonical form of complex hermitian matrix pencils, and then we may assume that $A_0 + tB_0$ coincides with a primitive block. The primitive blocks for this canonical form are given in Theorem 15.3.1; they are either real, in which case the statement of the lemma is trivial, or they have the form $\begin{bmatrix} 0 & X \\ \overline{X} & 0 \end{bmatrix}$, where X is a square-size complex matrix, in which case the congruence

$$\begin{bmatrix} 0 & I \\ I & 0 \end{bmatrix} \begin{bmatrix} 0 & X \\ \overline{X} & 0 \end{bmatrix} \begin{bmatrix} 0 & I \\ I & 0 \end{bmatrix} = \begin{bmatrix} 0 & \overline{X} \\ X & 0 \end{bmatrix}$$

completes the proof.   □

Proof of Theorem 8.5.2. Again, we prove only the "only if" part. Using the map $\widetilde{\omega}_n$ of (5.6.1) and the fact that $\widetilde{\omega}_n$ is a one-to-one unital homomorphism of real algebras that preserves the $*$ operation, we see that the complex hermitian $2m \times 2m$ matrix pencils

$$\begin{bmatrix} A + tB & 0 \\ 0 & \overline{A} + t\overline{B} \end{bmatrix} \quad \text{and} \quad \begin{bmatrix} A' + tB' & 0 \\ 0 & \overline{A'} + t\overline{B'} \end{bmatrix}$$

are C-congruent. Using Lemma 8.5.3, it follows that $\begin{bmatrix} A + tB & 0 \\ 0 & A + tB \end{bmatrix}$ is C-congruent to $\begin{bmatrix} A' + tB' & 0 \\ 0 & A' + tB' \end{bmatrix}$. Now apply Corollary 15.3.7.   □

8.6   EXPANSIVE AND PLUS-MATRICES: SINGULAR H

In this section we focus on H-expansive matrices. We fix a (not necessarily invertible) hermitian matrix $H \in \mathsf{H}^{n \times n}$, and let $[x, y] = \langle Hx, y \rangle = \langle x, Hy \rangle$.

Definition 8.6.1. A matrix $A \in \mathsf{H}^{n \times n}$ is called H-*expansive* if $[Ax, Ax] \geq [x, x]$ for all $x \in \mathsf{H}^{n \times 1}$ or, equivalently, if $A^* H A - H \geq 0$.

It turns out that the kernel Ker $A := \{x \in \mathsf{H}^{n \times 1} : Ax = 0\}$ of an H-expansive matrix A is H-negative (as long as H is invertible). To prove this, we need the following auxiliary result.

Lemma 8.6.2. *Let $A \in \mathsf{H}^{n \times n}$ be H-expansive. Then there exists a nonsingular matrix $S \in \mathsf{H}^{n \times n}$ such that*

$$S^* H S = H_1 \oplus (-I_{p_2}) \oplus 0_{p_3}, \quad S^* A^* H A S = M_1 \oplus 0_{p_2} \oplus 0_{p_3}, \quad (8.6.1)$$

where $M_1, H_1 \in \mathsf{H}^{p_1 \times p_1}$ are nonsingular and p_1, p_2, p_3 are nonnegative integers.

Proof. By Theorem 8.1.2 we may assume that H and $A^* H A$ have the forms

$$H = H_1 \oplus H_2 \oplus \cdots \oplus H_m, \quad A^* H A = M_1 \oplus M_2 \oplus \cdots \oplus M_m,$$

where M_j and H_j have the same size, H_1 and M_1 are nonsingular, and M_j and H_j, $j > 1$, are blocks of one of the following types:

type 1: $H_j = \varepsilon F_p$ and $M_j = \varepsilon F_p J_p(0)$ for some positive integer p, $\varepsilon = \pm 1$;

type 2: p nonnegative integer,

$$H_j = \begin{bmatrix} 0 & 0 & I_p \\ 0 & 0 & 0 \\ I_p & 0 & 0 \end{bmatrix}, \quad M_j = \begin{bmatrix} 0 & 0 & 0 \\ 0 & 0 & I_p \\ 0 & I_p & 0 \end{bmatrix} \in \mathsf{H}^{(2p+1) \times (2p+1)};$$

type 3: $H_j = \varepsilon F_p J_p(0)$ and $M_j = \varepsilon F_p$ for some positive integer p, $\varepsilon = \pm 1$.

Clearly, since A is H-expansive, each matrix $M_j - H_j$ is positive semidefinite. It is easy to check that this is possible if and only if $p = 1$ and $\varepsilon = -1$ if H_j and M_j are of type 1, $p = 1$ and $\varepsilon = 1$ if H_j and M_j are of type 3, and $p = 0$ if H_j and M_j are of type 2. But then, after eventually permuting some blocks, H and A^*HA have the forms

$$H = H_1 \oplus 0_{p_2} \oplus (-I_{p_3}) \oplus 0_{p_4}, \quad A^*HA = M_1 \oplus 0_{p_2} \oplus 0_{p_3} \oplus I_{p_4}.$$

Note that $M_1 - H_1$ is still positive semidefinite; thus, the number of positive eigenvalues of M_1 is larger or equal to the number of positive eigenvalues of H_1 (counted with algebraic multiplicities). See Theorem 5.13.1. From the well-known fact that the number of positive (resp. negative) eigenvalues of A^*HA is always less or equal to the number of positive (resp. negative) eigenvalues of H (Theorem 5.13.2), it follows that blocks of type 3 cannot occur and, hence, A^*HA and H have forms as in (8.6.1). $\qquad\square$

Corollary 8.6.3. *Let H be invertible and let $A \in \mathsf{H}^{n \times n}$ be H-expansive. Then* Ker A *is H-negative.*

Proof. If A is nonsingular, Ker A is H-negative by definition. Otherwise, let $y \in$ Ker A. Then $y \in$ Ker A^*HA. Since H is invertible, it follows immediately from Lemma 8.6.2 and Proposition 4.2.5(b) that $y^*Hy < 0$. $\qquad\square$

Proposition 8.6.4. *Let $A \in \mathsf{H}^{n \times n}$ be H-expansive. Then* Ker H *is A-invariant.*

Proof. Applying a transformation of the form $(A, H) \mapsto (S^{-1}AS, S^*HS)$, we may assume that H and A^*HA have the forms as in (8.6.1). Applying one more transformation on M_1 and H_1, we may furthermore assume that

$$H = I_{p_1} \oplus -I_{p_2} \oplus -I_{p_3} \oplus 0, \quad A^*HA = \begin{bmatrix} M_{11} & M_{12} \\ M_{12}^* & M_{22} \end{bmatrix} \oplus 0,$$

where p_1, p_2, p_3 are nonnegative integers, $M_{11} \in \mathsf{H}^{p_1 \times p_1}$, $M_{12} \in \mathsf{H}^{p_1 \times p_2}$, $M_{22} \in \mathsf{H}^{p_2 \times p_2}$. Let A be partitioned conformally:

$$A = \begin{bmatrix} A_{11} & A_{12} & A_{13} & A_{14} \\ A_{21} & A_{22} & A_{23} & A_{24} \\ A_{31} & A_{32} & A_{33} & A_{34} \\ A_{41} & A_{42} & A_{43} & A_{44} \end{bmatrix}.$$

With $A^*HA - H \geq 0$ also $M_{11} - I_{p_1} = A_{11}^*A_{11} - A_{21}^*A_{21} - A_{31}^*A_{31} - I_{p_1}$ must be positive semidefinite. This is possible only if $A_{11}^*A_{11}$ is positive definite, i.e., only if A_{11} is nonsingular. Next, we show $A_{14} = 0$. To see this, assume that A_{14} is not zero. Then there exists a matrix $P \in \mathsf{H}^{p_4 \times p_1}$, where $p_4 = n - p_1 - p_2 - p_3$, such that $A_{11} - PA_{14}$ is singular. (For example, if $v \in \mathsf{H}^{p_4}$ is such that $A_{14}v \neq 0$, choose P such that $PA_{14}v = A_{11}v$.) Applying the transformation $(A, H) \mapsto (P^{-1}AP, P^*HP)$ with

$$P = \begin{bmatrix} I_{p_1} & 0 & 0 & 0 \\ 0 & I_{p_2} & 0 & 0 \\ 0 & 0 & I_{p_3} & 0 \\ P & 0 & 0 & I_{p_4} \end{bmatrix},$$

we find that the $(1,1)$-block of $P^{-1}AP$ is $A_{11} - PA_{14}$, whereas $P^*HP = H$ and $P^*A^*HAP = A^*HA$. This contradicts the fact just mentioned that the $(1,1)$-block of $P^{-1}AP$ must be nonsingular. But $A_{14} = 0$ implies that the $(4,4)$-block of A^*HA has the form $-A_{24}^*A_{24} - A_{34}^*A_{34} = 0$. This is possible only if $A_{24} = 0$ and $A_{34} = 0$. Hence, Ker H is A-invariant.   □

The key result in this section is the following theorem.

Theorem 8.6.5. *Let $A \in \mathsf{H}^{n \times n}$ be H-expansive and let $\mathcal{M}_0 \subseteq \mathsf{H}^n$ be an A-invariant subspace which is H-nonnegative. Then there exists an H-nonnegative A-invariant subspace $\mathcal{M} \supseteq \mathcal{M}_0$ such that $\dim \mathcal{M} = \mathrm{In}_+(H) + \mathrm{In}_0(H)$.*

Proof. For the case when H is invertible, the result is proved in Theorem 6.2.6. Thus, consider the case that H is singular. Without loss of generality, we may assume that H has the form (6.3.1), i.e., $H = H_1 \oplus 0_\nu$, where $H_1 = I_p \oplus (-I_q)$. Then Proposition 8.6.4 implies that A takes the form

$$A = \begin{bmatrix} A_{11} & 0 \\ A_{21} & A_{22} \end{bmatrix},$$

where $A_{11} \in \mathsf{H}^{(p+q) \times (p+q)}$ is easily seen to be H_1-expansive. Represent $\mathcal{M}_0$ according to Lemma 6.3.1:

$$\mathcal{M}_0 = \mathrm{Im} \begin{bmatrix} Q_0 & 0 \\ Y_0 & X_0 \end{bmatrix}, \quad Q_0 = \begin{bmatrix} P_0 \\ K_0 \end{bmatrix},$$

where $P_0^*P_0 = I_p$, $\|K_0\| \leq 1$, $X_0^*X_0 = I$, and $X_0^*Y_0 = 0$. Then $\widetilde{\mathcal{M}}_0 = \mathrm{Im}\, Q_0$ is H_1-nonnegative and A_{11}-invariant. By the part already proved, there exists an $\mathrm{In}_+(H)$-dimensional, H_1-nonnegative, and A_{11}-invariant subspace $\widetilde{\mathcal{M}}$ that contains $\widetilde{\mathcal{M}}_0$. Let $\widetilde{\mathcal{M}} = \mathrm{Im}\, Q$ for some matrix Q of appropriately chosen dimension, i.e., $Q_0 = QW_0$ for some matrix W_0. Then $\mathcal{M} := \mathrm{Im}(Q \oplus I_\nu)$ is H-nonnegative and A-invariant. Furthermore, it contains $\mathcal{M}_0$ and has dimension $\mathrm{In}_+(H) + \mathrm{In}_0(H)$.   □

Obviously, every H-isometric matrix is H-expansive (recall that a matrix A is called H-*isometric* if $[Ax, Ax] = [x, x]$ for every $x \in \mathsf{H}^{n \times 1}$), Thus, as an important corollary of Theorem 8.6.5, we obtain the following.

Theorem 8.6.6. *Let $A \in \mathsf{H}^{n \times n}$ be an H-isometric matrix, and let $\mathcal{M}_0 \subseteq \mathsf{H}^n$ be an A-invariant H-nonnegative, resp. H-nonpositive, subspace. Then there exists an A-invariant H-nonnegative, resp. H-nonpositive, subspace $\mathcal{M}$ such that $\mathcal{M} \supseteq \mathcal{M}_0$ and $\dim \mathcal{M} = \mathrm{In}_+(H)$, resp. $\dim \mathcal{M} = \mathrm{In}_-(H)$.*

The part of Theorem 8.6.6 concerning H-nonpositive subspaces follows by noticing that A is also expansive with respect to $-H$ and applying Theorem 8.6.5 with H replaced by $-H$.

Recall that a matrix $A \in \mathsf{H}^{n \times n}$ is called an H-*plus-matrix* if $[Ax, Ax] \geq 0$ for every $x \in \mathsf{H}^{n \times 1}$ such that $[x, x] \geq 0$.

Theorem 8.6.7. *Let A be an H-plus-matrix such that the subspace $\mathrm{Ran}\, A$ is not H-nonnegative, and let $\mathcal{M}_0 \subseteq \mathsf{H}^{n \times n}$ be an A-invariant subspace which is H-nonnegative. Then there exists an H-nonnegative A-invariant subspace $\mathcal{M} \supseteq \mathcal{M}_0$ such that $\dim \mathcal{M} = \mathrm{In}_+(H) + \mathrm{In}_0(H)$.*

Proof. By Proposition 6.2.2, we have $[Ax, Ax] \geq \mu(A)[x, x]$ for all $x \in \mathsf{H}^{n \times 1}$, where $\mu(A) \geq 0$ is independent of x. Since Ran A is not H-nonnegative, it follows that $\mu(A) > 0$. Scaling A, if necessary, we can assume that $\mu(A) = 1$. Then A is H-expansive, and the result follows from Theorem 8.6.5. $\qquad\qquad\square$

Theorem 8.6.8. *Let $A \in \mathsf{H}^{n \times n}$ be an H-plus-matrix. Then there exists an H-nonnegative A-invariant subspace $\mathcal{M}$ such that* $\dim \mathcal{M} = \mathrm{In}_+(H) + \mathrm{In}_0(H)$.

Proof. By Proposition 6.2.2, we have $[Ax, Ax] \geq \mu(A)[x, x]$ for all $x \in \mathsf{H}^{n \times 1}$ and $\mu(A) \geq 0$. If $\mu(A) = 0$, then the range of A is an H-nonnegative subspace, and we are done by Theorem 4.2.6. Indeed, any maximal H-nonnegative subspace $\mathcal{M}$ containing the range of A is A-invariant. If $\mu(A) > 0$, apply Theorem 8.6.7 with $\mathcal{M}_0 = \{0\}$. $\qquad\qquad\square$

It is instructive to compare Theorem 6.2.6 and 8.6.8. The statement of Theorem 6.2.6 is stronger in the sense that it asserts that every H-nonnegative A-invariant subspace can be enlarged to an A-invariant subspace which is also maximal H-nonnegative, whereas Theorem 8.6.8 asserts merely existence of A-invariant subspaces which are maximal H-nonnegative. Of course, Theorem 6.2.6 has stronger hypothesis, namely, that H is invertible. It is an open question whether or not the statement of Theorem 6.2.6 holds true without the invertibility hypothesis (cf. Open Problem 6.2.8).

Definition 8.6.9. As in Section 6.4 (for the case of invertible H), we say that a matrix $A \in \mathsf{H}^{n \times n}$ is H-*expansive* if $[Ax, Ax] \geq [x, x]$ for every $x \in \mathsf{H}^{n \times 1}$. A matrix $B \in \mathsf{H}^{n \times n}$ is called H-*dissipative* if $\mathfrak{R}([Bx, x]) \leq 0$ for every $x \in \mathsf{H}^{n \times 1}$ or, equivalently, $HB + B^*H \leq 0$.

Lemma 6.4.5 remains valid also for singular (noninvertible) H. So, by using the linear fractional transformations of Section 6.4, we can obtain results on H-dissipative matrices from those for H-expansive matrices.

Theorem 8.6.10. *Let B be H-dissipative.*

(a) *Let $\mathcal{M}_0 \subseteq \mathsf{H}^{n \times 1}$ be a B-invariant H-nonnegative subspace. Then there exists a B-invariant maximal H-nonnegative subspace $\mathcal{M}$ such that $\mathcal{M} \supseteq \mathcal{M}_0$.*

(b) *Let $\mathcal{M}_0' \subseteq \mathsf{H}^{n \times 1}$ be a B-invariant H-nonpositive subspace. Then there exists a B-invariant maximal H-nonpositive subspace $\mathcal{M}'$ such that $\mathcal{M}' \supseteq \mathcal{M}_0'$.*

The proof is parallel to that of Theorem 6.4.6, using Theorem 8.6.5 instead of Corollary 6.4.3. Theorem 8.6.10 extends the result of Theorem 6.4.6 to the case of singular H.

8.7 EXERCISES

Ex. 8.7.1. Provide complete details for verification of the alternative version of Theorem 8.1.2, Part (b).

Ex. 8.7.2. Verify directly that the pair of matrices (A, B) of Example 8.3.5 cannot be simultaneously diagonalized.

Ex. 8.7.3. Describe the canonical form under congruence of the hermitian pencil $A + tB$ (Theorem 8.1.2) in each of the following situations:

(1) B is positive definite;

(2) B is positive semidefinite;

(3) A and B are positive definite;

(4) A and B are positive semidefinite;

(5) A and B each have only one negative eigenvalue (counted with multiplicities).

Ex. 8.7.4. What can you say about the canonical form (8.1.6) if it is known that the rank of $xA + yB$ is constant for all real x and y not both zero?

Ex. 8.7.5. Find the canonical forms under strict equivalence and under congruence of Theorem 8.1.2 for the following hermitian pencils:

$$\text{(a)} \quad \begin{bmatrix} I_k & tiF_k \\ -tiF_k & -I_k \end{bmatrix}, \quad \text{(b)} \quad F_6 + t \begin{bmatrix} 0 & 0 & j & 0 \\ 0 & 1 & 0 & 0 \\ -j & 0 & 0 & 0 \\ 0 & 0 & 0 & I_3 \end{bmatrix} \in \mathsf{H}^{6 \times 6}.$$

Ex. 8.7.6. Find all H-expansive, all H-dissipative, and all H-plus-matrices for $H = \begin{bmatrix} 1 & 0 \\ 0 & 0 \end{bmatrix}$.

Ex. 8.7.7. Let $H = I_p \oplus (-I_{n-p})$, and let $A = \operatorname{diag}(a_1, \ldots, a_n) \in \mathsf{H}^{n \times n}$ be a diagonal matrix.
   (a) Find for which values of $a_1, \ldots, a_n$ the matrix A is an H-plus-matrix.
   (b) For the values of $a_1, \ldots, a_n$ obtained in Part (a), find all A-invariant maximal H-nonnegative subspaces.

Ex. 8.7.8. Repeat Ex. 8.7.7 for $H = I_p \oplus (-I_q) \oplus 0_{n-p-q}$.

8.8   NOTES

The material of this chapter (except Sections 8.5 and 8.6) is adapted from Lancaster and Rodman [92], where the exposition is done for complex hermitian and real symmetric pencils. In the context of real and complex matrices, the canonical forms of pairs of hermitian/skewhermitian or symmetric/skewsymmetric matrices, have a long and distinguished history; we refer the reader to Lancaster and Rodman [92, 93], and especially to Thompson [151] for an extensive bibliography on the subject. Canonical forms for quaternion hermitian/skewhermitian pencils of matrices, as well as for symplectic and unitary matrices with respect to sesquilinear, symmetric, or skewsymmetric forms, have also been known in the literature for some time (see, e.g., Djoković [33] or Djoković et al. [34]) and recently have been subject to extensive renewed interest and interpretation: Horn and Sergeichuk [64], Sergeichuk [142, 143], Rodman [132, 133, 135, 134].

The alternative form (8.1.9) was used in Wall [155] (in the context of real symmetric matrix pencils).

Lemma 8.2.1 is attributed to Hua [67].

In the proof of Theorem 8.1.2 we use the standard approach, as in the proofs of Theorem 6.1 in Lancaster and Rodman [92] (for complex matrices), Theorem 5.1 in Lancaster and Rodman [93] (for real matrices), and Theorem 8.1(b) in Rodman [132].

Other characterizations of the positive semidefinite linear combination property and an alternative proof of Theorem 8.3.3 are given by Cheung et al. [26] in the context of selfadjoint operators on complex Hilbert spaces. See also Lancaster and Ye [95] and Rodman [136].

Theorem 8.3.4 in the context of complex matrices was proved by Finsler [43] and later reproved many times using various methods; see, e.g., Au-Yeung [7] and the survey by Uhlig [152]. The proof of Theorem 8.3.4 based on the convexity of the numerical range (for complex matrices) is due to Au-Yeung and Poon [10].

A proof of Theorem 8.3.6 for real and complex matrices is given in several texts, including Gantmacher and Krein [51], Gantmacher [48], and Franklin [45].

The material of Section 8.6 is taken from Mehl et al. [110] (where the presentation is in the framework of real and complex matrices).

Various classes of complex plus-matrices have been studied in van der Mee et al. [106]. A study of the corresponding classes of quaternion plus-matrices remains an open problem.

Chapter Nine

Skewhermitian and mixed pencils

Here we present canonical forms under strict equivalence and under congruence for pencils of quaternion matrices $A + tB$, where one of the matrices A and B is skewhermitian and the other can be hermitian or skewhermitian. We treat the case when both A and B are skewhermitian in Section 9.1 (this is the relatively easy case). The case when one of A and B is hermitian and the other is skewhermitian is treated in subsequent sections. Comparisons are made with the corresponding results for real and complex matrix pencils. We give an application to the canonical form of a quaternion matrix under congruence.

9.1 CANONICAL FORMS FOR SKEWHERMITIAN MATRIX PENCILS

Here, it turns out that for matrix pencils of type $A + tB$, where A and B are skewhermitian quaternion matrices, the canonical forms of strict equivalence and of simultaneous congruence are the same. The main result runs as follows:

Theorem 9.1.1. (i) *Every matrix pencil $A+tB$, where A and B are quaternion skewhermitian $n \times n$ matrices, is strictly equivalent to a pencil of skewhermitian matrices of the form*

$$
0_u \;\oplus\; \oplus_{j=1}^{p} \left(t \begin{bmatrix} 0 & 0 & F_{\varepsilon_j} \\ 0 & 0_1 & 0 \\ -F_{\varepsilon_j} & 0 & 0 \end{bmatrix} + \begin{bmatrix} 0 & F_{\varepsilon_j} & 0 \\ -F_{\varepsilon_j} & 0 & 0 \\ 0 & 0 & 0 \end{bmatrix} \right)
$$

$$
\oplus\; \oplus_{j=1}^{r} \begin{bmatrix} 0 & 0 & \cdots & 0 & 0 & \mathrm{i}\beta_j + t\mathrm{i} \\ 0 & 0 & \cdots & 0 & \mathrm{i}\beta_j + t\mathrm{i} & \mathrm{i} \\ 0 & 0 & \cdots & \mathrm{i}\beta_j + t\mathrm{i} & \mathrm{i} & 0 \\ \vdots & \vdots & \iddots & \vdots & \vdots & \vdots \\ \mathrm{i}\beta_j + t\mathrm{i} & \mathrm{i} & 0 & \cdots & 0 & 0 \end{bmatrix}_{k_j \times k_j}
$$

$$
\oplus\; \oplus_{j=1}^{q} \begin{bmatrix} 0 & 0 & \cdots & 0 & 0 & \mathrm{i} \\ 0 & 0 & \cdots & 0 & \mathrm{i} & t\mathrm{i} \\ 0 & 0 & \cdots & \mathrm{i} & t\mathrm{i} & 0 \\ \vdots & \vdots & \iddots & \vdots & \vdots & \vdots \\ \mathrm{i} & t\mathrm{i} & 0 & \cdots & 0 & 0 \end{bmatrix}_{\ell_j \times \ell_j} , \tag{9.1.1}
$$

where the positive integers, ε_j's, satisfy $\varepsilon_1 \le \cdots \le \varepsilon_p$ and the quaternions β_j's are such that the $\mathrm{i}\beta_j$'s all have zero real parts (the case when some or all the β_j's are zero is not excluded). The subscript in $[\,\cdot\,]_{p \times p}$ indicates the size, $p \times p$, of the matrix $[\,\cdot\,]$.

The form (9.1.1) *is uniquely determined by* $A + tB$ *up to arbitrary permutations of diagonal blocks in each of the two parts,*

$$\oplus_{j=1}^{r} \begin{bmatrix} 0 & 0 & \cdots & 0 & 0 & i\beta_j + ti \\ 0 & 0 & \cdots & 0 & i\beta_j + ti & i \\ 0 & 0 & \cdots & i\beta_j + ti & i & 0 \\ \vdots & \vdots & \cdot^{\cdot^{\cdot}} & \vdots & \vdots & \vdots \\ i\beta_j + ti & i & 0 & \cdots & 0 & 0 \end{bmatrix}_{k_j \times k_j} \tag{9.1.2}$$

and

$$\oplus_{j=1}^{q} \begin{bmatrix} 0 & 0 & \cdots & 0 & 0 & i \\ 0 & 0 & \cdots & 0 & i & ti \\ 0 & 0 & \cdots & i & ti & 0 \\ \vdots & \vdots & \cdot^{\cdot^{\cdot}} & \vdots & \vdots & \vdots \\ i & ti & 0 & \cdots & 0 & 0 \end{bmatrix}_{\ell_j \times \ell_j} ,$$

and up to replacement of β_j *in every block in* (9.1.2) *by a similar quaternion* γ_j *subject to the property that* $\Re(i\gamma_j) = 0$.

(ii) *Two quaternion matrix pencils,* $A_1 + tB_1$ *and* $A_2 + tB_2$, *with skewhermitian* A_1, B_1, A_2 *and* B_2 *are congruent if and only if they are strictly equivalent.*

Thus, the form (9.1.1) is a canonical form for a pair of quaternion skewhermitian matrices under simultaneous congruence as well as under strict equivalence.

Remark 9.1.2. Note that any nonzero quaternion with zero real part can be used instead of i in this theorem. Indeed, if $\alpha \in \mathsf{H} \setminus \{0\}$ has zero real part, then $\alpha = \lambda^* i \lambda$ for some nonzero quaternion λ (Theorem 2.2.6). Moreover,

$$\lambda^* (i\beta_j) \lambda = (\lambda^* i \lambda) (\lambda^{-1} \beta_j \lambda),$$

and if the real part of $i\beta_j$ is zero, then the real part of $\alpha \cdot \lambda^{-1} \beta_j \lambda = \lambda^* (i\beta_j) \lambda$ is zero as well (by the same Theorem 2.2.6).

For the proof of Theorem 9.1.1, it will be convenient to verify first the following independently interesting fact.

Proposition 9.1.3. *Let* $A, B \in \mathsf{H}^{m \times n}$. *The following statements are equivalent for the matrix pencil* $A + tB$:

(a) $A + tB$ *is strictly equivalent to* $-A^* - tB^*$;

(b) $A + tB$ *is strictly equivalent to a pencil* $A' + tB'$, *such that the quaternion matrices* A' *and* B' *are skewhermitian;*

(c) $m = n$, *and the left indices of* $A + tB$, *when arranged in nondecreasing order, coincide with the right indices of* $A + tB$, *also arranged in nondecreasing order.*

Proof. (b) $\Longrightarrow$ (a) is easy. If $A + tB = T(A' + tB')S$ for some invertible matrices S and T, with skewhermitian A' and B', then

$$-A^* - tB^* = -S^*(A'^* + tB'^*)T^* = S^*(A' + tB')T^* = S^*T^{-1}(A + tB)S^{-1}T^*.$$

(a) $\Longrightarrow$ (c). This follows from the uniqueness of the Kronecker form of $A + tB$ and the easily verified fact that the left (resp., right) indices of $A + tB$ are the right (resp., left) indices of $-A^* - tB^*$.

(c) $\implies$ (b). In view of the Kronecker form of $A + tB$, we need to consider only the following cases (leaving aside the trivial block $0_{u \times v}$, where we must have $u = v$):

(i) $A + tB = L_{\varepsilon \times (\varepsilon+1)} \oplus L_{\varepsilon \times (\varepsilon+1)}^T$;

(ii) $A + tB = I_k + t J_k(0)$;

(iii) $A + tB = t I_\ell + J_\ell(\alpha)$, where $\alpha \in \mathsf{H}$.

In Case (ii), let $A' + tB' = \mathrm{i} F_k(I_k + t J_k(0))$. In Case (iii), replacing α by a similar quaternion, we may assume without loss of generality that $\mathrm{i}\alpha$ has zero real part; then let $A' + tB' = \mathrm{i} F_\ell(t I_\ell + J_\ell(\alpha))$. In Case (i), use the equality

$$
\begin{bmatrix} I_{\varepsilon+1} & 0 \\ 0 & -F_\varepsilon \end{bmatrix} \left(t \begin{bmatrix} 0 & 0 & F_\varepsilon \\ 0 & 0_1 & 0 \\ -F_\varepsilon & 0 & 0 \end{bmatrix} + \begin{bmatrix} 0 & F_\varepsilon & 0 \\ -F_\varepsilon & 0_1 & 0 \\ 0 & 0 & 0 \end{bmatrix} \right) \begin{bmatrix} F_\varepsilon & 0 \\ 0 & I_{\varepsilon+1} \end{bmatrix}
$$

$$
= \begin{bmatrix} 0 & L_{\varepsilon \times (\varepsilon+1)} \\ L_{\varepsilon \times (\varepsilon+1)}^T & 0 \end{bmatrix},
$$

and we are done.   □

Proof of Theorem 9.1.1. Part (i) follows by applying the proof of the implication (c) $\implies$ (b) of Proposition 9.1.3 to the Kronecker form $A_0 + tB_0$ of the given skewhermitian matrix pencil $A + tB$. The uniqueness statement in (i) follows from the uniqueness properties of the Kronecker form $A_0 + tB_0$. Note that the Kronecker forms of the blocks

$$
M_\ell := \begin{bmatrix} 0 & 0 & \cdots & 0 & 0 & \mathrm{i} \\ 0 & 0 & \cdots & 0 & \mathrm{i} & t\mathrm{i} \\ 0 & 0 & \cdots & \mathrm{i} & t\mathrm{i} & 0 \\ \vdots & \vdots & \ddots & \vdots & \vdots & \vdots \\ \mathrm{i} & t\mathrm{i} & 0 & \cdots & 0 & 0 \end{bmatrix}_{\ell \times \ell}
$$

and

$$
N_k(\beta) := \begin{bmatrix} 0 & 0 & \cdots & 0 & 0 & \mathrm{i}\beta + t\mathrm{i} \\ 0 & 0 & \cdots & 0 & \mathrm{i}\beta + t\mathrm{i} & \mathrm{i} \\ 0 & 0 & \cdots & \mathrm{i}\beta + t\mathrm{i} & \mathrm{i} & 0 \\ \vdots & \vdots & \ddots & \vdots & \vdots & \vdots \\ \mathrm{i}\beta + t\mathrm{i} & \mathrm{i} & 0 & \cdots & 0 & 0 \end{bmatrix}_{k \times k}, \quad \beta \in \mathsf{H},
$$

are $I_\ell + t G_\ell$ and $t I_k + J_k(\beta)$, respectively.

For the proof of existence in Part (ii), we use the same approach as in the proof of Theorem 8.1.2. The proof boils down to the verification that each primitive block in (9.1.1) is congruent to its negative. This is easy: indeed, for the block

$$
K_{2\varepsilon+1} := t \begin{bmatrix} 0 & 0 & F_\varepsilon \\ 0 & 0 & 0 \\ -F_\varepsilon & 0 & 0 \end{bmatrix} + \begin{bmatrix} 0 & F_\varepsilon & 0 \\ -F_\varepsilon & 0 & 0 \\ 0 & 0 & 0 \end{bmatrix},
$$

we have

$$\begin{bmatrix} I_\varepsilon & 0 \\ 0 & -I_{\varepsilon+1} \end{bmatrix} K_{2\varepsilon+1} \begin{bmatrix} I_\varepsilon & 0 \\ 0 & -I_{\varepsilon+1} \end{bmatrix} = -K_{2\varepsilon+1};$$

for the block M_ℓ use $(-\mathrm{j}I)M_\ell(\mathrm{j}I) = -M_\ell$; finally, for the block $N_k(\beta)$, where the real part of $\mathrm{i}\beta$ is zero, we have

$$(-\gamma I)N_k(\beta)(\gamma I) = -N_k(\beta).$$

Here γ is any quaternion that satisfies the equalities $\gamma^2 = -1$, $\gamma \mathrm{i} \gamma = \mathrm{i}$, $\gamma(\mathrm{i}\beta)\gamma = \mathrm{i}\beta$. The existence of such γ is ensured by Ex. 2.7.14. Indeed, take γ to be a quaternion with zero real part of norm 1, which is orthogonal to both i and $\mathrm{i}\beta$ in the sense of (2.7.1).

Finally, the uniqueness statement in Part (ii) follows from the uniqueness of the Kronecker form and the Kronecker forms of the blocks M_ℓ and $N_k(\beta)$ given above.   □

9.2   COMPARISON WITH REAL AND COMPLEX SKEWHERMITIAN PENCILS

For the real skewsymmetric pencils we have the following.

Theorem 9.2.1. *Let $A+tB$, $A'+tB'$ be real $n \times n$ matrix pencils with skewsymmetric A, A', B, B'. Then the following statements are equivalent:*

(1) $A + tB$ *and* $A' + tB'$ *are* R-*congruent;*

(2) $A + tB$ *and* $A' + tB'$ *are* H-*congruent;*

(3) $A + tB$ *and* $A' + tB'$ *are* R-*strictly equivalent;*

(4) $A + tB$ *and* $A' + tB'$ *are* H-*strictly equivalent.*

The result is obtained by combining Theorems 7.6.3, 9.1.1, and 15.2.2.

The situation is somewhat more complicated for complex skewhermitian pencils: H-congruence generally does not imply C-congruence.

Theorem 9.2.2. *The following statements are equivalent for complex skewhermitian $n \times n$ matrix pencils $A + tB$ and $A' + tB'$:*

(a) $A + tB$ *and* $A' + tB'$ *are* C-*strictly equivalent;*

(b) $A + tB$ *and* $A' + tB'$ *are* H-*strictly equivalent;*

(c) $A + tB$ *and* $A' + tB'$ *are* H-*congruent.*

Proof. (b) $\Longleftrightarrow$ (c) by Theorem 9.1.1. The implication (a) $\Longrightarrow$ (b) is obvious. To prove (b) $\Longrightarrow$ (a), observe that by Theorem 15.3.1 the indices (partial multiplicities) of nonreal eigenvalues in the C-Kronecker form of $A + tB$ come in pairs, one for an eigenvalue β and the same one for the eigenvalue $\bar{\beta}$ (note that $A + tB$ has the same C-Kronecker form as the hermitian matrix pencil $\mathrm{i}A + t\mathrm{i}B$). Of course, this also holds for the pencil $A' + tB'$. Now, suppose (b) holds. By Theorem 7.6.3, the C-Kronecker form of $A + tB$ is obtained from the C-Kronecker form of $A' + tB'$ by replacing some Jordan blocks $J_m(\alpha)$, where α is a complex nonreal eigenvalue, with $J_m(\overline{\alpha})$. But since the indices of nonreal eigenvalues come in pairs (see the observation above), such replacements actually do not alter the C-Kronecker form of $A' + tB'$ (up to permutation of blocks allowed in the C-Kronecker form), and (a) follows. $\square$

Using the canonical form for complex hermitian matrix pencils $iA + tiB$ (Theorem 15.3.1), we have a more precise statement concerning H-congruence versus C-congruence for complex skewhermitian matrix pencils $A + tB$.

Theorem 9.2.3. *Let $A + tB$ be a skewhermitian $n \times n$ complex matrix pencil. Let $\lambda_1, \ldots, \lambda_u$ be all the distinct real eigenvalues of $A + tB$, and we include infinity in the set $\{\lambda_1, \ldots, \lambda_u\}$ if blocks of type $I + tJ_m(0)$ are present in the C-Kronecker form of $A + tB$. Let $p_{\lambda_j, s}$ be the number of Jordan blocks $J_s(\lambda_j)$ in the C-Kronecker form of $A + tB$, or the number of blocks $I + tJ_s(0)$ if $\lambda_j = \infty$. Then the set of all skewhermitian complex matrix pencils that are H-congruent to $A + tB$ consists of exactly p disjoint classes $\mathcal{F}_1, \ldots, \mathcal{F}_p$, each class consisting of mutually C-congruent skewhermitian complex matrix pencils, so that matrix pencils from different $\mathcal{F}_j$'s are not C-congruent. The integer p is given by*

$$p = \prod_{j=1}^{u} \left(\prod_{s=1}^{r_j} (p_{\lambda_j, s} + 1) \right). \tag{9.2.1}$$

Here r_j is the largest size of a Jordan block with eigenvalue λ_j, or the largest size of blocks $I + tJ_m(0)$ if $\lambda_j = \infty$, in the C-Kronecker form of $A + tB$.

The proof follows considerations analogous to those in the proof of Theorem 5.8.3. We leave the details to the reader.

Note also that the H-Kronecker form of $A + tB$ can be used in Theorem 9.2.3 in place of the C-Kronecker form.

The classes $\mathcal{F}_1, \ldots, \mathcal{F}_p$ can be identified in terms of the canonical form (15.3.1) of the complex hermitian matrix pencil $iA + tiB$. For notational simplicity, suppose $\lambda_1 = \infty$. Namely, for each pair of integers (j, s), where $s = 1, 2, \ldots, r_j$ and $j = 1, 2, \ldots, u$ such that $p_{\lambda_j, s} \neq 0$, select an integer $q_{\lambda_j, s}$ such that $0 \leq q_{\lambda_j, s} \leq p_{\lambda_j, s}$. We associate with these selections the form

$$K_0 \quad \oplus \quad \oplus_{\{s \,:\, p_{\lambda_1, s} \neq 0\}} \left(\underbrace{(F_s + tG_s) \oplus \cdots \oplus (F_s + tG_s)}_{q_{\lambda_1, s} \text{ times}} \right.$$

$$\oplus \underbrace{(-F_s - tG_s) \oplus \cdots \oplus (-F_s - tG_s)}_{p_{\lambda_1, s} - q_{\lambda_1, s} \text{ times}} \right)$$

$$\oplus \quad \oplus_{j=2}^{u} \oplus_{\{s \,:\, p_{\lambda_j, s} \neq 0\}} \left(\underbrace{((t - \lambda_j)F_s + G_s) \oplus \cdots \oplus ((t - \lambda_j)F_s + G_s)}_{q_{\lambda_j, s} \text{ times}} \right.$$

$$\oplus \underbrace{-((t - \lambda_j)F_s + G_s) \oplus \cdots \oplus -((t - \lambda_j)F_s + G_s)}_{p_{\lambda_j, s} - q_{\lambda_j, s} \text{ times}} \right), \tag{9.2.2}$$

where K_0 is the part of (15.3.1) that consists of blocks other than $(t+\alpha_j)F_{\ell_j}+G_{\ell_j}$ with real α_j and $F_{k_i}+tG_{k_i}$. Now, a class $\mathcal{F}_i$ consists exactly of those complex skewhermitian pencils $A'+tB'$ for which the canonical form of $iA'+tiB'$ under C-congruence is (9.2.2) for a fixed selection of the $q_{\lambda_j,s}$'s.

9.3 CANONICAL FORMS FOR MIXED PENCILS: STRICT EQUIVALENCE

In this and subsequent sections we study mixed pairs of matrices $A, B \in \mathsf{H}^{n \times n}$, where A is hermitian and B is skewhermitian or, equivalently, quaternion matrix pencils $A + tB$, where $A = A^*$ and $B = -B^*$, which will be called *hermitian/skewhermitian* . The case where A is skewhermitian and B is hermitian is easily reduced to the hermitian/skewhermitian pencils by interchanging the roles of A and B.

We start with description of primitive forms of hermitian/skewhermitian quaternion matrix pencils $A + tB$. In this description, we use the standard matrices

$$
F_m = \begin{bmatrix} 0 & \cdots & \cdots & 0 & 1 \\ \vdots & & & 1 & 0 \\ \vdots & & & & \vdots \\ 0 & 1 & & & \vdots \\ 1 & 0 & \cdots & \cdots & 0 \end{bmatrix}, \quad
G_m = \begin{bmatrix} F_{m-1} & 0 \\ 0 & 0 \end{bmatrix}, \quad
\widetilde{G}_m = \begin{bmatrix} 0 & 0 \\ 0 & F_{m-1} \end{bmatrix},
$$

and

$$
\Xi_m(\alpha) = \begin{bmatrix} 0 & & 0 & \cdots & 0 & \alpha \\ 0 & & 0 & \cdots & -\alpha & 0 \\ \vdots & & \vdots & \mathinner{\raise1pt\hbox{.}\mkern2mu\raise4pt\hbox{.}\mkern2mu\raise7pt\hbox{.}} & \vdots & \vdots \\ 0 & & (-1)^{m-2}\alpha & \cdots & 0 & 0 \\ (-1)^{m-1}\alpha & & 0 & \cdots & 0 & 0 \end{bmatrix} \in \mathsf{H}^{m \times m}, \quad \alpha \in \mathsf{H}. \quad (9.3.1)
$$

The notation "q-h-sk" points to the class of pencils under consideration: quaternion hermitian/skewhermitian.

(q-h-sk0)   a square-size zero matrix, i.e., $0_m + t0_m$.

(q-h-sk1) $G_{2\varepsilon+1} + t \begin{bmatrix} 0 & 0 & F_\varepsilon \\ 0 & 0_1 & 0 \\ -F_\varepsilon & 0 & 0 \end{bmatrix}.$

(q-h-sk2) $F_k + tiG_k.$

(q-h-sk3) $G_\ell + tiF_\ell.$

(q-h-sk4) $\begin{bmatrix} 0 & \alpha F_p + G_p \\ \alpha^* F_p + G_p & 0 \end{bmatrix} + t \begin{bmatrix} 0 & F_p \\ -F_p & 0 \end{bmatrix}$, where $\alpha \in \mathsf{H}$ has positive real part.

(q-h-sk5)   Let β be a nonzero quaternion with zero real part if m is even and a positive real number if m is odd. Then a primitive form is

$$
\begin{bmatrix}
0 & 0 & \cdots & 0 & 0 & \beta \\
0 & 0 & \cdots & 0 & -\beta & -1 \\
0 & 0 & \cdots & \beta & -1 & 0 \\
\vdots & \vdots & \ddots & \vdots & \vdots & \vdots \\
(-1)^{m-1}\beta & -1 & 0 & \cdots & 0 & 0
\end{bmatrix} + t\Xi_m(\mathrm{i}^m)
$$

$$
= \Xi_m(\beta) - \widetilde{G}_m + t\Xi_m(\mathrm{i}^m). \tag{9.3.2}
$$

The pencil in (q-h-sk5) is $m \times m$, where m is a positive integer. Note that the size $m \times m$ matrices

$$
\Xi_m(\beta) - \widetilde{G}_m \quad \text{and} \quad \Xi_m(\mathrm{i}^m)
$$

are hermitian and skewhermitian, respectively, for every m and every β subject to the conditions in (q-h-sk5).

The next theorem is the main result on strict equivalence of mixed hermitian/skewhermitian matrix pencils.

Theorem 9.3.1. *Every hermitian/skewhermitian quaternion matrix pencil $A + tB$ is strictly equivalent to a direct sum of blocks of types (q-h-sk0)–(q-h-sk5). In this direct sum, several blocks of the same type and of different sizes may be present. The direct sum is uniquely determined by A and B, up to an arbitrary permutation of blocks, up to replacement of α with a similar quaternion in any block of type (q-h-sk4), and up to replacement of β with a similar quaternion in any block of type (q-h-sk5) of even size.*

Note that G_m's in Theorem 9.3.1 can be replaced by $\widetilde{G}_m$'s, and any nonzero quaternion with zero real part can be used in place of i. Indeed, given $q \in \mathsf{H} \setminus \{0\}$ with $\mathfrak{R}(q) = 0$, we have the equality

$$
(\mathrm{diag}\,(s_1^*, \ldots, s_n^*))\,(G_n + t\mathrm{i}F_n)\,(\mathrm{diag}\,(s_1, \ldots, s_n)) = G_n + tqF_n,
$$

where $s_1, \ldots, s_n \in \mathsf{H} \setminus \{0\}$ are determined as follows: if n is odd, let $s_{(n+1)/2}$ be such that $s_{(n+1)/2}^* \mathrm{i} s_{(n+1)/2} = q$, and if

$$
s_{(n+1)/2}, s_{(n+1)/2-1}, \ldots, s_{(n+1)/2-k}, s_{(n+1)/2+1}, \ldots, s_{(n+1)/2+k}
$$

are already determined, where $k \in \{0, 1, \ldots, (n-1)/2 - 1\}$, we let

$$
s_{(n+1)/2-(k+1)} = \left(s_{(n+1)/2+k}^*\right)^{-1}, \qquad s_{(n+1)/2+(k+1)} = \left(s_{(n+1)/2-(k+1)}^* \mathrm{i}\right)q.
$$

In the case n is even, then we define $s_{n/2} = 1$, and if

$$
s_{n/2}, s_{n/2-1}, \ldots, s_{n/2-k}, s_{n/2+1}, \ldots, s_{n/2+k}
$$

are already determined, where $k \in \{0, 1, \ldots, n/2 - 1\}$, we let

$$
s_{n/2+(k+1)} = \left(s_{n/2-k}^* \mathrm{i}\right)^{-1}q, \qquad s_{n/2-(k+1)} = \left(s_{n/2+(k+1)}^*\right)^{-1}.
$$

Analogously, for a given $q \in \mathsf{H} \setminus \{0\}$ with $\mathfrak{R}(q) = 0$, a diagonal matrix T can be found such that

$$
T^*(F_n + t\mathrm{i}G_n)T = F_n + tqG_n. \tag{9.3.3}
$$

See Ex. 9.11.8.

Also, the matrix $\Xi_m(\beta) - \widetilde{G}_m$, where β is subject to the conditions in Theorem 9.3.1, can be replaced by any one of $\Xi_m(\beta) \pm G_m$ or $\Xi_m(\beta) + \widetilde{G}_m$.

The proof will be patterned on that of Theorem 9.1.1, Part (i). First we state and prove the analogue of Proposition 9.1.3.

Proposition 9.3.2. *Let* $A, B \in \mathsf{H}^{m \times n}$. *The following statements* (a), (b), *and* (c) *are equivalent for the matrix pencil* $A + tB$:

(a) $A + tB$ *is strictly equivalent to* $A^* - tB^*$;

(b) $A + tB$ *is strictly equivalent to a pencil* $A' + tB'$, *such that the quaternion matrices* A' *and* B' *are hermitian and skewhermitian, respectively;*

(c) (c1) $m = n$;

    (c2) *the left indices of* $A+tB$, *when arranged in nondecreasing order, coincide with the right indices, also arranged in nondecreasing order;*

    (c3) *for every eigenvalue* $\alpha \in \mathsf{H}$ *of* $A+tB$ *with nonzero real part, the quaternion* $-\alpha$ *is also an eigenvalue of* $A+tB$, *and the partial multiplicities of the eigenvalue* α, *arranged in nondecreasing order, coincide with those of the eigenvalue* $-\alpha$, *also arranged in nondecreasing order.*

Proof. The implication (b) $\Longrightarrow$ (a) follows analogously to the implication (b) $\Longrightarrow$ (a) in the proof of Proposition 9.1.3.

(a) $\Longrightarrow$ (c). Obviously, if (a) holds, then we must have $m = n$. Next observe that if (a) holds for $A + tB$, then (a) holds for every matrix pencil which is strictly equivalent to $A + tB$. Indeed, if

$$A_0 + tB_0 = S(A + tB)T, \quad A + tB = S_0(A^* - tB^*)T_0$$

for some matrices $A_0, B_0 \in \mathsf{H}^{n \times n}$ and some invertible matrices S, S_0, T, T_0, then

$$
\begin{aligned}
A_0^* - tB_0^* &= T^*(A^* - tB^*)S^* \\
&= T^*S_0^{-1}(A + tB)T_0^{-1}S^* = T^*S_0^{-1}S^{-1}(A_0 + tB_0)T^{-1}T_0^{-1}S^*.
\end{aligned}
$$

Therefore, without loss of generality, we may assume that $A+tB$ is in the Kronecker form (7.3.1). Now, the Kronecker form of

$$L_{\varepsilon \times (\varepsilon+1)}(-t)^* = \begin{bmatrix} 0_{1 \times \varepsilon} \\ I_\varepsilon \end{bmatrix} - t \begin{bmatrix} I_\varepsilon \\ 0_{1 \times \varepsilon} \end{bmatrix}$$

coincides with $L_{\varepsilon \times (\varepsilon+1)}(t)^T$ because of equality

$$\mathrm{diag}\,(1, -1, 1, \ldots, \pm 1) L_{\varepsilon \times (\varepsilon+1)}(t)^T \mathrm{diag}\,(-1, 1, -1, \ldots, \pm 1)$$

$$= L_{\varepsilon \times (\varepsilon+1)}(-t)^T. \tag{9.3.4}$$

This proves (c2) (in view of the uniqueness of the Kronecker form). Also, if $\alpha \in \mathsf{H}$ has nonzero real part, then the Kronecker form of $-tI_\ell + (J_\ell(\alpha))^*$ is $tI_\ell + J_\ell(-\alpha)$, and the matrix pencils $tI_\ell + J_\ell(\alpha)$ and $tI_\ell + J_\ell(-\alpha)$ are not strictly equivalent (since the quaternions α and $-\alpha$ ar not similar). This proves (c3), again by using the uniqueness of the Kronecker form.

(c) $\implies$ (b). Since both properties (c) and (b) are easily seen to be invariant under strict equivalence, it suffices to prove this implication for the case when $A+tB$ is given in a Kronecker form. In turn, we need to consider only the following two cases:

$$(1) \quad A + tB = L_{\varepsilon \times (\varepsilon+1)} \oplus L_{\varepsilon \times (\varepsilon+1)}^T;$$

$$(2) \quad A + tB = (tI_k + J_k(\alpha)) \oplus (tI_k + J_k(-\alpha)), \quad \Re(\alpha) \neq 0. \qquad (9.3.5)$$

In Case (1), use the equalities

$$\left(L_{\varepsilon \times (\varepsilon+1)}(t) \oplus \Lambda_{\varepsilon \times (\varepsilon+1)}(-t)^T\right) F_{2\varepsilon+1} = G_{2\varepsilon+1} + t \begin{bmatrix} 0 & 0 & F_\varepsilon \\ 0 & 0_1 & 0 \\ -F_\varepsilon & 0 & 0 \end{bmatrix},$$

where the $\varepsilon \times (\varepsilon+1)$ matrix pencil

$$\Lambda_{\varepsilon \times (\varepsilon+1)}(-t) := \begin{bmatrix} 1 & -t & 0 & \cdots & 0 \\ 0 & 1 & -t & \cdots & 0 \\ \vdots & & \ddots & \ddots & \vdots \\ 0 & 0 & \cdots & 1 & -t \end{bmatrix} \in \mathsf{H}^{\varepsilon \times (\varepsilon+1)}$$

is strictly equivalent to $L_{\varepsilon \times (\varepsilon+1)}(-t)$, which, in turn, is strictly equivalent to $L_{\varepsilon \times (\varepsilon+1)}(t)$ by virtue of (9.3.4). Indeed,

$$F_\varepsilon L_{\varepsilon \times (\varepsilon+1)}(-t) F_{\varepsilon+1} = \Lambda_{\varepsilon \times (\varepsilon+1)}(-t). \qquad (9.3.6)$$

In Case (2), note that the right-hand side of (9.3.5) is clearly strictly equivalent to

$$\begin{bmatrix} 0 & tI + J_(\alpha) \\ -tI - J_k(-\alpha) & 0 \end{bmatrix},$$

and furthermore $-J_k(-\alpha)$ is similar to $(J(\alpha))^*$. $\qquad \square$

Proof of Theorem 9.3.1. Just repeat the arguments of the proof of (c) $\implies$ (b) in Proposition 9.3.2. $\qquad \square$

9.4 CANONICAL FORMS FOR MIXED PENCILS: CONGRUENCE

We now state the main result on congruence of mixed hermitian/skewhermitian quaternion matrix pencils.

Theorem 9.4.1. *Let $A + tB$ be a quaternion hermitian/skewhermitian $n \times n$ matrix pencil. Then $A + tB$ is congruent to a hermitian/skewhermitian pencil of the form*

$$(A_0 + tB_0) \oplus \oplus_{j=1}^r \delta_j \left(F_{k_j} + tiG_{k_j}\right) \oplus \oplus_{i=1}^p \eta_i \left(G_{\ell_i} + tiF_{\ell_i}\right) \qquad (9.4.1)$$

$$\oplus \ \oplus_{u=1}^q \zeta_u \left(\begin{bmatrix} 0 & 0 & \cdots & 0 & 0 & \beta_u \\ 0 & 0 & \cdots & 0 & -\beta_u & -1 \\ 0 & 0 & \cdots & \beta_u & -1 & 0 \\ \vdots & \vdots & \ddots & \vdots & \vdots & \vdots \\ (-1)^{m_u-1}\beta_u & -1 & 0 & \cdots & 0 & 0 \end{bmatrix}_{m_u \times m_u} \right.$$

$$+ t\Xi_{m_u}(i^{m_u})\big). \qquad (9.4.2)$$

Here, $A_0 + tB_0$ is a direct sum of blocks of types (q-h-sk0), (q-h-sk1), (q-h-sk2) *for even k,* (q-h-sk3) *for odd ℓ, and* (q-h-sk4), *in which several blocks of the same type and of different and/or the same sizes may be present. Furthermore, the parameters of the blocks in* (9.4.1), (9.4.2) *are as follows*:

(1) *the k_j's are odd positive integers;*

(2) *the ℓ_i's are even positive integers;*

(3) *the β_u's are positive real numbers if m_u is odd, and β_u's are nonzero quaternions with zero real parts if m_u is even;*

(4) *the δ_j, η_i, and ζ_u are signs ± 1.*

The subscript in $[\,\cdot\,]_{x \times x}$ designates the size $x \times x$ of the matrix $[\,\cdot\,]$.

  The blocks in (9.4.1) *and* (9.4.2) *are uniquely determined by $A + tB$ up to an arbitrary permutation of constituent matrix pencil blocks in $A_0 + tB_0$, up to an arbitrary permutation of blocks in each of the three parts*

$$\oplus_{j=1}^{r} \delta_j \left(F_{k_j} + tiG_{k_j} \right), \qquad \oplus_{i=1}^{p} \eta_i \left(G_{\ell_i} + tiF_{\ell_i} \right),$$

and (9.4.2), *and up to the following replacements:*

(a) *α can be replaced with a similar quaternion in any block of type* (q-h-sk4);

(b) *β can be replaced with a similar quaternion in any block of type* (9.4.2) *of even size $m_u \times m_u$.*

  Conversely, any matrix pencil, which is congruent to a pencil of the form (9.4.1), (9.4.2) *is hermitian/skewhermitian.*

  Remark 9.4.2. The signs δ_j, η_i, and ζ_u form the *sign characteristic* of the hermitian/skewhermitian matrix pencil $A + tB$. Note that the blocks $F_k + tiG_k$, $G_k + tiF_k$, and

$$\Xi_{m_u}(\beta_u) - \widetilde{G}_{m_u} + t\Xi_{m_u}(\mathrm{i}^{m_u}), \tag{9.4.3}$$

where β_u is as in Theorem 9.4.1, are strictly equivalent to $I + tJ_k(0)$, $tI + J_\ell(0)$, and $tI + J_{m_u}(-\mathrm{i}^{m_u}\beta_u)$ (if m_u is odd), $tI + J_{m_u}(\mathrm{i}^{m_u}\beta_u)$ (if m_u is even), respectively. Thus, the sign characteristic associates a sign ± 1 to every odd partial multiplicity at infinity, to every even partial multiplicity at zero, and to every partial multiplicity of nonzero eigenvalues with zero real parts, for hermitian/skewhermitian matrix pencils.

  Remark 9.4.3. The strict equivalence in Remark 9.4.2 implies that $-\beta_u$ can be used in place of β_u in some of the blocks (9.4.2) with odd m_u. Indeed, (9.4.3) is strictly equivalent to $tI + J_{m_u}(-\mathrm{i}^{m_u}\beta_u)$, which is, in turn, strictly equivalent to $tI + J_{m_u}(\mathrm{i}^{m_u}\beta_u)$, as well as to (9.4.3) with β_u replaced by $-\beta_u$. Explicitly, for m_u odd and $\beta_u > 0$ we have

$$(\mathrm{diag}\,(-\mathrm{j}, \mathrm{j}, \ldots, -\mathrm{j}))(\Xi_{m_u}(\beta_u) - \widetilde{G}_{m_u} + t\Xi_{m_u}(\mathrm{i}^{m_u}))(\mathrm{diag}\,(\mathrm{j}, -\mathrm{j}, \ldots, \mathrm{j}))$$

$$= -(\Xi_{m_u}(-\beta_u) - \widetilde{G}_{m_u} + t\Xi_{m_u}(\mathrm{i}^{m_u})).$$

Thus, replacement of β_u by $-\beta_u$ will reverse the sign of ζ_u.

Remark 9.4.4. Noting the equalities

$$F_m(s_1 F_m + s_2 \widetilde{G}_m)F_m = s_1 F_m + s_2 G_m,$$

where s_1, s_2 are independent variables, we see that $\pm G_m$ and $\pm \widetilde{G}_m$ can be interchanged in Theorem 9.4.1. Also note that for m even, we have

$$\mathrm{diag}\,(1, -1, 1, \ldots, -1)(\Xi_m(\beta) - \widetilde{G}_m + t\Xi_m(\mathrm{i}^m))$$

$$\times \mathrm{diag}\,(1, -1, 1, \ldots, -1) = -(\Xi_m(\beta) + \widetilde{G}_m + t\Xi_m(\mathrm{i}^m)),$$

and for m odd,

$$\mathrm{diag}\,(1, -1, 1, \ldots, 1)(\Xi_m(\beta) - \widetilde{G}_m + t\Xi_m(\mathrm{i}^m))\mathrm{diag}\,(1, -1, 1, \ldots, 1)$$

$$= \Xi_m(\beta) + \widetilde{G}_m + t\Xi_m(\mathrm{i}^m).$$

Analogous equalities hold true for the blocks $F_k + ti G_k$ and $G_\ell + ti F_\ell$. Thus, $\pm G_m$, resp. $\pm \widetilde{G}_m$, can be replaced by $\mp G_m$, resp. $\mp \widetilde{G}_m$, in Theorem 9.4.1. However, upon such replacement, the sign in the sign characteristic of blocks of even size will reverse to its negative, and the sign of blocks of odd size will remain the same (cf. Remark 8.1.5).

Remark 9.4.5. The quaternion i can be replaced in (9.4.1) by any quaternion λ with zero real part and norm 1. Indeed, by Theorem 2.3.3 the real dimension of the solution set of $\mathrm{i}x - x\lambda = 0$ is equal to 2, so there exists $x \in \mathsf{H}$ such that $\mathrm{i}x - x\lambda = 0$ and x has zero real part and norm 1. Now, clearly,

$$(xI)^*(F_k + ti G_k)(xI) = (F_k + t\lambda G_k); \quad (xI)^*(G_\ell + ti F_\ell)(xI) = (G_\ell + t\lambda F_\ell).$$

The lengthy proof of Theorem 9.4.1 will be relegated to the next two sections. In this section we present only the next proposition, which justifies the statements concerning replacements of α's and β's, as stipulated in the theorem.

Proposition 9.4.6. (1) *If m is even and β_1, β_2 are similar nonzero quaternions with zero real part, then the two $m \times m$ hermitian/skewhermitian matrix pencils*

$$\Upsilon_j^{(1)} := \begin{bmatrix} 0 & 0 & \cdots & 0 & 0 & \beta_j + t \\ 0 & 0 & \cdots & 0 & -\beta_j - t & \pm 1 \\ 0 & 0 & \cdots & \beta_j + t & \pm 1 & 0 \\ \vdots & \vdots & \cdot^{\cdot^{\cdot}} & \vdots & \vdots & \vdots \\ -\beta_j - t & \pm 1 & 0 & \cdots & 0 & 0 \end{bmatrix}_{m \times m}, \quad j = 1, 2,$$

are congruent.

(2) *If α_1, α_2 are similar quaternions with nonzero real parts, then the four $2p \times 2p$ hermitian/skewhermitian matrix pencils*

$$\Upsilon^{(2)}(j, \pm) := \begin{bmatrix} 0 & \pm\alpha_j F_p + G_p \\ \pm\alpha_j^* F_p + G_p & 0 \end{bmatrix} + t \begin{bmatrix} 0 & F_p \\ -F_p & 0 \end{bmatrix}, \quad j = 1, 2,$$

are congruent.

Proof. For Part (a), notice that $\beta_1 = u^*\beta_2 u$ for some $u \in H$ with $u^*u = 1$ (see Theorem 2.2.6). Then $(u^*I_m)\Upsilon_2^{(1)}(uI_m) = \Upsilon_1^{(1)}$. For Part (b), let $q \in H \setminus \{0\}$ be such that $q\alpha_1 q^{-1} = \alpha_2$. Then

$$\begin{bmatrix} qI & 0 \\ 0 & (q^{-1})^*I \end{bmatrix} \Upsilon^{(2)}(1, \pm) \begin{bmatrix} q^*I & 0 \\ 0 & q^{-1}I \end{bmatrix} = \Upsilon^{(2)}(2, \pm).$$

Also, let $X \in \mathbb{R}^{p \times p}$ be an invertible matrix such that

$$X(-G_pF_p)X^{-1} = G_pF_p.$$

(The existence of X is guaranteed because both $-G_pF_p$ and G_pF_p are real $p \times p$ matrices of rank $p - 1$ such that $(-G_pF_p)^p = (G_pF_p)^p = 0$, and any two such matrices are similar because they have the same real Jordan form.) Then, with $Y = -F_p(X^{-1})^*F_p$, we have

$$\begin{bmatrix} 0 & X \\ Y & 0 \end{bmatrix} \Upsilon^{(2)}(j, \pm) \begin{bmatrix} 0 & Y^* \\ X^* & 0 \end{bmatrix} = \Upsilon^{(2)}(j, \mp), \quad j = 1, 2.$$

Statement (b) follows. $\qquad \square$

9.5 PROOF OF THEOREM 9.4.1: EXISTENCE

We start with a preliminary result.

Lemma 9.5.1. (1) *The block $F_k + tiG_k$ is congruent to its negative $-F_k - tiG_k$ if k is even.*

(2) *The block $F_k + tiG_k$ is not congruent to its negative if k is odd.*

(3) *The block $G_\ell + tiF_\ell$ is congruent to its negative $-G_\ell - tiF_\ell$ if ℓ is odd.*

(4) *The block $G_\ell + tiF_\ell$ is not congruent to its negative if ℓ is even.*

(5) *The block (q-h-sk5) is not congruent to its negative.*

(6) *Each of the blocks (q-h-sk1) and (q-h-sk4) is congruent to its negative.*

Proof. Parts (1) and (3) follow from the equalities

$$\text{diag}\,(j, -j, \ldots, j, -j)\,(F_k + tiG_k)\,\text{diag}\,(-j, j, \ldots, -j, j) = -(F_k + tiG_k) \qquad (9.5.1)$$

for even k and

$$\text{diag}\,(j, -j, \ldots, -j, j)\,(G_\ell + tiF_\ell)\,\text{diag}\,(-j, j, \ldots, j, -j) = -(G_\ell + tiF_\ell) \qquad (9.5.2)$$

for odd ℓ.

Part (2) is obvious because for odd k, the signature of F_k (as a hermitian matrix) is different from the signature of $-F_k$. Since the signature of G_ℓ is different from that of $-G_\ell$ for ℓ even, (4) follows as well.

For Part (6) observe the equalities

$$\begin{bmatrix} I_\varepsilon & 0 \\ 0 & -I_{\varepsilon+1} \end{bmatrix} \left(G_{2\varepsilon+1} + \begin{bmatrix} 0 & 0 & tF_\varepsilon \\ 0 & 0_1 & 0 \\ -tF_\varepsilon & 0 & 0 \end{bmatrix} \right) \begin{bmatrix} I_\varepsilon & 0 \\ 0 & -I_{\varepsilon+1} \end{bmatrix}$$

$$= - \left(G_{2\varepsilon+1} + \begin{bmatrix} 0 & 0 & tF_\varepsilon \\ 0 & 0_1 & 0 \\ -tF_\varepsilon & 0 & 0 \end{bmatrix} \right)$$

and

$$\begin{bmatrix} I_p & 0 \\ 0 & -I_p \end{bmatrix} \begin{bmatrix} 0 & \alpha F_p + G_p + tF_p \\ \overline{\alpha}F_p + G_p - tF_p & 0 \end{bmatrix} \begin{bmatrix} I_p & 0 \\ 0 & -I_p \end{bmatrix}$$

$$= \begin{bmatrix} 0 & -\alpha F_p - G_p - tF_p \\ -\overline{\alpha}F_p - G_p + tF_p & 0 \end{bmatrix}.$$

It remains to prove Part (5). If m is odd, then the hermitian matrix A given by (9.3.2) has signature different from that of $-A$, and we are done in this case. Suppose now m is even. It will be convenient to prove a lemma first.

Lemma 9.5.2. *Let β be a nonzero quaternion with zero real part. For m even, define the $m \times m$ matrices $\Phi_m(\beta)$ and $\Xi_m(1)$ by (1.2.7) and (1.2.6), respectively. If S and T are $m \times m$ quaternion matrices such that the equalities*

$$S\Phi_m(\beta) = -\Phi_m(\beta)T, \qquad S\Xi_m(1) = -\Xi_m(1)T \tag{9.5.3}$$

hold true, then the entries of S and T belong to the real subalgebra $\mathrm{Span}_{\mathsf{R}}\{1, \beta\}$.

Proof. We may assume without loss of generality, that $\beta = x\mathrm{i}$, where $x \in \mathsf{R} \backslash \{0\}$ (indeed, by Theorem 2.4.4 there exists an automorphism of H that maps any fixed nonzero quaternion with zero real part to a real multiple of i).
Letting

$$X_m := \begin{bmatrix} 0 & 0 & \cdots & 0 & 0 & 0 \\ 0 & 0 & \cdots & 0 & 0 & -1 \\ 0 & 0 & \cdots & 0 & -1 & 0 \\ \vdots & \vdots & \ddots & \vdots & \vdots & \vdots \\ 0 & -1 & 0 & \cdots & 0 & 0 \end{bmatrix},$$

and writing

$$S = S_0 + \mathrm{i}S_1 + \mathrm{j}S_2 + \mathrm{k}S_3, \qquad T = T_0 + \mathrm{i}T_1 + \mathrm{j}T_2 + \mathrm{k}T_3,$$

where $S_0, S_1, S_2, S_3, T_0, T_1, T_2, T_3 \in \mathsf{R}^{m \times m}$, the equalities (9.5.3) take the form

$$(S_0 + \mathrm{i}S_1 + \mathrm{j}S_2 + \mathrm{k}S_3)(\mathrm{i}x\Xi_m(1) + X_m)$$

$$= (-\mathrm{i}x\Xi_m(1) - X_m)(T_0 + \mathrm{i}T_1 + \mathrm{j}T_2 + \mathrm{k}T_3),$$

$$(S_0 + \mathrm{i}S_1 + \mathrm{j}S_2 + \mathrm{k}S_3)\Xi_m(1) = (-\Xi_m(1))(T_0 + \mathrm{i}T_1 + \mathrm{j}T_2 + \mathrm{k}T_3).$$

Equating the coefficients of j in both sides of each of these equations, and similarly for k, we obtain

$$\begin{aligned} S_2 X_m + x S_3 \Xi_m(1) &= x\Xi_m(1)T_3 - X_m T_2, \\ -x S_2 \Xi_m(1) + S_3 X_m &= -x\Xi_m(1)T_2 - X_m T_3, \end{aligned} \tag{9.5.4}$$

$$S_2 \Xi_m(1) = -\Xi_m(1)T_2, \qquad S_3 \Xi_m(1) = -\Xi_m(1)T_3. \tag{9.5.5}$$

Letting

$$\widetilde{X} := X_m \Xi_m(1)^{-1}, \qquad \widetilde{T}_2 = \Xi_m(1)T_2\Xi_m(1)^{-1}, \qquad \widetilde{T}_3 = \Xi_m(1)T_3\Xi_m(1)^{-1},$$

and using (9.5.5), the first equality in (9.5.4) gives

$$2xS_3 = -\widetilde{X}\widetilde{T}_2 - S_2\widetilde{X} = \widetilde{X}S_2 - S_2\widetilde{X}. \tag{9.5.6}$$

Analogously, the second equality in (9.5.4) gives

$$-2xS_2 = -S_3\widetilde{X} - \widetilde{X}\widetilde{T}_3 = -S_3\widetilde{X} + \widetilde{X}S_3. \tag{9.5.7}$$

Solve (9.5.6) for S_3 and substitute in (9.5.7):

$$-2xS_2 = \frac{1}{2x}\left(\widetilde{X}^2 S_2 - 2\widetilde{X}S_2\widetilde{X} + S_2\widetilde{X}^2\right).$$

Iterating this equality and using the easily verifiable fact that $\widetilde{X}$ is *nilpotent*, i.e., $\widetilde{X}^p = 0$ for some positive integer p, we obtain $S_2 = 0$, and then $S_3 = 0$. Then also $T_2 = T_3 = 0$, and the lemma is proved. $\qquad\square$

We now return to the proof of Part (5), the case of even m. Without loss of generality, we take β to be a nonzero real multiple of i: $\beta = ri$, where $r \in \mathsf{R} \setminus \{0\}$. Suppose $A + tB$ are given by (9.3.2), where m is even, and suppose, arguing by contradiction, that $A + tB$ is congruent (over the quaternions) to $-A - tB$. By Lemma 9.5.2, the complex matrix pencils $A + tB$ and $-A - tB$ are C-congruent. Then the complex hermitian matrix pencils $A + t(iB)$ and $-A + t(-iB)$ are also C-congruent. However, this is impossible. To see this, let

$$\widetilde{A} := (iB)^{-1}A = rI_m + D,$$

where the matrix D has $-i, i, -i, \ldots, i, -i$ on the first superdiagonal and zeros everywhere else. Clearly, the Jordan form of $\widetilde{A}$ is $J_m(r)$, and both $A + t(iB)$ and $-A + t(-i)B$ are C-strictly equivalent to $tI + J_m(r)$. By Theorem 15.3.1 in the Appendix, the canonical form of $A + t(iB)$ under C-congruence is $\pm((t+r)F_m + G_m)$; therefore, the canonical form of $-A + t(-i)B$ must be $\mp((t+r)F_m + G_m)$, and these canonical forms are not C-congruent by the same Theorem 15.3.1. $\qquad\square$

This concludes the proof of Lemma 9.5.1.

We now are able to prove the existence part of Theorem 9.4.1. First, consider the scalar case $n = 1$. If $a + tb$, where $a \in \mathsf{R}$, $\Re(b) = 0$, is a scalar hermitian/skewhermitian quaternion pencil, then we have the following three possibilities (excluding the trivial case $a = b = 0$):

1. $a = 0$. Then in view of Theorem 2.2.6, the pencil is congruent to ti.

2. $b = 0$. Obviously, the pencil is congruent to ± 1.

3. $a, b \neq 0$. Then by the same Theorem 2.2.6, $a + tb$ is congruent to a pencil of the form $\pm(\beta + ti)$ for some real positive β.

In all possibilities, the conditions of the form (9.4.1), (9.4.2) are fulfilled.

In the general case, the statement that every hermitian/skewhermitian quaternion matrix pencil is congruent to a pencil of the form (9.4.1), (9.4.2), subject to the conditions specified in Theorem 9.4.1, is proved by induction on the size of the matrices. In the proof, Lemma 9.5.1 and Theorem 5.11.1 are used.

Let $A + tB$ be an $n \times n$ hermitian/skewhermitian matrix pencil. 1. Using Theorem 9.3.1, we find invertible matrices $P \in \mathsf{H}^{n \times n}$ and $Q \in \mathsf{H}^{n \times n}$ such that

$$(A + tB)\,Q = P\,(A_0 + tB_0),\qquad(9.5.8)$$

where $A_0 + tB_0$ is a direct sum of blocks of the forms (q-h-sk0)–(q-h-sk5). We will show that $A + tB$ and $A_0 + tB_0$ are ϕ-congruent for some choice of the signs δ_j, κ_j, η_i, and ζ_u.

Without loss of generality, assume $Q = I$ (otherwise, consider $Q_\phi(A + tB)Q$ in place of $A + tB$). Then we have

$$\begin{aligned}
PA_0 &= A = A^* = (A_0)^* P^* = A_0 P^*,\\
PB_0 &= B = -B_0^* = -B_0^* P^* = B_0 P^*,
\end{aligned}\qquad(9.5.9)$$

and, therefore,

$$F(P)A_0 = A_0(F(P)),\quad F(P)B_0 = B_0(F(P))^*\qquad(9.5.10)$$

for every scalar polynomial $F(t)$ with real coefficients.

Several cases with respect to P may occur.

Case 9.5.3. *P has only real eigenvalues, there is only one distinct real eigenvalue of P, and this eigenvalue γ is positive.*

From the Jordan form of P it easily follows that for the polynomial with real coefficients $(t - \gamma^{-1})^n$, we have $(P^{-1} - \gamma^{-1})^n = 0$. By Lemma 8.2.1 there exists a polynomial $f_n(t)$ with real coefficients such that $t = (f_n(t))^2 \pmod{(t - \gamma^{-1})^n}$. Then $P^{-1} = (f_n(P^{-1}))^2$, and since P^{-1} itself is a polynomial of P with real coefficients, we have $P^{-1} = (f(P))^2$ for some polynomial $f(t)$ with real coefficients. Let $R = f(P)$, and use (9.5.10) to obtain

$$\begin{aligned}
R(A + tB)R^* &= RP(A_0 + tB_0)R^* = RP(A_0 + tB_0)(f(P))^*\\
&= f(P)Pf(P)(A_0 + tB_0) = A_0 + tB_0,
\end{aligned}\qquad(9.5.11)$$

and the existence of the form (9.4.1), (9.4.2) under congruence follows.

Case 9.5.4. *P has only real eigenvalues, there is only one distinct real eigenvalue of P, and this eigenvalue is negative.*

Arguing as in the preceding paragraph for $-P$ rather than P, it is found that $(-P)^{-1} = (f(-P))^2$ for some polynomial $f(t)$ with real coefficients. As in (9.5.11), it follows that

$$R(A + tB)R^* = -(A_0 + tB_0),$$

where $R = f(-P)$. Thus, to complete the proof of existence of the form (9.4.1) and (9.4.2), in Case 9.5.4, it remains to show that each of the blocks of types (q-h-sk1), (q-h-sk4), $F_p + t\beta G_p$ with p odd, and $G_p + t\beta F_p$ with p even, is congruent to its negative. This was shown in Lemma 9.5.1.

Case 9.5.5. *All eigenvalues of P have equal real parts, and their vector parts are nonzero and have equal euclidean lengths.*

Let $\beta' \in \mathsf{H}$ be an eigenvalue (necessarily nonreal) of P. Denote $\alpha = (\beta')^{-1}$. In view of the Jordan form of P, for some positive integer w (which may be taken the largest size of Jordan blocks in the Jordan form of P), we have $g(P^{-1}) = 0$, where $g(t) = ((t - \beta')(t - (\beta')^*))^w$ is a polynomial with real coefficients. Using Lemma 8.2.1 again, we find a polynomial $g_w(t)$ with real coefficients such that $t = (g_w(t))^2$ modulo a real polynomial multiple of $g(t)$). Then $P^{-1} = (g_w(P^{-1}))^2$. Since P is itself a matrix root of a polynomial with real coefficients, namely, $\widehat{g}(P) = 0$, where

$$\widehat{g}(t) = ((t - \alpha)(t - (\alpha^*)))^w,$$

it follows that P^{-1} is also a polynomial of P with real coefficients. Now argue as in Case 9.5.3.

Case 9.5.6. *All other possibilities (not covered in Cases 9.5.3, 9.5.4, and 9.5.5).*

In this case, using the Jordan form, P can be written in the form

$$P = S \operatorname{diag}(P_1, P_2, \ldots, P_r) S^{-1} \tag{9.5.12}$$

where S is an invertible quaternion matrix and $P_1, \ldots, P_r$ are matrices of sizes $n_1 \times n_1, \ldots, n_r \times n_r$, respectively, such that any pair of eigenvalues λ of P_j and μ of P_i $(i \neq j)$ satisfies the following condition:

$$\text{either } \Re(\lambda) \neq \Re(\mu) \text{ or } |\mathfrak{V}(\lambda)| \neq |\mathfrak{V}(\mu)|, \text{ or both.} \tag{9.5.13}$$

We also have $r \geq 2$ (the situations when $r = 1$ are covered in Cases 9.5.3–9.5.5). Substituting (9.5.12) in the equality $A + tB = P(A_0 + tB_0)$, we obtain

$$\left(P_1^{-1} \oplus \cdots \oplus P_r^{-1}\right)\left(\widetilde{A} + t\widetilde{B}\right) = \widetilde{A}_0 + t\widetilde{B}_0, \tag{9.5.14}$$

where

$$\begin{aligned}
\widetilde{A} &= S^{-1}A(S^*)^{-1}, & \widetilde{B} &= S^{-1}B(S^*)^{-1}, \\
\widetilde{A}_0 &= S^{-1}A_0(S^*)^{-1}, & \widetilde{B}_0 &= S^{-1}B_0(S^*)^{-1}.
\end{aligned}$$

Partition the matrix $\widetilde{A}$:

$$\widetilde{A} = [M_{ij}]_{i,j=1}^r,$$

where M_{ij} is of the size $n_i \times n_j$. Since $\widetilde{A}$ and $\widetilde{A}_0$ are hermitian, (9.5.14) implies the equality

$$P_i^{-1}M_{ij} = M_{ij}\left((P_j)^*\right)^{-1};$$

cf. (9.5.9). An elementary calculation shows that if (9.5.13) holds for two nonzero quaternions λ and μ, then the same property holds for the inverses λ^{-1} and μ^{-1}. Note also that in view of Corollary 5.7.2, $\nu \in \sigma(((P_j)^*)^{-1})$ if and only if $\nu \in \sigma(P_j^{-1})$. Thus, as a corollary of these remarks, we have

$$\sigma(P_i^{-1}) \cap \sigma(((P_j)^*)^{-1}) = \emptyset \quad \text{for} \quad i \neq j.$$

Now by Theorem 5.11.1 we have $M_{ij} = 0$ $(i \neq j)$. In other words,

$$\widetilde{A} = M_{11} \oplus \cdots \oplus M_{rr}.$$

Similarly, $\widetilde{B} = N_{11} \oplus \cdots \oplus N_{rr}$, where N_{ii} is of size $n_i \times n_i$.

Now, induction is used on the size of the matrix pencil $A + tB$ to complete the proof of existence part in Theorem 9.4.1. $\qquad\square$

9.6   PROOF OF THEOREM 9.4.1: UNIQUENESS

In this section we will make extensive use of the canonical form for real symmetric/skewsymmetric matrix pencils (Theorem 15.2.3 in the Appendix).

Let $A + tB$ be a hermitian/skewhermitian quaternion matrix pencil, and assume that it is congruent to two forms (9.4.1), (9.4.2). We have to prove that the two forms are obtained from each other by the permutations and replacements indicated in Theorem 9.4.1. These two forms are congruent; hence, a fortiori strictly equivalent. Therefore, by Theorem 9.3.1 we may assume that the two forms may differ only in the signs δ_j, η_k, and ζ_u. Thus, suppose that the hermitian/skewhermitian matrix pencil $A' + tB'$ given by (9.4.1), (9.4.2) is congruent to the pencil

$$
A'' + tB'' := (A_0 + tB_0) \oplus \oplus_{j=1}^{r} \delta'_j \left(F_{k_j} + tiG_{k_j} \right)
$$

$$
\oplus \quad \oplus_{i=1}^{p} \eta'_i \left(G_{\ell_i} + tiF_{\ell_i} \right) \tag{9.6.1}
$$

$$
\oplus \quad \oplus_{u=1}^{q} \zeta'_u \begin{bmatrix} & 0 & & 0 & \cdots \\ & 0 & & 0 & \cdots \\ & 0 & & 0 & \cdots \\ & \vdots & & \vdots & \cdot\cdot\cdot \\ (-1)^{m_u-1}(\beta_u + ti^{m_u}) & -1 & 0 & \end{bmatrix}
$$

$$
\begin{bmatrix} 0 & 0 & \beta_u + ti^{m_u} \\ 0 & -\beta_u - ti^{m_u} & -1 \\ \beta_u + ti^{m_u} & -1 & 0 \\ \vdots & \vdots & \vdots \\ \cdots & 0 & 0 \end{bmatrix} \tag{9.6.2}
$$

for some signs δ'_j, η'_i, and ζ'_u. We need to show that $A'' + tB''$ is obtained from $A' + tB'$ by a permutation of constituent blocks.

To this end we will apply the map χ to the pencils $A' + tB'$ and $A'' + tB''$. By properties (iii) and (iv) of $\chi_{n,n}$, we have

$$
\chi_{n,n} \left(A' + tB' \right) = \chi_{n,n} \left(A' \right) + t\chi_{n,n} \left(B' \right)
$$

is a real symmetric/skewsymmetric $4n \times 4n$ matrix pencil. We describe first the canonical forms of real symmetric/skewsymmetric pencils obtained from the primitive blocks (q-h-sk2), (q-h-sk3), and (q-h-sk5) by the map χ.

We will use the following real symmetric/skewsymmetric matrix pencil in the proof of the next lemma. Denote $\Xi_2 := \Xi_2(1) = \begin{bmatrix} 0 & 1 \\ -1 & 0 \end{bmatrix}$; then

$$
Q_{2m}(\nu) := \begin{bmatrix}
0 & 0 & \cdots & 0 & 0 & \nu\Xi_2^{m+1} \\
0 & 0 & \cdots & 0 & -\nu\Xi_2^{m+1} & -I_2 \\
0 & 0 & \cdots & \nu\Xi_2^{m+1} & -I_2 & 0 \\
\vdots & \vdots & \reflectbox{$\ddots$} & \vdots & \vdots & \vdots \\
(-1)^{m-1}\nu\Xi_2^{m+1} & -I_2 & 0 & \cdots & 0 & 0
\end{bmatrix}
$$

$$
+ \, t \begin{bmatrix}
0 & 0 & \cdots & 0 & \Xi_2^m \\
0 & 0 & \cdots & -\Xi_2^m & 0 \\
\vdots & \vdots & \reflectbox{$\ddots$} & \vdots & \vdots \\
0 & (-1)^{m-2}\Xi_2^m & \cdots & 0 & 0 \\
(-1)^{m-1}\Xi_2^m & 0 & \cdots & 0 & 0
\end{bmatrix}, \qquad \nu > 0. \qquad (9.6.3)
$$

The pencil in (9.6.3) is $2m \times 2m$, where m is a positive integer.

Lemma 9.6.1. (1) *The real symmetric/skewsymmetric matrix pencil $\chi_{k,k}\,(F_k + tiG_k)$, where k is odd, is R-congruent to*

$$
\left(F_k + t \begin{bmatrix} 0_1 & 0 & 0 \\ 0 & 0 & F_{\frac{k-1}{2}} \\ 0 & -F_{\frac{k-1}{2}} & 0 \end{bmatrix} \right)^{\oplus 4}.
$$

(2) *The real symmetric/skewsymmetric matrix pencil $\chi_{k,k}\,(F_k + tiG_k)$, where k is even, is R-congruent to*

$$
\left(F_{2k} + t \begin{bmatrix} 0_1 & 0 & 0 & 0 \\ 0 & 0 & 0 & F_{k-1} \\ 0 & 0 & 0_1 & 0 \\ 0 & -F_{k-1} & 0 & 0 \end{bmatrix} \right)^{\oplus 2}. \qquad (9.6.4)
$$

(3) *The real symmetric/skewsymmetric matrix pencil $\chi_{\ell,\ell}\,(G_\ell + tiF_\ell)$, where ℓ is even, is R-congruent to*

$$
\left(G_\ell + t \begin{bmatrix} 0 & F_{\frac{\ell}{2}} \\ -F_{\frac{\ell}{2}} & 0 \end{bmatrix} \right)^{\oplus 4}. \qquad (9.6.5)
$$

(4) *The real symmetric/skewsymmetric matrix pencil $\chi_{\ell,\ell}\,(G_\ell + tiF_\ell)$, where ℓ is odd, is R-congruent to*

$$
\begin{bmatrix} 0 & G_\ell + tF_\ell \\ G_\ell - tF_\ell & 0 \end{bmatrix}^{\oplus 2}. \qquad (9.6.6)
$$

(5) *The real symmetric/skewsymmetric matrix pencil*

$$
\chi_{m,m} \left(\begin{bmatrix}
0 & 0 & \cdots & 0 & 0 & \beta \\
0 & 0 & \cdots & 0 & -\beta & -1 \\
0 & 0 & \cdots & \beta & -1 & 0 \\
\vdots & \vdots & \reflectbox{$\ddots$} & \vdots & \vdots & \vdots \\
(-1)^{m-1}\beta & -1 & 0 & \cdots & 0 & 0
\end{bmatrix} + t\Xi_m(i^m) \right), \qquad (9.6.7)
$$

where m is odd and β is positive real number, is R-*congruent to*

$$(-1)^{(m+1)/2}\left(Q_{2m}(\beta)\oplus Q_{2m}(\beta)\right),$$

where $Q_{2m}(\beta)$ is defined by (9.6.3).

(6) *The real symmetric/skewsymmetric matrix pencil*

$$\chi_{m,m}\left(\begin{bmatrix} 0 & 0 & \cdots & 0 & 0 & \beta \\ 0 & 0 & \cdots & 0 & -\beta & -1 \\ 0 & 0 & \cdots & \beta & -1 & 0 \\ \vdots & \vdots & \ddots & \vdots & \vdots & \vdots \\ (-1)^{m-1}\beta & -1 & 0 & \cdots & 0 & 0 \end{bmatrix} + t\Xi_m(\mathrm{i}^m)\right), \qquad (9.6.8)$$

where m is even and β is a nonzero quaternion with zero real part, is R-*congruent to $(Q_{2m}(|\beta|))^{\oplus 2}$, where $Q_{2m}(|\beta|)$ is defined by* (9.6.3).

Proof. Part (1). It is easy to see that $\chi_{k,k}\left(F_k + tiG_k\right)$ is R-congruent, using a suitable permutation matrix for the congruence transformation, to

$$(F_k + t\operatorname{diag}(1,-1,1,-1,\ldots,1)G_k)^{\oplus 2}$$

$$\oplus(F_k + t\operatorname{diag}(-1,1,-1,1,\ldots,-1)G_k)^{\oplus 2}.$$

In turn,

$$\begin{bmatrix} 0 & 0 & \cdots & 0 & 0 & \epsilon_1 \\ 0 & 0 & \cdots & 0 & \epsilon_2 & 0 \\ 0 & 0 & \cdots & \epsilon_3 & 0 & 0 \\ \vdots & \vdots & \ddots & \vdots & \vdots & \vdots \\ \epsilon_k & & 0 & \cdots & 0 & 0 \end{bmatrix} (F_k + t\operatorname{diag}(1,-1,1,-1,\ldots,1)G_k)$$

$$\times \begin{bmatrix} 0 & 0 & \cdots & 0 & 0 & \epsilon_k \\ 0 & 0 & \cdots & 0 & \epsilon_{k-1} & 0 \\ 0 & 0 & \cdots & \epsilon_{k-2} & 0 & 0 \\ \vdots & \vdots & \ddots & \vdots & \vdots & \vdots \\ \epsilon_1 & & 0 & \cdots & 0 & 0 \end{bmatrix} = F_k + t\begin{bmatrix} 0_1 & 0 & 0 \\ 0 & 0 & F_{\frac{k-1}{2}} \\ 0 & -F_{\frac{k-1}{2}} & 0 \end{bmatrix},$$

where $\epsilon_j = \pm 1$ are found from the following system of equations:

$$\epsilon_1\epsilon_k = \epsilon_2\epsilon_{k-1} = \cdots = \epsilon_{(k-1)/2}\epsilon_{(k+1)/2} = 1, \qquad (9.6.9)$$

$$\epsilon_2\epsilon_k = -1, \quad \epsilon_3\epsilon_{k-1} = 1, \ldots, \quad \epsilon_{(k+1)/2}\epsilon_{(k+3)/2} = (-1)^{(k+3)/2}. \quad (9.6.10)$$

It is easy to see that the system (9.6.9), (9.6.10) has a unique solution once the value of $\epsilon_1 = \pm 1$ is arbitrarily assigned. Analogously, one proves that

$$F_k + t\operatorname{diag}(-1,1,-1,1,\ldots,-1)G_k$$

is R-congruent to $F_k + t\begin{bmatrix} 0_1 & 0 & 0 \\ 0 & 0 & F_{(k-1)/2} \\ 0 & -F_{(k-1)/2} & 0 \end{bmatrix}$.

Part (2). By Theorem 15.2.3, it suffices to prove that $\chi_{k,k}\left(F_k + tiG_k\right)$ is R-strictly equivalent to (9.6.4) (indeed, it follows from that theorem that a symmetric/

skewsymmetric real matrix pencil is R-strictly equivalent to (9.6.4) if and only if it is R-congruent to (9.6.4)). This is easily seen by using R-strict equivalence

$$\chi_{k,k}\,(F_k + tiG_k) \quad \mapsto \quad Q_1\chi_{k,k}\,(F_k + tiG_k)Q_2$$

with suitable permutation matrices Q_1 and Q_2.

Part (3). Applying a congruence with a suitable permutation matrix, it is easy to see that $\chi_{\ell,\ell}\,(G_\ell + tiF_\ell)$ is R-congruent to the matrix

$$
\begin{bmatrix}
0 & 0 & \cdots & 0 & I_2 & t\widetilde{\Xi}_2 \\
0 & 0 & \cdots & I_2 & t\widetilde{\Xi}_2 & 0 \\
0 & 0 & \cdots & t\widetilde{\Xi}_2 & 0 & 0 \\
\vdots & \vdots & \iddots & \vdots & \vdots & \vdots \\
I_2 & t\widetilde{\Xi}_2 & 0 & \cdots & 0 & 0 \\
t\widetilde{\Xi}_2 & 0 & 0 & \cdots & 0 & 0
\end{bmatrix}_{2\ell \times 2\ell}
$$

$$
\oplus
\begin{bmatrix}
0 & 0 & \cdots & 0 & I_2 & t\widetilde{\Xi}_2 \\
0 & 0 & \cdots & I_2 & t\widetilde{\Xi}_2 & 0 \\
0 & 0 & \cdots & t\widetilde{\Xi}_2 & 0 & 0 \\
\vdots & \vdots & \iddots & \vdots & \vdots & \vdots \\
I_2 & t\widetilde{\Xi}_2 & 0 & \cdots & 0 & 0 \\
t\widetilde{\Xi}_2 & 0 & 0 & \cdots & 0 & 0
\end{bmatrix}_{2\ell \times 2\ell}, \tag{9.6.11}
$$

where

$$\widetilde{\Xi}_2 := \Xi_2(-1) = \begin{bmatrix} 0 & -1 \\ 1 & 0 \end{bmatrix}. \tag{9.6.12}$$

The matrix (9.6.11) is of size $4\ell \times 4\ell$ and all zeros there are the zero 2×2 matrices. Denote by X either of the two equal $2\ell \times 2\ell$ blocks in (9.6.11). Let P be the $2\ell \times 2\ell$ permutation matrix whose rows, starting with the top row and going down, are

$$\mathbf{e}_1, \mathbf{e}_4, \mathbf{e}_5, \mathbf{e}_8, \mathbf{e}_9, \ldots, \mathbf{e}_{2\ell-4}, \mathbf{e}_{2\ell-3}, \mathbf{e}_{2\ell}, \mathbf{e}_2, \mathbf{e}_3, \ldots, \mathbf{e}_{2\ell-2}, \mathbf{e}_{2\ell-1},$$

where we have denoted by $\mathbf{e}_j$ the unit row vector with 1 in the jth position and zeros in all other positions. For typographical convenience, let

$$
\Psi_\ell =
\begin{bmatrix}
0 & 0 & \cdots & 0 & \widetilde{\Xi}_2 \\
0 & 0 & \cdots & \widetilde{\Xi}_2 & 0 \\
\vdots & \vdots & \iddots & \vdots & \vdots \\
0 & \widetilde{\Xi}_2 & \cdots & 0 & 0 \\
\widetilde{\Xi}_2 & 0 & \cdots & 0 & 0
\end{bmatrix}_{\ell \times \ell}.
$$

A computation shows that

$$PXP^* = (G_\ell + t\Psi_\ell) \oplus (G_\ell - t\Psi_\ell).$$

Next, observe that

$$\text{diag}\,(1, -1, 1, \ldots - 1) \cdot (G_\ell + t\Psi_\ell) \cdot \text{diag}\,(1, -1, 1, \ldots - 1) = G_\ell - t\Psi_\ell,$$

and, in turn,

$$\operatorname{diag}(\epsilon_1,\ldots,\epsilon_\ell)\cdot(G_\ell+t\Psi_\ell)\cdot\operatorname{diag}(\epsilon_1,\ldots,\epsilon_\ell)=G_\ell+t\begin{bmatrix}0 & F_{\ell/2}\\-F_{\ell/2} & 0\end{bmatrix},$$

where $\epsilon_1,\ldots,\epsilon_\ell$ are signs ±1 that are found from the following equations:

$$\begin{aligned}\epsilon_1\epsilon_\ell &= -1,\quad \epsilon_2\epsilon_{\ell-1}=1,\\ \epsilon_3\epsilon_{\ell-2} &= -1,\ldots,\epsilon_{\ell/2}\epsilon_{(\ell/2)+1}=(-1)^{\ell/2}, & (9.6.13)\\ \epsilon_1\epsilon_{\ell-1} &= \epsilon_2\epsilon_{\ell-2}=\cdots=\epsilon_{(\ell/2)-1}\epsilon_{(\ell/2)+1}=1. & (9.6.14)\end{aligned}$$

It is easy to see that the system (9.6.13), (9.6.14) has unique solution once the value of ϵ_1 is arbitrarily assigned.

Part (4). It suffices to prove that $\chi_{\ell,\ell}(G_\ell+tiF_\ell)$, where ℓ is odd, is R-strictly equivalent to (9.6.6) (indeed, by Theorem 15.2.3, a real symmetric/skewsymmetric matrix pencil is R-strictly equivalent to (9.6.6) if and only if it is R-congruent to (9.6.6)). Using the same transformation as in the proof of Part (3), we need to show that

$$\begin{bmatrix}0 & 0 & \cdots & 0 & I_2 & t\widetilde{\Xi}_2\\ 0 & 0 & \cdots & I_2 & t\widetilde{\Xi}_2 & 0\\ 0 & 0 & \cdots & t\widetilde{\Xi}_2 & 0 & 0\\ \vdots & \vdots & \iddots & \vdots & \vdots & \vdots\\ I_2 & t\widetilde{\Xi}_2 & 0 & \cdots & 0 & 0\\ t\widetilde{\Xi}_2 & 0 & 0 & \cdots & 0 & 0\end{bmatrix}_{2\ell\times2\ell},\quad\text{with}\quad \widetilde{\Xi}=\begin{bmatrix}0 & -1\\1 & 0\end{bmatrix},\quad(9.6.15)$$

is R-strictly equivalent to

$$\begin{bmatrix}0 & G_\ell+tF_\ell\\ G_\ell-tF_\ell & 0\end{bmatrix}. \qquad (9.6.16)$$

In fact, it is easy to see that the Kronecker form over the reals of both (9.6.15) and (9.6.16) is $(tI+J_\ell(0))\oplus(tI+J_\ell(0))$.

Part (5). Let $\Xi_2:=\begin{bmatrix}0 & 1\\-1 & 0\end{bmatrix}$. It will be convenient to denote also $\Upsilon:=\beta\Xi_2^{m+1}+t(-\Xi_2^m)$.

A congruence transformation with a permutation congruence matrix shows that (9.6.7) is R-congruent to

$$\begin{bmatrix}0 & 0 & \cdots & 0 & 0 & \Upsilon\\ 0 & 0 & \cdots & 0 & -\Upsilon & -I_2\\ 0 & 0 & \cdots & \Upsilon & -I_2 & 0\\ \vdots & \vdots & \iddots & \vdots & \vdots & \vdots\\ \Upsilon & -I_2 & 0 & \cdots & 0 & 0\end{bmatrix}^{\oplus2} \qquad (9.6.17)$$

if $m-1$ is not divisible by 4 and to

$$\begin{bmatrix}0 & 0 & \cdots & 0 & 0 & -\Upsilon\\ 0 & 0 & \cdots & 0 & \Upsilon & -I_2\\ 0 & 0 & \cdots & -\Upsilon & -I_2 & 0\\ \vdots & \vdots & \iddots & \vdots & \vdots & \vdots\\ -\Upsilon & -I_2 & 0 & \cdots & 0 & 0\end{bmatrix}^{\oplus2} \qquad (9.6.18)$$

if $m - 1$ is divisible by 4 (we assume here that m is odd). We see that the congruence transformation with the congruence matrix $F_2^{\oplus 2m}$ applied to the left-hand side of (9.6.17) yields the form (9.6.3) with $\nu = \beta$, whereas the the congruence transformation with the congruence matrix

$$(\text{diag}\,(F_2, -F_2, F_2, -F_2, \cdots, F_2))^{\oplus 2}$$

applied to the left-hand side of (9.6.18) yields the negative of (9.6.3) with $\nu = \beta$.

Part (6). Assume m is even. For an automorphism ψ of H, consider the matrix pencil

$$W_{4m}(\psi) := \chi_{m,m}\left(\begin{bmatrix} 0 & 0 & \cdots & 0 & 0 & \psi(\beta) \\ 0 & 0 & \cdots & 0 & -\psi(\beta) & -1 \\ 0 & 0 & \cdots & \psi(\beta) & -1 & 0 \\ \vdots & \vdots & \ddots & \vdots & \vdots & \vdots \\ -\psi(\beta) & -1 & 0 & \cdots & 0 & 0 \end{bmatrix} + t\Xi_m(\mathrm{i}^m)\right),$$

where β is a fixed nonzero quaternion with zero real part. Since $\psi(\beta)$ also has zero real part, clearly $W_{4m}(\psi)$ is a real symmetric/skewsymmetric pencil. By Proposition 3.3.2, $W_{4m}(\psi)$, for various ψ's, are all real unitarily similar to each other and, in particular, have the same canonical form under real congruence. By Theorem 2.2.6 we see that there exists an automorphism ψ of H such that $\psi(\beta) = q\mathrm{i}$ for some $q > 0$. Therefore, we need only to verify (6) for $\beta = q\mathrm{i}$, where $q > 0$; in fact, $q = |\beta|$. A congruence with a suitable permutation matrix shows that

$$\chi_{m,m}\left(\begin{bmatrix} 0 & 0 & \cdots & 0 & 0 & q\mathrm{i} \\ 0 & 0 & \cdots & 0 & -q\mathrm{i} & -1 \\ 0 & 0 & \cdots & q\mathrm{i} & -1 & 0 \\ \vdots & \vdots & \ddots & \vdots & \vdots & \vdots \\ -q\mathrm{i} & -1 & 0 & \cdots & 0 & 0 \end{bmatrix} + t\Xi_m(\mathrm{i}^m)\right)$$

is R-congruent to $(Q_{2m}(q))^{\oplus 2}$ if m is not divisible by 4 and is R-congruent to $(Q_{2m}(-q))^{\oplus 2}$ if m is divisible by 4. Here, $Q_{2m}(\pm q)$ is defined by (9.6.3). Finally, the equality

$$F_2^{\oplus m}(Q_{2m}(-q))F_2^{\oplus m} = Q_{2m}(q)$$

yields the desired R-congruence. $\square$

Now we can we easily complete the proof of the uniqueness part of Theorem 9.4.1. Indeed, if $A'' + tB''$ were not obtained from $A' + tB'$ by a permutation of constituent blocks, then by applying the map χ and using Lemma 9.6.1 we obtain a contradiction with the uniqueness (up to a permutation of constituent primitive blocks) of the canonical form of the real symmetric/skewsymmetric matrix pencil $\chi_{n,n}\,(A' + tB')$ (Theorem 15.2.3). $\square$

9.7 COMPARISON WITH REAL AND COMPLEX PENCILS: STRICT EQUIVALENCE

The canonical form for mixed hermitian/skewhermitian quaternion matrix pencils allows us to sort out the relations between R-congruence and H-congruence of mixed pencils of real matrices, and between C-congruence and H-congruence of mixed pencils of complex matrices.

For real matrices we have the following.

Theorem 9.7.1. *Let $A, B, A', B' \in \mathsf{R}^{m \times m}$ be such that*

$$A^T = A, \ (A')^T = A', \quad B^T = -B, \ (B')^T = -B'.$$

Then, the real matrix pencils $A + tB$ and $A' + tB'$ are H-congruent, resp. H-strictly equivalent, if and only if $A + tB$ and $A' + tB'$ are R-congruent, resp. R-strictly equivalent.

Proof. The statement on strict equivalence is a particular case of Theorem 7.6.3. For congruence, let an invertible matrix $S \in \mathsf{H}^{m \times m}$ be such that equality

$$S^*(A + tB)S = A' + tB' \tag{9.7.1}$$

holds, and write $S = S_0 + \mathrm{i}S_1 + \mathrm{j}S_2 + \mathrm{k}S_3$, where S_0, S_1, S_2, S_3 are real matrices. Then (9.7.1) takes the form (8.5.2), and the proof is completed as in the proof of Theorem 8.5.1. $\qquad\square$

The complex case is more involved. In this section we consider the strict equivalence relation. It will be convenient to work with hermitian matrix pencils $A + t\mathrm{i}B$, where $A = A^*, B = -B^* \in \mathsf{C}^{n \times n}$.

Theorem 9.7.2. *Let $A, A' \in \mathsf{C}^{n \times n}$ be complex hermitian matrices and let $B, B' \in \mathsf{C}^{n \times n}$ be complex skewhermitian matrices. Assume that the complex matrix pencils $A + tB$ and $A' + tB'$ are H-strictly equivalent and that the hermitian matrix pencil $A + t(\mathrm{i}B)$ is C-strictly equivalent to the form (15.3.1), where*

$$\alpha_1 = \alpha_2 = \cdots = \alpha_{q'} = 0, \quad \alpha_w \in \mathsf{R} \setminus \{0\} \text{ for } w = q' + 1, q' + 2, \ldots, q$$

for some q' (the case $q' = 0$ is not excluded), and where all the signs δ_j's and η_j's are taken to be equal to 1.

Then the hermitian pencil $A' + t(\mathrm{i}B')$ is C-strictly equivalent to a complex hermitian matrix pencil of the following form:

$$
0_u \quad \oplus \quad \left(t \begin{bmatrix} 0 & 0 & F_{\varepsilon_1} \\ 0 & 0 & 0 \\ F_{\varepsilon_1} & 0 & 0 \end{bmatrix} + G_{2\varepsilon_1 + 1} \right)
$$

$$
\oplus \cdots \oplus \left(t \begin{bmatrix} 0 & 0 & F_{\varepsilon_p} \\ 0 & 0 & 0 \\ F_{\varepsilon_p} & 0 & 0 \end{bmatrix} + G_{2\varepsilon_p + 1} \right)
$$

$$
\oplus \ (F_{k_1} + tG_{k_1}) \oplus \cdots \oplus (F_{k_r} + tG_{k_r})
$$

$$
\oplus \ (tF_{\ell_1} + G_{\ell_1}) \oplus \cdots \oplus \left(tF_{\ell_{q'}} + G_{\ell_{q'}} \right)
$$

$$
\oplus \ \left((t + \kappa'_{q'+1}\alpha_{q'+1}) F_{\ell_{q'+1}} + G_{\ell_{q'+1}} \right) \oplus \cdots \oplus \left((t + \kappa'_q \alpha_q) F_{\ell_q} + G_{\ell_q} \right)
$$

$$
\oplus \ \left(\begin{bmatrix} 0 & (t + \beta'_1)F_{m_1} \\ (t + \overline{\beta'_1})F_{m_1} & 0 \end{bmatrix} + \begin{bmatrix} 0 & G_{m_1} \\ G_{m_1} & 0 \end{bmatrix} \right)
$$

$$
\oplus \ \cdots \oplus \left(\begin{bmatrix} 0 & (t + \beta'_s)F_{m_s} \\ (t + \overline{\beta'_s})F_{m_s} & 0 \end{bmatrix} + \begin{bmatrix} 0 & G_{m_s} \\ G_{m_s} & 0 \end{bmatrix} \right), \tag{9.7.2}
$$

where

$$\text{for each } j = 1, \ldots, s, \text{ either } \beta'_j = \beta_j \text{ or } \beta'_j = -\overline{\beta_j}, \qquad (9.7.3)$$

$$\text{and } \kappa'_{q'+1}, \ldots, \kappa'_q \in \{1, -1\}. \qquad (9.7.4)$$

Conversely, suppose that $A' + tB'$ and $A'' + tB''$ are pencils of the form (9.7.2), with the parameters

$$\{\kappa'_{q'+1}, \ldots, \kappa'_q; \beta'_1, \ldots, \beta'_s\}$$

subject to (9.7.3) and (9.7.4) for $A' + tB'$, and with the corresponding parameters

$$\{\kappa''_{q'+1}, \ldots, \kappa''_q; \beta''_1, \ldots, \beta''_s\},$$

again subject to conditions analogous to (9.7.3) and (9.7.4) for $A'' + tB''$. Then the complex hermitian-skewhermitian matrix pencils $A' + t(-\mathsf{i})B'$ and $A'' + t(-\mathsf{i})B''$ are H-strictly equivalent.

We illustrate this theorem with an example.

Example 9.7.3. Let A, B, A', B' be as in Theorem 9.7.2. Let

$$A + tB = (t(-\mathsf{i}) + \alpha)F_\ell + G_\ell, \qquad \alpha \in \mathsf{R}.$$

Multiplying this equality on the right by $\mathsf{i}F_\ell$, we see that $A + tB$ has the C-Kronecker form $tI_\ell + J_\ell(\mathsf{i}\alpha)$. Since $A' + tB'$ has the same H-Kronecker form as $A + tB$, by Theorem 7.6.3 the C-Kronecker form of $A' + tB'$ is $tI_\ell + J_\ell(\mathsf{i}\alpha)$ or $tI_\ell + J_\ell(-\mathsf{i}\alpha)$. Therefore, $A' + t\mathsf{i}B'$ is C-strictly equivalent to $t\mathsf{i}I_\ell + J_\ell(\mathsf{i}\alpha)$ or to $t\mathsf{i}I_\ell + J_\ell(-\mathsf{i}\alpha)$. Multiplication on the right by $(-\mathsf{i})F_\ell$ yields C-strict equivalence of $A' + t\mathsf{i}B'$ to $(t + \alpha)F_\ell - \mathsf{i}G_\ell$ or to $(t - \alpha)F_\ell - \mathsf{i}G_\ell$. Finally, note the equalities

$$\text{diag}\,((-\mathsf{i})^j)_{j=0}^{\ell-1}\,((t \pm \alpha)F_\ell - \mathsf{i}G_\ell)\,\text{diag}\,((\mathsf{i})^{\ell-1-j})_{j=0}^{\ell-1} = (t \pm \alpha)F_\ell + G_\ell,$$

which produce the desired statement concerning C-strict equivalence of $A' + t\mathsf{i}B'$ according to Theorem 9.7.2.   □

We prove Theorem 9.7.2 in Section 9.9. Here, we indicate a corollary of this theorem describing the situations when strict equivalence of two hermitian/skewhermitian pencils is the same over H and over H. Recall that $-\alpha$ is the eigenvalue of a matrix pencil $tI + J_m(\alpha)$.

Corollary 9.7.4. *Let $A, B \in \mathsf{C}^{n \times n}$, where $A = A^*$, $B = -B^*$. Then the following two statements are equivalent:*

(1) *Every hermitian-skewhermitian complex matrix pencil $A' + tB'$, $A' = A'^*$, $B' = -B'^*$, which is H-strictly equivalent to $A + tB$ is also C-strictly equivalent to $A + tB$;*

(2) *$A + tB$ has no nonzero pure imaginary eigenvalues.*

The cases when zero is an eigenvalue of $A + tB$ and/or there are indices at infinity are not excluded in Corollary 9.7.4(b).

Proof. Observe that the complex matrix pencils $A + tB$ and $A + t\mathsf{i}B$ have the same indices at infinity and the same partial multiplicities at the eigenvalue zero, and $-\alpha \in \mathsf{C}$ is an eigenvalue of $A + tB$ if and only if $\mathsf{i}\alpha$ is an eigenvalue of $A + t\mathsf{i}B$; moreover the partial multiplicities of $A + tB$ at its eigenvalue $-\alpha$ coincide with the partial multiplicities of $A + tiB$ at its eigenvalue $i\alpha$. In view of this observation, the result of Corollary 9.7.4 is immediate from Theorem 9.7.2. $\square$

If the conditions of Corollary 9.7.4 do not hold, then we have more than one class of C-strict equivalence within the class of H-strict equivalence. We compute the precise number of such classes, as follows.

Let $A_0 + tB_0$ be a complex matrix pencil of hermitian matrices, and let (15.3.1) (with δ_i and η_j taken to be equal 1) be the canonical form of $A_0 + tB_0$ under C-strict equivalence. For every pair of nonzero real numbers $(\alpha, -\alpha)$, where α is taken to be positive, and every positive integer ℓ, let $s(\alpha, \ell)$ be the total number of blocks $(t + \alpha_j)F_{\ell_j} + G_{\ell_j}$ in (15.3.1) with $\alpha_j = \alpha$ or $\alpha_j = -\alpha$, and $\ell_j = \ell$. (We formally define $s(\alpha, \ell) = 0$ if neither α nor $-\alpha$ is an eigenvalue of $A_0 + tB_0$ or if at least one of $\alpha, -\alpha$ is an eigenvalue of $A_0 + tB_0$ but ℓ is not a partial multiplicity of $A_0 + tB_0$ at $\pm\alpha$.) Define

$$s(A_0 + tB_0) := \prod_{\alpha > 0} \prod_{\ell=1}^{\infty} (s(\alpha, \ell) + 1).$$

Theorem 9.7.5. *Let $A + tB$ be as in Theorem 9.7.2, and let*

$$\Omega(A + tB) := \{A' + tB' : A', B' \in \mathsf{C}, (A')^* = A', (B')^* = -B',$$
$$A' + tB' \text{ is } \mathsf{H} - \text{strictly equivalent to } A + tB\}$$

be the set of complex hermitian/skewhermitian pencils that are H-strictly equivalent to $A + tB$. Then $\Omega(A + tB)$ consists of exactly $s(A + tiB)$ disjoint classes so that the pencils in each class are all C-strictly equivalent and the pencils from different classes are not C-strictly equivalent to each other.

Proof. By Theorem 9.7.2 we need to show that $s(A + tiB)$ coincides with the number of mutually C-nonstrictly equivalent hermitian matrix pencils given by formula (9.7.2), where $\kappa'_{q'+1}, \ldots, \kappa'_q$ are arbitrary signs ± 1.

Denote by $\widetilde{A} + t\widetilde{B}$ the complex matrix pencil (9.7.2), where all signs κ'_j are taken to be equal to 1 and $\beta'_j = \beta_j$ for $j = 1, 2, \ldots, s$.

Fix a pair of nonzero real numbers $\alpha, -\alpha$, at least one of which is an eigenvalue of $\widetilde{A} + t\widetilde{B}$ and where $\alpha > 0$, and let

$$((t + \kappa'_1\alpha)F_\ell + G_\ell)\oplus \quad \cdots \quad \oplus((t + \kappa'_v\alpha)F_\ell + G_\ell)$$
$$\oplus((t + \kappa''_1(-\alpha))F_\ell + G_\ell)\oplus \quad \cdots \quad \oplus((t + \kappa''_w(-\alpha))F_\ell + G_\ell) \qquad (9.7.5)$$

be the part of (9.7.2) with blocks of size ℓ and eigenvalues $\pm\alpha$; for simplicity of notation, we have changed the subscripts of some of the κ'_j's in (9.7.2) as well as renamed some of them as κ''_j.

Suppose that exactly x of the signs $\kappa'_1, \ldots, \kappa'_v$ are -1 and exactly y of the signs $\kappa''_1, \ldots, \kappa''_w$ are -1, where $0 \leq x \leq v$; $0 \leq y \leq w$. Then the C-Kronecker form (as well as the H-Kronecker form) of the pencil (9.7.5) is

$$(U + tV)_{x,y} := \underbrace{(t + \alpha)F_\ell + G_\ell) \oplus \cdots \oplus (t + \alpha)F_\ell + G_\ell)}_{v - x + y \text{ times}}$$
$$\oplus \underbrace{(t - \alpha)F_\ell + G_\ell) \oplus \cdots \oplus (t - \alpha)F_\ell + G_\ell)}_{w - y + x \text{ times}}.$$

We see that if (x', y') is another selection of integers subject to $0 \leq x' \leq v$, $0 \leq y' \leq w$, then $(U + tV)_{x',y'}$ is C-strictly equivalent to $(U + tV)_{x,y}$ precisely when $x' - y' =$

$x - y$. Note that $x - y$ can take values in the set $\{-w, -w+1, \ldots, 0, 1, \ldots, v\}$, a total of $v + w + 1 = s(\alpha, \ell) + 1$ values. Thus, we obtain exactly $s(\alpha, \ell) + 1$ mutually C-nonstrictly equivalent matrix pencils from (9.7.5) by all possible selections of signs κ'_i and κ''_j. Putting this information together for all pairs $(\alpha, -\alpha)$, where $\alpha > 0$, and all integers ℓ, the result of Theorem 9.7.5 follows. $\qquad\square$

9.8 COMPARISON WITH COMPLEX PENCILS: CONGRUENCE

Here we consider comparison of H-congruence and C-congruence of mixed complex matrix pencils. The sign characteristic plays a key role here. As in Section 9.7, it will be convenient to work with hermitian matrix pencils $A + tiB$, where $A = A^*, B = -B^* \in \mathsf{C}^{n \times n}$.

Theorem 9.8.1. *Let $A, A' \in \mathsf{C}^{n \times n}$ be complex hermitian matrices and let $B, B' \in \mathsf{C}^{n \times n}$ be complex skewhermitian matrices. Assume that the complex matrix pencils $A + tB$ and $A' + tB'$ are H-congruent. Let (15.3.1) be the canonical form of the hermitian pencil $A + t(iB)$ under C-congruence, where*

$$\alpha_1 = \alpha_2 = \cdots = \alpha_{q'} = 0, \quad \alpha_w \in \mathsf{R} \setminus \{0\} \text{ for } w = q'+1, q'+2, \ldots, q$$

for some q' (the case $q' = 0$ is not excluded).

Then the canonical form of the hermitian pencil $A' + t(iB')$ under C-congruence has the following structure:

$$
\begin{aligned}
0 \ \oplus \ & \left(t \begin{bmatrix} 0 & 0 & F_{\varepsilon_1} \\ 0 & 0 & 0 \\ F_{\varepsilon_1} & 0 & 0 \end{bmatrix} + G_{2\varepsilon_1 + 1} \right) \oplus \cdots \oplus \left(t \begin{bmatrix} 0 & 0 & F_{\varepsilon_p} \\ 0 & 0 & 0 \\ F_{\varepsilon_p} & 0 & 0 \end{bmatrix} + G_{2\varepsilon_p + 1} \right) \\
\oplus \ & \delta'_1 \left(F_{k_1} + t G_{k_1} \right) \oplus \cdots \oplus \delta'_r \left(F_{k_r} + t G_{k_r} \right) \\
\oplus \ & \eta'_1 \left(t F_{\ell_1} + G_{\ell_1} \right) \oplus \cdots \oplus \eta'_{q'} \left(t F_{\ell_{q'}} + G_{\ell_{q'}} \right) \\
\oplus \ & \eta'_{q'+1} \left(\left(t + \kappa'_{q'+1} \alpha_{q'+1} \right) F_{\ell_{q'+1}} + G_{\ell_{q'+1}} \right) \\
\oplus \ & \cdots \oplus \eta'_q \left(\left(t + \kappa'_q \alpha_q \right) F_{\ell_q} + G_{\ell_q} \right) \\
\oplus \ & \left(\begin{bmatrix} 0 & (t + \beta'_1) F_{m_1} \\ (t + \overline{\beta'_1}) F_{m_1} & 0 \end{bmatrix} + \begin{bmatrix} 0 & G_{m_1} \\ G_{m_1} & 0 \end{bmatrix} \right) \\
\oplus \ & \cdots \oplus \left(\begin{bmatrix} 0 & (t + \beta'_s) F_{m_s} \\ (t + \overline{\beta'_s}) F_{m_s} & 0 \end{bmatrix} + \begin{bmatrix} 0 & G_{m_s} \\ G_{m_s} & 0 \end{bmatrix} \right),
\end{aligned}
\qquad (9.8.1)
$$

where for each $j = 1, \ldots, s$, either $\beta'_j = \beta_j$ or $\beta'_j = -\overline{\beta_j}$, and

$$\delta'_1, \ldots, \delta'_r; \quad \eta'_1, \ldots, \eta'_q; \quad \kappa'_{q'+1}, \ldots, \kappa'_q$$

are signs ± 1 subject to the following restrictions (1), (2), (3):

(1) $\delta'_j = \delta_j$ if k_j is odd $(j = 1, 2, \ldots, r)$.

(2) $\eta'_w = \eta_w$ if ℓ_w is even $(w = 1, 2, \ldots, q')$.

(3) *For every nonzero real eigenvalue α of $A + t(iB)$ and for every index ℓ of α, the following holds: if $q' + 1 \le w_1 < \cdots < w_k \le q$ are all the distinct integers*

between $q' + 1$ and q such that $\alpha_{w_j} = \alpha$ and $\ell_{w_j} = \ell$, $j = 1, 2, \ldots, k$, then there exists a permutation λ of $\{1, 2, \ldots, k\}$ such that the integers

$$\ell + \frac{1}{2}((1 + \kappa'_{w_{\lambda(j)}})\ell + \eta'_{w_{\lambda(j)}} - \eta_{w_j}) \tag{9.8.2}$$

are even for $j = 1, 2, \ldots, k$.

Conversely, suppose that $A' + tB'$ and $A'' + tB''$ are pencils of the form (9.8.1), with the parameters

$$\{\delta'_1, \ldots, \delta'_r; \eta'_1, \ldots, \eta'_q; \kappa'_{q'+1}, \ldots, \kappa'_q; \beta'_1, \ldots, \beta'_s\}$$

for $A' + tB'$ (subject to the restrictions (1), (2), and (3)), and with the corresponding parameters

$$\{\delta''_1, \ldots, \delta''_r; \eta''_1, \ldots, \eta''_q; \kappa''_{q'+1}, \ldots, \kappa''_q; \beta''_1, \ldots, \beta''_s\}, \quad \delta''_j, \ \eta''_k, \ \kappa''_\ell \in \{1, -1\},$$

for $A'' + tB''$ subject to the conditions:

(a) *$\delta''_j = \delta_j$ if k_j is odd $(j = 1, 2, \ldots, r)$;*

(b) *$\eta''_w = \eta_w$ if ℓ_w is even $(w = 1, 2, \ldots, q')$;*

(c) *for every nonzero real eigenvalue α of $A + t(\mathrm{i}B)$ and for every index ℓ of α, the following holds: if $w_1, \ldots, w_k$ are the distinct integers between $q' + 1$ and q such that $\alpha_{w_j} = \alpha$ and $\ell_{w_j} = \ell$, $j = 1, 2, \ldots, k$, then there exists a permutation λ of $\{1, 2, \ldots, k\}$ such that the integers*

$$\ell + \frac{1}{2}((1 + \kappa''_{w_{\lambda(j)}})\ell + \eta''_{w_{\lambda(j)}} - \eta_{w_j})$$

are even for $j = 1, 2, \ldots, k$;

(d) *for each $j = 1, \ldots, s$, either $\beta''_j = \beta_j$ or $\beta''_j = -\overline{\beta_j}$.*

Then the complex hermitian-skewhermitian matrix pencils $A' + t(-\mathrm{i})B'$ and $A'' + t(-\mathrm{i})B''$ are H-congruent.

Thus, in the structure (9.8.1) there are no restrictions on the signs δ'_j for even k_j and on the signs η'_w for odd ℓ_w $(w = 1, 2, \ldots, q)$. Note also that condition (3) determines uniquely the signs η'_w, $w = q' + 1, q' + 2, \ldots, q$, once the κ'_w's are assigned. More specifically, the integer (9.8.2) is even if and only if ℓ is even and $\eta'_{w_{\lambda(j)}} = \eta_{w_j}$ or ℓ is odd and $\eta'_{w_{\lambda(j)}} = \kappa'_{w_{\lambda(j)}}\eta_{w_j}$.

We give a proof of Theorem 9.8.1 in Section 9.9. Here, we indicate a corollary (the counterpart of Corollary 9.7.4) of Theorem 9.8.1 describing the situations when congruence of two hermitian/skewhermitian pencils is the same over H and over C. Recall that $-\alpha$ is the eigenvalue of a matrix pencil $tI + J_m(\alpha)$.

Corollary 9.8.2. *Let $A, B \in \mathsf{C}^{n \times n}$, where $A = A^*$, $B = -B^*$. Then the following two statements are equivalent:*

(1) *Every hermitian-skewhermitian complex matrix pencil $A' + tB'$, $A' = A'^*$, $B' = -B'^*$ which is H-congruent to $A + tB$ is also C-congruent to $A + tB$.*

(2) $A + tB$ *has no nonzero pure imaginary eigenvalues, the partial multiplicities at the zero eigenvalue are all even, and the indices at infinity are all odd.*[1]

Proof. In view of the observations made in the proof of Corollary 9.7.4, the result of Corollary 9.8.2 is immediate from Theorem 9.8.1. $\qquad\qquad\square$

We formulate the counterpart of Theorem 9.7.5 for C-congruence as an open problem.

Open Problem 9.8.3. *Let $A+tB$ be a hermitian/skewhermitian complex matrix pencil, and let*

$$\Omega_c(A + tB) \quad := \quad \{A' + tB' \,:\, A', B' \in \mathsf{C}, \ (A')^* = A', \ (B')^* = -B',$$
$$A' + tB' \text{ is } \mathsf{H}-\text{congruent to } A + tB\}$$

be the set of complex hermitian/skewhermitian pencils that are H-congruent to $A + tB$. Describe the disjoint classes in $\Omega_c(A + tB)$ so that the pencils in each class are all C-congruent to each other and the pencils from different classes are not C-congruent to each other. In particular, find a formula for the number of such classes in terms of the canonical form under C-congruence of the complex hermitian matrix pencil $A + t\mathrm{i}B$.

9.9   PROOFS OF THEOREMS 9.7.2 AND 9.8.1

It will be convenient to collect separately several facts that will be used in the proof.

Lemma 9.9.1. *In the statements* (C)–(G) *below, η, η', κ' are signs ± 1 and α is a nonzero real number. We have:*

(A) $F_k + t(-\mathrm{i})G_k$ *is H-congruent to $-F_k + t\mathrm{i}G_k$ for odd k;*

($\widehat{\text{A}}$) $F_k + t(-\mathrm{i})G_k$ *is H-congruent to $-F_k + t\mathrm{i}G_k$ for even k;*

(B) $t(-\mathrm{i})F_\ell + G_\ell$ *is not H-congruent to $t\mathrm{i}F_\ell - G_\ell$ for even ℓ;*

($\widehat{\text{B}}$) $t(-\mathrm{i})F_\ell + G_\ell$ *is H-congruent to $t\mathrm{i}F_\ell - G_\ell$ for odd ℓ;*

(C) *for $\eta' \neq \eta$ and $\kappa' = 1$, the pencil*

$$\eta' \left((-\mathrm{i}t + \kappa'\alpha) F_\ell + G_\ell \right) \tag{9.9.1}$$

   is not H-congruent to
$$\eta \left((-\mathrm{i}t + \alpha) F_\ell + G_\ell \right); \tag{9.9.2}$$

(D) *for $\kappa' = -1$, ℓ even, and $\eta' = \eta$, the pencil* (9.9.1) *is H-congruent to* (9.9.2);

(E) *for $\kappa' = -1$, ℓ even, and $\eta' = -\eta$, the pencil* (9.9.1) *is not H-congruent to* (9.9.2);

(F) *for $\kappa' = -1$, ℓ odd and $\eta' = -\eta$, the pencil* (9.9.1) *is H-congruent to* (9.9.2);

---

[1]We point out that Parts (2) in Corollaries 6.2 and 6.4 in Rodman [134] are misstated. To obtain correct statements, Part (2) in Corollary 6.2, resp. Corollary 6.4, should be replaced by statement (2) of Corollary 9.7.4, resp. Corollary 9.8.2.

(G) *for $\kappa' = -1$, ℓ odd and $\eta' = \eta$, (9.9.1) is not* H-*congruent to (9.9.2)*;

(H) *if $\beta \in C \setminus R$, then the matrix pencil*

$$\begin{bmatrix} 0 & (-\mathrm{i}t + \beta)F_m + G_m \\ (-\mathrm{i}t + \overline{\beta})F_m + G_m & 0 \end{bmatrix}$$

is H-*congruent to*

$$\begin{bmatrix} 0 & (-\mathrm{i}t - \overline{\beta})F_m + G_m \\ (-\mathrm{i}t - \beta)F_m + G_m & 0 \end{bmatrix}.$$

Proof. Many statements in Lemma 9.9.1 can be proved using Theorem 9.4.1 and Remark 9.4.2. We give a direct proof for most of them.

Statements (A) and (B) follow from the inertia theorem for hermitian quaternion matrices, because the signature of F_k for odd k is different from the signature of $-F_k$, and the signature of G_k for even k is different from that of $-G_k$. Statements $(\widehat{\mathrm{A}})$ and $(\widehat{\mathrm{B}})$ follow from the equalities

$$\mathrm{diag}\,(\mathrm{j}, -\mathrm{j}, \ldots, \mathrm{j}, -\mathrm{j})\,(F_k + t\mathrm{i}G_k)\,\mathrm{diag}\,(-\mathrm{j}, \mathrm{j}, \ldots, -\mathrm{j}, \mathrm{j}) = -(F_k + t\mathrm{i}G_k)$$

for even k and

$$\mathrm{diag}\,(\mathrm{j}, -\mathrm{j}, \ldots, -\mathrm{j}, \mathrm{j})\,(G_\ell + t\mathrm{i}F_\ell)\,\mathrm{diag}\,(-\mathrm{j}, \mathrm{j}, \ldots, \mathrm{j}, -\mathrm{j}) = -(G_\ell + t\mathrm{i}F_\ell)$$

for odd ℓ. Statement (C) is a consequence of Theorem 9.4.1; see also Remark 9.4.2. Indeed, it follows that any matrix pencil of mixed hermitian-skewhermitian quaternion matrices whose H-Kronecker form consists of exactly one Jordan block and the eigenvalue of this block is nonzero with zero real part is not H-congruent to its negative. For (D) note the equality (taking $\eta' = \eta = 1$ without loss of generality)

$$\mathrm{diag}\,(-\mathrm{j}, \mathrm{j}, \ldots, -\mathrm{j}, \mathrm{j})\,((-\mathrm{i}t - \alpha)F_\ell + G_\ell)\,\mathrm{diag}\,(\mathrm{j}, -\mathrm{j}, \ldots, \mathrm{j}, -\mathrm{j})$$

$$= ((-\mathrm{i}t + \alpha)F_\ell + G_\ell),$$

where $\alpha \in R \setminus \{0\}$ and ℓ is even.

Consider (E). By the already proved statement (D), the matrix pencils

$$\eta'\,((-\mathrm{i}t - \alpha)\,F_\ell + G_\ell) \quad \text{and} \quad \eta'\,((-\mathrm{i}t + \alpha)\,F_\ell + G_\ell)$$

are H-congruent; but

$$\eta'\,((-\mathrm{i}t + \alpha)\,F_\ell + G_\ell) = -\eta\,((-\mathrm{i}t + \alpha)\,F_\ell + G_\ell),$$

which is not H-congruent to (9.9.2) by statement (C). Statement (F) is proved by the equality

$$(\mathrm{diag}\,(\mathrm{j}, -\mathrm{j}, \mathrm{j}, \ldots, -\mathrm{j}, \mathrm{j}))\,((\mathrm{i}t + \alpha)F_\ell - G_\ell)\,(\mathrm{diag}\,(-\mathrm{j}, \mathrm{j}, -\mathrm{j}, \ldots, \mathrm{j}, -\mathrm{j}))$$

$$= (\mathrm{i}t + \alpha)F_\ell + G_\ell,$$

where ℓ is odd. For the proof of (G), observe that the real symmetric matrices $\eta'\kappa'\alpha F_\ell + \eta'G_\ell$ and $\eta\alpha F_\ell + \eta G_\ell$ (the values of the pencils (9.9.1) and (9.9.2) when

$t = 0$) have different inertia; therefore, these two matrices cannot be H-congruent. For statement (H), let α and γ be nonzero quaternions satisfying the properties

$$\alpha(-\mathrm{i})\alpha^{-1} = \mathrm{i}, \quad \alpha(-1)^m \gamma^* = 1,$$

and let

$$X := (\mathrm{diag}\,(\alpha, -\alpha, \ldots, (-1)^{m-1}\alpha)) \oplus (\mathrm{diag}\,(\gamma, -\gamma, \ldots, (-1)^{m-1}\gamma)) \in \mathsf{H}^{2m \times 2m}.$$

Then a straightforward computation shows that

$$X \begin{bmatrix} 0 & (-\mathrm{i}t + \beta)F_m + G_m \\ (-\mathrm{i}t + \overline{\beta})F_m + G_m & 0 \end{bmatrix} X^* =$$

$$\begin{bmatrix} 0 & (-\mathrm{i}t - \overline{\beta})F_m + G_m \\ (-\mathrm{i}t - \beta)F_m + G_m & 0 \end{bmatrix}.$$

This completes the proof of Lemma 9.9.1. □

Statements (C)–(G) of Lemma 9.9.1 can be conveniently summarized as follows.

Corollary 9.9.2. *For signs $\eta, \eta', \kappa' \in \{1, -1\}$, positive integer ℓ, and $\alpha \in \mathsf{R}\backslash\{0\}$, the matrix pencils (9.9.1) and (9.9.2) are H-congruent if and only if the integer*

$$\ell + \frac{1}{2}((1 + \kappa')\ell + \eta' - \eta) \tag{9.9.3}$$

is even, i.e., ℓ is even and $\eta' = \eta$, or ℓ is odd and $\eta' = \kappa'\eta$.

Proof of Theorem 9.8.1. The direct statement: The pencils $A + tB$ and $A' + tB'$ clearly have the same H-Kronecker form. Therefore, denoting by $A'_0 + tB'_0$ the C-Kronecker form of $A' + tB'$ and denoting by $A_0 + tB_0$ the C-Kronecker form of $A + tB$, we obtain, in view of Theorem 7.6.3, that $A'_0 + tB'_0$ follows from $A_0 + tB_0$ by replacing some blocks $J_{\ell_j}(\alpha_j)$ with $J_{\ell_j}(\overline{\alpha_j})$, for nonreal α_j. Note that $A'_0 + t(\mathrm{i}B'_0)$ is C-strictly equivalent to $A' + t(\mathrm{i}B')$, whereas $A_0 + t(\mathrm{i}B_0)$ is C-strictly equivalent to $A + t(\mathrm{i}B)$. Note also that

$$x = y + \mathrm{i}z, \ y, z \in \mathsf{R}, \ \text{is an eigenvalue of} \ A_0 + t(\mathrm{i}B_0)$$
$$\Longleftrightarrow \quad \mathrm{i}y - z \ \text{is an eigenvalue of} \ A_0 + tB_0$$
$$\Longleftrightarrow \quad \mathrm{i}y - z \ \text{and/or} \ -\mathrm{i}y - z \ \text{is an eigenvalue of} \ A'_0 + tB'_0$$
$$\Longleftrightarrow \quad -y + \mathrm{i}z \ \text{and/or} \ y + \mathrm{i}z \ \text{is an eigenvalue of} \ A'_0 + t(\mathrm{i}B'_0).$$

Taking into account Theorem 15.3.1, we see that the canonical form of the complex hermitian pencil $A' + t(iB')$ under C-congruence must have the following structure:

$$0_u \quad \oplus \quad \left(t \begin{bmatrix} 0 & 0 & F_{\varepsilon_1} \\ 0 & 0 & 0 \\ F_{\varepsilon_1} & 0 & 0 \end{bmatrix} + G_{2\varepsilon_1+1} \right)$$

$$\oplus \quad \cdots \quad \oplus \left(t \begin{bmatrix} 0 & 0 & F_{\varepsilon_p} \\ 0 & 0 & 0 \\ F_{\varepsilon_p} & 0 & 0 \end{bmatrix} + G_{2\varepsilon_p+1} \right)$$

$$\oplus \quad \delta_1'\left(F_{k_1} + tG_{k_1}\right) \oplus \cdots \oplus \delta_r'\left(F_{k_r} + tG_{k_r}\right)$$

$$\oplus \quad \eta_1'\left(tF_{\ell_1} + G_{\ell_1}\right) \oplus \cdots \oplus \eta_{q'}'\left(tF_{\ell_{q'}} + G_{\ell_{q'}}\right)$$

$$\oplus \quad \eta_{q'+1}'\left(\left(t + \eta_{q'+1}''\alpha_{q'+1}\right)F_{\ell_{q'+1}} + G_{\ell_{q'+1}}\right) \oplus \cdots \oplus \eta_q'\left(\left(t + \eta_q''\alpha_q\right)F_{\ell_q} + G_{\ell_q}\right)$$

$$\oplus \quad \left(\begin{bmatrix} 0 & (t+\beta_1')F_{m_1} \\ (t+\overline{\beta_1'})F_{m_1} & 0 \end{bmatrix} + \begin{bmatrix} 0 & G_{m_1} \\ G_{m_1} & 0 \end{bmatrix} \right)$$

$$\oplus \quad \cdots \quad \oplus \left(\begin{bmatrix} 0 & (t+\beta_s')F_{m_s} \\ (t+\overline{\beta_s'})F_{m_s} & 0 \end{bmatrix} + \begin{bmatrix} 0 & G_{m_s} \\ G_{m_s} & 0 \end{bmatrix} \right), \qquad (9.9.4)$$

where for each $j = 1, \ldots, s$, either $\beta_j' = \beta_j$ or $\beta_j' = -\overline{\beta_j}$, and

$$\delta_1', \ldots, \delta_r', \quad \eta_1', \ldots, \eta_q', \eta_{q'+1}'', \ldots, \eta_q''$$

are signs ± 1. Denote by $A_1 + tB_1$ the hermitian pencil (15.3.1), and by $A_1' + tB_1'$ the hermitian pencil (9.9.4). Then the pencils $A_1 + t(-i)B_1$ and $A_1' + t(-i)B_1'$ are H-congruent. In view of the canonical form for hermitian-skewhermitian quaternion pencils under H-congruence (Theorem 9.4.1), the direct statement of Theorem 9.8.1 follows from Lemma 9.9.1(A), (B), (H), and from Corollary 9.9.2.

The converse statement follows by analogous arguments using $(\widehat{A})$ and $(\widehat{B})$ of Lemma 9.9.1 and Corollary 9.9.2 again. $\square$

Proof of Theorem 9.7.2. The direct statement: Arguing as in the proof of the direct statement of Theorem 9.8.1, we see that the canonical form of the complex hermitian pencil $A' + t(iB')$ under C-congruence must have the same structure as in (9.9.4). Evidently, (9.9.4) is C-strictly equivalent to (9.7.2).

The converse statement: The H-strict equivalence of $A' + t(-i)B'$ and $A'' + t(-i)B''$ follows by observing that every pair of corresponding constituent blocks in $A' + t(-i)B'$ and $A'' + t(-i)B''$ has the same H-Kronecker form. $\square$

9.10 CANONICAL FORMS FOR MATRICES UNDER CONGRUENCE

Theorem 9.4.1, together with the obvious observation that every matrix $A \in \mathsf{H}^{n \times n}$ can be uniquely written in the form $A = F + G$, where $F = F^*$ and $G = -G^*$, leads to the following canonical form of square-size quaternion matrices under the congruence transformation $A \mapsto P^* A P$ with invertible matrix $P \in \mathsf{H}^{n \times n}$. Again, it will be convenient to list the primitive forms first; the prefix q-c stands for quaternion congruence.

(q-c0) $0_{m \times m}$.

(q-c1) $G_{2\varepsilon+1} + \begin{bmatrix} 0 & 0 & F_\varepsilon \\ 0 & 0_1 & 0 \\ -F_\varepsilon & 0 & 0 \end{bmatrix}$.

(q-c2) $F_k + iG_k$, where k is an even integer.

(q-c3) $G_\ell + iF_\ell$, where ℓ is an odd integer.

(q-c4) $\begin{bmatrix} 0 & (\alpha+1)F_p + G_p \\ (\alpha^*-1)F_p + G_p & 0 \end{bmatrix}$, where $\alpha \in \mathsf{H}$
          has positive real part.

Theorem 9.10.1. *Let $X \in \mathsf{H}^{n \times n}$. Then X is congruent to a matrix of the form*

$$A_0 \quad \oplus \quad \oplus_{j=1}^r \delta_j \left(F_{k_j} + iG_{k_j} \right) \quad \oplus \quad \oplus_{i=1}^p \eta_i \left(G_{\ell_i} + iF_{\ell_i} \right) \tag{9.10.1}$$

$$\oplus \quad \oplus_{u=1}^q \zeta_u \begin{bmatrix} 0 & 0 & \cdots \\ 0 & 0 & \cdots \\ 0 & 0 & \cdots \\ \vdots & \vdots & \iddots \\ (-1)^{m_u-1}(\beta_u + i^{m_u}) & -1 & 0 \end{bmatrix}$$

$$\begin{bmatrix} 0 & 0 & \beta_u + i^{m_u} \\ 0 & -\beta_u - i^{m_u} & -1 \\ \beta_u + i^{m_u} & -1 & 0 \\ \vdots & \vdots & \vdots \\ \cdots & 0 & 0 \end{bmatrix}_{m_u \times m_u} . \tag{9.10.2}$$

Here, A_0 is a direct sum of blocks of types (q-c0), (q-c1), (q-c2), (q-c3), and (q-c4), in which several blocks of the same type and of different and/or the same sizes may be present. Furthermore, the k_j's are odd positive integers, the ℓ_i's are even positive integers, the β_u's are positive real numbers if m_u is odd, the β_u's are nonzero quaternions with zero real parts if m_u is even, and the δ_j's and η_i's are signs 1 or -1.

The blocks in (9.10.1) and (9.10.2) are uniquely determined by X up to a permutation of constituent blocks in A_0, up to a permutation of blocks in each of the three parts

$$\oplus_{j=1}^r \delta_j \left(F_{k_j} + iG_{k_j} \right), \qquad \oplus_{i=1}^p \eta_i \left(G_{\ell_i} + iF_{\ell_i} \right),$$

and (9.10.2), up to replacements of α with a similar quaternion in any block of type (q-c4), and up to replacements of β with a similar quaternion in any block of type (9.10.2) of even size $m_u \times m_u$.

The uniqueness statement in Theorem 9.10.1 follows from the uniqueness statement in Theorem 9.8.1 upon the trivial observation that the correspondence between hermitian/skewhermitian pairs of matrices (F, G), $F, G \in \mathsf{H}^{n \times n}$, and matrices $A \in \mathsf{H}^{n \times n}$ given by $A = F + G$, is one-to-one and onto.

It is instructive to compare H-congruence of complex or real matrices with C-congruence or R-congruence. The following corollary is immediate from Theorem 8.5.1.

Corollary 9.10.2. *Two real $n \times n$ matrices A and B are* R-*congruent if and only if they are* H-*congruent; in other words, if $A = R^*BR$ for some invertible $R \in \mathsf{H}^{n \times n}$, then also $A = S^*BS$ for some invertible $S \in \mathsf{R}^{n \times n}$.*

An analogous result for complex matrices does not hold true. As a direct consequence of Corollary 9.8.2, we have the following.

Corollary 9.10.3. *A complex matrix $A \in \mathsf{C}^{n \times n}$ has the property that any complex matrix A' which is* H-*congruent to A, is also* C-*congruent to A, if and only if the complex hermitian pencil*

$$\frac{1}{2}(A + A^*) + ti\left(\frac{1}{2}(A - A^*)\right)$$

has no nonzero real eigenvalues, its indices at zero are all even, and its indices at infinity are all odd.

9.11   EXERCISES

Ex. 9.11.1. Provide details for the proof of Theorem 9.2.3.

Ex. 9.11.2. Show that a matrix $A \in \mathsf{H}^{n \times n}$ is nilpotent if and only if $\sigma(A) = \{0\}$.

Ex. 9.11.3. Describe the canonical form (9.1.1) of the skewhermitian pencil $A + tB$ under each of the following conditions:

(a) rank $(xA + yB)$ is constant for all $x, y \in \mathsf{R}$ not both zero;

(b) rank $A = 1$;

(c) rank $A = 2$.

Ex. 9.11.4. Describe the canonical form for strict equivalence of the hermitian/skewhermitian pencil $A + tB$ (Theorem 9.3.1), under each of the conditions of Ex. 9.11.3.

Ex. 9.11.5. Describe the canonical form under congruence of the hermitian/skewhermitian pencil $A + tB$ (Theorem 9.4.1), assuming one of the following hypotheses:

(a) A is positive definite;

(b) A is positive semidefinite;

(c) A has only one negative eigenvalue (counted with multiplicities).

Ex. 9.11.6. Find all possible canonical forms under congruence of quaternion matrices whose ranks do not exceed 2.

Ex. 9.11.7. Consider the following property of a hermitian/skewhermitian quaternion matrix pencil $A + tB$:

(A) *If $A' + tB'$ is a hermitian/skewhermitian matrix pencil that is strictly equivalent to $A + tB$, then in fact $A' + tB'$ is congruent to $A + tB$.*

(a) Identify matrix pencils with property (A) in terms of their canonical forms (9.4.1), (9.4.2).

(b) For every hermitian/skewhermitian matrix pencil $A + tB$ that does not have property (A), find the number of congruence classes within the set of hermitian/skewhermitian matrix pencils which are strictly equivalent to $A + tB$.

Ex. 9.11.8. Provide an explicit expression for the matrix T in (9.3.3).

9.12 NOTES

As indicated in the notes to the preceding chapter (see also references there), the canonical forms for skewhermitian and mixed (hermitian/skewhermitian) quaternion matrix pencils are known in the literature. Exposition in Sections 9.1 and 9.2, as well as that of Theorem 9.4.1 and its proof, follows Rodman [135].

Results of Sections 9.9 and 9.7 were proved in Rodman [134].

Canonical forms for complex matrices under congruence can be found in Horn and Sergeichuk [65, 66], Sergeichuk [141], and Lancaster and Rodman [93]. The canonical form of Theorem 9.10.1 is given in Rodman [135].

Congruences of real, complex, and quaternion matrices have been compared in Rodman [134].

Chapter Ten

Indefinite inner products: Conjugation

In this chapter we study indefinite inner products defined on $\mathsf{H}^{n \times 1}$ of the hermitian- and skewhermitian-types and matrices having symmetry properties with respect to one of these indefinite inner products. The hermitian-type inner product is a function

$$[\cdot, \cdot] : \mathsf{H}^{n \times 1} \times \mathsf{H}^{n \times 1} \longrightarrow \mathsf{H}$$

with the following properties:

(1) Linearity in the first argument: $[x_1 \alpha_1 + x_2 \alpha_2, y] = [x_1, y] \alpha_1 + [x_2, y] \alpha_2$ for all $x_1, x_2, y \in \mathsf{H}^{n \times 1}$ and all $\alpha_1, \alpha_2 \in \mathsf{H}$.

(2) Symmetry: $[x, y] = [y, x]^*$ for all $x, y \in \mathsf{H}^{n \times 1}$.

(3) Nondegeneracy: if $x_0 \in \mathsf{H}^{n \times 1}$ is such that $[x_0, y] = 0$ for all $y \in \mathsf{H}^{n \times 1}$, then $x_0 = 0$.

The skewhermitian-type inner product $[\cdot, \cdot]$ is defined by properties (1), (3), and

(2') antisymmetry: $[x, y] = -[y, x]^*$ for all $x, y \in \mathsf{H}^{n \times 1}$.

It follows from (1) and (2), or from (1) and (2'), that

$$[x, y_1 \alpha_1 + y_2 \alpha_2] = \alpha_1^* [x, y_1] + \alpha_2^* [x, y_2] \quad \text{for all} \quad x, y_1, y_2 \in \mathsf{H}^{n \times 1}, \quad \alpha_1, \alpha_2 \in \mathsf{H}.$$

Proposition 10.0.1. $[\cdot, \cdot]$ *is an inner product of hermitian-type, resp. skewhermitian-type, if and only if there exists a hermitian, resp. skewhermitian, invertible* $n \times n$ *matrix* H *such that*

$$[x, y] = \langle Hx, y \rangle = \pm \langle x, Hy \rangle = y^* H x, \quad \text{for all} \quad x, y \in \mathsf{H}^{n \times 1},$$

with the sign $+$*, resp.* $-$*. Such a matrix* H *is uniquely determined by the inner product.*

Proof. The "if" part is straightforward, as is the uniqueness of H. If $[\cdot, \cdot]$ is an inner product of hermitian or skewhermitian-type, then for every fixed $y \in \mathsf{H}^{n \times 1}$, the map $x \mapsto [x, y]$ is (quaternion) linear; hence, there exists $z \in \mathsf{H}^{n \times 1}$ such that $[x, y] = \langle x, z \rangle$ for all $x \in \mathsf{H}^{n \times 1}$. One verifies that the map $y \mapsto z$ is linear, so $z = Gy$ for some linear transformation G on $\mathsf{H}^{n \times 1}$, which we represent as a matrix (with respect to the standard basis $e_1, \ldots, e_n$). Now let $H = G^*$. It is easy to see that H is invertible and hermitian, resp. skewhermitian, if the inner product is of hermitian-type, resp. skewhermitian-type. $\square$

The matrix H of Proposition 10.0.1 is said to be *associated* with the indefinite iner product $[\cdot, \cdot]$. In the sequel, we will often work with associated matrices rather than directly with indefinite inner products.

We develop canonical forms for $(H,^*)$-hermitian and $(H,^*)$-skewhermitian matrices, with respect to hermitian-type indefinite inner products, as well as for $(H,^*)$-Hamiltonian and $(H,^*)$-skew-Hamiltonian matrices, with respect to skewhermitian-type inner products. These forms are based on the canonical forms given in Chapters 8 and 9. Applications are given to invariant semidefinite and neutral subspaces and to boundedness properties of systems of linear differential equations with certain symmetries.

10.1 H-HERMITIAN AND H-SKEWHERMITIAN MATRICES

Let $[\cdot,\cdot]$ be an inner product on $\mathsf{H}^{n\times 1}$ of hermitian type. A matrix $A \in \mathsf{H}^{n\times n}$ is said to be *selfadjoint*, resp. *skewadjoint*, with respect to $[\cdot,\cdot]$ if $[Ax, y] = [x, Ay]$, resp. $[Ax, y] = -[x, Ay]$, for all $x, y \in \mathsf{H}^{n\times 1}$. If $H = H^*$ is the matrix associated with $[\cdot,\cdot]$, then A is selfadjoint, resp. skewadjoint, with respect to $[\cdot,\cdot]$ if and only if $HA = A^*H$, resp. $HA = -A^*H$, i.e., A is $(H,^*)$-hermitian, resp. $(H,^*)$-skewhermitian.

In this chapter, we abbreviate "$(H,^*)$-hermitian" and "$(H,^*)$-skewhermitian" as "H-hermitian" and "H-skewhermitian," respectively.

Note that A is H-hermitian, resp. H-skewhermitian, if and only if $S^{-1}AS$ is S^*HS-hermitian, resp. H-skewhermitian, for any invertible $S \in \mathsf{H}^{n\times n}$.

The canonical forms of H-hermitian and H-skewhermitian matrices are given as follows. We denote by $\mathsf{C}_{+,0}$ the set of complex numbers with positive imaginary parts.

Theorem 10.1.1. *Let $H \in \mathsf{H}^{n\times n}$ be hermitian and invertible, and let A be H-hermitian. Denote by $d_p(A, \lambda)$ the number of Jordan blocks in the Jordan form of A of size p corresponding to the set of mutually similar eigenvalues that contains the eigenvalue λ of A. Then for every $p \geq 0$ and for every nonreal eigenvalue λ of A, the number $d_p(A, \lambda)$ is even, and there exists an invertible quaternion matrix S such that*

$$S^*HS \;=\; \oplus_{\lambda\in\sigma(A)\cap\mathsf{C}_{+,0}} \oplus_{p\geq 1} \oplus_{k=1}^{\frac{1}{2}d_p(A,\lambda)} \oplus F_{2p}$$

$$\oplus \;\; \oplus_{\lambda\in\sigma(A)\cap\mathsf{R}} \oplus_{p\geq 1} \oplus_{k=1}^{d_p(A,\lambda)} \; \eta_{k,\lambda}^{(p)}F_p, \tag{10.1.1}$$

$$S^{-1}AS \;=\; \oplus_{\lambda\in\sigma(A)\cap\mathsf{C}_{+,0}} \oplus_{p\geq 1} \oplus_{k=1}^{\frac{1}{2}d_p(A,\lambda)} \begin{bmatrix} J_p(\lambda) & 0_p \\ 0_p & J_p(\bar\lambda) \end{bmatrix}$$

$$\oplus \;\; \oplus_{\lambda\in\sigma(A)\cap\mathsf{R}} \oplus_{p\geq 1} \oplus_{k=1}^{d_p(A,\lambda)} \; J_p(\lambda), \tag{10.1.2}$$

where $\eta_{k,\lambda}^{(p)}$ are signs ± 1.

The form $(10.1.1), (10.1.2)$ is uniquely determined by the pair (H, A), up to an arbitrary permutation of diagonal blocks in each of the two parts

$$\oplus_{\lambda\in\sigma(A)\cap\mathsf{C}_{+,0}} \oplus_{p\geq 1} \oplus_{k=1}^{\frac{1}{2}d_p(A,\lambda)} \oplus \; F_{2p}$$

and

$$\oplus_{\lambda\in\sigma(A)\cap\mathsf{R}} \oplus_{p\geq 1} \oplus_{k=1}^{d_p(A,\lambda)} \eta_{k,\lambda}^{(p)}F_p$$

in S^*HS and, simultaneously, in the same permutation of the blocks in $S^{-1}AS$.

Conversely, if H and A are given by formulas $(10.1.1), (10.1.2)$ *for some invertible S, then H is hermitian and invertible and A is H-hermitian.*

The proof is obtained by applying Theorem 8.1.2 to the pair of hermitian matrices (HA, H). The converse statement is easily verified.

We note that the canonical form $(10.1.2)$ of the pair (A, H) involves only real and complex matrices. Using Remark 8.1.8, an alternative form can be given entirely in terms of real matrices. Another alternative form can be given using Remark 8.1.6.

The signs $\eta_{k,\lambda}^{(p)}$ constitute the *sign characteristic* of the H-hermitian matrix A. Thus, the sign characteristic assigns a sign ± 1 to every partial multiplicity of every real eigenvalue of A.

Remark 10.1.2. It is well known, and found in many sources in literature (see, e.g., Gohberg et al. [52, 53]), that the result of Theorem 10.1.1 is applicable in the context of complex matrices as well. It can be obtained by applying Theorem 15.3.1 to the pair of complex hermitian matrices (HA, H).

We now give a proof of Proposition 8.1.7 based on the form $(10.1.2)$.

Proof of Proposition 8.1.7. Note that the matrix $A := H_1^{-1} H_2$ is H_1-hermitian. A straightforward computation shows that A is a lower triangular Toeplitz matrix with $p_k^{(2)}/p_k^{(1)}$ on the main diagonal and x on the first subdiagonal. Thus, the Jordan form of A is $J_k(p_k^{(2)}/p_k^{(1)})$. Now use the canonical form $(10.1.2)$ of the H_1-hermitian matrix A. The canonical form yields $(8.1.11)$. The sign η can be determined using the following rule, which may be applied in the case the H-hermitian $k \times k$ matrix $A \in \mathsf{H}^{k \times k}$ has only one Jordan block (necessarily corresponding to a real eigenvalue) in its Jordan form. Namely, if $y_1, y_2, \ldots, y_k$ is a chain consisting of an eigenvector y_1 and associated generalized eigenvectors $y_2, \ldots, y_k$ of A, then the number $y_k^* H y_1 = y_1^* H y_k$ is real and nonzero and its sign coincides with the sign η in the sign characteristic of the pair (A, H). (This rule is a particular case of what is known as the second description of the sign characteristic of the H-hermitian matrix A, for complex matrices; see Gohberg et al. [53, Sections 5.8, 6.3] or Gohberg et al. [52, Section 3.9].) Applying the rule results in formula $(8.1.12)$. $\square$

In the canonical form for H-skewhermitian matrices, we will use the standard matrix

$$\Xi_n(\mathsf{i}^{n-1}) := \begin{bmatrix} 0 & 0 & 0 & \cdots & 0 & 0 & \mathsf{i}^{n-1} \\ 0 & 0 & 0 & \cdots & 0 & -\mathsf{i}^{n-1} & 0 \\ 0 & 0 & 0 & \cdots & \mathsf{i}^{n-1} & 0 & 0 \\ \vdots & \vdots & & \cdot^{\cdot^{\cdot}} & \vdots & \vdots & \vdots \\ 0 & (-1)^{n-2}\mathsf{i}^{n-1} & 0 & \cdots & 0 & 0 & 0 \\ (-1)^{n-1}\mathsf{i}^{n-1} & 0 & 0 & \cdots & 0 & 0 & 0 \end{bmatrix}.$$

Note that $\Xi_n(\mathsf{i}^{n-1})$ is hermitian for every positive integer n.

Theorem 10.1.3. *Let $H \in \mathsf{H}^{n \times n}$ be hermitian and invertible, and let $A \in \mathsf{H}^{n \times n}$ be H-skewhermitian: $HA = -A^* H$. Then there is an invertible quaternion matrix W such that $W^{-1}AW$ and $W^* HW$ are block diagonal matrices*

$$W^{-1}AW = A_1 \oplus \cdots \oplus A_s, \quad W^* HW = H_1 \oplus \cdots \oplus H_s, \tag{10.1.3}$$

where each diagonal block (A_α, H_α) is of one of the following four types:

(i)

$$A_\alpha \;=\; J_{2n_1+1}(0) \oplus J_{2n_2+1}(0) \oplus \cdots \oplus J_{2n_p+1}(0),$$

$$H_\alpha \;=\; \kappa_1 \Xi_{2n_1+1}(\mathrm{i}^{2n_1}) \oplus \kappa_2 \Xi_{2n_2+1}(\mathrm{i}^{2n_2}) \oplus \cdots \oplus \kappa_p \Xi_{2n_p+1}(\mathrm{i}^{2n_p}),$$

where n_j's are nonnegative integers and κ_j's are signs ± 1;

(ii)

$$A_\alpha \;=\; J_{2m_1}(0) \oplus \cdots \oplus J_{2m_q}(0),$$
$$H_\alpha \;=\; \Xi_{2m_1}(\mathrm{i}^{2m_1-1}) \oplus \cdots \oplus \Xi_{2m_q}(\mathrm{i}^{2m_q-1}),$$

where m_j's are positive integers;

(iii)

$$A_\alpha \;=\; J_{k_1}(a+\mathrm{i}b) \oplus -(J_{k_1}(a+\mathrm{i}b))^*$$
$$\oplus \cdots \oplus J_{k_u}(a+\mathrm{i}b) \oplus -(J_{k_u}(a+\mathrm{i}b))^*, \qquad (10.1.4)$$

$$H_\alpha \;=\; \begin{bmatrix} 0 & I_{k_1} \\ I_{k_1} & 0 \end{bmatrix} \oplus \cdots \oplus \begin{bmatrix} 0 & I_{k_u} \\ I_{k_u} & 0 \end{bmatrix}, \qquad (10.1.5)$$

where a and b are real numbers such that $a > 0$ and $b \geq 0$, and the numbers a and b, the total number $2u$ of Jordan blocks, and the sizes $k_1, \ldots, k_u$ may depend on (A_α, H_α);

(iv)

$$A_\alpha \;=\; J_{h_1}(\mathrm{i}b) \oplus J_{h_2}(\mathrm{i}b) \oplus \cdots \oplus J_{h_t}(\mathrm{i}b),$$

$$H_\alpha \;=\; \eta_1 \Xi_{h_1}(\mathrm{i}^{h_1-1}) \oplus \cdots \oplus \eta_t \Xi_{h_t}(\mathrm{i}^{h_t-1}),$$

where b is a positive real number and $\eta_1, \ldots, \eta_t$ are signs ± 1. Again, the parameters $b, t, h_1, \ldots, h_t$ and $\eta_1, \ldots, \eta_t$ may depend on the particular diagonal block (A_α, H_α).

The form (10.1.3) is uniquely determined by the pair (A, H), up to an arbitrary permutation of the constituent pairs of blocks $(A_1, H_1), \ldots, (A_s, H_s)$.

Conversely, if for a pair of quaternion matrices (A, H), formula (10.1.3) holds for some invertible quaternion matrix W, then H is invertible hermitian and A is H-skewhermitian.

The proof of Theorem 10.1.3 is obtained by applying Theorem 9.4.1 to the hermitian/skewhermitian matrix pencil $H + tHA$ and taking advantage of the assumption that H is invertible.

We have a *sign characteristic* for an H-skewhermitian matrix A: a sign ± 1 is attached to every partial multiplicity of A corresponding to nonzero eigenvalues with zero real parts and to odd partial multiplicities corresponding to the eigenvalue zero.

The number i can be replaced in Theorem 10.1.3 by any quaternion γ with zero real part and length one; indeed, to verify that, note the equalities

$$(-\lambda I)\Xi_n(\mathrm{i}^{n-1})(\lambda I) \;=\; \Xi_n((\lambda^{-1}\mathrm{i}\lambda)^{n-1}),$$
$$(-\lambda I)J_n(\mathrm{i}b)(\lambda I) \;=\; J_n((\lambda^{-1}\mathrm{i}\lambda)b),$$

where $\lambda \in H$ has zero real part and length one and $b \in R$, and note also that every $\gamma \in H$ with zero real part and length one can be represented in the form $\gamma = \lambda^* i \lambda$, for some $\lambda \in H$ with zero real part and length one (Theorem 2.5.1(c)).

It is instructive to compare blocks of type (ii) with the corresponding blocks of H_α-hermitian complex matrices. If A_α and H_α are $n \times n$ complex matrices such that H_α is invertible and iA_α is H_α-hermitian and the Jordan form of iA_α is $J_{2m_1}(0) \oplus \cdots \oplus J_{2m_q}(0)$, then by the canonical form for H_α-hermitian matrices over C (use Theorem 15.3.1 for the pair of complex hermitian matrices H_α and $H_\alpha(iA_\alpha)$), there exists a complex invertible matrix W such that

$$\begin{aligned} W^{-1}(iA_\alpha)W &= J_{2m_1}(0) \oplus \cdots \oplus J_{2m_q}(0), \\ W^* H_\alpha W &= \tau_1 F_{2m_1} \oplus \cdots \oplus \tau_q F_{2m_q}, \end{aligned}$$

where $\tau_1, \ldots, \tau_q$ are signs ± 1. Another transformation with the invertible complex matrix

$$W_1 := \oplus_{j=1}^q \left(\operatorname{diag}\left(1, i, i^2, \ldots, i^{2m_j - 1}\right) \right)$$

shows that

$$W_1^{-1} W^{-1} A_\alpha W W_1 = J_{2m_1}(0) \oplus \cdots \oplus J_{2m_q}(0),$$

$$W_1^* W^* H_\alpha W W_1 = \tau_1 \Xi_{2m_1}(i^{2m_1 - 1}) \oplus \cdots \oplus \tau_q \Xi_{2m_q}(i^{2m_q - 1}).$$

The signs here are uniquely defined by the pair (A_α, H_α), up to a permutation of signs corresponding to the Jordan blocks of equal sizes. However, over the quaternions the signs are immaterial because there is an invertible quaternion matrix T_{2m} such that

$$T_{2m}^{-1} J_{2m}(0) T_{2m} = J_{2m}(0), \quad T_{2m}^*(-\Xi_{2m}(i^{2m-1})) T_{2m} = \Xi_{2m}(i^{2m-1}).$$

Indeed, $T_{2m} = jI_{2m}$ will do.

Open Problem 10.1.4. *Develop canonical forms for pairs of matrices (A, H), where $H = H^* \in H^{n \times n}$ and $HA = A^* H$. The matrix H is not assumed to be invertible.*

Same problem with respect to pairs of matrices (A, H), where H is hermitian and $HA = -A^* H$, or where $H = -H$ and $HA = \pm A^* H$.

Some results in the direction of Open Problem 10.1.4, in the context of complex H-hermitian matrices, are found in Mehl and Rodman [111].

10.2 INVARIANT SEMIDEFINITE SUBSPACES

In this section we assume that $H = H^* \in H^{n \times n}$ and H is invertible. Our main result here (the next theorem) asserts the possibility of extending invariant semidefinite subspaces to subspaces that are simultaneously invariant and maximal semidefinite. The invariance here refers to H-hermitian or H-skewhermitian matrices.

Theorem 10.2.1. *Let $A \in H^{n \times n}$ be H-hermitian or H-skewhermitian. Suppose subspace $\mathcal{M} \subseteq H^{n \times 1}$ is H-nonnegative, resp. H-nonpositive, and A-invariant. Then there is an A-invariant H-nonnegative, resp. H-nonpositive, subspace $\mathcal{N} \subseteq H^{n \times 1}$ that contains $\mathcal{M}$ and has dimension*

$$\dim_H \mathcal{N} = p, \qquad resp. \quad \dim_H \mathcal{N} = q,$$

where $p = \operatorname{In}_+(H)$, $q = \operatorname{In}_-(H)$.

Proof. We consider the case of H-nonnegative subspaces (for the H-nonpositive subspaces apply the result for $-H$ in place of H).

We use induction on the size n of the matrices H and A, and on the dimension of $\mathcal{M}$. The case $n = 1$ is trivial, and the case $\mathcal{M} = \{0\}$ follows by constructions (which are standard for complex H-hermitian matrices; see, e.g., Gohberg et al. [53, 52]) using the canonical forms (10.1.2) and (10.1.3).

We provide details of these constructions. Suppose A is H-hermitian. We can assume without loss of generality, that H and A are given by the right-hand sides of (10.1.2); indeed, a subspace $\mathcal{M}$ is A-invariant H-nonnegative (or H-nonpositive) if and only if $S^{-1}(\mathcal{M})$ is $S^{-1}AS$-invariant S^*HS-nonnegative (or S^*HS-nonpositive). Select the vectors $e_1, \ldots, e_p$ corresponding to each pair $F_{2p}, \begin{bmatrix} J_p(\lambda) & 0_p \\ 0_p & J_p(\lambda^*) \end{bmatrix}$ (λ nonreal) of $2p \times 2p$ blocks, and select the vectors $e_1, \ldots, e_s$ corresponding to each pair $\eta_{k,\lambda}^{(p)} F_p, J_p(\lambda)$ (λ real) of $p \times p$ blocks as follows:

$$
s = \begin{cases}
\dfrac{p}{2} & \text{if } p \text{ is even,} \\[2ex]
\dfrac{p+1}{2} & \text{if } p \text{ is odd and } \eta_{k,\lambda}^{(p)} = 1, \\[2ex]
\dfrac{p-1}{2} & \text{if } p \text{ is odd and } \eta_{k,\lambda}^{(p)} = -1.
\end{cases}
$$

Observe that

$$
\text{In}_+(F_m) = \text{In}_-(F_m) = \frac{m}{2} \quad \text{if } m \text{ is even,}
$$

and

$$
\text{In}_+(F_m) = \frac{m+1}{2}, \quad \text{In}_-(F_m) = \frac{m-1}{2} \quad \text{if } m \text{ is odd.}
$$

Using this observation, it is easy to see that the selected vectors span (over H) an A-invariant H-nonnegative subspace of dimension p.

Suppose now A is H-skewhermitian, and we may (and do) assume that A and H are given by the right-hand sides of (10.1.3). Then we select vectors as follows. For the block of type (ii), select

$$
e_1, e_2, \ldots, e_{m_1}, e_{2m_1+1}, e_{2m_1+2} \ldots, e_{2m_1+m_2},
$$

$$
\ldots, e_{2m_1+\cdots+2m_{q-1}+1}, e_{2m_1+\cdots+2m_{q-1}+2}, \ldots, e_{2m_1+\cdots+2m_{q-1}+m_q}.
$$

For the block of type (iii) select

$$
e_1, e_2, \ldots, e_{k_1}, e_{2k_1+1}, e_{2k_1+2} \ldots, e_{2k_1+k_2},
$$

$$
\ldots, e_{2k_1+\cdots+2k_{u-1}+1}, e_{2k_1+\cdots+2k_{u-1}+2}, \ldots, e_{2k_1+\cdots+2k_{u-1}+k_u}.
$$

To make a selection of vectors in the blocks of type (i) and (iv), we first note that

$$
\text{In}_+\left(\Xi_n(\mathrm{i}^{n-1})\right) = \frac{n+1}{2}, \qquad \text{In}_-\left(\Xi_n(\mathrm{i}^{n-1})\right) = \frac{n-1}{2} \tag{10.2.1}
$$

for n odd; in particular, the $((n+1)/2, (n+1)/2)$th entry of $\Xi_n(\mathrm{i}^{n-1})$ is equal to 1, and

$$
\text{In}_+\left(\Xi_n(\mathrm{i}^{n-1})\right) = \text{In}_-\left(\Xi_n(\mathrm{i}^{n-1})\right) = \frac{n}{2} \tag{10.2.2}
$$

for n even. Now, for each pair $(J_{2n_j+1}(0), \kappa_j \Xi_{2n_j+1}(i^{2n_j}))$ of blocks as in (i), select $e_1, e_2, \ldots, e_{n_j+1}$ if $\kappa_j = 1$ and select $e_1, e_2, \ldots, e_{n_j}$ if $\kappa_j = -1$. Finally, for a pair of blocks $(J_{h_j}(ib), \eta_j \Xi_{h_j}(i^{h_j-1}))$, as in (iv), we select $e_1, e_2, \ldots, e_{h_j/2}$ if h_j is even, select $e_1, e_2, \ldots, e_{(h_j+1)/2}$ if h_j is odd and $\eta_j = 1$, and select $e_1, e_2, \ldots, e_{(h_j-1)/2}$ if h_j is odd and $\eta_j = -1$. The subspace spanned by all the selected vectors is A-invariant H-nonnegative and has dimension $p = \text{In}_+(H)$.

Now proceed to the general case of the proof of Theorem 10.2.1. Using Lemma 4.5.1 and the notation of that lemma, we may assume that

$$\mathcal{M} = \text{Span}_\mathsf{H}\{e_1, \ldots, e_{n_2+n_3}\},$$

$$H = \begin{bmatrix} 0 & 0 & I_{n_2} & 0 \\ 0 & I_{n_3} & 0 & 0 \\ I_{n_2} & 0 & 0 & 0 \\ 0 & 0 & 0 & H_0 \end{bmatrix}, \quad A = \begin{bmatrix} A_{22} & A_{23} & A_{24} & A_{25} \\ A_{32} & A_{33} & A_{34} & A_{35} \\ 0 & 0 & A_{44} & A_{45} \\ 0 & 0 & A_{54} & A_{55} \end{bmatrix}, \quad (10.2.3)$$

where $H_0 = H_0^*$, $A_{pp} \in \mathsf{H}^{n_p \times n_p}$ for $p = 2, 3, 4, 5$, and $n_2 + n_3 + n_4 + n_5 = n$. Note that we must have $n_1 = 0$ because H is assumed to be invertible.

The condition $(HA)^* = \pm(HA)$ gives

$$A_{32} = 0, \quad A_{35} = 0, \quad A_{45} = 0,$$

and $(H_0 A_{55})^* = \pm H_0 A_{55}$. Let

$$I_{p_0} \oplus -I_{q_0} \quad (10.2.4)$$

be the canonical form of H_0 as in Theorem 4.1.6 (note that H_0 is invertible because H is). By the induction hypothesis there exists an A_{55}-invariant H_0-nonnegative subspace $\mathcal{M}_0$ of dimension p_0. Then the subspace

$$\begin{bmatrix} \mathsf{H}^{n_2} \\ 0 \\ 0 \\ 0 \end{bmatrix} \dotplus \begin{bmatrix} 0 \\ \mathsf{H}^{n_3} \\ 0 \\ 0 \end{bmatrix} \dotplus \begin{bmatrix} 0 \\ 0 \\ 0 \\ \mathcal{M}_0 \end{bmatrix}$$

is clearly A-invariant H-nonnegative and has dimension $n_2 + n_3 + p_0$. It is easy to see that $n_2 + n_3 + p_0 = \text{In}_+(H)$. $\qquad\square$

Remark 10.2.2. It turns out that for an H-skewhermitian matrix A, there exists an A-invariant maximal H-nonnegative (resp. H-nonpositive) subspace with additional properties regarding location of the spectrum. Namely, let Λ be a set of eigenvalues of A such that

$$\lambda \in \Lambda, \quad \mu \in \mathsf{H}, \quad \mu \text{ similar to } \lambda \quad \Longrightarrow \quad \mu \in \Lambda \quad (10.2.5)$$

and

$$\lambda \in \Lambda \quad \Longrightarrow \quad -\lambda \notin \Lambda. \quad (10.2.6)$$

For example, the set of all eigenvalues of A with positive real parts satisfies these conditions. Let $\mathcal{M}_\Lambda$ be the sum of root subspaces of A corresponding to the eigenvalues in Λ. Then there exists an A-invariant maximal H-nonnegative subspace that contains $\mathcal{M}_\Lambda$. Likewise, there exists an A-invariant maximal H-nonpositive subspace that contains $\mathcal{M}_\Lambda$.

To see that, we may assume that A and H are given by the right-hand sides of (10.1.3). The canonical form of Theorem 10.1.3 (see Part (iii) there) shows that if Λ and $\mathcal{M}_\Lambda$ are as above, then $\mathcal{M}$ is H-neutral. Now apply Theorem 10.2.1.

It is not generally true that every A-invariant H-positive (or H-negative) subspace is contained in an A-invariant H-positive (or H-negative) of dimension p (or q) as in Theorem 13.3.1, where A is H-hermitian or H-skewhermitian. See Exs. 10.10.1, 10.10.2, and 10.10.3.

10.3 INVARIANT LAGRANGIAN SUBSPACES I

Of particular interest here are Lagrangian subspaces.

Definition 10.3.1. Given a hermitian or skewhermitian matrix $G \in \mathsf{H}^{n \times n}$, a subspace $\mathcal{M} \subseteq \mathsf{H}^{n \times 1}$ is said to be *G-Lagrangian* if it is G-neutral and has dimension $n/2$.

Clearly this definition makes sense only if n is even. As it follows from Theorems 4.2.6 and 4.2.7 and assuming n is even, G-Lagrangian subspaces exist if and only if

$$\min\{\text{In}_+(G), \text{In}_-(G)\} + \text{In}_0(G) \geq \frac{n}{2} \tag{10.3.1}$$

if G is hermitian, and

$$\lfloor \frac{\text{rank} \, G}{2} \rfloor + (n - \text{rank} \, G) \geq \frac{n}{2}$$

if G is skewhermitian. The latter inequality can be rewritten in the form

$$\lfloor \frac{\text{rank} \, G + 1}{2} \rfloor \leq \frac{n}{2}. \tag{10.3.2}$$

We will use the notion of G-Lagrangian subspaces for invertible G (and even n); in this case, (10.3.2) is always satisfied, and (10.3.1) becomes

$$\text{In}_+(G) = \text{In}_-(G) = \frac{n}{2}. \tag{10.3.3}$$

The following property of Lagrangian subspaces will be very useful.

Proposition 10.3.2. *Let*

$$G = \pm G^* = G_1 \oplus G_2 \oplus \cdots \oplus G_k \in \mathsf{H}^{n \times n},$$

where $G_j \in \mathsf{H}^{n_j \times n_j}$, $j = 1, 2, \ldots, k$, and where n is even. Assume G is invertible. Then a subspace $\mathcal{M} \subseteq \mathsf{H}^{n \times 1}$ of the form

$$\mathcal{M} = \mathcal{M}_1 \oplus \mathcal{M}_2 \oplus \cdots \oplus \mathcal{M}_k := \left\{ \begin{bmatrix} x_1 \\ x_2 \\ \vdots \\ x_k \end{bmatrix} : x_1 \in \mathcal{M}_1, \ldots, x_k \in \mathcal{M}_k \right\}$$

is G-Lagrangian if and only if each $\mathcal{M}_j$ is G_j-Lagrangian, for $j = 1, 2, \ldots, k$ (in particular, each n_j must be even).

Proof. First note that each G_j is invertible since G is. Suppose first G is hermitian; then each G_j is hermitian. If $\mathcal{M}$ is G-Lagrangian, then each $\mathcal{M}_j$ is

G_j-neutral, and because G_j is invertible, $\dim_{\mathsf{H}} (\mathcal{M}_j) \le n_j/2$ (Theorem 4.2.6(e)). On the other hand,

$$\sum_{j=1}^{k} \dim_{\mathsf{H}} (\mathcal{M}_j) = \dim_{\mathsf{H}} (\mathcal{M}) = \frac{n}{2}, \qquad (10.3.4)$$

so we must have the equality $\dim_{\mathsf{H}} (\mathcal{M}_j) = n_j/2$, and $\mathcal{M}_j$ is G_j-Lagrangian. Conversely, if each $\mathcal{M}_j$ is G_j-Lagrangian, then clearly $\mathcal{M}$ is G-neutral, and the first equality in (10.3.4) shows that $\mathcal{M}$ is G-Lagrangian.

If G is skewhermitian, the proof is similar (Ex. 10.10.5). $\square$

If A is H-hermitian or H-skewhermitian, then A does not necessarily have an invariant H-Lagrangian subspace even when (10.3.3) is fulfilled. A necessary and sufficient condition for this to happen is given by the next theorem.

Theorem 10.3.3. *Let $H \in \mathsf{H}^{n \times n}$ be hermitian and invertible.*

(a) *If $A \in \mathsf{H}^{n \times n}$ is H-hermitian, then there exists an A-invariant H-Lagrangian subspace if and only if for every real eigenvalue λ of A (if any), the number of odd partial multiplicities corresponding to λ is even, and exactly half of them have sign -1 (the other half having the sign 1) in the sign characteristic of (A, H).*

(b) *If $A \in \mathsf{H}^{n \times n}$ is H-skewhermitian, then there exists an A-invariant H-Lagrangian subspace if and only if for every eigenvalue λ of A with zero real part (if any), the number of odd partial multiplicities corresponding to λ is even, and exactly half of them have sign -1 (the other half having the sign 1) in the sign characteristic of (A, H).*

Proof. Note that for any invertible $S \in \mathsf{H}^{n \times n}$, the subspace $\mathcal{M}$ is H-Lagrangian if and only if $S^{-1}(\mathcal{M})$ is S^*HS-Lagrangian. Therefore, without loss of generality, we may (and do) assume that the pair (A, H) is given by the canonical form of Theorem 10.1.1 (if A is H-hermitian) or Theorem 10.1.3 (if A is H-skewhermitian).

Proof of Part (a). In view of Proposition 10.3.2 and the fact that every A-invariant subspace is the sum of its intersections with the root subspaces of A (Proposition 5.1.4), we need only consider the situation when $\mathsf{H}^{n \times 1}$ is a root subspace for A. Two cases can occur. (1) The eigenvalues of A (all similar to each other) are nonreal; (2) the sole eigenvalue λ of A is real. In the first case, the condition in Part (a) is vacuous, and an A-invariant H-Lagrangian subspace always exists. Indeed, as in the proof of Theorem 10.2.1, select the vectors $e_1, \ldots, e_p$ corresponding to each pair F_{2p}, $\begin{bmatrix} J_p(\lambda) & 0_p \\ 0_p & J_p(\lambda^*) \end{bmatrix}$ of $2p \times 2p$ blocks. Then the subspace spanned by the selected vectors is A-invariant and H-Lagrangian.

Suppose now Case (2) holds true. We prove the "if" direction. Rearranging the blocks (if necessary) in the canonical form, we assume that

$$H = (F_{p_1} \oplus -F_{p_2}) \oplus \cdots \oplus (F_{p_{2u-1}} \oplus -F_{p_{2u}})$$

$$\oplus \eta_{2u+1} F_{p_{2u+1}} \oplus \cdots \oplus \eta_{2u+v} F_{p_{2u+v}},$$

$$A = J_{p_1}(\lambda) \oplus \cdots \oplus J_{p_{2u+v}}(\lambda),$$

where $p_1, \ldots, p_{2u}$ are odd and $p_{2u+1}, \ldots, p_{2u+v}$ are even, and the η_j's are signs ± 1. (The cases when $u = 0$ or $v = 0$ are not excluded.) Select the following vectors in $\mathsf{H}^{n \times 1}$:

$$e_1, \ldots, e_{(p_1-1)/2}, e_{(p_1+1)/2} + e_{p_1+(p_2+1)/2}, e_{p_1+1}, \ldots, e_{p_1+(p_2-1)/2}, \ldots, \quad (10.3.5)$$

$$e_{p_1+\cdots+p_{2j}+1}, \ldots, e_{p_1+\cdots+p_{2j}+(p_{2j+1}-1)/2}, \quad (10.3.6)$$

$$e_{p_1+\cdots+p_{2j}+(p_{2j+1}+1)/2} + e_{p_1+\cdots+p_{2j}+p_{2j+1}+(p_{2j+2}+1)/2}, \quad (10.3.7)$$

$$e_{p_1+\cdots+p_{2j}+p_{2j+1}+1}, \ldots, e_{p_1+\cdots p_{2j}+p_{2j+1}+(p_{2j+2}-1)/2} \quad (10.3.8)$$

for $j = 1, 2, \ldots, u-1$, and

$$e_{w+1}, \ldots, e_{w+p_{2u+1}/2}, e_{w+p_{2u+1}+1}, \ldots, e_{w+p_{2u+1}+p_{2u+2}/2}, \ldots, \quad (10.3.9)$$

$$e_{w+p_{2u+1}+\cdots+p_{2u+v-1}+1}, \ldots, e_{w+p_{2u+1}+\cdots+p_{2u+v-1}+p_{2u+v}/2}, \quad (10.3.10)$$

where $w = p_1 + \cdots + p_{2u}$. By inspection, we see that the selected vectors (10.3.5), (10.3.6), (10.3.8), (10.3.9), and (10.3.10) span an A-invariant H-Lagrangian subspace.

For the "only if" direction, let

$$H = \oplus_{j=1}^r \eta_j F_{p_j}, \quad A = \oplus_{j=1}^r J_{p_j}(\lambda), \quad (10.3.11)$$

where $\eta_j = \pm 1$. If there exists an A-invariant H-Lagrangian subspace, then, in particular, n is even and we must have $\mathrm{In}_+ (H) = \mathrm{In}_- (H) = n/2$ (cf. (10.3.3)). It is easily seen that for H given by (10.3.11),

$$\mathrm{In}_+ (H) - \mathrm{In}_- (H) = \sum_{\{j \,:\, p_j \text{ is odd}\}} \eta_j,$$

so the condition in Theorem 10.3.3(a) follows.

The proof of Part (b) follows the same pattern as that of Part (a), using equalities (10.2.1), (10.2.2), and the fact that the $((n+1)/2, (n+1)/2)$th entry of $\Xi_n(i^{n-1})$ is equal to 1 if n is odd (Ex. 10.10.6). $\qquad\square$

Remark 10.3.4. For any H-skewhermitian matrix A, if there exists an A-invariant H-Lagrangian subspace, then there exists such subspace with additional properties regarding location of the spectrum, as in Remark 10.2.2. Namely, let Λ be a set of eigenvalues of A with the properties (10.2.5), (10.2.6). Then there exists an A-invariant H-Lagrangian subspace $\mathcal{M}$ that contains the sum of root subspaces for A corresponding to the eigenvalues in Λ.

To prove that, it suffices to consider the case when the set Λ is maximal (by inclusion) subject to conditions (10.2.5), (10.2.6). In this case, it turns out that the restriction $A|_{\mathcal{M}}$ has no eigenvalues with nonzero real parts besides those in Λ. To construct such a subspace $\mathcal{M}$, in view of the canonical form (10.1.3), it suffices to consider two cases: (1) all eigenvalues of A have zero real parts; (2) A and H are given by the right-hand sides of (10.1.4) and (10.1.5), respectively. In the first case Λ is vacuous, so any A-invariant H-Lagrangian subspace $\mathcal{M}$ will do. In the second case, suppose for example that $a + ib \in \Lambda$, $-a + ib \notin \Lambda$ (if $a + ib \notin \Lambda$, $-a + ib \in \Lambda$, the consideration is completely analogous). Select the following vectors:

$$e_1, e_2, \ldots, e_{k_1}, e_{2k_1+1}, \ldots, e_{2k_1+k_2}, \ldots,$$

$$e_{2k_1+2k_2+\cdots+2k_{u-1}+1}, e_{2k_1+2k_2+\cdots+2k_{u-1}+2}, \ldots, e_{2k_1+2k_2+\cdots+2k_{u-1}+k_u}.$$

The selected vectors span in $\mathsf{H}^{n \times 1}$, $n = 2k_1 + \cdots + 2k_u$, a subspace $\mathcal{M}$ which is A-invariant H-Lagrangian, and $\sigma(A|_{\mathcal{M}})$ consists of all quaternions similar to $a + ib$.

10.4   DIFFERENTIAL EQUATIONS I

In this section we study the system of differential equations with constant coefficients

$$A_\ell x^{(\ell)}(t) + A_{\ell-1} x^{(\ell-1)}(t) + \cdots + A_1 x'(t) + A_0 x(t) = 0, \quad t \in \mathsf{R}, \qquad (10.4.1)$$

where $A_\ell, \ldots, A_1, A_0 \in \mathsf{H}^{n \times n}$, and $x(t)$ is an unknown ℓ times continuously differentiable $\mathsf{H}^{n \times 1}$-valued function of the real independent variable t. It will be assumed in addition that A_k is hermitian if k is odd, A_k is skewhermitian if k is even, and A_ℓ is invertible. As in Section 5.14, consider the companion matrix of the system (10.4.1):

$$C = \begin{bmatrix} 0 & I_n & 0 & \ldots & 0 \\ 0 & 0 & I_n & \ldots & 0 \\ \vdots & \vdots & \vdots & \ddots & \vdots \\ 0 & 0 & 0 & \ldots & I_n \\ -A_\ell^{-1} A_0 & -A_\ell^{-1} A_1 & -A_\ell^{-1} A_2 & \ldots & -A_\ell^{-1} A_{\ell-1} \end{bmatrix} \in \mathsf{H}^{\ell n \times \ell n}. \quad (10.4.2)$$

Define also

$$G := \begin{bmatrix} A_1 & A_2 & A_3 & \ldots & A_\ell \\ -A_2 & -A_3 & \ldots & -A_\ell & 0_n \\ A_3 & A_4 & \ldots & 0_n & 0_n \\ -A_4 & \ldots & \ldots & 0_n & 0_n \\ \vdots & \cdot^{\cdot^{\cdot}} & \cdot^{\cdot^{\cdot}} & \cdot^{\cdot^{\cdot}} & \vdots \\ (-1)^{\ell-1} A_\ell & 0_n & 0_n & \ldots & 0_n \end{bmatrix} \in \mathsf{H}^{n\ell \times n\ell}. \qquad (10.4.3)$$

Clearly, G is hermitian and invertible, and

$$(\mathrm{In}_+(G), \mathrm{In}_-(G)) = \begin{cases} (\dfrac{n\ell}{2}, \dfrac{n\ell}{2}) & \text{if } \ell \text{ is even,} \\[2mm] \left(\dfrac{n(\ell-1)}{2} + \mathrm{In}_-(A_\ell), \dfrac{n(\ell-1)}{2} + \mathrm{In}_+(A_\ell) \right) \\ \quad \text{if } \ell \text{ is odd}, \dfrac{\ell-1}{2} \text{ is odd,} \\[2mm] \left(\dfrac{n(\ell-1)}{2} + \mathrm{In}_+(A_\ell), \dfrac{n(\ell-1)}{2} + \mathrm{In}_-(A_\ell) \right) \\ \quad \text{if } \ell \text{ is odd}, \dfrac{\ell-1}{2} \text{ is even.} \end{cases} \qquad (10.4.4)$$

To verify (10.4.4), consider a family of hermitian matrices

$$G_\varepsilon = \begin{bmatrix} \varepsilon A_1 & \varepsilon A_2 & \varepsilon A_3 & \ldots & A_\ell \\ -\varepsilon A_2 & -\varepsilon A_3 & \ldots & -A_\ell & 0_n \\ \varepsilon A_3 & \varepsilon A_4 & \ldots & 0_n & 0_n \\ -\varepsilon A_4 & \ldots & \ldots & 0_n & 0_n \\ \vdots & \cdot^{\cdot^{\cdot}} & \cdot^{\cdot^{\cdot}} & \cdot^{\cdot^{\cdot}} & \vdots \\ (-1)^{\ell-1} A_\ell & 0_n & 0_n & \ldots & 0_n \end{bmatrix}, \quad 0 \le \varepsilon \le 1. \qquad (10.4.5)$$

Clearly, G_ε is hermitian and invertible for all $\varepsilon \in [0,1]$. By the continuous dependence of eigenvalues on the entries of the matrix (Theorem 5.2.5), the inertia of G_ε

is constant (i.e., independent of ε). It is not difficult to see that the inertia of G_0 is given by the right-hand side of (10.4.4) (Ex. 10.10.4).

The key observation is that the matrix GC is skewhermitian; in other words, C is G-skewhermitian. The equality $GC = -C^*G$ is verified by direct computation.

Definition 10.4.1. We say that the system (10.4.1) is *forward*, resp. *backward*, *bounded* if all solutions are bounded as $t \longrightarrow +\infty$, resp. $t \longrightarrow -\infty$. The system (10.4.1) is said to be *bounded* if all solutions are bounded on the real line.

Theorem 10.4.2. *The following four statements are equivalent for the system* (10.4.1), *where A_k is hermitian if k is odd, A_k is skewhermitian if k is even, and A_ℓ is invertible*:

(a) *the system is forward bounded*;

(b) *the system is backward bounded*;

(c) *the system is bounded*;

(d) *all eigenvalues of C have zero real parts, and for every eigenvalue the geometric multiplicity coincides with the algebraic multiplicity.*

Proof. Clearly (c) implies both (a) and (b). The equivalence of (c) and (d) follows from Theorem 5.14.2(b) and from the analogue of Theorem 5.14.2(b) for backward stability (Ex. 5.16.17). It remains to prove that (a) or (b) implies (d). We show that (a) implies (d); the proof of the statement that (b) implies (d) is completely analogous. If (a) holds true, then by Theorem 5.14.2(b) all eigenvalues of C have nonpositive real parts. But then the canonical form (Theorem 10.1.3) shows that the G-skewhermitian matrix C cannot have eigenvalues with nonzero real parts. Equality of geometric and algebraic multiplicities for every eigenvalue follows from the same Theorem 5.14.2(b). □

In applications, it is often desirable to know if a given system is not only bounded, but all nearby systems are also bounded. This concept of stable boundedness is formally defined as follows.

Definition 10.4.3. A system of differential equations (10.4.1) is said to be *stably bounded* if there exists $\varepsilon > 0$ such that every system of differential equations

$$A'_\ell x^{(\ell)}(t) + A'_{\ell-1} x^{(\ell-1)}(t) + \cdots + A'_1 x'(t) + A'_0 x(t) = 0, \quad t \in \mathsf{R}, \qquad (10.4.6)$$

is bounded, provided the coefficients $A'_j \in \mathsf{H}^{n \times n}$, $j = 0, 1, \ldots, \ell$, satisfy the following conditions:

(1) If k is odd, then A'_k is hermitian.

(2) If k is even, then A'_k is skewhermitian.

(3) The inequalities

$$\|A'_j - A_j\| < \varepsilon, \quad \text{for } j = 0, 1, 2, \ldots, \ell,$$

hold true.

Note that A'_ℓ will be invertible if $\|A'_\ell - A_\ell\| < \varepsilon$ holds and ε is small enough. Indeed, this follows, for example, from the continuity of the determinant function and Theorem 5.9.2(2).

In particular, a stable bounded system is bounded. A sufficient condition for stable boundedness is given next.

Theorem 10.4.4. *Assume that every root subspace of the companion matrix C is either G-positive or G-negative. Then the system* (10.4.1) *is stably bounded.*

Proof. We prove Theorem 10.4.4 in two steps.

Step 1. We show that (10.4.1) is bounded. Indeed, the canonical form of Theorem 10.1.3 applied to the pair (G, C), where C is G-skewhermitian, shows that all eigenvalues of C have zero real parts (otherwise, the corresponding eigenvectors are G-neutral) and for every eigenvalue the geometric multiplicity coincides with the algebraic multiplicity (otherwise, there is a G-neutral eigenvector). Then the boundedness of (10.4.1) follows from Theorem 10.4.2.

Step 2. We prove the stable boundedness of (10.4.1). Since every root subspace $\mathcal{M}$ of C is either G-positive or G-negative, the same property holds for any subspace sufficiently close to $\mathcal{M}$ (in the gap metric). Indeed, we have $|x^*Gx| > 0$ for every nonzero $x \in \mathcal{M}$, and therefore, by compactness of the unit sphere $S_\mathcal{M}$ in $\mathcal{M}$, the inequality

$$\min_{x \in S_\mathcal{M}} \{|x^*Gx|\} \geq \varepsilon > 0$$

holds true. Now suppose a subspace $\mathcal{X} \subseteq \mathsf{H}^{n \times 1}$ is close to $\mathcal{M}$:

$$\theta(\mathcal{X}, \mathcal{M}) = \|P_\mathcal{X} - P_\mathcal{M}\| \leq \delta,$$

where $\delta > 0$ is small. Take $x \in \mathcal{X}$, $\|x\| = 1$. Then

$$\|x - P_\mathcal{M}x\| = \|P_\mathcal{X}x - P_\mathcal{M}x\| \leq \delta, \tag{10.4.7}$$

so for $y := P_\mathcal{M}x$ we have $\|y\| \geq 1 - \delta$ and

$$|y^*Gy| = (1-\delta)^2 \left|(y/(1-\delta))^* G(y/(1-\delta))\right| \geq (1-\delta)^2\varepsilon.$$

On the other hand,

$$x^*Gx = (x-y)^*G(x-y) + y^*G(x-y) + (x-y)^*Gy + y^*Gy;$$

hence,

$$|x^*Gx - y^*Gy| \leq \delta^2\|G\| + 2\delta\|G\|\|y\| \leq \delta^2\|G\| + 2\delta(1+\delta)\|G\|,$$

where (10.4.7) was used. Thus,

$$|x^*Gx| \geq (1-\delta)^2\varepsilon - (\delta^2\|G\| + 2\delta(1+\delta)\|G\|),$$

which can be made $\geq \varepsilon/2$ if δ is sufficiently small. This shows that $\mathcal{X}$ is also G-positive (or G-negative, as the case may be).

Let λ_0 be the eigenvalue of C in the closed upper half of the complex plane C_+ such that the root subspace $\mathcal{M}$ corresponds to λ. The C-invariant subspace $\mathcal{M}$, being a root subspace, is a Lipschitz function (Theorem 5.2.5), so every nearby matrix C' has an invariant subspace $\mathcal{M}'$ close to $\mathcal{M}$, and every such subspace is either G-positive or G-negative by the observation in the preceding paragraph. Moreover, in fact, $\mathcal{M}'$ is the sum of root subspaces for C' corresponding to the eigenvalues in C_+ which are close to λ, as indicated in the same Theorem 5.2.5. It follows that every root subspace of C' is either G-positive or G-negative. By Step 1, taking C' to be the companion matrix of a system (10.4.6), we obtain that (10.4.6) is bounded, if (1), (2), and (3) are satisfied with sufficiently small $\varepsilon > 0$.   □

As the proof of Theorem 10.4.4 shows, if (10.4.1) is stably bounded, then, for sufficiently small ε, every system (10.4.6) satisfying (1), (2), and (3) is also stably bounded (not only bounded as required by the definition of stable boundedness).

We do not know whether or not the condition of Theorem 10.4.4 is also necessary.

Open Problem 10.4.5. *Prove or disprove that if the system of differential equations (10.4.1) is stably bounded, then every root subspace of C is either G-negative or G-positive.*

The solution of Open Problem 10.4.5 is affirmative if $\ell = 1$, as given in the next theorem.

Theorem 10.4.6. *The system of first order differential equations*

$$A_1 x'(t) + A_0 x(t) = 0, \tag{10.4.8}$$

where

$$A_1 = A_1^* \in \mathsf{H}^{n \times n}, \qquad A_0 = -A_0^* \in \mathsf{H}^{n \times n},$$

and A_1 is invertible, is stably bounded if and only if every root subspace of $A_1^{-1} A_0$ is either A_1-positive or A_1-negative.

Proof. The "if" part is contained in Theorem 10.4.4. To prove the "only if" part, assume that there is a root subspace of $A_1^{-1} A_0$ that is not A_1-positive or A_1-negative, and we will show that the system (10.4.8) is not stably bounded. In view of Theorem 10.4.2 we may (and do) also assume that all eigenvalues of $A_1^{-1} A_0$ have zero real parts and for every eigenvalue of $A_1^{-1} A_0$, the geometric multiplicity is equal to its algebraic multiplicity. Taking A_1 and $A_1^{-1} A_0$ as in the right-hand side of (10.1.3), we see that these matrices have the following form:

$$A_1 = \eta_1 \oplus \cdots \oplus \eta_t \oplus \kappa_1 \oplus \cdots \oplus \kappa_p, \qquad A_1^{-1} A_0 = (ib_1) \oplus \cdots \oplus (ib_t) \oplus 0_p,$$

where $b_1, \ldots, b_t$ are positive real numbers and the κ_i's, η_j's are signs ± 1. The condition that not all root subspaces of $A_1^{-1} A_0$ are A_1-positive or A_1-negative amounts to the following: not all the κ_i's are the same, or $b_{j_1} = b_{j_2}$, $\eta_{j_1} \neq \eta_{j_2}$ for some indices j_1, j_2 ($1 \leq j_1 < j_2 \leq t$). For notational simplicity suppose $b_1 = b_2$, $\eta_1 = -\eta_2 = 1$. Then for nonzero real number y, the matrix

$$X := A_1^{-1} A_0 + \left(\begin{bmatrix} 0 & y \\ y & 0 \end{bmatrix} \oplus 0_{t+p-2} \right)$$

is A_1-skewhermitian, close to $A_1^{-1} A_0$ (if y is close to zero), and has eigenvalues $ib_1 \pm y$ with nonzero real parts. Thus, the system $A_1 x'(t) + (A_1 X) x(t) = 0$ is not bounded, and (10.4.8) is not stably bounded.   □

10.5   HAMILTONIAN, SKEW-HAMILTONIAN MATRICES: CANONICAL FORMS

In this section $H \in \mathsf{H}^{n \times n}$ is an invertible skewhermitian matrix. Recall that $A \in \mathsf{H}^{n \times n}$ is said to be $(H,^*)$-*Hamiltonian* if $A^*H = -HA$, and $(H,^*)$-*skew-Hamiltonian* if $A^*H = HA$. In this chapter, we write, for short, H-Hamiltonian and H-skew-Hamiltonian instead of $(H,^*)$-Hamiltonian and $(H,^*)$-skew-Hamiltonian, respectively.

We begin with canonical forms. The Hamiltonian case is treated in the next theorem.

Theorem 10.5.1. *Let $H \in \mathsf{H}^{n \times n}$ be an invertible skewhermitian matrix, and let $X \in \mathsf{H}^{n \times n}$ be H-Hamiltonian. Then for some invertible quaternion matrix S, the matrices S^*HS and $S^{-1}XS$ have, simultaneously, the following form:*

$$S^*HS = \oplus_{j=1}^r \eta_j \Xi_{\ell_j}(\mathsf{i}^{\ell_j}) \oplus \oplus_{v=1}^s \begin{bmatrix} 0 & F_{p_v} \\ -F_{p_v} & 0 \end{bmatrix} \oplus \oplus_{u=1}^q \zeta_u \Xi_{m_u}(\mathsf{i}^{m_u}), \quad (10.5.1)$$

$$S^{-1}XS = \oplus_{j=1}^r J_{\ell_j}(0) \oplus \oplus_{v=1}^s \begin{bmatrix} -J_{p_v}(\overline{\alpha_v}) & 0 \\ 0 & J_{p_v}(\alpha_v) \end{bmatrix} \oplus \oplus_{u=1}^q J_{m_u}(\gamma_u), \quad (10.5.2)$$

where η_j, ζ_u are signs ± 1 with the additional condition that $\eta_j = 1$ if ℓ_j is odd, the complex numbers $\alpha_1, \ldots, \alpha_s$ have positive real parts, and the complex numbers $\gamma_1, \ldots, \gamma_q$ are nonzero with zero real parts.

The form (10.5.1), (10.5.2) is uniquely determined by the pair (X, H), up to an arbitrary simultaneous permutation of primitive blocks in each of the three parts,

$$\left(\oplus_{j=1}^r \eta_j \Xi_{\ell_j}(\mathsf{i}^{\ell_j}), \ \oplus_{j=1}^r J_{\ell_j}(0) \right),$$

$$\left(\oplus_{v=1}^s \begin{bmatrix} 0 & F_{p_v} \\ -F_{p_v} & 0 \end{bmatrix}, \ \oplus_{v=1}^s \begin{bmatrix} -J_{p_v}(\overline{\alpha_v}) & 0 \\ 0 & J_{p_v}(\alpha_v) \end{bmatrix} \right),$$

and

$$\left(\oplus_{u=1}^q \zeta_u \Xi_{m_u}(\mathsf{i}^{m_u}), \ \oplus_{u=1}^q J_{m_u}(\gamma_u) \right),$$

and up to replacements of some α_k's and some γ_j's with their complex conjugates.

Conversely, if H, X have the forms (10.5.1), (10.5.2), then H is invertible skewhermitian, and X is H-Hamiltonian.

Remarks:

(1) A more general form than (10.5.1), (10.5.2) can be given (see Rodman [135]). Namely, the α_k'a and γ_j's are allowed to be quaternions (under additional restriction that $\mathsf{i}\gamma_j$ is real if m_j is odd), and the uniqueness is up to replacements in some primitive blocks, simultaneously in (10.5.1) and (10.5.2) of the α_k's and the γ_j's with similar quaternions. In this connection, note the formulas

$$S^{-1}(\mathsf{i}J_m(\lambda))S = J_m(\mathsf{i}\lambda),$$

$$S^*(\mathsf{i}F_m)S = \begin{cases} -\Xi_m(\mathsf{i}^m) & \text{if } m \text{ is even}, \\ \Xi_m(\mathsf{i}^m) & \text{if } m \text{ is odd}, \end{cases} \quad (10.5.3)$$

where λ is real, and $S = \text{diag}\,(1, -\mathsf{i}, (-\mathsf{i})^2, \ldots, (-\mathsf{i})^{m-1})$.

(2) The formulas

$$(jI_{2p_\nu})^{-1} \begin{bmatrix} -J_{p_\nu}(\overline{\alpha_\nu}) & 0 \\ 0 & J_{p_\nu}(\alpha_\nu) \end{bmatrix} jI_{2p_\nu} = \begin{bmatrix} -J_{p_\nu}(\alpha_\nu) & 0 \\ 0 & J_{p_\nu}(\overline{\alpha_\nu}) \end{bmatrix},$$

$$(jI_{2p_\nu})^* \begin{bmatrix} 0 & F_{p_\nu} \\ -F_{p_\nu} & 0 \end{bmatrix} jI_{2p_\nu} = \begin{bmatrix} 0 & F_{p_\nu} \\ -F_{p_\nu} & 0 \end{bmatrix}$$

make explicit the replacement of α_ν by its complex conjugate.

(3) The following formulas make explicit the replacement of γ_u by its complex conjugate:

$$S^{-1} J_{m_u}(-\gamma_u) S = J_{m_u}(\gamma_u),$$

$$S^*(\Xi_{m_u}(i^{m_u})) S = \begin{cases} -\Xi_{m_u}(i^{m_u}) & \text{if } m_u \text{ is odd,} \\ \Xi_{m_u}(i^{m_u}) & \text{if } m_u \text{ is even,} \end{cases}$$

where $S = jI$. Note that under this replacement ζ_u reverses to its negative if m_u is odd and remains invariant if m_u is even.

(4) The *sign characteristic* of an H-Hamiltonian matrix A assigns a sign ± 1 to every even partial multiplicity of A corresponding to the zero eigenvalue and to all partial multiplicities of every nonzero eigenvalue with zero real part.

Proof of Theorem 10.5.1. Note that H is skewhermitian and HA is hermitian. We use the canonical form for the hermitian/skewhernitian pencil $HA + tH$ under congruence given in Theorem 9.4.1. Because H is invertible, blocks of the forms (q-h-sk0), (q-h-sk1), (q-h-sk2) cannot appear. We consider separately each of the blocks (q-h-sk3), (q-h-sk4), (q-h-sk5).

For a block of type (q-h-sk3), we have

$$H = \eta i F_\ell, \qquad A = H^{-1}(HA) = (\eta i F_\ell)^{-1} \cdot \eta G_\ell = -i J_\ell(0)^T, \qquad (10.5.4)$$

where the sign η is always $+1$ if ℓ is odd. We can easily transform, using the transformation

$$(H, A) \mapsto (W^* H W, W^{-1} A W) \quad \text{for some invertible} \quad W \in \mathsf{H}^{\ell \times \ell}, \qquad (10.5.5)$$

the pair (H, A) given by (10.5.4) to the pair $(\eta \Xi_\ell(i^\ell), J_\ell(0))$ as required in Theorem 10.5.1. Indeed, let

$$W = \begin{bmatrix} 0 & 0 & \cdots & 0 & i^{\ell-1} \\ 0 & 0 & \cdots & i^{\ell-2} & 0 \\ \vdots & \vdots & \ddots & \vdots & \vdots \\ 1 & 0 & \cdots & 0 & 0 \end{bmatrix},$$

then a calculation shows that the equalities

$$W^*(i F_\ell) W = \Xi_\ell(i^\ell), \qquad W^{-1}(-i J_\ell(0)^T) W = J_\ell(0)$$

hold true.

For block of type (q-h-sk4), we have $H = \begin{bmatrix} 0 & F_p \\ -F_p & 0 \end{bmatrix}$, and

$$
\begin{aligned}
A &= H^{-1}(HA) = \begin{bmatrix} 0 & F_p \\ -F_p & 0 \end{bmatrix}^{-1} \begin{bmatrix} 0 & \alpha F_p + G_p \\ \alpha^* F_p + G_p & 0 \end{bmatrix} \\
&= \begin{bmatrix} -J_p(\overline{\alpha})^T & 0 \\ 0 & J_p(\alpha)^T \end{bmatrix},
\end{aligned}
$$

where $\alpha \in \mathsf{H}$ has positive real part, and we may take α to be a complex number. Applying a suitable transformation of type (10.5.5), we obtain the pair

$$
\begin{bmatrix} 0 & F_p \\ -F_p & 0 \end{bmatrix}, \quad \begin{bmatrix} -J_p(\overline{\alpha}) & 0 \\ 0 & J_p(\alpha) \end{bmatrix},
$$

again as required in Theorem 10.5.1.

Finally, for a block of type (q-h-sk5), we have

$$
H = \zeta \Xi_m(\mathrm{i}^m), \qquad HA = \zeta(\Xi_m(\beta) - \widetilde{G}_m), \qquad \zeta = \pm 1, \tag{10.5.6}
$$

where β is a nonzero quaternion with zero real part if m is even and β is a positive real number if m is odd. We may (and do) assume that, in fact, β is a complex number, and if necessary we may replace β with $-\beta$ (see Remark 9.4.3). It will be convenient to use the following notation:

$$
\Gamma_m(\lambda) := \begin{bmatrix} 0 & \lambda & 0 & \dots & 0 \\ 0 & 0 & -\lambda & \dots & 0 \\ \vdots & \vdots & \ddots & \ddots & \vdots \\ 0 & 0 & 0 & \dots & (-1)^m \lambda \\ 0 & 0 & 0 & \dots & 0 \end{bmatrix} \in \mathsf{H}^{m \times m}, \qquad \lambda \in \mathsf{H}.
$$

A straightforward computation shows that, for H and HA given in (10.5.6), we have

$$
\begin{aligned}
A &= \beta I_m + \Gamma_m(1) && \text{if } m \text{ is even, divisible by 4;} \\
A &= -\beta I_m + \Gamma_m(-1) && \text{if } m \text{ is even, not divisible by 4;} \\
A &= -\mathrm{i}\beta I_m + \Gamma_m(\mathrm{i}) && \text{if } m \text{ is odd, } m-1 \text{ divisible by 4;} \\
A &= \mathrm{i}\beta I_m + \Gamma_m(-\mathrm{i}) && \text{if } m \text{ is odd, } m-1 \text{ not divisible by 4.}
\end{aligned}
$$

In each of these cases, the pair (H, A) can be easily transformed (using the transformation (10.5.5)) to the form

$$
(\widetilde{\zeta}\Xi_m(\mathrm{i}^m), J_m(\gamma)), \qquad \gamma \in \mathsf{C} \setminus \{0\}, \quad \mathfrak{R}(\gamma) = 0, \quad \widetilde{\zeta} = \pm 1.
$$

Indeed, for m even we have

$$
(\operatorname{diag}(I_2, -I_2, \dots, (-1)^{m/2-1} I_2)) \, \Xi_m(\mathrm{i}^m)
$$

$$
\cdot (\operatorname{diag}(I_2, -I_2, \dots, (-1)^{m/2-1} I_2)) = (-1)^{m/2-1} \Xi_m(\mathrm{i}^m),
$$

$$
(\operatorname{diag}(I_2, -I_2, \dots, (-1)^{m/2-1} I_2)) (\pm(\beta I_m + \Gamma_m(1)))
$$

$$
\cdot (\operatorname{diag}(I_2, -I_2, \dots, (-1)^{m/2-1} I_2)) = \pm J_m(\beta);
$$

if m is even but not divisible by 4, a further transformation is needed:

$$(\text{diag}\,(1,-1,1,\ldots,-1))\,\Xi_m(\text{i}^m)\,(\text{diag}\,(1,-1,1,\ldots,-1)) = -\Xi_m(\text{i}^m),$$
$$(\text{diag}\,(1,-1,1,\ldots,-1))\,(-J_m(\beta))\,(\text{diag}\,(1,-1,1,\ldots,-1)) = J_m(-\beta).$$

For m odd,

$$(\text{diag}\,(1,\text{i},1,\text{i},\ldots,1))\,\Xi_m(\text{i}^m)\,(\text{diag}\,(1,-\text{i},1,\ldots,1)) = \Xi_m(\text{i}^m),$$

$$(\text{diag}\,(1,\text{i},1,\text{i},\ldots,1))\,(\pm(-\text{i}\beta I_m + \Gamma_m(\text{i}))(\text{diag}\,(1,-\text{i},1,-\text{i},\ldots,1))$$
$$= \pm J_m(-\text{i}\beta),$$

and for the case when $m-1$ is not divisible by 4, a further transformation is needed, as in the case m is even, not divisible by 4. $\quad\square$

Next, consider H-skew-Hamiltonian matrices. The canonical form runs as follows.

Theorem 10.5.2. *Let $H \in \mathsf{H}^{n\times n}$ be skewhermitian and invertible, and let A be H-skew-Hamiltonian. Then there exists an invertible matrix $S \in \mathsf{H}^{n\times n}$ such that*

$$S^{-1}AS = J_{\ell_1}(\beta_1) \oplus \cdots \oplus J_{\ell_q}(\beta_q), \quad S^*HS = \text{i}F_{\ell_1} \oplus \cdots \oplus \text{i}F_{\ell_q}, \qquad (10.5.7)$$

where $\beta_1,\ldots,\beta_q \in \mathsf{H}$ are such that $\text{i}\beta_j$ have zero real parts for $j = 1,2,\ldots,q$.

The form (10.5.7) *is uniquely determined by the pair (A,H), up to a simultaneous permutation of the pairs of corresponding primitive blocks in $S^{-1}AS$ and S^*HS, and up to replacement of each β_j by a similar quaternion β'_j subject to the condition that $\text{i}\beta'_j$ has zero real part.*

Conversely, if A and H are given by the right-hand sides of the equalities in (10.5.7)*, then A is H-skew-Hamiltonian.*

Note that the canonical form for a pair (A, H), where $H = -H^*$ is invertible and A is H-skew-Hamiltonian, does not involve sign characteristic. Also, in contrast with many other canonical forms in the book, the form (10.5.7) cannot be given generally speaking in terms of real and complex matrices. For example, one can take $\beta_1,\ldots,\beta_q \in \text{Span}_\mathsf{R}\{1,\text{j}\}$ or, alternatively $\beta_1,\ldots,\beta_q \in \text{Span}_\mathsf{R}\{1,\text{k}\}$.

Proof. We follow the same approach as in the proof of Theorem 10.5.1. Here H and HA are both skewhermitian. The canonical form for the skewhermitian matrix pencil $HA+tH$ (Theorem 9.1.1) allows us to consider each primitive block separately. Taking into account that H is invertible, we have to consider only the case when

$$H = \text{i}F_k, \qquad HA = \begin{bmatrix} 0 & 0 & \cdots & 0 & 0 & \text{i}\beta \\ 0 & 0 & \cdots & 0 & \text{i}\beta & \text{i} \\ 0 & 0 & \cdots & \text{i}\beta & \text{i} & 0 \\ \vdots & \vdots & \ddots & \vdots & \vdots & \vdots \\ \text{i}\beta & \text{i} & 0 & \cdots & 0 & 0 \end{bmatrix}_{k\times k},$$

where $\beta \in \mathsf{H}$ is such that $\text{i}\beta$ has zero real part. Thus,

$$A = H^{-1}(HA) = J_k(\beta),$$

and so the pair (H, A) is in the form as required in Theorem 10.5.2. $\quad\square$

Corollary 10.5.3.   (a) *A matrix* $A \in \mathsf{H}^{n \times n}$ *is similar to an H-Hamiltonian matrix for some invertible skewhermitian H if and only its Jordan structure is symmetric with respect to negation: if $\lambda \in \sigma(A)$ and λ has nonzero real part, then also $-\lambda \in \sigma(A)$, and the partial multiplicities of A at λ coincide with those at $-\lambda$.*

(b) *Any matrix $A \in \mathsf{H}^{n \times n}$ is similar to an H-skew-Hamiltonian matrix for some invertible skewhermitian H.*

Recall that partial multiplicities of A at its eigenvalue λ also include, by definition, all $\mu \in \mathsf{H}$ similar to λ.

The corollary is immediate upon inspection of the canonical forms (10.5.2) and (10.5.7).

10.6   INVARIANT LAGRANGIAN SUBSPACES II

We assume in this section, as before, that $H \in \mathsf{H}^{n \times n}$ is skewhermitian and invertible and assume in addition that n is even. Recall that a subspace $\mathcal{M} \subseteq \mathsf{H}^{n \times 1}$ is said to be *H-Lagrangian* if $\mathcal{M}$ is H-neutral and $\dim \mathcal{M} = n/2$. By Theorem 4.2.7(2) (or (10.3.2)) Lagrangian subspaces always exist.

In this section we give criteria (based on the canonical forms of Section 10.5) for existence of invariant Lagrangian subspaces for H-Hamiltonian and H-skew-Hamiltonian matrices.

Theorem 10.6.1. (a) *Let $X \in \mathsf{H}^{n \times n}$ be H-Hamiltonian. Then there exists an X-invariant H-Lagrangian subspace if and only if every root subspace of X corresponding to eigenvalues with zero real part (including the zero eigenvalue if X is not invertible) has even (quaternion) dimension.*

(b) *Let $X \in \mathsf{H}^{n \times n}$ be H-skew-Hamiltonian. Then there exists an X-invariant H-Lagrangian subspace if and only if every root subspace for X is even dimensional.*

Proof. We start with the proof of Part (b).

The condition is obviously necessary in view of the canonical form (10.5.7) for the pair (X, H) and Proposition 10.3.2. Now suppose that every root subspace of X is even dimensional. We may assume that X and H are given by the right-hand sides of the formulas in (10.5.7):

$$X = J_{\ell_1}(\beta_1) \oplus \cdots \oplus J_{\ell_q}(\beta_q), \quad H = \mathsf{i}F_{\ell_1} \oplus \cdots \oplus \mathsf{i}F_{\ell_q},$$

where $\beta_1, \ldots, \beta_q \in \mathrm{Span}_\mathsf{R} \{1, \mathsf{j}\}$. Replacing some of the β_j's by similar quaternions, we further assume that

$$\beta_j = a_j + \mathsf{j}b_j, \quad j = 1, 2, \ldots, q, \quad \text{where } a_j, b_j \in \mathsf{R}, \ b_j \geq 0.$$

By Proposition 10.3.2 it suffices to consider the case when X has only one root subspace; in other words, given the assumptions on the β_j's, the equalities $\beta_1 = \cdots = \beta_q$ hold true. By our hypothesis, the number of odd indices (counted with repetitions) among $\ell_1, \ldots, \ell_q$ is even. Rearranging blocks (if necessary) in (5.3.7), we let $\ell_1, \ldots, \ell_{2u}$ be odd and $\ell_{2u+1}, \ldots, \ell_q$ be even for some nonnegative integer u. Noting that $\mathrm{Span}_\mathsf{H} \begin{bmatrix} 1 \\ \mathsf{j} \end{bmatrix}$ is $\begin{bmatrix} \mathsf{i} & 0 \\ 0 & \mathsf{i} \end{bmatrix}$-neutral, we select the following vectors in $\mathsf{H}^{n \times 1}$:

$$e_1, \ldots, e_{(\ell_1 - 1)/2}, e_{(\ell_1 + 1)/2} + \mathsf{j}e_{\ell_1 + (\ell_2 + 1)/2}, e_{\ell_1 + 1}, \ldots, e_{\ell_1 + (\ell_2 - 1)/2}, \tag{10.6.1}$$

$$e_{\ell_1+\cdots+\ell_{2j}+1}, \ldots, e_{\ell_1+\cdots+\ell_{2j}+\cdots+(\ell_{2j+1}-1)/2}, \tag{10.6.2}$$

$$e_{\ell_1+\cdots+\ell_{2j}+\cdots+(\ell_{2j+1}+1)/2} + \mathrm{j}e_{\ell_1+\cdots+\ell_{2j}+\ell_{2j+1}+(\ell_{2j+2}+1)/2}, \tag{10.6.3}$$

$$e_{\ell_1+\cdots+\ell_{2j}+\ell_{2j+1}+1}, \ldots, e_{\ell_1+\cdots+\ell_{2j}+\ell_{2j+1}+(\ell_{2j+2}-1)/2}, \tag{10.6.4}$$

for $j = 1, 2, \ldots, u-1$, and, setting $w = \ell_1 + \cdots + \ell_{2u}$,

$$e_{w+1}, \ldots, e_{w+\ell_{2u+1}/2}, e_{w+\ell_{2u+1}+1}, \ldots, e_{w+\ell_{2u+1}+\ell_{2u+2}/2}, \ldots, \tag{10.6.5}$$

$$e_{w+\ell_{2u+1}+\cdots+\ell_{q-1}+1}, e_{w+\ell_{2u+1}+\cdots+\ell_{q-1}+2}, \ldots, e_{w+\ell_{2u+1}+\cdots+\ell_{q-1}+\ell_q/2}. \tag{10.6.6}$$

The selected vectors span an X-invariant H-Lagrangian subspace in $\mathsf{H}^{n\times1}$.

Proof of Part (a). As in Part (b), by Proposition 10.3.2 and the canonical form in Theorem 10.5.1, the conditions in Part (a) are necessary for existence of X-invariant H-Lagrangian subspaces. To prove sufficiency of these conditions, in view of Theorem 10.5.1 we may (and do) assume that X and H have one of the following three forms:

(1)

$$H = \oplus_{v=1}^s \begin{bmatrix} 0 & F_{p_v} \\ -F_{p_v} & 0 \end{bmatrix}, \qquad X = \oplus_{v=1}^s \begin{bmatrix} -J_{p_v}(\overline{\alpha}) & 0 \\ 0 & J_{p_v}(\alpha) \end{bmatrix},$$

where $\alpha \in \mathsf{C}$ has positive real part;

(2)

$$H = \oplus_{j=1}^v \Xi_{\ell_j}(\mathrm{i}^{\ell_j}) \oplus \oplus_{j=v+1}^q \eta_j \Xi_{\ell_j}(\mathrm{i}^{\ell_j}), \qquad X = \oplus_{j=1}^q J_{\ell_j}(0),$$

where $\ell_1, \ldots, \ell_v$ are odd, $\ell_{v+1}, \ldots, \ell_q$ are even, and the η_j's are signs ±1;

(3)

$$H = \oplus_{j=1}^q \zeta_j \Xi_{m_j}(\mathrm{i}^{m_j}), \qquad X = \oplus_{j=1}^q J_{m_j}(\gamma),$$

where the ζ_j's are signs ±1, and $\gamma \in \mathsf{C}$ has zero real part and positive imaginary part.

In Case (1), an X-invariant H-Lagrangian subspace always exists. To construct such a subspace, select

$$e_1, e_2, \ldots, e_{p_1}, e_{2p_1+1}, e_{2p_1+2}, \ldots, e_{2p_1+p_2}, \ldots,$$

$$e_{2p_1+\cdots+2p_{v-1}+1}, \ldots, e_{2p_1+\cdots+2p_{v-1}+p_v}$$

and span the subspace in $\mathsf{H}^{n\times1}$ by the selected vectors.

Assume Case (2) holds and assuming that the conditions of Theorem 10.6.1 hold true, i.e., $v = 2u$ is even, we select the vectors as in (10.6.1)–(10.6.6). Then the subspace spanned by the selected vectors in $\mathsf{H}^{n\times1}$ is X-invariant and H-Lagrangian.

Finally, assume (3) holds. For simplicity of notation, suppose that $m_1, \ldots m_v$ are odd and $m_{v+1}, \ldots, m_q$ are even. Since we assume that the conditions of Theorem 10.6.1(a) hold true, $v = 2u$ is even. Now select the following vectors in $\mathsf{H}^{n\times1}$:

$$e_1, \ldots, e_{(m_1-1)/2}, e_{(m_1+1)/2} + z_1 e_{m_1+(m_2+1)/2}, e_{m_1+1}, \ldots, e_{m_1+(m_2-1)/2},$$

$$e_{m_1+\cdots+m_{2j}+1}, \ldots, e_{m_1+\cdots+m_{2j}+\cdots+(m_{2j+1}-1)/2},$$

$$e_{m_1+\cdots+m_{2j}+\cdots+(m_{2j+1}+1)/2} + z_{j+1} e_{m_1+\cdots+m_{2j}+m_{2j+1}+(m_{2j+2}+1)/2},$$

$$e_{m_1+\cdots+m_{2j}+m_{2j+1}+1}, \ldots, e_{m_1+\cdots+m_{2j}+m_{2j+1}+(m_{2j+2}-1)/2}$$

for $j = 1, 2, \ldots, u-1$, where $z_j = 1$ if $\zeta_{2j-1} = -\zeta_{2j}$ and $z_j = \mathrm{j}$ if $\zeta_{2j-1} = \zeta_{2j}$, and, setting $w = \ell_1 + \cdots + \ell_{2u}$,

$$e_{w+1}, \ldots, e_{w+ml_{2u+1}/2}, e_{w+m_{2u+1}+1}, \ldots, e_{w+m_{2u+1}+m_{2u+2}/2}, \ldots,$$

$$e_{w+m_{2u+1}+\cdots+m_{q-1}+1}, e_{w+m_{2u+1}+\cdots+m_{q-1}+2}, \ldots, e_{w+m_{2u+1}+\cdots+m_{q-1}+m_q/2}.$$

We verify, using Proposition 10.3.2, that the selected vectors span an X-invariant H-Lagrangian subspace. $\square$

10.7   EXTENSION OF SUBSPACES

Throughout this section we fix a skewhermitian matrix $H \in \mathsf{H}^{n \times n}$ and an H-Hamiltonian matrix $A \in \mathsf{H}^{n \times n}$. Thus, HA is hermitian, and we can consider the classes of HA-nonnegative, of HA-nonnegative, of maximal HA-nonnegative, and of maximal HA-nonnegative subspaces. The main results of this section have to do with extensions of A-invariant H-neutral subspaces to A-invariant maximal HA-semidefinite subspaces.

Remark 10.7.1. Note that for an A-invariant subspace $\mathcal{M}$, if $\mathcal{M}$ is H-neutral, then $\mathcal{M}$ is also HA-neutral. The converse generally speaking is false (take $A = 0$ to provide a counterexample). However, if A is invertible, then an A-invariant subspace $\mathcal{M}$ is H-neutral if and only if it is HA-neutral. Indeed, suppose $\mathcal{M}$ is HA-neutral; then $x^* HAy = 0$ for all $x, y \in \mathcal{M}$. Now take y such that $Ay = x$ to obtain that $x^* Hx = 0$ for all $x \in \mathcal{M}$, and it follows that $\mathcal{M}$ is H-neutral (cf. the polarization identity (3.5.1)).

Theorem 10.7.2. Let $\mathcal{N} \subseteq \mathsf{H}^{n \times 1}$ be an A-invariant H-neutral subspace. Then there exist A-invariant subspaces $\mathcal{L}_+$ and $\mathcal{L}_-$ such that each of them contains $\mathcal{N}$ and $\mathcal{L}_+$ is maximal HA-nonnegative, whereas $\mathcal{L}_-$ is maximal HA-nonpositive.

Remark 10.7.3. We compute the dimensions of the maximal HA-nonnegative and maximal HA-nonpositive subspaces in Theorem 10.7.2. Inspection of the canonical form of the pair (H, A) as in Theorem 10.5.1 reveals that

$$\mathrm{In}_+ (HA) - \mathrm{In}_- (HA) = \sum_{\{u \,:\, m_u \text{ is odd}\}} (-\zeta_u \cdot \mathrm{sign}(\Im \gamma_u))$$

$$+ \sum_{\{j \,:\, \ell_j \text{ is even}\}} \eta_j \qquad (10.7.1)$$

and

$$\mathrm{In}_0 (HA) = r.$$

So, denoting by Υ the right-hand side of (10.7.1), we have

$$\mathrm{In}_\pm (HA) = \frac{1}{2} (\pm \Upsilon + n - r),$$

and the dimension of a maximal HA-nonnegative, resp. maximal HA-nonpositive, subspace is equal to $(1/2)\,(\Upsilon + n + r)$, resp. $(1/2)\,(-\Upsilon + n + r)$.

We mention in passing that Theorem 10.7.2 also remains valid for singular H; in other words, for every pair of quaternion matrices H and A such that $H = -H^*$ and $HA = -A^*H$. Indeed, if H is singular, then applying a transformation

$$A, \ H \quad \mapsto \quad S^{-1}AS, \ S^*HS \tag{10.7.2}$$

for a suitable invertible matrix S, we may assume that H has the block form $H = \begin{bmatrix} H_1 & 0 \\ 0 & 0 \end{bmatrix}$, where H_1 is invertible and skewhermitian. Because of the equality $HA = -A^*H$, we obtain that A has the conformally partitioned block form $A = \begin{bmatrix} A_{11} & 0 \\ A_{21} & A_{22} \end{bmatrix}$, where A_{11} is H_1-Hamiltonan. Without loss of generality, we may assume that $\mathcal{N} \supseteq \operatorname{Ker} H$ (otherwise replace $\mathcal{N}$ with $\mathcal{N} + \operatorname{Ker} H$); then the statement of Theorem 10.7.2 for A and H is easily reduced to the situation where A and H are replaced by A_{11} and H_1, respectively.

The subspaces $\mathcal{L}_\pm$ of Theorem 10.7.2 may have additional spectral properties. Since A is an H-Hamiltonian matrix, the canonical form 10.5.2 shows that A is similar to $-A$, and, therefore, the set of eigenvalues of A is symmetric relative to negation: if $\lambda \in \sigma(A)$, then $-\lambda \in \sigma(A)$.

Definition 10.7.4. A set of eigenvalues $\mathcal{S} \in \mathsf{H}$ of A will be called a *c-set* if the following four conditions are fulfilled:

(1) The eigenvalues in $\mathcal{S}$ all have nonzero real parts.

(2) If $\lambda_0 \in \mathcal{S}$, then all quaternions similar to λ_0 are also in $\mathcal{S}$.

(3) If $\lambda_0 \in \mathcal{S}$, then $-\lambda_0 \notin \mathcal{S}$.

(4) $\mathcal{S}$ is maximal (in the sense of sets containment) set of eigenvalues of A that satisfies conditions (1), (2), and (3).

This terminology is borrowed from Gohberg et al. [52]. For example, the set of all eigenvalues of A with positive real parts is a *c*-set. It may happen (when all eigenvalues of A have zero real parts) that any *c*-set is empty.

Theorem 10.7.5. *Under the hypotheses of Theorem 10.7.2, assume in addition that the set $\mathcal{S}_0$ of eigenvalues with nonzero real parts of the restriction $A|_{\mathcal{N}}$ is such that*

$$\lambda_0 \in \mathcal{S}_0 \quad \Longrightarrow \quad -\lambda_0 \notin \mathcal{S}_0.$$

Then for every c-set $\mathcal{S}$ such that $\mathcal{S} \supseteq \mathcal{S}_0$, there exist subspaces $\mathcal{L}_\pm$ as in Theorem 10.7.2, with the additional property that $\mathcal{S}$ coincides with the set of eigenvalues with nonzero real parts of $A|_{\mathcal{L}_\pm}$.

The cases when $\mathcal{N} = \{0\}$ and/or when $\mathcal{S}_0 = \emptyset$ are not excluded in Theorems 10.7.2 and 10.7.5.

The case when A is singular presents additional difficulty largely due to the fact that the spectrum of A is symmetric with respect to negation, and zero is the fixed point of this symmetry.

Open Problem 10.7.6. *Are the results of Theorem 10.7.2 and 10.7.5 valid under the weaker hypothesis that $\mathcal{N}$ is A-invariant HA-neutral (rather than A-invariant H-neutral)?*

As follows from Remark 10.7.1, the solution to Open Problem 10.7.6 is affirmative if A is invertible.

The proofs of Theorems 10.7.2 and 10.7.5 are given in the next section.

10.8 PROOFS OF THEOREMS 10.7.2 AND 10.7.5

We start with a lemma.

Lemma 10.8.1. *Let $Z \in \mathsf{H}^{m \times m}$ be a hermitian matrix partitioned as follows:*

$$Z = \begin{bmatrix} 0 & 0 & Q_1^* \\ 0 & Q_2 & K_1^* \\ Q_1 & K_1 & K_2 \end{bmatrix},$$

where the $p \times q$ block Q_1 is right-invertible (thus, $p \leq q$). Then

$$\text{In}_\pm(Z) + \text{In}_0(Z) = q + \text{In}_\pm(Q_2) + \text{In}_0(Q_2). \tag{10.8.1}$$

Proof. Let $Q_1^{[-1]}$ be a right inverse of Q_1, and let

$$X = -K_1^* \left(Q_1^{[-1]} \right)^*, \qquad Y = -\frac{1}{2} K_2^* \left(Q_1^{[-1]} \right)^*.$$

Then

$$\begin{bmatrix} I & 0 & 0 \\ X & I & 0 \\ Y & 0 & I \end{bmatrix} Z \begin{bmatrix} I & X^* & Y^* \\ 0 & I & 0 \\ 0 & 0 & I \end{bmatrix} = \begin{bmatrix} 0 & 0 & Q_1^* \\ 0 & Q_2 & 0 \\ Q_1 & 0 & 0 \end{bmatrix}.$$

In turn, let $Q_1 = S\,[I_p\ 0]\,T$, where S and T are invertible, be a rank decomposition of Q_1 (Proposition 3.2.5), then

$$\begin{bmatrix} T^* & 0 \\ 0 & S \end{bmatrix} \begin{bmatrix} 0 & 0 & I_p \\ 0 & 0 & 0_{(q-p)\times p} \\ I_p & 0_{p\times(q-p)} & 0 \end{bmatrix} \begin{bmatrix} T & 0 \\ 0 & S^* \end{bmatrix} = \begin{bmatrix} 0 & Q_1^* \\ Q_1 & 0 \end{bmatrix}.$$

Thus,

$$\text{In}_\pm(Z) + \text{In}_0(Z) = \text{In}_\pm(Q_2) + \text{In}_0(Q_2) + \text{In}_\pm \begin{bmatrix} 0 & Q_1^* \\ Q_1 & 0 \end{bmatrix} + \text{In}_0 \begin{bmatrix} 0 & Q_1^* \\ Q_1 & 0 \end{bmatrix}$$

$$= \text{In}_\pm(Q_2) + \text{In}_0(Q_2)$$

$$+ \text{In}_\pm \begin{bmatrix} 0 & 0 & I_p \\ 0 & 0 & 0_{(q-p)\times p} \\ I_p & 0_{p\times(q-p)} & 0 \end{bmatrix} + \text{In}_0 \begin{bmatrix} 0 & 0 & I_p \\ 0 & 0 & 0_{(q-p)\times p} \\ I_p & 0_{p\times(q-p)} & 0 \end{bmatrix},$$

and

$$\text{In}_\pm \begin{bmatrix} 0 & 0 & I_p \\ 0 & 0 & 0_{(q-p)\times p} \\ I_p & 0_{p\times(q-p)} & 0 \end{bmatrix} + \text{In}_0 \begin{bmatrix} 0 & 0 & I_p \\ 0 & 0 & 0_{(q-p)\times p} \\ I_p & 0_{p\times(q-p)} & 0 \end{bmatrix}$$

is easily seen to be equal to q.   □

Proof of Theorems 10.7.2 and 10.7.5. We prove these results only for HA-nonnegative subspaces (for nonpositive subspaces the proof is analogous, or else use $-H$ in place of H).

The canonical form of the pair of matrices (A, H) (with invertible H) under the transformations (10.7.2) given in Theorem 10.5.1 allows us to reduce the proofs to separate consideration of two cases: (1) A is invertible; (2) A is nilpotent.

Assume first that A is invertible. Let $\widehat{H} = HA$. Then $\widehat{H}$ is hermitian and invertible, and A is $\widehat{H}$-skewhermitian: $\widehat{H}A = -A^*\widehat{H}$. The subspace $\mathcal{N}$ is easily seen to be $\widehat{H}$-neutral. By Theorem 10.2.1 and Remark 10.2.2, there exist $\widehat{H}$-nonnegative, resp. $\widehat{H}$-nonpositive, subspaces which are A-invariant and have the required properties of dimension and location of the spectrum as required in Theorems 10.7.2 and 10.7.5.

Thus, it remains to prove Theorems 10.7.2 and 10.7.5 for nilpotent A. Note that in this case both $\mathcal{S}_0$ and $\mathcal{S}$ of Theorem 10.7.5 are vacuous, and in fact Theorem 10.7.5 coincides with Theorem 10.7.2. Therefore, we will prove Theorem 10.7.2 assuming A is nilpotent.

Consider the subspace

$$\mathcal{N}^{[\perp]} := \{x \in \mathsf{H}^{n \times 1} \,|\, x^* H y = 0 \quad \text{for all} \quad y \in \mathcal{N}\},$$

the H-orthogonal companion of $\mathcal{N}$. As $\mathcal{N}$ is H-neutral, we have $\mathcal{N} \subseteq \mathcal{N}^{[\perp]}$. Since A is H-Hamiltonian and $\mathcal{N}$ is A-invariant, the subspace $\mathcal{N}^{[\perp]}$ is easily seen to be A-invariant as well. Assuming $\mathcal{N} \neq \mathcal{N}^{[\perp]}$, choose an (ordered) euclidean orthonormal basis

$$(y_1, \ldots, y_n) \tag{10.8.2}$$

in $\mathsf{H}^{n \times 1}$ so that the first vectors in (10.8.2) form a basis of $\mathcal{N}$, the next vectors in (10.8.2) form a basis of the euclidean orthogonal complement of $\mathcal{N}$ in $\mathcal{N}^{[\perp]}$, and the remaining vectors in (10.8.2) form a basis of the euclidean orthogonal complement of $\mathcal{N}^{[\perp]}$ in $\mathsf{H}^{n \times 1}$. With respect to the basis (10.8.2), A has a block form

$$A = \begin{bmatrix} A_{11} & A_{12} & A_{22} \\ 0 & A_{22} & A_{23} \\ 0 & 0 & A_{33} \end{bmatrix},$$

and the corresponding representation of H is

$$H = [y_i^* H y_j]_{i,j=1}^n = \begin{bmatrix} 0 & 0 & H_{13} \\ 0 & H_{22} & H_{23} \\ -H_{13}^* & -H_{23}^* & H_{33} \end{bmatrix}.$$

The matrix H_{22} is skewhermitian and A_{22} is H_{22}-Hamiltonian and nilpotent.

If $\dim(\mathrm{Ker}\,(A_{22})) \geq 2$, then by Lemma 4.4.2 there exists a nonzero $x_0 \in \mathrm{Ker}\,(A_{22})$ which is H_{22}-neutral. Then the subspace $\mathcal{N} + \mathrm{Span}_{\mathsf{H}}\{x_0\}$ is clearly A-invariant and H-neutral. If $\dim(\mathrm{Ker}\,(A_{22})) = 1$ but $A_{22} \neq 0$, then the canonical form (10.5.1), (10.5.2) shows that we may suppose that

$$A_{22} = J_\ell(0), \qquad H_{22} = \pm \Xi_\ell(\mathrm{i}^\ell),$$

where $\ell \geq 2$. Then the subspace $\mathcal{N} + \mathrm{Span}_{\mathsf{H}}\{x_1\}$ with x_1 an eigenvector of A_{22}, is A-invariant and H-neutral.

We repeat the above procedure with $\mathcal{N}$ replaced with $\mathcal{N} + \mathrm{Span}_H \{x_0\}$ or $\mathcal{N} + \mathrm{Span}_H \{x_1\}$ as the case may be. Eventually, we reduce the proof to the case when

$$\mathcal{N} = \mathcal{N}^{[\perp]} \quad \text{or} \quad \dim_H \mathcal{N} = \dim_H \mathcal{N}^{[\perp]} - 1. \tag{10.8.3}$$

Case 10.8.2. *Assume* $\mathcal{N} = \mathcal{N}^{[\perp]}$.

Since

$$\dim \mathcal{N} = n - \dim \mathcal{N}^{[\perp]}$$

(this equality holds because $\mathcal{N}^{[\perp]}$ coincides with the euclidean orthogonal complement to the $\dim \mathcal{N}$-dimensional subspace $H(\mathcal{N})$), we have

$$\dim \mathcal{N} = \dim \mathcal{N}^{[\perp]} = \frac{n}{2}, \tag{10.8.4}$$

and n is necessarily even.

Choosing a euclidean orthogonal basis in $H^{n \times 1}$ such that the first part of its elements form a basis in $\mathcal{N}$, we represent A and H in the form

$$A = \begin{bmatrix} B_{11} & B_{12} \\ 0 & B_{22} \end{bmatrix}, \quad B_{ij} \in H^{n/2 \times n/2}, \quad H = \begin{bmatrix} 0 & H_1 \\ -H_1^* & H_2 \end{bmatrix}. \tag{10.8.5}$$

Here

$$\mathcal{N} = \mathrm{Span}_H \{e_1, \ldots, e_{n/2}\},$$

the matrix H_1 is invertible (because H is so), and H_2 is skewhermitian. Applying a transformation (10.7.2) with $S = \begin{bmatrix} I & W_1 \\ 0 & W_2 \end{bmatrix}$, for suitable W_1 and W_2, we may (and do) assume that in fact

$$H = \begin{bmatrix} 0 & I \\ -I & 0 \end{bmatrix}. \tag{10.8.6}$$

Then, since A is H-Hamiltonian, we have

$$A = \begin{bmatrix} B_{11} & B_{12} \\ 0 & -B_{11}^* \end{bmatrix}, \quad B_{12} \text{ hermitian.} \tag{10.8.7}$$

Next, with A and H given by (10.8.6) and (10.8.7), we apply a transformation (10.7.2) with S of the form $S = \begin{bmatrix} U & 0 \\ 0 & V \end{bmatrix}$, where the invertible matrices U and V are chosen so that $U^*V = I$ and

$$U^{-1}B_{11}U = \begin{bmatrix} 0_{r \times r} & 0 \\ C_1 & C_2 \end{bmatrix}, \quad r = \dim \left(\mathrm{Ker}\, B_{11} \right).$$

The matrix $[C_1 \quad C_2]$ is clearly right-invertible, the matrix H given by (10.8.6) is fixed under this transformation, whereas the transformed matrix A (which will be again denoted by A) is of the form

$$A = \begin{bmatrix} 0 & 0 & D_1 & D_2 \\ C_1 & C_2 & D_3 & D_4 \\ 0 & 0 & 0 & -C_1^* \\ 0 & 0 & 0 & -C_2^* \end{bmatrix}.$$

Thus,

$$HA = \begin{bmatrix} 0 & 0 & 0 & -C_1^* \\ 0 & 0 & 0 & -C_2^* \\ 0 & 0 & -D_1 & -D_2 \\ -C_1 & -C_2 & -D_3 & -D_4 \end{bmatrix}.$$

Since HA is hermitian, we have $D_1 = D_1^*$, $D_4 = D_4^*$, $D_3 = D_2^*$. Let $\mathcal{M}_+$ be a maximal $(-D_1)$-nonnegative subspace, and let

$$\mathcal{L}_+ = \mathcal{N} + \begin{bmatrix} 0_{n/2 \times 1} \\ \mathcal{M}_+ \\ 0_{(n/2-r) \times 1} \end{bmatrix}.$$

By Lemma 10.8.1,

$$\dim \mathcal{L}_+ = \text{In}_+ (HA) + \text{In}_0 (HA).$$

Also, $\mathcal{L}_+$ is clearly HA-nonnegative and A-invariant. This concludes the proof of Theorems 10.7.2 and 10.7.5 in Case 10.8.2.

Case 10.8.3. *Assume* $\dim_H \mathcal{N} = \dim_H \mathcal{N}^{[\perp]} - 1$.

Arguing as in Case 10.8.2, we see that n is odd and $\dim_H \mathcal{N} = (n-1)/2$. Choosing a euclidean orthogonal basis in $H^{n \times 1}$ such that the first $(n-1)/2$ of its elements form a basis in $\mathcal{N}$ and the first $(n+1)/2$ of its elements form a basis in $\mathcal{N}^{[\perp]}$, we represent A and H in the form

$$A = \begin{bmatrix} B_{11} & B_{12} & B_{13} \\ 0 & 0_{1 \times 1} & B_{23} \\ 0 & 0 & B_{33} \end{bmatrix}, \quad \text{where } B_{11}, B_{33} \in H^{(n-1)/2 \times (n-1)/2},$$

$$H = \begin{bmatrix} 0 & 0 & H_{13} \\ 0 & H_{22} & H_{23} \\ -H_{13}^* & -H_{23}^* & H_{33} \end{bmatrix}, \quad \text{where } H_{22} = -H_{22}^* \in H \setminus \{0\}.$$

Here

$$\mathcal{N} = \text{Span}_H \{e_1, \ldots, e_{(n-1)/2}\}, \quad \mathcal{N}^{[\perp]} = \text{Span}_H \{e_1, \ldots, e_{(n+1)/2}\},$$

and the matrix H_{13} is invertible. Applying the transformations as in Case 10.8.2, we suppose that the matrices A and H have the following form, conformally partitioned:

$$A = \begin{bmatrix} 0 & 0 & D_1 & D_2 & D_3 \\ C_1 & C_2 & D_4 & D_5 & D_6 \\ 0 & 0 & 0_{1 \times 1} & D_7 & D_8 \\ 0 & 0 & 0 & 0 & -C_1^* \\ 0 & 0 & 0 & 0 & -C_2^* \end{bmatrix},$$

$$H = \begin{bmatrix} 0 & 0 & 0 & I & 0 \\ 0 & 0 & 0 & 0 & I \\ 0 & 0 & H_{22} & H_{231} & H_{232} \\ -I & 0 & -H_{231}^* & 0 & 0 \\ 0 & -I & -H_{232}^* & 0 & 0 \end{bmatrix}.$$

Here $[C_1 \ C_2] \in \mathsf{H}^{q \times (n-1)/2}$ is right-invertible ($q := \operatorname{rank} B_{1,1}$). Due to HA being hermitian, we have $-D_1 = (H_{22}D_7)^*$, and

$$HA = \begin{bmatrix} 0 & 0 & 0 & 0 & -C_1^* \\ 0 & 0 & 0 & 0 & -C_2^* \\ 0 & 0 & 0 & H_{22}D_7 & -D_4^* \\ 0 & 0 & (H_{22}D_7)^* & G & \star \\ -C_1 & -C_2 & -D_4 & \star & \star \end{bmatrix},$$

where $G = G^*$ and where we denote by $\star$ block entries of no immediate interest. Let

$$\widetilde{A} = \begin{bmatrix} 0_{1 \times 1} & D_7 \\ 0 & 0 \end{bmatrix}, \qquad \widetilde{H} = \begin{bmatrix} 0_{1 \times 1} & H_{22}D_7 \\ (H_{22}D_7)^* & G \end{bmatrix}.$$

If $\mathcal{M}$ is an $\widetilde{A}$-invariant maximal $\widetilde{H}$-nonnegative subspace, then, in view of Lemma 10.8.1, the subspace $\mathcal{N} + \begin{bmatrix} 0_{(n-1)/2 \times 1} \\ \mathcal{M} \\ 0_{q \times 1} \end{bmatrix}$ is A-invariant and maximal HA-nonnegative, and we are done.

So it remains to prove existence of an $\widetilde{A}$-invariant maximal $\widetilde{H}$-nonnegative subspace. If $D_7 = 0$, then any maximal $\widetilde{H}$-nonnegative subspace will do. If $D_7 \neq 0$, then for some invertible S we have $D_7 S = [1 \ 0 \ \dots \ 0]$. We may replace $\widetilde{A}$ and $\widetilde{H}$ by

$$\widehat{A} := \begin{bmatrix} 1 & 0 \\ 0 & S^{-1} \end{bmatrix} \widetilde{A} \begin{bmatrix} 1 & 0 \\ 0 & S \end{bmatrix} = \begin{bmatrix} 0_{1 \times 1} & 1 & 0 \\ 0 & 0_{1 \times 1} & 0 \\ 0 & 0 & 0 \end{bmatrix}$$

and

$$\widehat{H} := \begin{bmatrix} 1 & 0 \\ 0 & S^* \end{bmatrix} \widetilde{H} \begin{bmatrix} 1 & 0 \\ 0 & S \end{bmatrix} = \begin{bmatrix} 0_{1 \times 1} & H_{22} & 0 \\ H_{22}^* & G_{11} & G_{21}^* \\ 0 & G_{21} & G_{22} \end{bmatrix},$$

where $\begin{bmatrix} G_{11} & G_{21}^* \\ G_{21} & G_{22} \end{bmatrix} = S^*GS$, respectively. Let $\mathcal{M}_0$ be a maximal G_{22}-nonnegative subspace; then

$$\mathcal{M} := \operatorname{Span}_{\mathsf{H}} \{e_1\} + \begin{bmatrix} 0_{2 \times 1} \\ \mathcal{M}_0 \end{bmatrix}$$

is clearly $\widehat{A}$-invariant and $\widehat{H}$-nonnegative. We show that $\mathcal{M}$ is actually maximal $\widehat{H}$-nonnegative.

Observe the equality

$$\begin{bmatrix} 1 & 0 & 0 \\ 0 & 1 & 0 \\ -G_{21}H_{22}^{-1} & 0 & I \end{bmatrix} \widehat{H} \begin{bmatrix} 1 & 0 & -(H_{22}^*)^{-1}G_{21}^* \\ 0 & 1 & 0 \\ 0 & 0 & I \end{bmatrix} = \begin{bmatrix} 0 & H_{22} & 0 \\ H_{22}^* & G_{11} & 0 \\ 0 & 0 & G_{22} \end{bmatrix}.$$

Therefore (using the inertia theorem, and cf. (4.2.5)),

$$\begin{aligned} \operatorname{In}_+ \left(\widehat{H}\right) + \operatorname{In}_0 \left(\widehat{H}\right) &= \operatorname{In}_+ \left(\begin{bmatrix} 0 & H_{22} \\ H_{22}^* & G_{11} \end{bmatrix}\right) + \operatorname{In}_0 \left(\begin{bmatrix} 0 & H_{22} \\ H_{22}^* & G_{11} \end{bmatrix}\right) \\ &\quad + \operatorname{In}_+ (G_{22}) + \operatorname{In}_0 (G_{22}). \end{aligned}$$

But it is easily seen that

$$\text{In}_+ \left(\begin{bmatrix} 0 & H_{22} \\ H_{22}^* & G_{11} \end{bmatrix} \right) = 1, \quad \text{In}_0 \left(\begin{bmatrix} 0 & H_{22} \\ H_{22}^* & G_{11} \end{bmatrix} \right) = 0,$$

so

$$\text{In}_+ (\widehat{H}) + \text{In}_0 (\widehat{H}) = \text{In}_+ (G_{22}) + \text{In}_0 (G_{22}) + 1 = \dim \mathcal{M}$$

(where we have used Theorem 4.2.6(a)), and our claim follows.

10.9   DIFFERENTIAL EQUATIONS II

Consider the system of differential equations with constant coefficients

$$A_\ell x^{(\ell)}(t) + A_{\ell-1} x^{(\ell-1)}(t) + \cdots + A_1 x'(t) + A_0 x(t) = 0, \quad t \in \mathsf{R}, \qquad (10.9.1)$$

where $A_\ell, \ldots, A_1, A_0 \in \mathsf{H}^{n \times n}$, and $x(t)$ is an unknown ℓ times continuously differentiable $\mathsf{H}^{n \times 1}$-valued function of the real independent variable t. We assume in addition that A_k is skewhermitian if k is odd, A_k is hermitian if k is even, and A_ℓ is invertible.

The treatment of (10.9.1) is parallel to that of (10.4.1) in Section 10.4. Letting C and G be defined by (10.4.2) and (10.4.3), respectively, we see that G is skewhermitian and invertible and C is G-Hamiltonian.

The analogue of Theorem 10.4.2 holds true.

Theorem 10.9.1. *The following four statements are equivalent for the system* (10.9.1):

(a) *The system is forward bounded.*

(b) *The system is backward bounded.*

(c) *The system is bounded.*

(d) *All eigenvalues of C have zero real parts, and for every eigenvalue the geometric multiplicity coincides with the algebraic multiplicity.*

The proof is essentially the same as that of Theorem 10.4.2, using the canonical form of Theorem 10.5.1.

Stable boundedness of a system of differential equation (10.9.1) is defined analogously to that of (10.4.1).

Definition 10.9.2. System (10.9.1) is said to be *stably bounded* if there exists $\varepsilon > 0$ such that every system of differential equations

$$A_\ell' x^{(\ell)}(t) + A_{\ell-1}' x^{(\ell-1)}(t) + \cdots + A_1' x'(t) + A_0' x(t) = 0, \quad t \in \mathsf{R}, \qquad (10.9.2)$$

is bounded, provided the coefficients $A_j' \in \mathsf{H}^{n \times n}$, $j = 0, 1, \ldots, \ell$, satisfy the following conditions:

(1) If k is odd, then A_k' is skewhermitian.

(2) If k is even, then A_k' is hermitian.

(3) The inequalities

$$\|A'_j - A_j\| < \varepsilon, \quad \text{for} \quad j = 0, 1, 2, \dots, \ell,$$

hold true.

Theorem 10.9.3. *Assume that every root subspace of the companion matrix C is either GC-positive or GC-negative (note that GC is a hermitian matrix). Then system (10.9.1) is stably bounded.*

The proof is essentially the same as that of Theorem 10.4.4. Note that the hypotheses of Theorem 10.9.3 imply, in particular, that C is invertible and has only eigenvalues with zero real parts (cf. the canonical form of Theorem 10.5.1).

In contrast with Theorem 10.4.4, the condition of Theorem 10.9.3 is not necessary for stable boundedness, as the scalar example $A_1 = i$, $A_0 = 0$ shows. The scalar equation $ix' = 0$ is obviously stably bounded, but $A_1^{-1}A_0 = 0$ is not positive or negative definite. However, the question remains for *invertible* matrices C.

Open Problem 10.9.4. *Suppose the system (10.9.1) is stably bounded, and assume in addition that the companion matrix C is invertible. Does it follow that every root subspace of C is either GC-positive or GC-negative?*

The answer is affirmative for first order systems.

Theorem 10.9.5. *The system of first order differential equations*

$$A_1 x'(t) + A_0 x(t) = 0, \tag{10.9.3}$$

where

$$A_1 = -A_1^* \in \mathsf{H}^{n \times n}, \qquad A_0 = A_0^* \in \mathsf{H}^{n \times n},$$

and A_1, A_0 are invertible, is stably bounded if and only if every root subspace of $A_1^{-1}A_0$ is either A_0-positive or A_0-negative.

Proof. The "if" part is contained in Theorem 10.9.3. To prove the "only if" part, assume that there is a root subspace of $A_1^{-1}A_0$ which is not A_1-positive or A_1-negative, and we will show that the system (10.9.3) is not stably bounded. In view of Theorem 10.9.1 we may assume that all eigenvalues of $A_1^{-1}A_0$ have zero real parts, and for every eigenvalue of $A_1^{-1}A_0$ its geometric multiplicity coincides with the algebraic multiplicity. Using the canonical form of Theorem 10.5.1, we further assume that A_1 and $A_1^{-1}A_0$ have the form

$$A_1 = \oplus_{u=1}^{q} \zeta_u i, \quad A_1^{-1}A_0 = \oplus_{u=1}^{q} \gamma_u,$$

where γ_u's are complex numbers with zero real parts and positive imaginary parts (see remark (3) after Theorem 10.5.1). The condition that there is a root subspace of $A_1^{-1}A_0$ which is not A_1-positive or A_1-negative means that $\gamma_{u_1} = \gamma_{u_2}$ and $\zeta_{u_1} \neq \zeta_{u_2}$ for some indices u_1 and u_2. For notational simplicity suppose that $\gamma_1 = \gamma_2 = ia$, where $a > 0$, and $\zeta_1 = 1 = -\zeta_2$. Let y be a nonzero real number, and let

$$\widetilde{A}_0 := \begin{bmatrix} -a & y \\ y & a \end{bmatrix} \oplus \zeta_3 i \gamma_3 \oplus \cdots \oplus \zeta_q i \gamma_q.$$

Then $A_1^{-1}\widetilde{A}_0$ has eigenvalues $-ia \pm y$ with nonzero real parts, and $\lim_{y \to 0} \widetilde{A}_0 = A_0$. By Theorem 10.9.1, the system $A_1 x'(t) + \widetilde{A}_0 x(t) = 0$ is not bounded; hence, (10.9.3) is not stably bounded. $\square$

10.10   EXERCISES

Ex. 10.10.1. In Exs. 10.10.1, 10.10.2, and 10.10.3, $H \in \mathsf{H}^{n \times n}$ is assumed to be hermitian and invertible. Give examples of $(H,^*)$-hermitian and $(H,^*)$-skewhermitian matrices for which there does not exist an A-invariant H-positive (or H-negative) subspace of dimension $\mathrm{In}_+(H)$ (or $\mathrm{In}_-(H)$).

Ex. 10.10.2. Show that if A is $(H,^*)$-hermitian and has only real eigenvalues with algebraic multiplicity equal to the geometric multiplicity for every eigenvalue, then every A-invariant H-positive (or H-negative) subspace admits extension to an A-invariant H-positive (or H-negative) subspace of dimension $\mathrm{In}_+(H)$ (or $\mathrm{In}_-(H)$).

Ex. 10.10.3. State and prove the statement analogous to Ex. 10.10.2 for $(H,^*)$-skewhermitian matrices.

Ex. 10.10.4. For G_ε defined by (10.4.2), verify that the inertia of G_0 is given by

$$
(\mathrm{In}_+(G_0), \mathrm{In}_-(G_0)) = \begin{cases} \left(\dfrac{n\ell}{2}, \dfrac{n\ell}{2}\right) & \text{if } \ell \text{ is even,} \\[2mm] \left(\dfrac{n(\ell-1)}{2} + \mathrm{In}_-(A_\ell), \dfrac{n(\ell-1)}{2} + \mathrm{In}_+(A_\ell)\right) \\ \text{if } \ell \text{ is odd, } \dfrac{\ell-1}{2} \text{ is odd,} \\[2mm] \left(\dfrac{n(\ell-1)}{2} + \mathrm{In}_+(A_\ell), \dfrac{n(\ell-1)}{2} + \mathrm{In}_-(A_\ell)\right) \\ \text{if } \ell \text{ is odd, } \dfrac{\ell-1}{2} \text{ is even.} \end{cases}
$$

$$(10.10.1)$$

Ex. 10.10.5. Prove, with full detail, Proposition 10.3.2 for the case of skewhermitian G.

Ex. 10.10.6. Provide details for the proof of Theorem 10.3.3, Part (b).

Ex. 10.10.7. Let $H \in \mathsf{H}^{n \times n}$ be hermitian of balanced inertia: n is even and $\mathrm{In}_+ H = \mathrm{In}_- H = n/2$. Show that if A is $(H,^*)$-hermitian and has at most one real eigenvalue (perhaps of high multiplicity), then there exists an A-invariant H-Lagrangian subspace.

Ex. 10.10.8. Let H be as in Ex. 10.10.7. Show that if A is $(H,^*)$-skewhermitian and has at most one similarity class of eigenvalues with zero real parts (perhaps of high multiplicity), then there exists an A-invariant H-Lagrangian subspace.

Ex. 10.10.9. Show that a matrix $A \in \mathsf{H}^{n \times n}$ is $(H,^*)$-hermitian for some invertible hermitian matrix H if and only if A is similar (over H) to a real matrix.

Ex. 10.10.10. Prove that the following statements are equivalent for $A \in \mathsf{H}^{n \times n}$:

(a) A is $(H,^*)$-skewhermitian for some invertible hermitian matrix H;

(b) A is $(G,^*)$-Hamiltonian for some invertible skewhermitian matrix G;

(c) for every pair $(\lambda, -\lambda)$ of eigenvalues of A with nonzero real parts, the partial multiplicities corresponding to the eigenvalues similar to λ coincide with the partial multiplicities corresponding to the eigenvalues similar to $-\lambda$;

(d) A is similar to $-A$.

Hint: Use the Jordan form of A and the canonical forms of Theorems 10.1.3 and 10.5.1.

Ex. 10.10.11. Show that every matrix $A \in \mathsf{H}^{n \times n}$ is $(H,^*)$-skew-Hamiltonian for some skewhermitian invertible matrix H.

Ex. 10.10.12. Consider three matrix pairs:

$$\text{(a)} \qquad A_1 = \begin{bmatrix} \lambda_1 & \alpha \\ 0 & \lambda_2 \end{bmatrix}, \quad H_1 = \begin{bmatrix} 0 & 1 \\ 1 & 0 \end{bmatrix}, \quad \lambda_1, \lambda_2, \alpha \in \mathsf{H};$$

$$\text{(b)} \qquad A_2 = \begin{bmatrix} \lambda_1 & 0 \\ 0 & \lambda_2 \end{bmatrix}, \quad H_2 = \begin{bmatrix} 1 & 0 \\ 0 & -1 \end{bmatrix}, \quad \lambda_1, \lambda_2 \in \mathsf{H};$$

$$\text{(c)} \qquad A_3 = \begin{bmatrix} \lambda_1 & \alpha_1 & \alpha_2 \\ 0 & \lambda_2 & \alpha_3 \\ 0 & 0 & \lambda_3 \end{bmatrix}, \quad H_3 = \begin{bmatrix} 0 & 0 & 1 \\ 0 & 1 & 0 \\ 1 & 0 & 0 \end{bmatrix},$$

where $\lambda_j, \alpha_j \in \mathsf{H}$ for $j = 1, 2, 3$. When is A_j $(H_j,^*)$-hermitian for $j = 1, 2, 3$? If it is, find the canonical form of the pair (A_j, H_j).

Ex. 10.10.13. Find the sign characteristic for the pairs in Ex. 10.10.12, assuming that A_j is $(H_j,^*)$-hermitian.

Ex. 10.10.14. Determine which matrices A_j $(j = 1, 2, 3)$ given in Ex. 10.10.12 are $(H,^*)$-skewhermitian, and in case they are, find the canonical form of the pair (A_j, H_j).

Ex. 10.10.15. Describe the possible canonical forms of Theorem 10.1.1 if it is known that the matrix H:

(1) is positive definite;

(2) has only one negative eigenvalue (counted with multiplicities);

(3) has only two negative eigenvalues (counted with multiplicities).

Ex. 10.10.16. Repeat Ex. 10.10.15, but now with regard to the canonical forms for $(H,^*)$-skewhermitian matrices given in Theorem 10.1.3.

Ex. 10.10.17. Describe the possible canonical forms of $(H,^*)$-Hamiltonian matrices (Theorem 10.5.1) in each of the following cases:

(1) The hermitian matrix HA is positive semidefinite.

(2) HA is positive definite.

(3) HA has only one negative eigenvalue (counted with multiplicities).

Ex. 10.10.18. Show that under the hypotheses of Theorem 10.9.3, every system (10.9.2) is stably bounded, provided conditions (1), (2), (3) in Section 10.9 are satisfied and ε is sufficiently small.

Ex. 10.10.19. Let $H = -H^* \in \mathsf{H}^{n \times n}$ be invertible, and let A, A' be two similar $(H,^*)$-skew-Hamiltonian matrices. Show that there exists an $(H,^*)$-symplectic matrix S such that $S^{-1} A' S = A$.

Ex. 10.10.20. Let H be as in Ex. 10.10.19.

(a) Show that the result analogous to Ex. 10.10.19 does not generally hold for $(H,^*)$-Hamiltonian matrices.

(b) Find a condition on the Jordan structure of an $(H,^*)$-Hamiltonian matrix A that would ensure that for every $(H,^*)$-Hamiltonian matrix A' which is similar to A there exists an $(H,^*)$-symplectic matrix S such that $S^{-1}A'S = A$.

Ex. 10.10.21. Let $H \in \mathsf{H}^{n \times n}$ be hermitian and invertible. Is the result analogous to Ex. 10.10.19 valid for $(H,^*)$-hermitian matrices? $(H,^*)$-skewhermitian matrices? If not, find conditions on the Jordan structure of an $(H,^*)$-hermitian, resp. $(H,^*)$-skewhermitian, matrix A that would guarantee that for every $(H,^*)$-hermitian, resp. $(H,^*)$-skewhermitian, matrix A' which is similar to A, the similarity matrix between A and A' can be chosen to be $(H,^*)$-unitary.

Ex. 10.10.22. The system of differential equations (10.4.1) is said to be *polynomially bounded* if there exist positive integer m and $M_1, M_2 > 0$ such that $\|x(t)\| \le M_1|t|^m + M_2$ for every solution $x(t)$ of (10.4.1). State and prove a criterion for polynomial boundedness in terms of the companion matrix.

Ex. 10.10.23. Repeat Ex. 10.10.22 for the system (10.9.1).

10.11 NOTES

Selfadjoint matrices and operators with respect to real or complex indefinite inner product have been extensively studied, both in the finite dimensional and the infinite dimensional settings, and the literature here is voluminous. Among numerous books and papers on the subject, we mention here only Gohberg et al. [52, 53], which are dedicated to the finite dimensional aspects of the theory.

The main results of this chapter (Theorems 10.1.1, 10.1.3, 10.5.1, 10.5.2) are well known in the literature; see, e.g., Djoković et al. [34]. For Theorem 10.1.1, see also Karow [78] and Sergeichuk [142]. For comparison with the formulation in Djoković et al., note that the case we are interested in corresponds to the choice $D = H$, $\epsilon = 1$, $\rho = -1$, with A replaced by A^*, in Djocović et al. Also note that Djoković et al. work with lower triangular rather than upper triangular Jordan blocks.

Canonical forms for real and complex Hamiltonian matrices are found in many sources: Burgoyne and Cushman [23], Faßbender et al. [41], Lin et al. [101], Mehrmann and Xu [113], and Mehl et al. [109].

The presentation of Theorems 10.5.1 and 10.5.2 here is based on Rodman [135].

In the context of real and complex matrices, invariant Lagrangian subspaces have been the subject of intensive studies. We refer the reader to Freiling et al. [46], Mehl et al. [107], and Ran and Rodman [125, 126] for more information and in-depth treatment of real and complex invariant Lagrangian subspaces.

For real matrices, Theorems 10.7.2 and 10.7.5 were proved in Rodman [129, 130].

In connection with Theorem 10.7.2, note that invariant neutral subspaces (under the additional assumption that H is invertible) have been studied in Lancaster and Rodman [89].

Formulas similar to (10.8.1) were given in Theorem 2.1 of Alpay and Dym [3].

Ex. 10.10.12 and 10.10.13 are taken from Gohberg et al. [53], where they are stated in the context of complex matrices.

The exposition in Section 10.8 follows, for the most part, Rodman [130], where it was carried out for real matrices.

Chapter Eleven

Matrix pencils with symmetries: Nonstandard involution

In this chapter the subject matter involves quaternion matrix pencils or, equivalently, pairs of quaternion matrices, with symmetries with respect to a fixed nonstandard involution ϕ. Here, we provide canonical forms for ϕ-hermitian pencils, i.e., pencils of the form $A + tB$, where A and B are both ϕ-hermitian. We also provide canonical forms for ϕ-skewhermitian pencils. The canonical forms in question are with respect to either strict equivalence of pencils or to simultaneous ϕ-congruence of matrices. Applications are made to joint ϕ-numerical ranges of two ϕ-skewhermitian matrices and to the corresponding joint ϕ-numerical cones.

We fix a nonstandard involution ϕ throughout this chapter and a quaternion $\beta(\phi)$ such that $\phi(\beta(\phi)) = -\beta(\phi)$ and $|\beta(\phi)| = 1$.

11.1 CANONICAL FORMS FOR ϕ-HERMITIAN PENCILS

Consider a matrix pencil $A + tB$, $A, B \in \mathsf{H}^{n \times n}$.

Definition 11.1.1. The matrix pencil $A + tB$ is said to be ϕ-hermitian if $A_\phi = A$ and $B_\phi = B$.

Definition 11.1.2. Two matrix pencils $A + tB$ and $A' + tB'$ are said to be ϕ-congruent if

$$A' + tB' = S_\phi(A + tB)S$$

for some invertible $S \in \mathsf{H}^{n \times n}$.

Clearly, the ϕ-congruence of matrix pencils is an equivalence relation, and ϕ-congruent matrix pencils are strictly equivalent. However, in general, strictly equivalent matrix pencils need not be ϕ-congruent. Moreover, any pencil that is ϕ-congruent to a ϕ-hermitian matrix pencil is itself ϕ-hermitian.

We now present the canonical forms for ϕ-hermitian matrix pencils under strict equivalence and ϕ-congruence. It turns out that these forms are the same.

Theorem 11.1.3. (a) *Every ϕ-hermitian matrix pencil $A + tB$ is strictly equivalent to a ϕ-hermitian matrix pencil of the form*

$$0_u \ \oplus \left(t \begin{bmatrix} 0 & 0 & F_{\varepsilon_1} \\ 0 & 0 & 0 \\ F_{\varepsilon_1} & 0 & 0 \end{bmatrix} + G_{2\varepsilon_1+1} \right) \oplus \cdots \oplus \left(t \begin{bmatrix} 0 & 0 & F_{\varepsilon_p} \\ 0 & 0 & 0 \\ F_{\varepsilon_p} & 0 & 0 \end{bmatrix} + G_{2\varepsilon_p+1} \right)$$

$$\oplus (F_{k_1} + tG_{k_1}) \oplus \cdots \oplus (F_{k_r} + tG_{k_r})$$

$$\oplus ((t + \alpha_1) F_{\ell_1} + G_{\ell_1}) \oplus \cdots \oplus ((t + \alpha_q) F_{\ell_q} + G_{\ell_q}). \tag{11.1.1}$$

Here, $\varepsilon_1 \leq \cdots \leq \varepsilon_p$ and $k_1 \leq \cdots \leq k_r$ are positive integers, and $\alpha_j \in \mathrm{Inv}\,(\phi)$.

The form (11.1.1) is uniquely determined by A and B up to an arbitrary permutation of the diagonal blocks in the part

$$((t + \alpha_1) F_{\ell_1} + G_{\ell_1}) \oplus \cdots \oplus ((t + \alpha_q) F_{\ell_q} + G_{\ell_q}) \tag{11.1.2}$$

and up to replacement of α_j in each block $(t + \alpha_j) F_{\ell_j} + G_{\ell_j}$ with a quaternion $\beta_j \in \mathrm{Inv}\,(\phi)$ such that

$$\mathfrak{R}(\alpha_j) = \mathfrak{R}(\beta_j) \quad \text{and} \quad |\mathfrak{V}(\alpha_j)| = |\mathfrak{V}(\beta_j)|. \tag{11.1.3}$$

(b) *Every ϕ-hermitian matrix pencil $A + tB$ is ϕ-congruent to a ϕ-hermitian matrix pencil of the form* (11.1.1), *with the same uniqueness conditions as in the part* (a).

(c) *ϕ-hermitian matrix pencils are ϕ-congruent if and only if they are strictly equivalent.*

Note that in (11.1.1) one can replace each G_v by $\widetilde{G}_v$ using the formula $\widetilde{G}_v = F_v G_v F_v$.

Proof. We prove Theorem 11.1.3 by reduction to Theorem 9.1.1.

Proof of (a). Let $A + tB$ be a ϕ-hermitian matrix pencil and let $\widehat{A} = -\beta(\phi)A$, $\widehat{B} = -\beta(\phi)B$. By Proposition 4.1.7 the pencil $\widehat{A} + t\widehat{B}$ is skewhermitian. By Theorem 9.1.1, there exist invertible matrices S and T such that

$$S^*(\widehat{A} + t\widehat{B})T = W,$$

where W is the canonical from (9.1.1) and where we replace i with $-\beta(\phi)$ (see Remark 9.1.2). Then $S^*(A + tB)T = \beta(\phi)W$, again by Proposition 4.1.7, and it is easily seen that $\beta(\phi)W$ is strictly equivalent to (11.1.1) (with G_v replaced by $\widetilde{G}_v$). The uniqueness statement in (a) follows from that in Theorem 9.1.1(i).

Part (b) obviously follows from (a) and (c), so it remains to prove (c). Let $A + tB$ and $A_1 + tB_1$ be two strictly equivalent ϕ-hermitian $n \times n$ matrix pencils, so

$$S_\phi(A + tB)T = A_1 + tB_1, \quad S, T \in \mathsf{H}^{n \times n} \quad \text{are invertible}.$$

Letting

$$\widehat{A} = -\beta(\phi)A, \quad \widehat{B} = -\beta(\phi)B, \quad \widehat{A}_1 = -\beta(\phi)A_1, \quad \widehat{B}_1 = -\beta(\phi)B_1,$$

by Proposition 4.1.7 we have

$$S^*(\widehat{A} + t\widehat{B})T = \widehat{A}_1 + t\widehat{B}_1.$$

Now by Theorem 9.1.1 the skewhermitian matrix pencils $\widehat{A} + t\widehat{B}$ and $\widehat{A}_1 + t\widehat{B}_1$ are congruent, so

$$V^*(\widehat{A} + t\widehat{B})V = \widehat{A}_1 + t\widehat{B}_1, \quad V \in \mathsf{H}^{n \times n} \quad \text{is invertible}.$$

Applying Proposition 4.1.7 once more, we get

$$V_\phi(A + tB)V = A_1 + tB_1,$$

and (c) is proved. $\square$

11.2 CANONICAL FORMS FOR ϕ-SKEWHERMITIAN PENCILS

Definition 11.2.1. A matrix pencil $A+tB$ is called ϕ-*skewhermitian* if $A_\phi = -A$ and $B_\phi = -B$.

In this section we consider ϕ-skewhermitian matrix pencils and their canonical forms.

We start with the canonical form.

Theorem 11.2.2. *Fix $\beta(\phi) \in \mathsf{H}$ such that $\phi(\beta(\phi)) = -\beta(\phi)$, $|\mathfrak{V}(\beta(\phi))| = 1$.*

(a) *Every ϕ-skewhermitian matrix pencil $A + tB$ is strictly equivalent to a ϕ-skew- hermitian matrix pencil of the form*

$$
0_u \;\oplus\; \left(t\begin{bmatrix} 0 & 0 & F_{\varepsilon_1} \\ 0 & 0 & 0 \\ -F_{\varepsilon_1} & 0 & 0 \end{bmatrix} + \begin{bmatrix} 0 & F_{\varepsilon_1} & 0 \\ -F_{\varepsilon_1} & 0 & 0 \\ 0 & 0 & 0 \end{bmatrix} \right)
$$
$$
\oplus\; \cdots \;\oplus\; \left(t\begin{bmatrix} 0 & 0 & F_{\varepsilon_p} \\ 0 & 0 & 0 \\ -F_{\varepsilon_p} & 0 & 0 \end{bmatrix} + \begin{bmatrix} 0 & F_{\varepsilon_p} & 0 \\ -F_{\varepsilon_p} & 0 & 0 \\ 0 & 0 & 0 \end{bmatrix} \right)
$$
$$
\oplus\; (\beta(\phi)F_{k_1} + t\beta(\phi)G_{k_1}) \oplus \cdots \oplus (\beta(\phi)F_{k_r} + t\beta(\phi)G_{k_r})
$$
$$
\oplus\; ((t+\gamma_1)\beta(\phi)F_{m_1} + \beta(\phi)G_{m_1}) \oplus \cdots \oplus \left((t+\gamma_p)\beta(\phi)F_{m_p} + \beta(\phi)G_{m_p} \right)
$$
$$
\oplus\; \left((t+\alpha_1)\begin{bmatrix} 0 & F_{\ell_1} \\ -F_{\ell_1} & 0 \end{bmatrix} + \begin{bmatrix} 0 & G_{\ell_1} \\ -G_{\ell_1} & 0 \end{bmatrix} \right)
$$
$$
\oplus\; \cdots \;\oplus\; \left((t+\alpha_q)\begin{bmatrix} 0 & F_{\ell_q} \\ -F_{\ell_q} & 0 \end{bmatrix} + \begin{bmatrix} 0 & G_{\ell_q} \\ -G_{\ell_q} & 0 \end{bmatrix} \right). \tag{11.2.1}
$$

Here, $\varepsilon_1 \leq \cdots \leq \varepsilon_p$ and $k_1 \leq \cdots \leq k_r$ are positive integers, $\alpha_1, \ldots, \alpha_q \in \mathrm{Inv}\,(\phi)\setminus\mathsf{R}$, and $\gamma_1, \ldots, \gamma_p$ are real.

The form (11.2.1) is uniquely determined by A and B up to an arbitrary permutation of the diagonal blocks in each of the parts

$$
\oplus_{j=1}^p \left((t+\gamma_j)\beta(\phi)F_{m_j} + \beta(\phi)G_{m_j} \right)
$$

and

$$
\oplus_{j=1}^q \left((t+\alpha_j)\begin{bmatrix} 0 & F_{\ell_j} \\ -F_{\ell_j} & 0 \end{bmatrix} + \begin{bmatrix} 0 & G_{\ell_j} \\ -G_{\ell_j} & 0 \end{bmatrix} \right) \tag{11.2.2}
$$

and up to replacement of α_j in each block

$$
(t+\alpha_j)\begin{bmatrix} 0 & F_{\ell_j} \\ -F_{\ell_j} & 0 \end{bmatrix} + \begin{bmatrix} 0 & G_{\ell_j} \\ -G_{\ell_j} & 0 \end{bmatrix} \tag{11.2.3}
$$

with a quaternion $\beta_j \in \mathrm{Inv}\,(\phi)$, which is similar to α_j.

(b) *Every ϕ-skewhermitian matrix pencil $A + tB$ is ϕ-congruent to a matrix* *pencil (which is also ϕ-skewhermitian) of the form*

$$0_u \;\oplus\; \left(t \begin{bmatrix} 0 & 0 & F_{\varepsilon_1} \\ 0 & 0 & 0 \\ -F_{\varepsilon_1} & 0 & 0 \end{bmatrix} + \begin{bmatrix} 0 & F_{\varepsilon_1} & 0 \\ -F_{\varepsilon_1} & 0 & 0 \\ 0 & 0 & 0 \end{bmatrix} \right)$$

$$\oplus \;\cdots\; \oplus \left(t \begin{bmatrix} 0 & 0 & F_{\varepsilon_p} \\ 0 & 0 & 0 \\ -F_{\varepsilon_p} & 0 & 0 \end{bmatrix} + \begin{bmatrix} 0 & F_{\varepsilon_p} & 0 \\ -F_{\varepsilon_p} & 0 & 0 \\ 0 & 0 & 0 \end{bmatrix} \right)$$

$$\oplus \quad \delta_1 \left(\beta(\phi) F_{k_1} + t\beta(\phi) G_{k_1} \right) \oplus \cdots \oplus \delta_r \left(\beta(\phi) F_{k_r} + t\beta(\phi) G_{k_r} \right)$$

$$\oplus \quad \eta_1 \left((t+\gamma_1)\beta(\phi) F_{m_1} + \beta(\phi) G_{m_1} \right)$$

$$\oplus \quad \cdots \oplus \eta_p \left((t+\gamma_p)\beta(\phi) F_{m_p} + \beta(\phi) G_{m_p} \right)$$

$$\oplus \quad \left((t+\alpha_1) \begin{bmatrix} 0 & F_{\ell_1} \\ -F_{\ell_1} & 0 \end{bmatrix} + \begin{bmatrix} 0 & G_{\ell_1} \\ -G_{\ell_1} & 0 \end{bmatrix} \right)$$

$$\oplus \quad \cdots \oplus \left((t+\alpha_q) \begin{bmatrix} 0 & F_{\ell_q} \\ -F_{\ell_q} & 0 \end{bmatrix} + \begin{bmatrix} 0 & G_{\ell_q} \\ -G_{\ell_q} & 0 \end{bmatrix} \right). \qquad (11.2.4)$$

Here ε_i, k_j, α_m and γ_s are as in Part (a), and $\delta_1,\ldots,\delta_r$ and $\eta_1,\ldots,\eta_p$ are signs ± 1.

   The form (11.2.4) under ϕ-congruence is uniquely determined by A and B up to an arbitrary permutation of the diagonal blocks in each of the parts

$$\oplus_{j=1}^p \left(\eta_j \left((t+\gamma_j)\beta(\phi) F_{m_j} + \beta(\phi) G_{m_j} \right) \right)$$

and (11.2.2) and up to replacement of α_j in each block (11.2.3) with a similar quaternion $\beta_j \in \mathrm{Inv}\,(\phi)$.

   In connection with the form (11.2.4), note the following ϕ-congruence:

$$\begin{bmatrix} \beta(\phi) I_\varepsilon & 0 \\ 0 & I_{\varepsilon+1} \end{bmatrix} \left(t \begin{bmatrix} 0 & 0 & F_\varepsilon \\ 0 & 0 & 0 \\ -F_\varepsilon & 0 & 0 \end{bmatrix} + \begin{bmatrix} 0 & F_\varepsilon & 0 \\ -F_\varepsilon & 0 & 0 \\ 0 & 0 & 0 \end{bmatrix} \right)$$

$$\cdot \begin{bmatrix} -\beta(\phi) I_\varepsilon & 0 \\ 0 & I_{\varepsilon+1} \end{bmatrix} = \beta(\phi) \left(t \begin{bmatrix} 0 & 0 & F_\varepsilon \\ 0 & 0 & 0 \\ F_\varepsilon & 0 & 0 \end{bmatrix} + \begin{bmatrix} 0 & F_\varepsilon & 0 \\ F_\varepsilon & 0 & 0 \\ 0 & 0 & 0 \end{bmatrix} \right). \qquad (11.2.5)$$

   The signs δ_i and η_j of (11.2.4) form the *sign characteristic* of the ϕ-skewhermitian pencil $A + tB$. In view of the equalities

$$(\beta(\phi) F_{k_i} + t\beta(\phi) G_{k_i})(-\beta(\phi) F_{k_i}) = I + t J_{k_i}(0) \qquad (11.2.6)$$

and

$$(t\beta(\phi) F_{m_j} + \gamma_j \beta(\phi) F_{m_j} + \beta(\phi) G_{m_j})(-\beta(\phi) F_{m_j}) \;=\; tI + J_{m_j}(\gamma_j),$$
$$\gamma_j \in \mathsf{R}, \qquad (11.2.7)$$

the sign characteristic assigns a sign ± 1 to every index corresponding to a real eigenvalue of $A + tB$, as well as to every index at infinity. The signs are uniquely determined by $A+tB$, up to permutations of signs that correspond to equal indices of the same real eigenvalue and permutations of signs that correspond to the equal indices at infinity.

The proof of this theorem will be given in the next section.

In contrast with ϕ-hermitian matrix pencils, strictly equivalent ϕ-skewhermitian matrix pencils need not be ϕ-congruent.

Theorem 11.2.3. *Let $A + tB$ be a ϕ-skewhermitian $n \times n$ quaternion matrix pencil. Then the strict equivalence class of $A + tB$ consists of not more than 2^n ϕ-congruence classes. The upper bound 2^n is attained for*

$$A + tB = \mathrm{diag}\left((t + \gamma_1)\beta(\phi), (t + \gamma_2)\beta(\phi), \ldots, (t + \gamma_n)\beta(\phi)\right),$$

where $\gamma_1, \ldots, \gamma_n$ are distinct real numbers.

Proof. Assume that $A + tB$ is strictly equivalent to the form (11.2.1), and assume that in the form (11.2.1) there are exactly s distinct blocks of the types $\beta(\phi)F_{k_i} + t\beta(\phi)G_{k_i}$ and $(t + \gamma_i)\beta(\phi)F_{m_i} + \beta(\phi)G_{m_i}$. Denote these s distinct blocks by $K_1, \ldots, K_s$, and further assume that the block K_j appears in (11.2.1) exactly v_j times, for $j = 1, 2, \ldots, s$. Define the integer

$$q := \prod_{j=1}^{s}(v_j + 1).$$

Theorem 11.2.2 shows that the strict equivalence class of $A + tB$ contains exactly q ϕ-congruence classes. So, the maximal number of ϕ-congruence classes in a strict equivalence class of a ϕ-skewhermitian $n \times n$ quaternion matrix pencil is equal to

$$\max\{\prod_{j=1}^{s}(v_j + 1)\},$$

where the maximum is taken over all tuples of positive integers $(v_1, \ldots, v_s)$ such that $v_1 + \cdots + v_s \leq n$. It is easy to see that the maximum is achieved for $s = n$ and $v_1 = \cdots = v_n = 1$ (use the elementary inequality $v + 1 < (v - u + 1)(u + 1)$ for every pair of positive integers u, v such that $1 \leq u \leq v - 1$). $\qquad\square$

A criterion for ϕ-skewhermitian matrix pencils with the property that their strict equivalence to other ϕ-skewhermitian matrix pencils implies ϕ-congruence is given next.

Theorem 11.2.4. *Let $A + tB$ be a ϕ-skewhermitian matrix pencil. Assume that the following property holds:*

$$\mathrm{rank}\,(A + tB) = \mathrm{rank}\,A = \mathrm{rank}\,B \quad \text{for all} \quad t \in \mathsf{R}. \qquad (11.2.8)$$

Then a ϕ-skewhermitian matrix pencil $A' + tB'$ is ϕ-congruent to $A + tB$ if and only if $A' + tB'$ is strictly equivalent to $A + tB$.

Conversely, if a ϕ-skewhermitian matrix pencil $A + tB$ has the property that every ϕ-skewhermitian matrix pencil that is strictly equivalent to $A + tB$ is actually ϕ-congruent to $A + tB$, then (11.2.8) holds.

The proof of this theorem follows easily from Theorem 11.2.2 by inspection of (11.2.4). Indeed, property (11.2.8) is equivalent to the absence of blocks

$$\delta_j(\beta(\phi)F_{k_j} + t\beta(\phi)G_{k_j}) \quad \text{and} \quad \eta_j((t + \gamma_j)\beta(\phi)F_{m_j} + \beta(\phi)G_{m_j})$$

in (11.2.4).

11.3   PROOF OF THEOREM 11.2.2

Part (a) follows from Theorem 8.1.2 by using Proposition 4.1.7 (cf. the proof of Theorem 11.1.3). For that purpose, we need to verify, apart from evident verifications, that the matrix pencils

$$\beta(\phi)W_1(t) := \beta(\phi) \left[\begin{array}{cc} 0 & (t+\gamma)F_m + G_m \\ (t+\gamma^*)F_m + G_m & 0 \end{array} \right], \qquad \gamma \in \mathsf{H} \setminus \mathsf{R},$$

and

$$W_2(t) := \left[\begin{array}{cc} 0 & (t+\alpha)F_m + G_m \\ (t+\alpha)(-F_m) + (-G_m) & 0 \end{array} \right], \qquad \alpha \in \mathrm{Inv}\,(\phi) \setminus \mathsf{R},$$

are strictly equivalent. This is easy:

$$\left[\begin{array}{cc} r_2^{-1} & 0 \\ 0 & -r_1^{-1} \end{array} \right] W_1(t) \left[\begin{array}{cc} r_1 & 0 \\ 0 & r_2 \end{array} \right] = W_2(t),$$

where $r_1, r_2 \in \mathsf{H} \setminus \{0\}$ are such that

$$r_1^{-1}\gamma^* r_1 = r_2^{-1}\gamma r_2 = \alpha.$$

Note that for every $\gamma \in \mathsf{H} \setminus \mathsf{R}$, one can choose r_1 so that $r_1^{-1}\gamma^* r_1 \in \mathrm{Inv}\,(\phi) \setminus \mathsf{R}$.

Consider now the existence part of (b). We use again Proposition 4.1.7 and take advantage of Theorem 8.1.2. Thus, we need only to verify that each constituent primitive block in (8.1.6), when multiplied by $\beta(\phi)$ on the left, is ϕ-congruent to a constituent primitive block in (11.2.4). For this purpose, we may assume that all signs in (8.1.6) are $+1$'s. The required verification is carried by the formulas (11.2.5) and

$$\left[\begin{array}{cc} 1 & 0 \\ 0 & \beta(\phi) \end{array} \right] \cdot \beta(\phi) \left[\begin{array}{cc} 0 & (t+\gamma)F_m + G_m \\ (t+\gamma^*)F_m + G_m & 0 \end{array} \right]$$

$$\cdot \left[\begin{array}{cc} 1 & 0 \\ 0 & -\beta(\phi) \end{array} \right] = \left[\begin{array}{cc} 0 & (t+\alpha)F_m + G_m \\ -(t+\alpha)F_m - G_m & 0 \end{array} \right],$$

where γ is chosen in $\mathrm{Inv}\,(\phi) \setminus \mathsf{R}$, and $\alpha = \gamma^*$.

A proof of uniqueness in Part (b) can be also done using Proposition 4.1.7.

In what follows, we offer a direct and independent proof of the uniqueness part of Theorem 11.2.2(b). The proof is modeled after the proof of uniqueness of the canonical form for pairs of complex Hermitian matrices (see, e.g., Section 8 in Lancaster and Rodman [92, Section 8], or Thompson [151]).

We introduce the following terminology.

Definition 11.3.1. Let $A_1 + tB_1$ be a ϕ-skewhermitian pencil of the form (11.2.4). Then we say that

$$\delta_1 \left(\beta(\phi)F_{k_1} + t\beta(\phi)G_{k_1} \right) \oplus \cdots \oplus \delta_r \left(\beta(\phi)F_{k_r} + t\beta(\phi)G_{k_r} \right)$$

is the $\beta(\phi)$-part of $A_1 + tB_1$, and for every real γ,

$$\oplus_{\{j:\gamma_j=\gamma\}} \left(\eta_j \left((t+\gamma_j)\beta(\phi)F_{m_j} + \beta(\phi)G_{m_j} \right) \right)$$

is the γ-part of $A_1 + tB_1$.

We will write $A_1 + tB_1$ in the following form (perhaps, after permutation of primitive blocks):

$$
\begin{aligned}
A_1 + tB_1 \;=\; & 0_u \oplus (A_{1,\flat} + tB_{1,\flat}) \\
& \oplus (A_{1,\sharp} + tB_{1,\sharp}) \oplus (A_{1,\beta(\phi)} + tB_{1,\beta(\phi)}) \\
& \oplus \oplus_{j=1}^{m} (A_{1,\gamma_j} + tB_{1,\gamma_j}),
\end{aligned}
\tag{11.3.1}
$$

where $A_{1,\beta(\phi)} + tB_{1,\beta(\phi)}$ is the $\beta(\phi)$-part of $A_1 + tB_1$, the block $A_{1,\gamma_j} + tB_{1,\gamma_j}$ is the γ_j-part of $A_1 + tB_1$ (the numbers $\gamma_1, \ldots, \gamma_m$ are real, distinct, and arranged in increasing order: $\gamma_1 < \cdots < \gamma_m$), the part $A_{1,\flat} + tB_{1,\flat}$ consists of a direct sum of blocks

$$
t \begin{bmatrix} 0 & 0 & F_{\varepsilon_i} \\ 0 & 0 & 0 \\ -F_{\varepsilon_i} & 0 & 0 \end{bmatrix} + \begin{bmatrix} 0 & F_{\varepsilon_i} & 0 \\ -F_{\varepsilon_i} & 0 & 0 \\ 0 & 0 & 0 \end{bmatrix}, \quad i = 1, 2, \ldots, p,
$$

and the part $A_{1,\sharp} + tB_{1,\sharp}$ consists of the blocks

$$
(t + \alpha_i) \begin{bmatrix} 0 & F_{\ell_i} \\ -F_{\ell_i} & 0 \end{bmatrix} + \begin{bmatrix} 0 & G_{\ell_i} \\ -G_{\ell_i} & 0 \end{bmatrix}, \quad i = 1, 2, \ldots, q,
$$

where $\alpha_i \in \mathrm{Inv}\,(\phi) \setminus \mathsf{R}$.

Definition 11.3.2. We say that the form (11.3.1), with the indicated properties, is an *arranged form*.

Lemma 11.3.3. *Let two ϕ-skewhermitian matrix pencils $A_1 + tB_1$ and $A_2 + tB_2$ be given in the arranged forms (11.3.1) and*

$$
\begin{aligned}
A_2 + tB_2 \;=\; & 0_{u'} \oplus (A_{2,\flat} + tB_{2,\flat}) \oplus (A_{2,\sharp} + tB_{2,\sharp}) \oplus (A_{2,\beta(\phi)} + tB_{2,\beta(\phi)}) \\
& \oplus \oplus_{j=1}^{m'} (A_{2,\gamma_j'} + tB_{2,\gamma_j'}).
\end{aligned}
$$

Assume that $A_1 + tB_1$ and $A_2 + tB_2$ are ϕ-congruent. Then

$$
u = u', \quad m = m', \quad \gamma_j = \gamma_j', \quad \text{for } j = 1, 2, \ldots, m,
\tag{11.3.2}
$$

the ϕ-skewhermitian matrix pencils $A_{1,\flat} + tB_{1,\flat}$ and $A_{2,\flat} + tB_{2,\flat}$ are ϕ-congruent, the ϕ-skewhermitian matrix pencils $A_{1,\sharp} + tB_{1,\sharp}$ and $A_{2,\sharp} + tB_{2,\sharp}$ are ϕ-congruent, the $\beta(\phi)$-parts $A_{1,\beta(\phi)} + tB_{1,\beta(\phi)}$ and $A_{2,\beta(\phi)} + tB_{2,\beta(\phi)}$ are ϕ-congruent, and for each γ_j, the γ_j-parts $A_{1,\gamma_j} + tB_{1,\gamma_j}$ and $A_{2,\gamma_j} + tB_{2,\gamma_j}$ are ϕ-congruent.

Proof. Equalities (11.3.2) follow from the uniqueness of the Kronecker form of $A_1 + tB_1$ (which is the same as the Kronecker form of $A_2 + tB_2$). For the same reason, $A_{1,\flat} + tB_{1,\flat}$ is strictly equivalent to $A_{2,\flat} + tB_{2,\flat}$, and, therefore, permuting if necessary the blocks, we may (and will) assume that

$$
A_{1,\flat} + tB_{1,\flat} = A_{2,\flat} + tB_{2,\flat}.
$$

Analogously, we may assume that

$$
A_{1,\sharp} + tB_{1,\sharp} = A_{2,\sharp} + tB_{2,\sharp}.
$$

Also, the $\beta(\phi)$-parts of $A_1 + tB_1$ and $A_2 + tB_2$ are strictly equivalent, as well as the γ_j-parts of $A_1 + tB_1$ and $A_2 + tB_2$, for every fixed γ_j.

For uniformity of notation, we let

$$M_0 + tN_0 := 0_u,$$

$$M_i + tN_i \quad := \quad t \begin{bmatrix} 0 & 0 & F_{\varepsilon_i} \\ 0 & 0 & 0 \\ -F_{\varepsilon_i} & 0 & 0 \end{bmatrix} + \begin{bmatrix} 0 & F_{\varepsilon_i} & 0 \\ -F_{\varepsilon_i} & 0 & 0 \\ 0 & 0 & 0 \end{bmatrix},$$

$$i = 1, 2, \ldots, p, \tag{11.3.3}$$

$$M_{p+i} + tN_{p+i} \quad := \quad \oplus_{s=1}^{w_i} \left((t + \alpha_i) \begin{bmatrix} 0 & F_{\ell_{i,s}} \\ -F_{\ell_{i,s}} & 0 \end{bmatrix} + \begin{bmatrix} 0 & G_{\ell_{i,s}} \\ -G_{\ell_{i,s}} & 0 \end{bmatrix} \right),$$

$$i = 1, 2, \ldots, q, \tag{11.3.4}$$

where $\alpha_1, \ldots, \alpha_q \in \mathrm{Inv}\,(\phi) \setminus \mathsf{R}$ are mutually nonsimilar;

$$\begin{aligned}
M_{p+q+1} + tN_{p+q+1} \quad &:= \quad A_{1,\beta(\phi)} + tB_{1,\beta(\phi)}; \\
M'_{p+q+1} + tN'_{p+q+1} \quad &:= \quad A_{2,\beta(\phi)} + tB_{2,\beta(\phi)}; \\
M_{p+q+1+j} + tN_{p+q+1+j} \quad &:= \quad A_{1,\gamma_j} + tB_{1,\gamma_j}, \quad j = 1, 2, \ldots, m; \\
M'_{p+q+1+j} + tN'_{p+q+1+j} \quad &:= \quad A_{2,\gamma_j} + tB_{2,\gamma_j}, \quad j = 1, 2, \ldots, m.
\end{aligned}$$

Let $n_j \times n_j$ be the size of $M_j + tN_j$, for $j = 0, 1, \ldots, p+q+1+m$ (note that $n_j \times n_j$ is also the size of $M'_j + tN'_j$, for $j = p+q+1, p+q+2, \ldots, p+q+1+m$). Write

$$\begin{aligned}
T \left(\oplus_{j=0}^{p+q+1+m} (M_j + tN_j) \right) \quad &= \quad \left(\oplus_{j=0}^{p+q} (M_j + tN_j) \right) \\
&\oplus \oplus_{j=p+q+1}^{p+q+1+m} (M'_j + tN'_j) \, S, \tag{11.3.5} \\
S \quad &= \quad T_\phi^{-1},
\end{aligned}$$

for some invertible T, and partition T and S conformally with (11.3.5):

$$T = [T^{(ij)}]_{i,j=0}^{p+q+1+m}, \quad S = [S^{(ij)}]_{i,j=0}^{p+q+1+m},$$

where $T^{(ij)}$ and $S^{(ij)}$ are $n_i \times n_j$. We then have, from (11.3.5):

$$T^{(ij)} M_j = M_i S^{(ij)}; \quad T^{(ij)} N_j = N_i S^{(ij)} \quad (i, j = 0, \ldots, p+q+1+m), \tag{11.3.6}$$

where in the right-hand sides, M_i and N_i are replaced by M'_i and N'_i, respectively, for $i = p+q+1, p+q+2, \ldots, p+q+m+1$. In particular,

$$T^{(0j)} M_j = 0, \quad T^{(0j)} N_j = 0 \quad (j = 0, \ldots, p+q+1+m),$$

which immediately implies the equalities $T^{(0j)} = 0$ for $j = p+1, \ldots, p+q+1+m$. Since

$$\mathrm{Ran}\, M_j + \mathrm{Ran}\, N_j = \mathsf{H}^{n_j}, \qquad (j = 1, \ldots, p),$$

we also obtain $T^{(0j)} = 0$ for $j = 1, \ldots, p$. Therefore, in view of the equality $S = T_\phi^{-1}$, also $S^{(i0)} = 0$ for $i = 1, \ldots, p+q+1+m$. Now, clearly, the matrices

$$\widetilde{T} := \left[T^{(ij)} \right]_{i,j=1}^{p+q+1+m} \quad \text{and} \quad \widetilde{S} := \left[S^{(ij)} \right]_{i,j=1}^{p+q+1+m} = (\widetilde{T}_\phi)^{-1} \tag{11.3.7}$$

are invertible, and we have

$$\widetilde{T}\left(\oplus_{j=1}^{p+q+1+m}\left(M_j+tN_j\right)\right)$$

$$= \left(\oplus_{j=1}^{p+q}\left(M_j+tN_j\right)\oplus\oplus_{j=p+q+1}^{p+q+1+m}\left(M_j'+tN_j'\right)\right)\widetilde{S}. \qquad (11.3.8)$$

Next, consider equalities (11.3.6) for $i,j=p+1,\ldots,p+q+1+m$ and $i\neq j$ (these hypotheses on i and j will be assumed throughout the present paragraph). If $i,j\neq p+q+1$, then N_i and N_j are both invertible, and using (11.3.6) we obtain

$$T^{(ij)}M_j = M_iS^{(ij)} = M_iN_i^{-1}T^{(ij)}N_j; \qquad (11.3.9)$$

hence,

$$T^{(ij)}M_jN_j^{-1} = M_iN_i^{-1}T^{(ij)}$$

(if $i>p+q+1$, then we use M_i' and N_i' in place of M_i and N_i, respectively, in the right-hand side of (11.3.9)). A computation shows that

$$M_jN_j^{-1} = \oplus_{s=1}^{w_{j-p}}\left(J_{\ell_{j-p,s}}(\alpha_{j-p})\oplus J_{\ell_{j-p,s}}(\alpha_{j-p})\right),\quad\text{for}\quad j=p+1,\ldots,p+q,$$

and $M_jN_j^{-1} = M_j'(N_j')^{-1}$ is a Jordan matrix with eigenvalue $\gamma_{j-p-q-1}$ for $j=p+q+2,\ldots,p+q+1+m$. By Theorem 5.11.1, we obtain

$$T^{(ij)}=0\quad\text{for}\ i,j\in\{p+1,\ldots,p+q,p+q+2,\ldots,p+q+1+m\},\ i\neq j.$$

If $i\neq p+q+1$ but $j=p+q+1$, then N_i and M_j are invertible, and (11.3.6) leads to

$$\begin{aligned}T^{(ij)} &= M_iS^{(ij)}M_j^{-1} = M_iN_i^{-1}T^{(ij)}N_jM_j^{-1}\\ &= (M_iN_i^{-1})^2T^{(ij)}(N_jM_j^{-1})^2\\ &= \cdots = (M_iN_i^{-1})^wT^{(ij)}(N_jM_j^{-1})^w = 0\end{aligned} \qquad (11.3.10)$$

for sufficiently large positive integer w, because $N_jM_j^{-1}$ is easily seen to be nilpotent. (If $i>p+q+1$, then we replace M_i and N_i by M_i' and N_i', respectively, in (11.3.10).) Finally, if $i=p+q+1$ and $j\neq p+q+1$, then, analogously, (11.3.6) gives

$$\begin{aligned}T^{(ij)} &= N_i'S^{(ij)}N_j^{-1} = N_i'(M_i')^{-1}T^{(ij)}M_jN_j^{-1}\\ &= (N_i'(M_i')^{-1})^2T^{(ij)}(M_jN_j^{-1})^2\\ &= \cdots = (N_i'(M_i')^{-1})^wT^{(ij)}(N_jM_j^{-1})^w,\end{aligned} \qquad (11.3.11)$$

and since $N_i'(M_i')^{-1}$ is nilpotent, the equality $T^{(ij)}=0$ follows. Thus, $T^{(ij)}=0$ for $i,j=p+1,\ldots,p+q+1+m$ and $i\neq j$. Analogously, we obtain the equalities $S^{(ij)}=0$ for $i,j=p+1,\ldots,p+q+1+m$ and $i\neq j$.

Now consider $T^{(ji)}$ and $S^{(ij)}$ for $i=1,\ldots,p$ and $j=p+1,\ldots,p+q+1+m$ (these hypotheses on i and j will be assumed throughout the current paragraph). Assume first $j=p+q+1$. Then (11.3.6) gives

$$T^{(i,p+q+1)}M_{p+q+1} = M_iS^{(i,p+q+1)},\quad T^{(i,p+q+1)}N_{p+q+1} = N_iS^{(i,p+q+1)},$$

and therefore

$$M_i S^{(i,p+q+1)} M_{p+q+1}^{-1} N_{p+q+1} = N_i S^{(i,p+q+1)}. \tag{11.3.12}$$

Using the form (11.3.3) of M_i and N_i, and equating the bottom rows in the left- and right-hand sides of (11.3.12), we find that the first row of $S^{(i,p+q+1)}$ is zero. Then consideration of the next to the bottom row of (11.3.12) implies that the second row of $S^{(i,p+q+1)}$ is zero. Continuing in this fashion it is found that the top ε_i rows of $S^{(i,p+q+1)}$ consist of zeros. Applying a similar argument to the equality

$$T^{(p+q+1,i)} N_i = N'_{p+q+1}((M'_{p+q+1})^{-1} T^{(p+q+1,i)} M_i),$$

it is found that the first ε_i columns of $T^{(p+q+1,i)}$ consist of zeros. Now assume $j \neq p+q+1$. Then (11.3.6) gives

$$T^{(ij)} M_j = M_i S^{(ij)}; \quad T^{(ij)} N_j = N_i S^{(ij)},$$

and consequently

$$N_i S^{(ij)} N_j^{-1} M_j = M_i S^{(ij)}. \tag{11.3.13}$$

In view of the form (11.3.3) of M_i and N_i, the $(\varepsilon_i + 1)$th row in the left-hand side of (11.3.13) is zero. But $(\varepsilon_i + 1)$th row in the right-hand side is the negative of the ε_ith row of $S^{(ij)}$. So, the ε_ith row of $S^{(ij)}$ is equal to zero. By considering successively the $(\varepsilon_i + 2)$th, $(\varepsilon_i + 3)$th, etc., rows on both sides of (11.3.13), it is found that the first ε_i rows of $S^{(ij)}$ are zeros. Next, use the equalities

$$T^{(ji)} M_i = M_j S^{(ji)}, \quad T^{(ji)} N_i = N_j S^{(ji)},$$

with M_j, N_j replaced by M'_j, N'_j, respectively, if $j > p+q+1$, to obtain

$$T^{(ji)} M_i = M_j S^{(ji)} = M_j (N_j^{-1} T^{(ji)} N_i).$$

Arguing as above, it follows that the first ε_i columns of $T^{(ji)}$ consist of zeros.

It has been shown that matrix $\widetilde{T}$ has the following form:

$$\widetilde{T} = \begin{bmatrix} T^{(11)} & & \cdots & T^{(1p)} & \\ \vdots & & \vdots & \vdots & \\ T^{(p1)} & & \cdots & T^{(pp)} & \\ \hline 0_{n_{p+1} \times \epsilon_1} & * & \cdots & 0_{n_{p+1} \times \epsilon_p} & * \\ 0_{n_{p+2} \times \epsilon_1} & * & \cdots & 0_{n_{p+2} \times \epsilon_p} & * \\ \vdots & & \vdots & \vdots & \\ 0_{n_{p+q+1+m} \times \epsilon_1} & * & \cdots & 0_{n_{p+q+1+m} \times \epsilon_p} & * \end{bmatrix}$$

(p leftmost block columns), and

$$\widetilde{T} = \begin{bmatrix} T^{(1,p+1)} & T^{(1,p+2)} & \cdots & T^{(1,p+q+1+m)} \\ \vdots & \vdots & \vdots & \vdots \\ T^{(p,p+1)} & T^{(p,p+2)} & \cdots & T^{(p,p+q+1+m)} \\ T^{(p+1,p+1)} & 0 & \cdots & 0 \\ 0 & T^{(p+2,p+2)} & \cdots & 0 \\ \vdots & \vdots & \vdots & \vdots \\ 0 & 0 & \cdots & T^{(p+q+1+m,p+q+1+m)} \end{bmatrix}$$

$(q + 1 + m$ rightmost block columns). Let $W = T^{-1}$, and partition conformally with (11.3.1):

$$W = \left[W^{(ij)}\right]_{i,j=0}^{p+q+1+m}, \qquad \widetilde{W} := \widetilde{T}^{-1} = \left[W^{(ij)}\right]_{i,j=1}^{p+q+1+m}.$$

In view of (11.3.8) we have

$$\widetilde{W}\left(\oplus_{j=1}^{p+q}\left(M_j + tN_j\right) \oplus \oplus_{j=p+q+1}^{p+q+1+m}\left(M_j' + tN_j'\right)\right)$$

$$= \left(\oplus_{j=1}^{p+q+1+m}\left(M_j + tN_j\right)\right)\widetilde{S}^{-1}. \qquad (11.3.14)$$

A similar argument shows that for $i = 1,\ldots,p$ and $j = p+1,\ldots,p+q+1+m$, the first ε_i columns of $W^{(ji)}$ are zeros. Also, (11.3.14) implies $W^{(ij)} = 0$ for $i,j = p+1,\ldots,p+q+1+m$ and $i \neq j$, which is proved exactly in the same way as $T^{(ij)} = 0$ $(i,j = p+1,\ldots,p+q+1+m, i \neq j)$ was proved using (11.3.8).

Denoting by $\widetilde{W}^{(ik)}$ the matrix formed by the $(\varepsilon_k + 1)$ rightmost columns of $W^{(ik)}$ and by $\widetilde{S}^{(kj)}$ the matrix formed by the $(\varepsilon_k + 1)$ bottom rows of $S^{(kj)}$ (here $i,j = p+1,\ldots,p+q+1+m; k = 1,\ldots,p$), we have

$$W^{(ik)}(M_k + tN_k)S^{(kj)} = \begin{bmatrix} 0 & \widetilde{W}^{(ik)} \end{bmatrix} \begin{bmatrix} 0_{\varepsilon_k} & * \\ * & 0_{\varepsilon_k+1} \end{bmatrix} \begin{bmatrix} 0 \\ \widetilde{S}^{(kj)} \end{bmatrix} = 0, \qquad (11.3.15)$$

where the form (11.3.3) of M_k and N_k was used.

Hence, using the properties of $W^{(ij)}$ and $S^{(ij)}$ verified above, the following equalities are obtained, where $i,j \in \{p+1,\ldots,p+q+1+m\}$:

$$\sum_{k=1}^{p+q} W^{(ik)}(M_k + tN_k)S^{(kj)} + \sum_{k=p+q+1}^{p+q+1+m} W^{(ik)}\left(M_k' + tN_k'\right)S^{(kj)} = 0$$

if $i \neq j$, and

$$\sum_{k=1}^{p+q} W^{(ik)}(M_k + tN_k)S^{(kj)} + \sum_{k=p+q+1}^{p+q+1+m} W^{(ik)}\left(M_k' + tN_k'\right)S^{(kj)}$$

$$= W^{(ii)}\left(M_i + tN_i\right)S^{(ii)} \qquad (11.3.16)$$

if $i = j$. (For $i = p+q+1,\ldots,p+q+1+m$, replace $W^{(ii)}\left(M_i + tN_i\right)S^{(ii)}$ with $W^{(ii)}\left(M_i' + tN_i'\right)S^{(ii)}$ in the right-hand side of (11.3.16).) We now obtain, by a computation starting with (11.3.14), where (11.3.16) is used:

$$\text{diag}\left[M_i + tN_i\right]_{i=p+q+1}^{p+q+1+m}$$

$$= \left[\sum_{k=1}^{p+q} W^{(ik)}(M_k + tN_k)S^{(kj)}\right.$$

$$\left. + \sum_{k=p+q+1}^{p+q+1+m} W^{(ik)}\left(M_k' + tN_k'\right)S^{(kj)}\right]_{i,j=p+q+1}^{p+q+1+m}$$

$$= \text{diag}\left[W^{(ii)}\right]_{i=p+q+1}^{p+q+1+m} \text{diag}\left[M_i' + tN_i'\right]_{i=p+q+1}^{p+q+1+m} \text{diag}\left[S^{(ii)}\right]_{i=p+q+1}^{p+q+1+m}.$$

$$(11.3.17)$$

Note that the matrix diag $[M_i + tN_i]_{i=p+q+1}^{p+q+1+m}$ is clearly invertible for some real value of t; hence, we see from (11.3.17) that the matrices $W^{(ii)}$ and $S^{(ii)}$ are invertible ($i = p + q + 1, \ldots, p + q + 1 + m$). Since also $\widetilde{S} = \widetilde{W}_\phi$ (see equality (11.3.7)), the equality (11.3.17) shows that for every $i = p+q+1, \ldots, p+q+1+m$, the ϕ-skewhermitian matrix pencils $M_i + tN_i$ and $M_i' + tN_i'$ are ϕ-congruent, as required. $\qquad\square$

We now return to the proof of uniqueness in Theorem 11.2.2(b). Let $A + tB$ be a ϕ-skewhermitian matrix pencil that is congruent to two forms (11.2.4). The uniqueness part of Theorem 11.2.2(a) guarantees that apart from permutations of blocks, these two forms can possibly differ only in the signs δ_j and η_k. Lemma 11.3.3 allows us to reduce the proof to cases when either only blocks of the form

$$\delta_1 \left(\beta(\phi)F_{k_1} + t\beta(\phi)G_{k_1}\right) \oplus \cdots \oplus \delta_r \left(\beta(\phi)F_{k_r} + t\beta(\phi)G_{k_r}\right) \qquad (11.3.18)$$

are present or only blocks of the form

$$\eta_1 \left((t + \gamma)\beta(\phi)F_{m_1} + \beta(\phi)G_{m_1}\right)$$

$$\oplus \cdots \oplus \eta_p \left((t + \gamma)\beta(\phi)F_{m_p} + \beta(\phi)G_{m_p}\right), \quad \gamma \in \mathsf{R} \qquad (11.3.19)$$

are present. We now consider each of these two cases separately.

We start with the form (11.3.19). So it is assumed that a ϕ-skewhermitian matrix pencil $A+tB$ is ϕ-congruent to (11.3.19), as well as to (11.3.19) with possibly different signs $\widetilde{\eta}_j$. We have to prove that, in fact, the form

$$\widetilde{\eta}_1 \left((t + \gamma)\beta(\phi)F_{m_1} + \beta(\phi)G_{m_1}\right)$$

$$\oplus \cdots \oplus \widetilde{\eta}_p \left((t + \gamma)\beta(\phi)F_{m_p} + \beta(\phi)G_{m_p}\right), \qquad (11.3.20)$$

is obtained from (11.3.19) after a permutation of blocks. Write

$$T \left(\oplus_{j=1}^p \eta_j \left((t + \gamma)\beta(\phi)F_{m_j} + \beta(\phi)G_{m_j}\right)\right)$$

$$= \left(\oplus_{j=1}^p \widetilde{\eta}_j \left((t + \gamma)\beta(\phi)F_{m_j} + \beta(\phi)G_{m_j}\right)\right) S, \qquad (11.3.21)$$

where T is an invertible quaternion matrix and $S = (T_\phi)^{-1}$, and partition

$$T = [T_{ij}]_{i,j=1}^p, \quad S = [S_{ij}]_{i,j=1}^p,$$

where T_{ij} and S_{ij} are $m_i \times m_j$. Then

$$T_{ij} \left(\eta_j \beta(\phi)F_{m_j}\right) = \left(\widetilde{\eta}_i \beta(\phi)F_{m_i}\right) S_{ij},$$
$$T_{ij} \left(\eta_j \left(\gamma\beta(\phi)F_{m_j} + \beta(\phi)G_{m_j}\right)\right) = \left(\widetilde{\eta}_i \left(\gamma\beta(\phi)F_{m_i} + \beta(\phi)G_{m_i}\right)\right) S_{ij},$$

for $i, j = 1, 2, \ldots, p$; therefore,

$$T_{ij}(\eta_j(\gamma\beta(\phi)F_{m_j} + \beta(\phi)G_{m_j}))$$

$$= (\widetilde{\eta}_i(\gamma\beta(\phi)F_{m_i} + \beta(\phi)G_{m_i}))\widetilde{\eta}_i(\beta(\phi))^{-1}F_{m_i}T_{ij}(\eta_j\beta(\phi)F_{m_j}). \qquad (11.3.22)$$

Equality (11.3.22) implies

$$T_{ij}V_j = V_iT_{ij}, \qquad (11.3.23)$$

where

$$V_i = (\gamma F_{m_i} + G_{m_i}) F_{m_i}^{-1} = J_{m_i}(\gamma),$$

the $m_i \times m_i$ Jordan block with eigenvalue γ. Proposition 5.4.2 shows that T_{ij} has the form

$$T_{ij} = \begin{bmatrix} 0 & \widetilde{T}_{ij} \end{bmatrix} \quad (\text{if } m_i \leq m_j) \tag{11.3.24}$$

or

$$T_{ij} = \begin{bmatrix} \widetilde{T}_{ij} \\ 0 \end{bmatrix} \quad (\text{if } m_i \geq m_j), \tag{11.3.25}$$

where $\widetilde{T}_{ij}$ is an upper triangular Toeplitz matrix of size $\min(m_i, m_j) \times \min(m_i, m_j)$. Permuting blocks in (11.3.19) if necessary, it can be assumed that the sizes m_j are arranged in nondecreasing order. Let $u < v$ be indices such that

$$m_i < m_{u+1} = m_{u+2} = \cdots = m_v < m_j \quad \text{for all } i \leq u \text{ and for all } j > v.$$

Now for $u < i \leq v$, $u < k \leq v$, in view of (11.3.21) we obtain the following equality, where δ_{ik} is the Kronecker symbol, i.e., $\delta_{ik} = 1$ if $i = k$ and $\delta_{ik} = 0$ if $i \neq k$:

$$\delta_{ik}\widetilde{\eta}_i\beta(\phi)F_{m_i} = \sum_{j=1}^{p} T_{ij}\left(\eta_j\beta(\phi)F_{m_j}\right)(T_{kj})_\phi = \sum_{j=1}^{u} T_{ij}(\eta_j\beta(\phi)F_{m_j})(T_{kj})_\phi$$

$$+ \sum_{j=u+1}^{v} T_{ij}(\eta_j\beta(\phi)F_{m_j})(T_{kj})_\phi + \sum_{j=v+1}^{p} T_{ij}(\eta_j\beta(\phi)F_{m_j})(T_{kj})_\phi. \tag{11.3.26}$$

In view of (11.3.25) the lower-left corner in the first sum in the right-hand side of (11.3.26) is zero. Using (11.3.24), it is easily verified that the lower-left corner in the third sum is also zero. The lower-left corner in the second sum in the right-hand side of (11.3.26) is equal to

$$\sum_{j=u+1}^{v} t_{ij}\eta_j\beta(\phi)(t_{kj})_\phi,$$

where t_{ij} is the entry on the main diagonal of T_{ij}. Thus,

$$\delta_{ik}\widetilde{\eta}_i\beta(\phi) = \sum_{j=u+1}^{v} t_{ij}\eta_j\beta(\phi)(t_{kj})_\phi.$$

It follows that

$$\widetilde{\eta}_{u+1}\beta(\phi) \oplus \cdots \oplus \widetilde{\eta}_v\beta(\phi)$$

$$= [t_{ik}]_{i,k=u+1}^{v} \left(\eta_{u+1}\beta(\phi) \oplus \cdots \oplus \eta_v\beta(\phi)\right)\right) \left([t_{ik}]_{i,k=u+1}^{v}\right)_\phi. \tag{11.3.27}$$

Now Theorem 4.1.2(b) guarantees that the two systems of signs $\{\eta_{u+1}, \ldots, \eta_v\}$ and $\{\widetilde{\eta}_{u+1}, \ldots, \widetilde{\eta}_v\}$ have the same number of +1s (and also the same number of −1s). Thus, within each set of blocks of equal size m_j, the number of η_j's which are equal to +1 (resp. to −1) coincides with the number of $\widetilde{\eta}_j$'s which are equal to +1 (resp. to −1). This shows that (11.3.20) is indeed obtained from (11.3.19) after a permutation of blocks.

Finally, assume that $A + tB$ is ϕ-congruent to (11.3.18), and also ϕ-congruent to

$$\widetilde{\delta}_1 \left(\beta(\phi) F_{k_1} + t\beta(\phi) G_{k_1} \right) \oplus \cdots \oplus \widetilde{\delta}_r \left(\beta(\phi) F_{k_r} + t\beta(\phi) G_{k_r} \right)$$

with possibly different signs $\widetilde{\delta}_j$, $j = 1, 2, \ldots, r$. Arguing as in the preceding case, we obtain the equalities

$$T_{ij}(\delta_j \beta(\phi) F_{k_j}) = (\widetilde{\delta}_i \beta(\phi) F_{k_i}) S_{ij}, \qquad T_{ij}\left(\delta_j \beta(\phi) G_{k_j} \right) = \left(\widetilde{\delta}_i \beta(\phi) G_{k_i} \right) S_{ij}.$$

The proof that the form

$$\oplus_{j=1}^{r} \widetilde{\delta}_j \left(F_{k_j} + G_{k_j} \right)$$

is obtained from (11.3.18) after a permutation of blocks, proceeds from now on in the same way (letting $\gamma = 0$) as the proof that (11.3.19) and (11.3.20) are the same up to a permutation of blocks.

The proof of the uniqueness part of Theorem 11.2.2(b) is complete. $\square$

11.4   NUMERICAL RANGES AND CONES

Let ϕ be a nonstandard involution.

Definition 11.4.1. For a pair of ϕ-skewhermitian $n \times n$ quaternionic matrices (A, B), we define the ϕ-*numerical range*

$$W_\phi(A, B) := \{(x_\phi Ax, x_\phi Bx) : x \in \mathsf{H}^{n \times 1}, \ \|x\| = 1\} \subseteq \mathsf{H}^2$$

and the ϕ-*numerical cone*

$$C_\phi(A, B) := \{(x_\phi Ax, x_\phi Bx) : x \in \mathsf{H}^{n \times 1}\} \subseteq \mathsf{H}^2.$$

Since $\phi(x_\phi Ax) = -x_\phi Ax$, we clearly have that

$$W_\phi(A, B) \subseteq C_\phi(A, B) \subseteq \{(y_1 \beta(\phi), y_2 \beta(\phi)) : y_1, y_2 \in \mathsf{R}\}.$$

Proposition 11.4.2. *The ϕ-numerical range $W_\phi(A, B)$ and cone $C_\phi(A, B)$ are convex.*

Indeed, the convexity of $W_\phi(A, B)$ is proved in Theorem 3.7.13. Then the convexity of $C_\phi(A, B)$ follows without difficulty (Ex. 11.5.1).

We present a result that characterizes the situations when the ϕ-numerical cone is contained in a half-plane bounded by a line passing through the origin. We identify here $\mathsf{R}^2 \beta(\phi)$ with R^2.

Theorem 11.4.3. *The following statements are equivalent for a pair of ϕ-skewhermitian $n \times n$ matrices (A, B):*

(1) *$C_\phi(A, B)$ is contained in a half-plane bounded by a line passing through the origin;*

(2) *the pencil $A + tB$ is ϕ-congruent to a pencil of the form $\beta(\phi)A' + t\beta(\phi)B'$, where A' and B' are real symmetric matrices such that some linear combination $(\sin \mu)A' + (\cos \mu)B'$, $0 \leq \mu < 2\pi$ is positive semidefinite.*

Proof. Observe that implication $(2) \Longrightarrow (1)$ is evident.

We prove $(1) \Longrightarrow (2)$. Since $C_\phi(A, B) = C_\phi(S_\phi A S, S_\phi B S)$ for any invertible quaternion matrix S, we may (and do) assume without loss of generality, that the pair (A, B) is given in the canonical form of Theorem 11.2.2(b).

It will be convenient to consider particular blocks first.

Claim. *Let*

$$A_0 = \begin{bmatrix} 0 & \alpha F_\ell + G_\ell \\ -\alpha F_\ell - G_\ell & 0 \end{bmatrix}, \quad B_0 = \begin{bmatrix} 0 & F_\ell \\ -F_\ell & 0 \end{bmatrix},$$

where $\alpha \in \mathrm{Inv}\,(\phi) \setminus \mathsf{R}$. Then

$$C_\phi(A_0, B_0) = \mathsf{R}^2 \beta(\phi). \tag{11.4.1}$$

To verify (11.4.1), first observe that adding a real nonzero multiple of B_0 to A_0 does not alter the property (11.4.1). Thus, we may assume that the real part of α is zero. Next, replacing ϕ by a similar nonstandard involution, if necessary, we may also assume that $\beta(\phi) = \mathsf{k}$ and $\alpha = a\mathsf{i} + b\mathsf{j}$ for some real a and b not both zero. It will be proved that

$$(\mathsf{R}\mathsf{k}, 0) \subseteq \Omega_0, \tag{11.4.2}$$

where

$$\Omega_0 := \left\{ \left([\phi(y) \ 0 \ \dots \ 0 \ \phi(z)] A_0 \begin{bmatrix} y \\ 0 \\ \vdots \\ 0 \\ z \end{bmatrix}, [\phi(y) \ 0 \ \dots \ 0 \ \phi(z)] B_0 \begin{bmatrix} y \\ 0 \\ \vdots \\ 0 \\ z \end{bmatrix} \right) \right\},$$

with $x, y \in \mathsf{H}$ arbitrary, and

$$(0, \mathsf{R}\mathsf{k}) \subseteq \Omega_0. \tag{11.4.3}$$

In view of the convexity of $C_\phi(A_0, B_0)$, this will suffice to prove (11.4.1). Write

$$\begin{bmatrix} y \\ z \end{bmatrix} = x_1 + x_2\mathsf{i} + x_3\mathsf{j} + x_4\mathsf{k}, \quad x_1, x_2, x_3, x_4 \in \mathsf{R}^{2\times 1}.$$

Then

$$[\phi(y) \ 0 \ \dots \ 0 \ \phi(z)] B_0 [y \ 0 \ \dots \ 0 \ z]^T$$

$$= (x_1^T + x_2^T\mathsf{i} + x_3^T\mathsf{j} - x_4^T\mathsf{k}) \begin{bmatrix} 0 & 1 \\ -1 & 0 \end{bmatrix} (x_1 + x_2\mathsf{i} + x_3\mathsf{j} + x_4\mathsf{k}),$$

which, in turn, is equal to

$$\mathsf{k} \begin{bmatrix} x_1^T & x_2^T & x_3^T & x_4^T \end{bmatrix} \begin{bmatrix} 0 & 0 & 0 & 0 & 0 & 0 & 0 & 1 \\ 0 & 0 & 0 & 0 & 0 & 0 & -1 & 0 \\ 0 & 0 & 0 & 0 & 0 & 1 & 0 & 0 \\ 0 & 0 & 0 & 0 & -1 & 0 & 0 & 0 \\ 0 & 0 & 0 & -1 & 0 & 0 & 0 & 0 \\ 0 & 0 & 1 & 0 & 0 & 0 & 0 & 0 \\ 0 & -1 & 0 & 0 & 0 & 0 & 0 & 0 \\ 1 & 0 & 0 & 0 & 0 & 0 & 0 & 0 \end{bmatrix} \begin{bmatrix} x_1 \\ x_2 \\ x_3 \\ x_4 \end{bmatrix}. \tag{11.4.4}$$

Analogously,

$$[\phi(y) \ 0 \ \ldots \ 0 \ \phi(z)]A_0[y \ 0 \ \ldots \ 0 \ z]^T$$

$$= (x_1^T + x_2^T\mathsf{i} + x_3^T\mathsf{j} - x_4^T\mathsf{k}) \begin{bmatrix} 0 & a\mathsf{i} + b\mathsf{j} \\ -a\mathsf{i} - b\mathsf{j} & 0 \end{bmatrix} (x_1 + x_2\mathsf{i} + x_3\mathsf{j} + x_4\mathsf{k}),$$

which is equal to

$$\mathsf{k} \begin{bmatrix} x_1^T & x_2^T & x_3^T & x_4^T \end{bmatrix} \begin{bmatrix} 0 & 0 & 0 & -b & 0 & a & 0 & 0 \\ 0 & 0 & b & 0 & -a & 0 & 0 & 0 \\ 0 & b & 0 & 0 & 0 & 0 & 0 & -a \\ -b & 0 & 0 & 0 & 0 & 0 & a & 0 \\ 0 & -a & 0 & 0 & 0 & 0 & 0 & -b \\ a & 0 & 0 & 0 & 0 & 0 & b & 0 \\ 0 & 0 & 0 & a & 0 & b & 0 & 0 \\ 0 & 0 & -a & 0 & -b & 0 & 0 & 0 \end{bmatrix}$$

$$\cdot \begin{bmatrix} x_1 \\ x_2 \\ x_3 \\ x_4 \end{bmatrix}. \tag{11.4.5}$$

Assuming $x_3 = x_4 = 0$ (if $b \neq 0$) or $x_2 = x_4 = 0$ (if $a \neq 0$) in (11.4.4) and (11.4.5) shows the inclusion (11.4.2), and assuming $x_1 = x_4 = 0$ shows (11.4.3).

Suppose now that statement (1) holds. In view of the claim and taking advantage of formula (11.2.5), we may further assume that $A = \widetilde{A}\beta(\phi)$ and $B = \widetilde{B}\beta(\phi)$, where $\widetilde{A}$ and $\widetilde{B}$ are real symmetric $n \times n$ matrices. Now, clearly, statement (2) follows. $\qquad\square$

Other criteria for Theorem 11.4.3(2) to hold true, in terms of the pair of real symmetric matrices (A', B'), can be given using results of Theorem 11.3 in Lancaster and Rodman [92] and Theorem 5 in Cheung et al. [26].

Theorem 11.4.4. *Let (A, B) be a pair of ϕ-skewhermitian (quaternion) $n \times n$ matrices. Then $C_\phi(A, B)$ is contained in a half-plane bounded by a line passing through the origin if and only if the pencil $A + tB$ is ϕ-congruent to a pencil of the form $\beta(\phi)A' + t\beta(\phi)B'$, where A' and B' are real symmetric matrices that satisfy either one of the following two equivalent conditions:*

(a) *For every matrix $X \in \mathsf{R}^{n \times 2}$ such that $X^T X = I_2$, it holds that some nontrivial real linear combination of $X^T A'X$ and $X^T B'X$ is positive semidefinite.*

(b) *For every $x \in \mathsf{R}^{n \times 1}$ such that $\langle A'x, x \rangle = \langle B'x, x \rangle = 0$, the two vectors $A'x$ and $B'x$ are R-linearly dependent, and at least one of the following conditions (i) and (ii) fails:*

   (i) *$\operatorname{Ker} A' = \operatorname{Ker} B'$ has (real) dimension $n - 2$.*

   (ii) *The dimension of $\operatorname{Ran}(aA' + bB')$ is equal to 2 for every pair $a, b \in \mathsf{R}$ not both zero.*

11.5   EXERCISES

Ex. 11.5.1. Prove that $C_\phi(A, B)$ is convex.

Ex. 11.5.2. Find the canonical form under ϕ-congruence of Theorem 11.1.3 for the following ϕ-hermitian pencils:

$$(a) \quad \begin{bmatrix} 0 & \beta(\phi) \\ -\beta(\phi) & 0 \end{bmatrix} + t \begin{bmatrix} 1 & 1 \\ 1 & 1 \end{bmatrix};$$

$$(b) \quad \begin{bmatrix} 0 & 0 & \beta(\phi) \\ 0 & 0 & 0 \\ -\beta(\phi) & 0 & 0 \end{bmatrix} + t \begin{bmatrix} q & q & q \\ q & q & q \\ q & q & q \end{bmatrix},$$

where $q \in H \setminus \{0\}$ is such that $\beta(\phi)q = -q\beta(\phi)$.

Ex. 11.5.3. Characterize (in terms of structure of the Kronecker form) those ϕ-hermitian matrix pencils $A + tB$ that are:

(a) ϕ-congruent to real symmetric matrix pencils;

(b) ϕ-congruent to matrix pencils of the form $\beta(\phi)A + t\beta(\phi)B$, where A and B are real skewsymmetric matrices.

Ex. 11.5.4. Find all possible canonical forms of Theorem 11.1.3 if it is known that $\text{rank}\,(xA + yB) \leq 2$ for all real x, y.

Ex. 11.5.5. Find all possible canonical forms of Theorem 11.1.3 if it is known that $\text{rank}\,(xA + yB) = 3$ for all real x, y not both zero.

Ex. 11.5.6. Characterize those ϕ-skewhermitian matrix pencils $A + tB$ that are:

(a) ϕ-congruent to real skewsymmetric matrix pencils;

(b) ϕ-congruent to matrix pencils of the form $\beta(\phi)A + t\beta(\phi)B$, where A and B are real symmetric matrices.

Ex. 11.5.7. Find the canonical form under ϕ-congruence of Theorem 11.2.2 for the following ϕ-skewhermitian pencils:

$$\begin{bmatrix} 0 & \beta(\phi) \\ \beta(\phi) & 0 \end{bmatrix} + t \begin{bmatrix} a\beta(\phi) & q \\ -q & 0 \end{bmatrix},$$

where q is a nonzero quaternion such that $\beta(\phi)q = -q\beta(\phi)$, and a is a real parameter.

Ex. 11.5.8. For each ϕ-skewhermitian matrix pencil $A + tB$ of Ex. 11.5.6, verify whether or not the ϕ-numerical cone $C_\phi(A, B)$ is contained in a half-space bounded by a line passing through the origin.

Ex. 11.5.9. Let $A + tB$ be a ϕ-skewhermitian $n \times n$ matrix pencil such that A is invertible and the following property holds true:

(A) Every ϕ-skewhermitian matrix pencil which is strictly equivalent to $A + tB$ is, in fact, ϕ-congruent to $A + tB$.

Show that then there exists $\varepsilon > 0$ such that every ϕ-skewhermitian $n \times n$ matrix pencil $A' + tB'$ with $\|A' - A\| + \|B' - B\| < \varepsilon$ also has the property (A). (Cf. Theorem 11.2.4.)

Ex. 11.5.10. Find whether or not the result of Ex. 11.5.9 is valid if:

(a) the condition that A is invertible is omitted;

(b) the condition that A is invertible is replaced by the condition that B is invertible;

(c) the condition that A is invertible is replaced by the condition that $B + tA$ is invertible for some real t.

Ex. 11.5.11. Let $H \in \mathsf{H}^{n \times n}$ be ϕ-skewhermitian, not necessarily invertible. State and prove analogues of Theorems 8.6.5 and 8.6.8 for (H, ϕ)-expansive and (H, ϕ)-plus-matrices.

Ex. 11.5.12. Provide details of the derivation of Theorem 11.2.2 part (a) from Theorem 8.1.2, by taking advantage of Proposition 4.1.7. Hint: See the proof of Theorem 11.1.3.

11.6 NOTES

The contents of this chapter are based on Rodman [132]. In particular, formulations of the main results and the proofs are taken from that paper. The results of Theorems 11.1.3 and 11.2.2 (for the case when at least one of the matrices A and B is invertible) are found in Djoković et al. [34], for example.

Chapter Twelve

Mixed matrix pencils: Nonstandard involutions

The canonical forms of mixed quaternion matrix pencils, i.e., such that one of the two matrices is ϕ-hermitian and the other is ϕ-skewhermitian, are also studied here with respect to simultaneous ϕ-congruence. Other canonical forms of mixed matrix pencils are developed with respect to strict equivalence. As an application, we provide canonical forms of quaternion matrices under ϕ-congruence.

As in the preceding chapter, we fix a nonstandard involution ϕ throughout this chapter and a quaternion $\beta(\phi)$ such that $\phi(\beta(\phi)) = -\beta(\phi)$ and $|\beta(\phi)| = 1$.

12.1 CANONICAL FORMS FOR ϕ-MIXED PENCILS: STRICT EQUIVALENCE

Definition 12.1.1. A matrix pencil $A + tB$, where $A, B \in \mathsf{H}^{n \times n}$, is said to be *$\phi$-hermitian-skewhermitian*, in short *ϕ-hsk*, if $A_\phi = A$ and $B_\phi = -B$.

In this section we formulate the canonical form for ϕ-hsk matrix pencils under strict equivalence.

We start with a list of primitive forms of ϕ-hsk pencils. We will use the matrices $\Xi_m(\beta(\phi))$ defined in (1.2.6). Thus, $(\Xi_m(\beta(\phi)))_\phi = \Xi_m(\beta(\phi))$ if m is even, and $(\Xi_m(\beta(\phi)))_\phi = -\Xi_m(\beta(\phi))$ if m is odd. The equality

$$\begin{bmatrix} 1 & 1 \\ 1 & -1 \end{bmatrix} \begin{bmatrix} 0 & \beta(\phi) \\ \beta(\phi) & 0 \end{bmatrix} \begin{bmatrix} 1 & 1 \\ 1 & -1 \end{bmatrix} = \begin{bmatrix} 2\beta(\phi) & 0 \\ 0 & -2\beta(\phi) \end{bmatrix}$$

shows that the $\beta(\phi)$-signature (cf. Theorem 4.1.2(b)) of $\Xi_m(\beta(\phi))$ for odd m is equal to

$$\left(\frac{m-1}{2}, \frac{m+1}{2}, 0 \right) \quad \text{if } m = 4k + 3, \, k \text{ nonnegative integer} \tag{12.1.1}$$

and to

$$\left(\frac{m+1}{2}, \frac{m-1}{2}, 0 \right) \quad \text{if } m = 4k + 1, \, k \text{ nonnegative integer.} \tag{12.1.2}$$

We consider the following primitive ϕ-hsk pencils; the designation (q-ϕ-h-sk) stands for quaternion, ϕ, hermitian, skewhermitian.

(q-ϕ-h-sk0)   a square-size zero matrix.

(q-ϕ-h-sk1) $G_{2\varepsilon+1} + t \begin{bmatrix} 0 & 0 & F_\varepsilon \\ 0 & 0_1 & 0 \\ -F_\varepsilon & 0 & 0 \end{bmatrix}$, ε positive integer.

(q-ϕ-h-sk2) $F_k + t\beta(\phi)G_k$, k positive integer.

(q-ϕ-h-sk3) $\quad G_k + t\beta(\phi)F_k$, $\quad k$ positive integer.

(q-ϕ-h-sk4)

$$\begin{bmatrix} 0 & \alpha F_{\frac{\ell}{2}} + G_{\frac{\ell}{2}} + tF_{\frac{\ell}{2}} \\ \alpha F_{\frac{\ell}{2}} + G_{\frac{\ell}{2}} - tF_{\frac{\ell}{2}} & 0 \end{bmatrix},$$

where ℓ is even and $\alpha \in \mathrm{Inv}(\phi)$, $\Re(\alpha) > 0$.

(q-ϕ-h-sk5)

$$\begin{bmatrix} -\Xi_{s-1}(\beta(\phi)) & 0 \\ 0 & 0 \end{bmatrix} + \begin{bmatrix} 0 & 0 & \dots & 0 & \rho \\ 0 & 0 & \dots & -\rho & 0 \\ \vdots & \vdots & \ddots & \vdots & \vdots \\ 0 & -\rho & \dots & 0 & 0 \\ \rho & 0 & \dots & 0 & 0 \end{bmatrix} + t\Xi_s(\beta(\phi)), \qquad (12.1.3)$$

where s is odd and ρ is real positive.

(q-ϕ-h-sk6)

$$\Xi_s(\beta(\phi)) +$$
$$t\left(\begin{bmatrix} -\Xi_{s-1}(\beta(\phi)) & 0 \\ 0 & 0 \end{bmatrix} + \begin{bmatrix} 0 & 0 & \dots & 0 & \rho \\ 0 & 0 & \dots & -\rho & 0 \\ \vdots & \vdots & \ddots & \vdots & \vdots \\ 0 & \rho & \dots & 0 & 0 \\ -\rho & 0 & \dots & 0 & 0 \end{bmatrix} \right), \qquad (12.1.4)$$

where s is even and ρ is real positive.

The matrices $\widetilde{G}_k$ and $\widetilde{G}_{\ell/2}$ can be used in (q-ϕ-h-sk2), (q-ϕ-h-sk3), and (q-ϕ-h-sk4) in place of G_k and $G_{\ell/2}$, respectively.

We also remark that the block (q-ϕ-h-sk4) is ϕ-congruent to (q-ϕ-h-sk4), in which α is replaced by $-\alpha$. Thus, in (q-ϕ-h-sk4) one may replace the condition $\Re(\alpha) > 0$ with the condition $\Re(\alpha) < 0$. Indeed, we have

$$\begin{bmatrix} 0 & Y^T \\ X^T & 0 \end{bmatrix} \begin{bmatrix} 0 & \alpha F_{\frac{\ell}{2}} + G_{\frac{\ell}{2}} + tF_{\frac{\ell}{2}} \\ \alpha F_{\frac{\ell}{2}} + G_{\frac{\ell}{2}} - tF_{\frac{\ell}{2}} & 0 \end{bmatrix} \begin{bmatrix} 0 & X \\ Y & 0 \end{bmatrix}$$
$$= \begin{bmatrix} 0 & -\alpha F_{\frac{\ell}{2}} + G_{\frac{\ell}{2}} + tF_{\frac{\ell}{2}} \\ -\alpha F_{\frac{\ell}{2}} + G_{\frac{\ell}{2}} - tF_{\frac{\ell}{2}} & 0 \end{bmatrix}, \qquad (12.1.5)$$

where X and Y are invertible real $\ell/2 \times \ell/2$ matrices such that

$$Y^T(-G_{\frac{\ell}{2}}F_{\frac{\ell}{2}})(Y^T)^{-1} = G_{\frac{\ell}{2}}F_{\frac{\ell}{2}} \quad \text{and} \quad X = -F_{\ell/2}(Y^T)^{-1}F_{\ell/2}.$$

Such real matrices X and Y obviously exist because $GF = J_{\ell/2}(0)$ and $J_{\ell/2}(0)$ is similar to its negative. The verification of equality (12.1.5) is straightforward.

It will be convenient to denote

$$\Psi_s(\alpha) := \begin{bmatrix} -\Xi_{s-1}(\beta(\phi)) & 0 \\ 0 & 0 \end{bmatrix} + \begin{bmatrix} 0 & 0 & \dots & 0 & \alpha \\ 0 & 0 & \dots & -\alpha & 0 \\ \vdots & \vdots & \ddots & \vdots & \vdots \\ 0 & (-1)^{s-2}\alpha & \dots & 0 & 0 \\ (-1)^{s-1}\alpha & 0 & \dots & 0 & 0 \end{bmatrix},$$

where $\alpha \in \mathsf{H} \setminus \{0\}$. Thus, the primitive form (q-ϕ-h-sk5) is $\Psi_s(\rho) + t\Xi_s\left(\beta(\phi)\right)$, and the primitive form (q-ϕ-h-sk6) is $\Xi_s\left(\beta(\phi)\right) + t\Psi_s(\rho)$.

Theorem 12.1.2. *Every ϕ-hsk matrix pencil $A + tB$ is strictly equivalent to a ϕ-hsk matrix pencil that is a direct sum of blocks of types (q-ϕ-h-sk0)–(q-ϕ-h-sk6).*

The direct sum is uniquely determined by A and B up to an arbitrary permutation of the diagonal blocks and up to a replacement in each block of type (q-ϕ-h-sk4), the quaternion α with a similar quaternion $\alpha' \in \mathrm{Inv}\,(\phi)$.

12.2   PROOF OF THEOREM 12.1.2

We start with preliminary considerations.

Let $q = (q_1, q_2, q_3)$ be a units triple. Write $X \in \mathsf{H}^{m \times n}$ in the form

$$X = X_{11} + q_1 X_{12} + (X_{21} + q_1 X_{22})q_2,$$

where $X_{11}, X_{12}, X_{21}, X_{22} \in \mathsf{R}^{m \times n}$. Then we define

$$\widetilde{\omega}_{m,n}^{(q)}(X) := \begin{bmatrix} X_{11} + \mathrm{i}X_{12} & X_{21} + \mathrm{i}X_{22} \\ -X_{21} + \mathrm{i}X_{22} & X_{11} - \mathrm{i}X_{12} \end{bmatrix} \in \mathbb{C}^{2m \times 2n}. \qquad (12.2.1)$$

We will often write $\widetilde{\omega}^{(q)}(X)$ for $\widetilde{\omega}_{m,n}^{(q)}(X)$, with m, n understood form context.

In complete analogy with the map $\widetilde{\omega}_n$ of (5.6.1), the map $\widetilde{\omega}^{(q)}(\cdot)$ is a unital homomorphism.

Proposition 12.2.1. *The map $\widetilde{\omega}^{(q)}(\cdot)$ is one-to-one and has the properties:*

$$\widetilde{\omega}^{(q)}(aX + bY) = a\widetilde{\omega}^{(q)}(X) + b\widetilde{\omega}^{(q)}(Y), \quad \forall \ X, Y \in \mathsf{H}^{m \times n}, \quad a, b \in \mathsf{R};$$

$$\widetilde{\omega}^{(q)}(XY) = \widetilde{\omega}^{(q)}(X)\widetilde{\omega}^{(q)}(Y), \quad \forall \ X \in \mathsf{H}^{m \times n}, \ Y \in \mathsf{H}^{n \times p};$$

$$\widetilde{\omega}^{(q)}(X^*) = (\widetilde{\omega}^{(q)}(X))^*, \quad \forall \ X \in \mathsf{H}^{m \times n}; \qquad \widetilde{\omega}^{(q)}(I) = I.$$

In particular, $\widetilde{\omega}^{(q)}$ is a homomorphism of real algebras on $\mathsf{H}^{n \times n}$.

As in Theorem 5.7.1, we have the following.

Theorem 12.2.2. *If*

$$J_{m_1}(\alpha_1) \oplus \cdots \oplus J_{m_r}(\alpha_r), \qquad \alpha_1 = a_1 + \mathrm{i}b_1, \ldots, \alpha_r = a_r + \mathrm{i}b_r \in \mathsf{C}, \qquad (12.2.2)$$

is a Jordan form of $A \in \mathsf{H}^{n \times n}$, then

$$\begin{bmatrix} J_{m_1}(\alpha_1) & 0 \\ 0 & J_{m_1}(\overline{\alpha_1}) \end{bmatrix} \oplus \cdots \oplus \begin{bmatrix} J_{m_r}(\alpha_r) & 0 \\ 0 & J_{m_r}(\overline{\alpha_r}) \end{bmatrix}$$

is the complex Jordan form of $\widetilde{\omega}^{(q)}(A)$.

For the proof, use the Jordan form (12.2.2) with $\alpha_j \in \mathrm{Span}_{\mathsf{R}}\{1, q_1\}$, for $j = 1, 2, \ldots, r$.

We note the following connection between the action of a nonstandard involution and the complex representation $\widetilde{\omega}^{(q)}$.

Proposition 12.2.3. *Let ϕ be a nonstandard involution, and let $q = (q_1, q_2, q_3)$ be a units triple such that*

$$\mathrm{Inv}\,(\phi) = \mathrm{Span}_{\mathsf{R}}\{1, q_1, q_3\}.$$

Then for the map $\widetilde{\omega}^{(q)}(\cdot)$ given by (12.2.1), we have

$$\widetilde{\omega}^{(q)}(X_\phi) = (\widetilde{\omega}^{(q)}(X))^T. \tag{12.2.3}$$

The proof is by a straightforward verification.

In the next lemma, the properties of the Kronecker form of a mixed pencil are given.

Lemma 12.2.4. *Let $A + tB \in \mathsf{H}^{n \times n}$, where A is ϕ-hermitian and B is ϕ-skewhermitian. Then:*

(1) *the left indices of the pencil $A + tB$, arranged in the nondecreasing order, coincide with its right indices, also arranged in nondecreasing order;*

(2) *for every eigenvalue α with nonzero real part of $A + tB$, the indices of $A + tB$ that correspond to α are paired with the indices of $A + tB$ that correspond to $-\alpha$. In other words, if α is an eigenvalue of $A + tB$ with a nonzero real part, then $-\alpha$ is also an eigenvalue of $A + tB$, and for every positive integer k, the number of blocks $tI_{\ell_j} + J_{\ell_j}(\alpha_j)$ in the Kronecker form of $A + tB$ for which $\ell_j = k$ and α_j is similar to α coincides with the number of blocks $tI_{\ell_j} + J_{\ell_j}(\alpha_j)$ for which $\ell_j = k$ and α_j is similar to $-\alpha$.*

Proof. Let (7.3.1) be the Kronecker form of $A + tB$; thus

$$\begin{aligned}
S(A + tB)T \;=\; & 0_{u \times v} \oplus L_{\varepsilon_1 \times (\varepsilon_1 + 1)} \oplus \cdots \oplus L_{\varepsilon_p \times (\varepsilon_p + 1)} \\
& \oplus L^T_{\eta_1 \times (\eta_1 + 1)} \oplus \cdots \oplus L^T_{\eta_q \times (\eta_q + 1)} \\
& \oplus (I_{k_1} + tJ_{k_1}(0)) \oplus \cdots \oplus (I_{k_r} + tJ_{k_r}(0)) \\
& \oplus (tI_{\ell_1} + J_{\ell_1}(\alpha_1)) \oplus \cdots \oplus (tI_{\ell_s} + J_{\ell_s}(\alpha_s)) \quad (12.2.4)
\end{aligned}$$

for some invertible matrices S and T. Let $q := (q_1, q_2, q_3)$ be a units triple such that $\mathrm{Inv}\,(\phi) = \mathrm{Span}_{\mathsf{R}}\{1, q_1, q_3\}$. We may (and do) assume without loss of generality, that $\alpha_1, \ldots, \alpha_s \in \mathrm{Span}_{\mathsf{R}}\{1, q_1\}$. We now apply the map $\widetilde{\omega}^{(q)}$ (defined by (12.2.1)) to the equality (12.2.4). Using the definition of the map $\widetilde{\omega}^{(q)}$ and Theorem 12.2.2, after a permutation of blocks and similarity with a complex similarity matrix (if necessary), we obtain that the complex pencil $\widetilde{\omega}^{(q)}(A) + t\widetilde{\omega}^{(q)}(B)$ is strictly equivalent (over C) to the following complex pencil:

$$\begin{aligned}
0_{2u \times 2v} \quad \oplus \quad & \oplus_{j=1}^p \left(L_{\varepsilon_j \times (\varepsilon_j + 1)} \oplus L_{\varepsilon_j \times (\varepsilon_j + 1)} \right) \\
\oplus \quad & \oplus_{j=1}^q \left(L^T_{\eta_j \times (\eta_j + 1)} \oplus L^T_{\eta_j \times (\eta_j + 1)} \right) \\
\oplus \quad & \oplus_{j=1}^r \left((I_{k_j} + tJ_{k_j}(0)) \oplus (I_{k_j} + tJ_{k_j}(0)) \right) \\
\oplus \quad & \oplus_{j=1}^s \left((tI_{\ell_j} + J_{\ell_j}(\alpha_j)) \oplus (tI_{\ell_j} + J_{\ell_j}(\overline{\alpha_j})) \right).
\end{aligned}$$

By Proposition 12.2.3, the matrix $\widetilde{\omega}^{(q)}(A)$ is (complex) symmetric, and the matrix $\widetilde{\omega}^{(q)}(B)$ is skewsymmetric. By Theorem 15.3.6 the left and right indices of $\widetilde{\omega}^{(q)}(A) + t\widetilde{\omega}^{(q)}(B)$ coincide, and for every nonzero eigenvalue $\alpha \in \mathsf{C}$ of $\widetilde{\omega}^{(q)}(A) + t\widetilde{\omega}^{(q)}(B)$, the complex number $-\alpha$ is also an eigenvalue of $\widetilde{\omega}^{(q)}(A) + t\widetilde{\omega}^{(q)}(B)$, and the indices of α as the eigenvalue of $\widetilde{\omega}^{(q)}(A) + t\widetilde{\omega}^{(q)}(B)$ coincide with those of $-\alpha$. The result of Lemma 12.2.4 now follows. □

We are now ready to prove Theorem 12.1.2. Indeed, in view of Lemma 12.2.4 we may assume that the Kronecker form of the ϕ-hsk pencil $A + tB$ has the following form:

$$
\begin{aligned}
0_{u \times u} \quad &\oplus \quad \oplus_{j=1}^{p} \left(L_{\varepsilon_j \times (\varepsilon_j + 1)} \oplus L_{\varepsilon_j \times (\varepsilon_j + 1)}^T \right) \oplus \oplus_{j=1}^{r} \left((I_{k_j} + t J_{k_j}(0)) \right. \\
&\oplus \quad \oplus_{j=1}^{m} \left(t I_{\ell_j} + J_{\ell_j}(\alpha_j) \right) \\
&\oplus \quad \oplus_{j=m+1}^{s} \left((t I_{\ell_j} + J_{\ell_j}(\alpha_j)) \oplus (t I_{\ell_j} + J_{\ell_j}(-\alpha_j)) \right). \quad (12.2.5)
\end{aligned}
$$

Here $\alpha_1, \ldots, \alpha_s \in \text{Inv}(\phi)$; the real parts of $\alpha_1, \ldots, \alpha_m$ are equal to zero; and the real parts of $\alpha_{m+1}, \ldots, \alpha_s$ are nonzero. It remains to show that each block in (12.2.5) is strictly equivalent to one of the blocks (q-ϕ-h-sk0)–(q-ϕ-h-sk6). This is trivial for $0_{u \times u}$. For (q-ϕ-h-sk1), we have

$$
\left(L_{\varepsilon \times (\varepsilon + 1)}(t) \oplus \Lambda_{\varepsilon \times (\varepsilon + 1)}(-t)^T \right) F_{2\varepsilon + 1} = G_{2\varepsilon + 1} + t \begin{bmatrix} 0 & 0 & F_\varepsilon \\ 0 & 0_1 & 0 \\ -F_\varepsilon & 0 & 0 \end{bmatrix}.
$$

Here the $\varepsilon \times (\varepsilon + 1)$ matrix pencil

$$
\Lambda_{\varepsilon \times (\varepsilon + 1)}(-t) := \begin{bmatrix} 1 & -t & 0 & \cdots & 0 \\ 0 & 1 & -t & \cdots & 0 \\ \vdots & & \ddots & \ddots & \vdots \\ 0 & 0 & \cdots & 1 & -t \end{bmatrix} \in \mathsf{H}^{\varepsilon \times (\varepsilon + 1)}
$$

is strictly equivalent to $L_{\varepsilon \times (\varepsilon + 1)}(-t)$, which, in turn, is strictly equivalent to the pencil $L_{\varepsilon \times (\varepsilon + 1)}(t)$. Indeed,

$$
F_\varepsilon L_{\varepsilon \times (\varepsilon + 1)}(-t) F_{\varepsilon + 1} = \Lambda_{\varepsilon \times (\varepsilon + 1)}(-t),
$$

and

$$
\text{diag}(-1, 1, -1, \ldots, \pm 1) L_{\varepsilon \times (\varepsilon + 1)}(t) \text{diag}(1, -1, 1, \ldots, \pm 1)
$$

$$
= L_{\varepsilon \times (\varepsilon + 1)}(-t).
$$

The block $I_k + t J_k(0)$ is strictly equivalent to (q-ϕ-h-sk2):

$$
\left(\text{diag}(1, \beta(\phi)^{-1}, \ldots, \beta(\phi)^{-k+1}) \right) (I_k + t J_k(0)) F_k
$$

$$
\cdot \left(\text{diag}(\beta(\phi)^{k-1}, \beta(\phi)^{k-2}, \ldots, 1) \right) = F_k + t\beta(\phi) G_k.
$$

Analogously, the block $t I_\ell + J_\ell(0)$ is strictly equivalent to a block (q-ϕ-h-sk3) (of the same size $\ell \times \ell$):

$$
\left(\text{diag}(1, (-\beta(\phi))^{-1}, \ldots, (-\beta(\phi))^{-\ell+1}) \right) (t I_\ell + J_\ell(0)) \beta(\phi) F_\ell
$$

$$
\cdot \left(\text{diag}((-\beta(\phi))^{\ell-1}, (-\beta(\phi))^{\ell-2}, \ldots, 1) \right) = \beta(\phi) t F_\ell + G_\ell.
$$

If the real part of α is nonzero, then $(tI_{\ell/2} + J_{\ell/2}(\alpha)) \oplus (tI_{\ell/2} + J_{\ell/2}(-\alpha))$ is strictly equivalent to (q-ϕ-h-sk4):

$$\left(\begin{bmatrix} 0 & \alpha F_{\ell/2} + G_{\ell/2} \\ \alpha F_{\ell/2} + G_{\ell/2} & 0 \end{bmatrix} + t \begin{bmatrix} 0 & F_{\ell/2} \\ -F_{\ell/2} & 0 \end{bmatrix} \right)$$

$$\cdot \begin{bmatrix} 0 & -F_{\ell/2} \\ F_{\ell/2} & 0 \end{bmatrix} = \begin{bmatrix} tI + J_{\ell/2}(\alpha) & 0 \\ 0 & tI - J_{\ell/2}(\alpha) \end{bmatrix},$$

and note that $-J_{\ell/2}(\alpha)$ is similar to $J_{\ell/2}(-\alpha)$. Finally, consider a block $tI_s + J_s(\alpha)$, where $\alpha \in \mathrm{Inv}\,(\phi) \setminus \{0\}$ and the real part of α is zero. We show that for odd s this block is strictly equivalent to (q-ϕ-h-sk5), and for even s it is strictly equivalent to (q-ϕ-h-sk6). Indeed, $(tI_s + J_s(\alpha))\Xi_s\,(\beta(\phi))$ (for s odd) is equal to the matrix

$$\Psi_s(\alpha\beta(\phi)) + t\Xi_s\,(\beta(\phi)).$$

Note that $\alpha\beta(\phi) \in \mathrm{Inv}\,(\phi) \setminus \{0\}$; hence, the transformation

$$\Psi_s(\alpha\beta(\phi)) + t\Xi_s\,(\beta(\phi)) \quad \longrightarrow \quad \phi(\omega)I \cdot (\Psi_s(\alpha\beta(\phi)) + t\Xi_s\,(\beta(\phi))) \cdot \omega I$$

for a suitable $\omega \in \mathsf{H}$, $|\omega| = 1$, yields the form (12.1.3); see Corollary 4.1.4. For the case when s is even, we first observe that $(J_s(\alpha))^{-1}$ is similar to $J_s(\alpha^{-1})$, and, therefore, $tI_s + J_s(\alpha)$ is strictly equivalent to $I_s + tJ_s(\alpha^{-1})$. (Note that $\alpha^{-1} \in \mathrm{Inv}\,(\phi) \setminus \{0\}$, and the real part of α^{-1} is zero as well.) Now, as in the case of odd s, a suitable transformation of $(I_s + J_s(\alpha^{-1}))\Xi_s\,(\beta(\phi))$ yields the form (12.1.3). This completes the proof of the existence part of Theorem 12.1.2.

The proof of uniqueness of Theorem 12.1.2 follows from the uniqueness of the Kronecker form of the ϕ-hsk pencil $A + tB$, and from the proof (given above) that each block in (12.2.5) is strictly equivalent to one, in fact exactly one, of the blocks (q-ϕ-h-sk0)–(q-ϕ-h-sk6).

12.3   CANONICAL FORMS OF ϕ-MIXED PENCILS: CONGRUENCE

We state the main result on the canonical form of ϕ-hermitian-skewhermitian matrix pencils under (simultaneous) ϕ-conruence.

Theorem 12.3.1. *Every ϕ-hsk matrix pencil $A + tB$ is ϕ-congruent to a ϕ-hsk matrix pencil of the form*

$$\begin{aligned} (A_0 + tB_0) \quad &\oplus \quad \oplus_{j=1}^{p} \varepsilon_j (F_{k_j} + t\beta(\phi)G_{k_j}) \oplus \oplus_{j=1}^{q} \kappa_j (G_{\ell_j} + t\beta(\phi)F_{\ell_j}) \\ &\oplus \quad \oplus_{j=1}^{r} \delta_j (\Psi_{s_j}(\nu_j) + t\Xi_{s_j}\,(\beta(\phi))) \\ &\oplus \quad \oplus_{j=1}^{m} \mu_j (\Xi_{w_j}\,(\beta(\phi)) + t\Psi_{w_j}(\tau_j)), \end{aligned}$$

where the parameters have the following properties:

(1) *The integers k_j's are even, ℓ_j's are odd, s_j's are odd, and w_j's are even.*

(2) *The ε_j's, κ_j's, δ_j's, and μ_j's are signs ± 1.*

(3) *The ν_j and τ_j are positive reals.*

(4) *The part $A_0 + tB_0$ is block diagonal with the diagonal blocks consisting of blocks of the types (q-ϕ-h-sk0), (q-ϕ-h-sk1), (q-ϕ-h-sk4), (q-ϕ-h-sk2) of odd sizes, and (q-ϕ-h-sk3) of even sizes.*

The form (12.3.1) is uniquely determined by the pencil $A + tB$, up to permutations of diagonal blocks in each of the five parts

$$A_0 + tB_0, \quad \oplus_{j=1}^{p} \varepsilon_j (F_{k_j} + t\beta(\phi)G_{k_j}), \quad \oplus_{j=1}^{q} \kappa_j (G_{\ell_j} + t\beta(\phi)F_{\ell_j}),$$

$$\oplus_{j=1}^{r} \delta_j (\Psi_{s_j}(\nu_j) + t\Xi_{s_j}(\beta(\phi))), \quad \text{and} \quad \oplus_{j=1}^{m} \mu_j (\Xi_{w_j}(\beta(\phi)) + t\Psi_{w_j}(\tau_j)),$$

and up to a replacement in each block of type (q-ϕ-h-sk4) the quaternion α with a similar quaternion $\alpha' \in \mathrm{Inv}(\phi)$.

Conversely, if a matrix pencil $A + tB$ is ϕ-congruent to a pencil of the form (12.3.1), then $A + tB$ is ϕ-hsk.

A direct and independent proof of Theorem 12.3.1, with full detail, is given in Rodman [133]. It follows a general outline of the proof of Theorem 11.2.2(b) (see also Theorem 8.1 in Rodman [132]); this, in turn, is based on a similar approach for real and complex matrix pencils with symmetries that was used in many sources; see, e.g., Thompson [151] and Lancaster and Rodman [92, 89]. The proof in Rodman [133] is rather lengthy, so we will not reproduce it here. Instead, we will present in the next section a partial proof (of existence) based on reduction to Theorem 9.4.1.

Remark 12.3.2. The proofs of Theorems 12.1.2 (Section 12.2) and 12.3.1 (as presented in Rodman [133]) will show that in the form (12.3.1), the blocks of types (q-ϕ-h-sk5) and (q-ϕ-h-sk6) may be replaced by more general types, as follows: Fix $\alpha_0, \alpha_0' \in \mathbb{H} \setminus \{0\}$ such that $\phi(\alpha_0) = \alpha_0$ and $\phi(\alpha_0') = \alpha_0'$. Then (12.3.1) can be replaced by the form

$$
\begin{aligned}
(A_0 + tB_0) \quad &\oplus \quad \oplus_{j=1}^{p} \varepsilon_j (F_{k_j} + t\beta(\phi)G_{k_j}) \oplus \oplus_{j=1}^{q} \kappa_j (G_{\ell_j} + t\beta(\phi)F_{\ell_j}) \\
&\oplus \quad \oplus_{j=1}^{r} \delta_j (\Psi_{s_j}(x_j\alpha_0) + t\Xi_{s_j}(\beta(\phi))) \\
&\oplus \quad \oplus_{j=1}^{m} \mu_j (\Xi_{w_j}(\beta(\phi)) + t\Psi_{w_j}(y_j\alpha_0')), \quad (12.3.1)
\end{aligned}
$$

where x_j's and y_j's are real positive numbers, and all other parameters are as in Theorems 12.1.2 and 12.3.1. The form (12.3.1) is unique up to a permutation of diagonal blocks. In particular, one can take $\alpha_0 = \alpha_0' = -1$ in (12.3.1), which amounts to the requirement that ρ is real negative in (q-ϕ-h-sk5) and (q-ϕ-h-sk6), rather than real positive.

Remark 12.3.3. Note that for even ℓ, the pencil $G_\ell + t\beta(\phi)F_\ell$ is ϕ-congruent to its negative, $-G_\ell + t\beta(\phi)(-F_\ell)$. Indeed, if $q \in \mathrm{Inv}(\phi)$ is such that $q^2 = -1$, then $q\beta(\phi) = -\beta(\phi)q$, and

$$(\mathrm{diag}\,(q, -q, \ldots, -q))(G_\ell + tF_\ell)(\mathrm{diag}\,(q, -q, \ldots, -q)) = -G_\ell - t\beta(\phi)F_\ell.$$

Observe that the forms (q-ϕ-h-sk5) and (q-ϕ-h-sk6) involve only quaternions that are real linear combinations of 1 and $\beta(\phi)$. This circumstance allows one to compare these forms with the canonical forms of complex hermitian-skewhermitian matrix pencils by means of the real linear map

$$\chi : \mathrm{Span}_{\mathbb{R}}\{1, \beta(\phi)\} \quad \longrightarrow \quad \mathbb{C}, \quad \chi(1) = 1, \quad \chi(\beta(\phi)) = \mathrm{i}. \quad (12.3.2)$$

Lemma 12.3.4. (a) *Consider the ϕ-hsk matrix pencil $\Psi_s(\rho) + t\Xi_s(\beta(\phi))$, where ρ be a nonzero real number and s is odd. Then there exists an invertible matrix S with entries in $\mathrm{Span}_R\{1, \beta(\phi)\}$ such that*

$$S_\phi(\Psi_s(\rho) + t\Xi_s(\beta(\phi)))S = \pm(F_s + t\beta(\phi)F_sJ_s(\rho^{-1})).$$

The sign $\pm$ depends only on s and on the sign (positive or negative) of ρ.

(b) *For the ϕ-hsk matrix pencil $\Xi_s(\beta(\phi)) + t\Psi_s(\rho)$, where ρ is a nonzero real number and s is even, there exists an invertible matrix $\widetilde{S}$ with entries in the algebra $\mathrm{Span}_R\{1, \beta(\phi)\}$ such that*

$$\widetilde{S}_\phi(\Xi_s(\beta(\phi)) + t\Psi_s(\rho))\widetilde{S} = \pm(F_s + t\beta(\phi)F_sJ_s(-\rho)), \qquad (12.3.3)$$

where the sign $\pm$ depends only on s and on the sign of ρ.

Proof. We will prove (b) only, the proof of (a) being completely analogous. Define the complex matrices H and G by the equalities

$$H = \chi(\Xi_s(\beta(\phi))), \quad iG = \chi(\Psi_s(\rho)),$$

with the map χ applied entrywise. Then H and G are hermitian and invertible. Moreover, $H^{-1}G$ is similar to $J_s(-\rho)$. Now Theorem 15.3.1 yields existence of an invertible complex matrix W such that

$$W^*(H + tiG)W = \pm(F_s + tiF_sJ_s(-\rho)). \qquad (12.3.4)$$

Letting $\widetilde{S} = \chi^{-1}(W)$, formula (12.3.3) follows. To see that the sign in (12.3.4) depends only on s and on the sign of ρ, we appeal to the perturbation theory of sign characteristic of a pair of complex or real invertible hermitian matrices (see Gohberg et al. [53, Theorem 5.9.1] and Rodman [131]). $\qquad\square$

Using Lemma 12.3.4 and Remark 12.3.2, the main result of Theorem 12.3.1 may be reformulated as follows.

Theorem 12.3.5. *Every ϕ-hsk matrix pencil $A + tB$ is ϕ-congruent to a ϕ-hsk matrix pencil of the form*

$$
\begin{aligned}
(A_0 + tB_0) \quad &\oplus \quad \oplus_{j=1}^p \varepsilon_j(F_{k_j} + t\beta(\phi)G_{k_j}) \\
&\oplus \quad \oplus_{j=1}^q \kappa_j(G_{\ell_j} + t\beta(\phi)F_{\ell_j}) \\
&\oplus \quad \oplus_{j=1}^r \delta_j(F_{s_j} + t\beta(\phi)F_{s_j}J_{s_j}(\tau_j)), \qquad (12.3.5)
\end{aligned}
$$

where the parameters have the following properties:

(1) *The integers k_j's are even, and ℓ_j's are odd.*

(2) *The ε_j's, κ_j's, δ_j's are signs ± 1.*

(3) *τ_j are positive reals.*

(4) *The part $A_0 + tB_0$ is block diagonal with the diagonal blocks consisting of blocks of the types (q-ϕ-h-sk0), (q-ϕ-h-sk1), (q-ϕ-h-sk4), (q-ϕ-h-sk2) of odd sizes, and (q-ϕ-h-sk3) of even sizes.*

The uniqueness properties of the form (12.3.5) are the same as those of (5.9.1).

The signs ε_j's, κ_j's, δ_j's in (12.3.5) (or in (12.3.1)) constitute the *sign characteristic* of the ϕ-hsk pencil $A + tB$. Thus, the sign characteristic assigns a $\pm$ sign to every Jordan block in the Kronecker form of $A + tB$ with a nonzero eigenvalue having zero real part, to every Jordan block of $A + tB$ with zero eigenvalue and odd size, and to every block corresponding to infinity of even size. The uniqueness statements of Theorems 12.3.1 and 12.3.5 lead naturally to equivalence of sign characteristics: two sign characteristics are said to be *equivalent* if for every eigenvalue $\tau\beta(\phi)$ of $A + tB$, $0 \le \tau \le \infty$, and for every positive integer k, the number of signs $+1$ (or, equivalently, of signs -1) associated with the blocks of size k and eigenvalue $\tau\beta(\phi)$ is the same in both sign characteristics (k is assumed to be odd if $\tau = 0$ and even if $\tau = \infty$). We obtain from Theorems 12.3.1 and 12.3.5 that *two ϕ-hsk matrix pencils are ϕ-congruent if and only if they are strictly equivalent and have equivalent sign characteristics.*

12.4   PROOF OF THEOREM 12.3.1

We prove here the existence part of Theorem 12.3.1 only and refer the reader to Rodman [133] for a full detailed proof of the theorem.

Let $A + tB$ be a ϕ-hsk (quaternion) matrix pencil. Set

$$\widetilde{A} := -\beta(\phi)A, \qquad \widetilde{B} := -\beta(\phi)B.$$

By Proposition 4.1.7, the matrix pencil $\widetilde{B} + t\widetilde{A}$ is hermitian/skewhermitian. Let

$$Y := S^*(\widetilde{B} + t\widetilde{A})S \tag{12.4.1}$$

be a canonical form of $\widetilde{B} + t\widetilde{A}$ as in Theorem 9.4.1, and let $\widetilde{Y} := S^*(\widetilde{A} + t\widetilde{B})S$ be the form obtained from (12.4.1) with the roles of $\widetilde{A}$ and $\widetilde{B}$ interchanged. Then by Proposition 4.1.7,

$$\beta(\phi)\widetilde{Y} = S_\phi(A + tB)S,$$

and all that remains is to prove that the constituent blocks of $\beta(\phi)\widetilde{Y}$ are ϕ-congruent to the corresponding blocks in (9.4.1), (9.4.2), or (12.3.5). Ignoring the trivial zero block, this boils down to the following lemma.

Lemma 12.4.1. (1) *The matrix pencils*

$$A_1 + tB_1 := \beta(\phi)\begin{bmatrix} 0 & 0 & F_\varepsilon \\ 0 & 0_1 & 0 \\ -F_\varepsilon & 0 & 0 \end{bmatrix} + t\beta(\phi)G_{2\epsilon+1}$$

and

$$A_2 + tB_2 := G_{2\varepsilon+1} + t\begin{bmatrix} 0 & 0 & F_\varepsilon \\ 0 & 0_1 & 0 \\ -F_\varepsilon & 0 & 0 \end{bmatrix}$$

are ϕ-congruent.

(2) *The matrix pencils*

$$\beta(\phi)iG_k + t\beta(\phi)F_k \quad \text{and} \quad \pm(G_k + t\beta(\phi)F_k)$$

are ϕ-congruent if k is even, and

$$\beta(\phi)\mathrm{i}G_k + t\beta(\phi)F_k \quad \text{and} \quad G_k + t\beta(\phi)F_k$$

are ϕ-congruent if k is odd.

(3) *The matrix pencils*

$$\beta(\phi)\mathrm{i}F_\ell + t\beta(\phi)G_\ell \quad \text{and} \quad \pm (F_\ell + t\beta(\phi)G_\ell)$$

are ϕ-congruent if ℓ is odd, and

$$\beta(\phi)\mathrm{i}F_\ell + t\beta(\phi)G_\ell \quad \text{and} \quad F_\ell + t\beta(\phi)G_\ell$$

are ϕ-congruent if ℓ is even.

(4) *For every $\alpha \in \mathsf{H}$ with positive real part, the matrix pencils*

$$\beta(\phi) \begin{bmatrix} 0 & F_p \\ -F_p & 0 \end{bmatrix} + t\beta(\phi) \begin{bmatrix} 0 & \alpha F_p + G_p \\ \alpha^* F_p + G_p & 0 \end{bmatrix}$$

and

$$\begin{bmatrix} 0 & \widetilde{\alpha}F_p + G_p + tF_p \\ \widetilde{\alpha}F_p + G_p - tF_p & 0 \end{bmatrix}$$

are ϕ-congruent for some $\widetilde{\alpha} \in \mathrm{Inv}\,(\phi)$, which also has positive real part.

(5) *The matrix pencils*

$$A_3 + tB_3 \ := \ \beta(\phi)$$

$$\cdot \left(\Xi_m(\mathrm{i}^m) + t \begin{bmatrix} 0 & 0 & \dots & 0 & \beta \\ 0 & 0 & \dots & -\beta & -1 \\ \vdots & \vdots & \ddots & \vdots & \vdots \\ 0 & (-1)^{m-2}\beta & -1 & \dots & 0 \\ (-1)^{m-1}\beta & -1 & \dots & 0 & 0 \end{bmatrix} \right) \tag{12.4.2}$$

and

$$\delta(F_m + t\beta(\phi)F_m J_m(\tau))$$

are ϕ-congruent for some choice of $\delta = \pm 1$. Here $\beta > 0$ if m is odd, $\beta \in \mathsf{H}$ is nonzero with zero real part if m is even, and τ is a positive number in both cases.

The proof of (5) will show that $\delta = -1$ if m is even or if m is odd and $m - 1$ not divisible by 4 and $\delta = 1$ otherwise.

Proof. Proof of (1). We identify $\mathrm{Span}_\mathsf{R}\,\{1, \beta(\phi)\}$ with C, via identification of $\beta(\phi)$ with the complex imaginary unit. Then ϕ acts on $\mathrm{Span}_\mathsf{R}\,\{1, \beta(\phi)\}$ as complex conjugation, and both matrix pencils in (1) are complex hermitian/skewhermitian. Since both pencils have the same Kronecker form

$$\begin{bmatrix} 0 & L_{\epsilon \times (\epsilon+1)}(t) \\ L_{\epsilon \times (\epsilon+1)}(t) & 0 \end{bmatrix},$$

it follows from Theorem 15.3.1 (applied to the pencils $A_1 - t\beta(\phi)B_1$ and $A_2 - t\beta(\phi)B_2$) that the matrix pencils in (1) are C-congruent and, therefore, ϕ-congruent.

Proof of (2). In view of Remark 9.4.5, we may replace i by any quaternion λ with zero real part and norm 1. (In this respect note that $\beta(\phi)\alpha \in \mathrm{Inv}\,(\phi)$ for all

$\alpha \in \mathsf{H}$ with zero real part.) So we replace i with $\beta(\phi)$. Identifying $\mathrm{Span}_\mathsf{R}\{1, \beta(\phi)\}$ with C, and using Theorem 15.3.1 as in the proof of (1), we see that

$$\beta(\phi)\mathsf{i}G_k + t\beta(\phi)F_k \quad \text{and} \quad G_k + t\beta(\phi)F_k$$

are ϕ-congruent. If, in addition, k is even, then $G_k + t\beta(\phi)F_k$ is ϕ-congruent to its negative by Remark 12.3.3.

The proof of (3) is similar to that of (2), using the equality (analogous to Remark 12.3.3)

$$\mathrm{diag}\,(q, -q, \ldots, q)(F_\ell + t\beta(\phi)G_\ell)\mathrm{diag}\,(-q, q, \ldots, -q) = -(F_\ell + t\beta(\phi)G_\ell)$$

for an odd size ℓ, where $q \in \mathsf{H}$ is such that $\phi(q) = q$, $q^2 = -1$, and $q\beta(\phi)q = \beta(\phi)$.

Proof of (4). As it follows from the uniqueness statement of Theorem 9.4.1, α can be replaced by any similar quaternion. Thus, we may assume that $\alpha \in \mathrm{Span}_\mathsf{R}\{1, \beta(\phi)\}$. As before, we identify $\mathrm{Span}_\mathsf{R}\{1, \beta(\phi)\}$ with C and reduce the proof to C-congruence of complex hermitian/skewhermitian matrix pencils or, equivalently, (upon replacing t with $t\beta(\phi)$) to C-congruence of complex hermitian pencils (Theorem 15.3.1).

Proof of (5). Consider two cases separately: (a) m is even; (b) m is odd. Suppose first m is even. In view of Theorem 9.4.1, β can be replaced by any similar quaternion. So, we take $\beta = \pm\tau\beta(\phi)$, where $\tau > 0$ and where we take the sign $-$ if m is divisible by 4 and the sign $+$ if m is not divisible by 4. Let $(\beta(\phi), q_2, q_3)$ be a units triple. Then the following equalities hold true:

$$\mathrm{diag}\,(q_2, -q_3, \ldots, q_2, -q_3)(A_3 + tB_3)\mathrm{diag}\,(q_2, -q_3, \ldots, q_2, -q_3)$$

$$= -(F_m + t\beta(\phi)F_m J_m(\tau))$$

if m is divisible by 4 and

$$\mathrm{diag}\,(q_2, q_3, \ldots, q_2, q_3)(A_3 + tB_3)\mathrm{diag}\,(q_2, q_3, \ldots, q_2, q_3)$$

$$= -(F_m + t\beta(\phi)F_m J_m(\tau))$$

if m is not divisible by 4. This verifies the required ϕ-congruence.

Next, suppose m is odd. Note that one can replace i in (12.4.2) by any quaternion λ similar to α; indeed, if $\lambda = q^*\mathsf{i}q$ for some $q \in \mathsf{H}$ with $|q| = 1$, and denoting for convenience the matrix $\beta(\phi)^{-1}B_3$ by Υ, we have

$$(q^*I)(\Xi_m(\mathsf{i}^m) + t\Upsilon)(qI) = \Xi_m(\lambda^m) + t\Upsilon.$$

Thus, upon replacing i with $\pm\beta(\phi)$, where the sign chosen so that $\beta(\phi)(\pm\beta(\phi))^m = 1$, the matrix pencil $A_3 + tB_3$ takes the form

$$A_3 + tB_3 = \Xi(1) + t\beta(\phi)\Upsilon.$$

We now identify $\mathrm{Span}_\mathsf{R}\{1, \beta(\phi)\}$ with C and observe that the Kronecker form over C of $A_3 + tB_3$ is $tI + J_m(\alpha)$, where $\alpha := (\beta(\phi)\beta)^{-1}$, and that of $F_m + t\beta(\phi)F_m J_m(\tau)$ is $tI + J_m(\alpha')$, where $\alpha' := (\beta(\phi)\tau)^{-1}$. Thus, taking $\tau = \beta$, we see by Theorem 15.3.1 that the matrix pencil $A_3 + tB_3$ is ϕ-congruent over C (which amounts to C-congruence) to either $F_m + t\beta(\phi)F_m J_m(\tau)$ or to $-(F_m + t\beta(\phi)F_m J_m(\tau))$. In fact, using the canonical form of ϕ-skewsymmetric matrices (Theorem 4.1.2), one can easily see that the sign here must be -1 if $m - 1$ is not divisible by 4 and $+1$ otherwise. $\square$

12.5   STRICT EQUIVALENCE VERSUS ϕ-CONGRUENCE

In this section we develop some applications of the canonical form of Theorem 12.3.1 regarding the relations between strict equivalence and ϕ-congruence for ϕ-hsk matrix pencils.

Clearly, ϕ-congruent ϕ-hsk matrix pencils are strictly equivalent, but the converse is generally false. It turns out that every strict equivalence class contains only finitely many ϕ-congruent classes (when restricted to ϕ-hsk pencils), and the number of these can be identified in terms of the Kronecker form of the pencils.

Theorem 12.5.1. *Let $A+tB$ be a ϕ-hsk matrix pencil, where $A, B \in \mathsf{H}^{n \times n}$. Let*

$$\lambda_1 = 0, \quad \lambda_2 = \infty, \quad \lambda_3 \notin \{0, \infty\}, \dots, \lambda_r \notin \{0, \infty\}$$

be all the distinct eigenvalues of $A + tB$, including the zero eigenvalue and infinity, if applicable, having the following properties:

(1) *The real parts of $\lambda_3, \dots, \lambda_r$ are zeros.*

(2) *The norms of the vector parts $|\mathfrak{V}(\lambda_3)|, \dots, |\mathfrak{V}(\lambda_r)|$ are all distinct.*

Let $k_{1,1} < \cdots < k_{1,p_1}$ be the distinct odd indices of the eigenvalue 0 of $A + tB$, let $k_{2,1} < \cdots < k_{2,p_2}$ be the distinct even indices at infinity of $A + tB$, and for $j = 3, 4, \dots, r$, let $k_{j,1} < \cdots < k_{j,p_j}$ be the distinct indices of the eigenvalue λ_j. Furthermore, assume that the index $k_{j,m}$ appears $q_{j,m}$ times in the Kronecker form of $A + tB$; i.e., there are exactly $q_{j,m}$ blocks $tI_{k_{j,m}} + J_{k_{j,m}}(\lambda_j)$ in the Kronecker form of $A + tB$ ($k_{j,m}$ is odd if $j = 1$) for $m = 1, 2, \dots, p_j$ and $j = 1, 3, 4, \dots, r$, and there are exactly $q_{1,m}$ blocks $I_{k_{1,m}} + tJ_{k_{1,m}}(0)$ of an even size $k_{1,m}$. Then there exist

$$w := \prod_{j=1}^{r} \prod_{m=1}^{p_j} (q_{j,m} + 1) \tag{12.5.1}$$

ϕ-hsk mutually pairwise non-ϕ-congruent pencils $A_1 + tB_1, \dots, A_w + tB_w$, each of which is strictly equivalent to $A + tB$. Moreover, there do not exist $w + 1$ ϕ-hsk mutually pairwise non-ϕ-congruent pencils, each of which is strictly equivalent to $A + tB$.

The proof follows easily from Theorem 12.3.1; indeed, for fixed j and m, as in Theorem 12.5.1, there are exactly $q_{j,m} + 1$ mutually nonequivalent ways to assign the signs corresponding to the blocks of size $k_{j,m}$ with eigenvalue λ_j in the sign characteristic of $A + tB$.

Two corollaries of Theorem 12.5.1 are worthy of separate statements.

Corollary 12.5.2. *A ϕ-hsk quaternion matrix pencil $A+tB$ has the property that every ϕ-hsk quaternion pencil that is strictly equivalent to $A + tB$ is automatically ϕ-congruent to $A + tB$ if and only if $A + tB$ has no even indices at infinity, no odd indices corresponding to the zero eigenvalue, and no nonzero eigenvalues having zero real parts.*

Corollary 12.5.3. *The maximal number of elements in a set of $n \times n$ quaternion matrix pencils which are strictly equivalent to each other, but mutually pairwise non-ϕ-congruent, is equal to 2^n.*

For the proof of Corollary 12.5.3 observe that for a fixed n, the maximal value of w in (12.5.1) is 2^n, which is attained for any ϕ-hsk pencil $A + tB$ with n nonreal eigenvalues with distinct norms of their vector parts.

12.6 CANONICAL FORMS OF MATRICES UNDER ϕ-CONGRUENCE

As another application of Theorem 12.3.1, we derive a canonical form of matrices $A \in \mathsf{H}^{n \times n}$ under the ϕ-congruence relation; recall that two matrices $X, Y \in \mathsf{H}^{n \times n}$ are called ϕ-*congruent* if $Y = S_\phi X S$ for some invertible $S \in \mathsf{H}^{n \times n}$. Using the decomposition

$$A = B + C, \quad B = B_\phi, \quad C = -C_\phi,$$

where

$$B = \frac{A_\phi + A}{2}, \quad C = \frac{-A_\phi + A}{2},$$

the problem reduces to the problem of a canonical form of the ϕ-hsk pencil $B + tC$.

It will be convenient to work with suitably modified blocks (q-ϕ-h-sk1)–(q-ϕ-h-sk6).

(q-ϕ-h-sk1$'$)

$$G_{2\varepsilon+1} + \begin{bmatrix} 0 & 0 & F_\varepsilon \\ 0 & 0_1 & 0 \\ -F_\varepsilon & 0 & 0 \end{bmatrix}, \quad \varepsilon \text{ positive integer.}$$

(q-ϕ-h-sk2$'$) $\quad F_k + \beta(\phi)G_k, \quad k$ positive integer.

(q-ϕ-h-sk3$'$) $\quad G_k + \beta(\phi)F_k, \quad k$ positive integer.

(q-ϕ-h-sk4$'$)

$$\begin{bmatrix} 0 & (1+\alpha)F_{\frac{\ell}{2}} + G_{\frac{\ell}{2}} \\ (-1+\alpha)F_{\frac{\ell}{2}} + G_{\frac{\ell}{2}} & 0 \end{bmatrix},$$

where ℓ is even, $\alpha \in \mathrm{Inv}\,(\phi)$, $\Re(\alpha) > 0$.

(q-ϕ-h-sk5$'$)

$$\Lambda_s(\rho) := \begin{bmatrix} 0 & 0 & \cdots & 0 & -\beta(\phi) & \Upsilon \\ 0 & 0 & \cdots & \beta(\phi) & -\Upsilon & 0 \\ \vdots & \vdots & \reflectbox{$\ddots$} & \reflectbox{$\ddots$} & \vdots & \vdots \\ (-1)^{s-1}\beta(\phi) & (-1)^{s-2}\Upsilon & \cdots & 0 & 0 & 0 \\ (-1)^{s-1}\Upsilon & 0 & \cdots & 0 & 0 & 0 \end{bmatrix}, \quad (12.6.1)$$

where the size of the matrix is $s \times s$, ρ is real positive, and s may be even or odd; we have denoted here $\Upsilon := \rho + \beta(\phi)$.

In the forms (q-ϕ-h-sk1$'$)–(q-ϕ-h-sk5$'$) modifications are possible, analogous to the modifications in the forms (q-ϕ-h-sk1)–(q-ϕ-h-sk6) discussed in Section 12.3.

Invoking Theorem 12.3.1, the following result is obtained.

Theorem 12.6.1. *Every matrix $A \in \mathsf{H}^{m \times m}$ is ϕ-congruent to a matrix in the form*

$$0_{u \times u} \quad \oplus \quad (A_0 + tB_0) \oplus \oplus_{j=1}^{p} \varepsilon_j(F_{k_j} + \beta(\phi)G_{k_j})$$
$$\oplus \quad \oplus_{j=1}^{q} \kappa_j(G_{\ell_j} + \beta(\phi)F_{\ell_j}) \oplus \oplus_{j=1}^{r} \delta_j \Lambda_{s_j}(\rho_j), \quad (12.6.2)$$

where the parameters have the following properties:

(1) *the integers k_j's are even, and the ℓ_j's are odd;*

(2) *the ε_j's, κ_j's, and δ_j's are signs ± 1;*

(3) *the ρ_j's are positive reals;*

(4) *the part $A_0 + tB_0$ is block diagonal with the diagonal blocks consisting of blocks of the types (q-ϕ-h-sk1'), (q-ϕ-h-sk4'), (q-ϕ-h-sk2') of odd sizes, and (q-ϕ-h-sk3') of even sizes.*

The form (12.6.2) is uniquely determined by the pencil A, up to a permutation of the diagonal blocks in each of the four parts

$$A_0 + tB_0, \quad \oplus_{j=1}^{p} \varepsilon_j (F_{k_j} + \beta(\phi)G_{k_j}), \quad \oplus_{j=1}^{q} \kappa_j (G_{\ell_j} + \beta(\phi)F_{\ell_j}),$$

$$\text{and} \quad \oplus_{j=1}^{r} \delta_j \Lambda_{s_j}(\rho_j),$$

and up to a replacement, in each block of type (q-ϕ-h-sk4'), the quaternion α with a similar quaternion $\alpha' \in \text{Inv}(\phi)$.

An alternative statement can be given using Theorem 12.3.5 rather than Theorem 12.3.1. The only change in Theorem 12.6.1 would be that the block $\Lambda_{s_j}(\rho_j)$ is replaced with $F_{s_j} + \beta(\phi)F_{s_j}J_{s_j}(\rho_j)$.

12.7 COMPARISON WITH REAL AND COMPLEX MATRICES

We compare the strict equivalence and ϕ-congruence relations for real, complex, and quaternion matrix pencils and begin with real matrix pencils. It will be efficient to deal with several symmetries of real matrix pencils at once.

Definition 12.7.1. We say that complex matrix pencils $A + tB$ and $A' + tB'$ are $(\mathsf{C}, ^T)$-*congruent* if $A + tB = S^T(A' + tB')S$ for some invertible complex matrix S.

Theorem 12.7.2. *Fix $\eta = \pm 1$, $\tau = \pm 1$. Let $A, B, A', B' \in \mathsf{R}^{m \times m}$ be such that*

$$A^T = \eta A, \ (A')^T = \eta A', \quad B^T = \tau B, \ (B')^T = \tau B'. \tag{12.7.1}$$

Then statements (i), (ii), and (iii) are equivalent.

(i) *The matrix pencils $A + tB$ and $A' + tB'$ are R-strictly equivalent.*

(ii) *The matrix pencils $A + tB$ and $A' + tB'$ are $(\mathsf{C}, ^T)$-congruent.*

(iii) *The matrix pencils $A + tB$ and $A' + tB'$ are ϕ-congruent for some nonstandard involution ϕ, equivalently, for all nonstandard involutions ϕ.*

Proof. Clearly, (iii) implies that $A + tB$ and $A' + tB'$ are H-strictly equivalent, which, in turn, yields (i) (in view of Theorem 7.6.3). To see that (ii) implies (iii), identify C with $\text{Span}_{\mathsf{R}}\{1, q\}$, where $q \in \mathsf{H} \setminus \{0\}$ has zero real part and $\phi(q) = q$.

It remains to prove that (i) implies (ii). So, assume (i) holds. The canonical forms of symmetric, skewsymmetric, or mixed symmetric/skewsymmetric real matrix pencils under R-strict equivalence and under R-congruence are given in Theorems 15.2.1, 15.2.2, and 15.2.3. It follows from these theorems that there exist real invertible matrices S and T and pairs of matrices

$$A_j, B_j \in \mathsf{R}^{n_j \times n_j}, \quad n_1 + \cdots + n_p = m,$$

such that

$$S^T(A + tB)S = \oplus_{j=1}^{p}\delta_j(A_j + tB_j),$$
$$T^T(A' + tB')T = \oplus_{j=1}^{p}\xi_j(A_j + tB_j),$$

where for each j, the pair (A_j, B_j) satisfies

$$A_j^T = \eta A_j, \quad B_j^T = \tau B_j,$$

and $\delta_1, \ldots, \delta_p, \xi_1, \ldots, \xi_p$ are signs ± 1. It suffices to show that

$$\oplus_{j=1}^{p}\delta_j(A_j + tB_j) \quad \text{and} \quad \oplus_{j=1}^{p}\xi_j(A_j + tB_j)$$

are $(\mathsf{C},^T)$-congruent. This is easy:

$$\oplus_{j=1}^{p}\delta_j(A_j + tB_j) = \left(\oplus_{j=1}^{p}W_j\right)\left(\oplus_{j=1}^{p}\xi_j(A_j + tB_j)\right)\left(\oplus_{j=1}^{p}W_j\right),$$

where $W_j = I_{n_j}$ if $\delta_j = \xi_j$ and $W_j = \mathrm{i}I_{n_j}$ if $\delta_j \neq \xi_j$. $\qquad\square$

Consider now comparison with pairs of complex matrices. In the rest of this section we identify (as usual) C with $\mathrm{Span}_\mathsf{R}\{1, \mathrm{i}\} \subset \mathsf{H}$. and we assume that the nonstandard involution ϕ is such that $\beta(\phi) \in \mathrm{Span}_\mathsf{R}\{\mathrm{j}, \mathrm{k}\}$.

We obviously have the following.

Proposition 12.7.3. *If two complex matrix pencils $A + tB$ and $A' + tB'$ are $(\mathsf{C},^T)$-congruent, then they are also ϕ-congruent (over H).*

A key result of this section states that for complex pencils of symmetric or skewsymmetric matrices, H-strict equivalence is the same as ϕ-congruence.

Theorem 12.7.4. *Fix $\eta = \pm 1$, $\tau = \pm 1$. Let $A, B, A', B' \in \mathsf{C}^{m \times m}$ be such that*

$$A^T = \eta A, \quad (A')^T = \eta A', \quad B^T = \tau B, \quad (B')^T = \tau B'.$$

Then the matrix pencils $A + tB$ and $A' + tB'$ are H-strictly equivalent if and only if they are ϕ-congruent.

The lengthy proof of Theorem 12.7.4 is relegated to the next section.

A complete characterization of canonical forms of complex symmetric matrix pencils that are ϕ-congruent to a fixed complex symmetric matrix pencil in a canonical form is given in the next theorem.

Theorem 12.7.5. *Let $A + tB$ be a complex symmetric matrix pencil, and let*

$$0_u \quad \oplus \quad \left(t\begin{bmatrix} 0 & 0 & F_{\varepsilon_1} \\ 0 & 0 & 0 \\ F_{\varepsilon_1} & 0 & 0 \end{bmatrix} + G_{2\varepsilon_1 + 1}\right)$$

$$\oplus \quad \cdots \quad \oplus \left(t\begin{bmatrix} 0 & 0 & F_{\varepsilon_p} \\ 0 & 0 & 0 \\ F_{\varepsilon_p} & 0 & 0 \end{bmatrix} + G_{2\varepsilon_p + 1}\right)$$

$$\oplus \quad (F_{k_1} + tG_{k_1}) \oplus \cdots \oplus (F_{k_r} + tG_{k_r})$$

$$\oplus \quad ((t + \alpha_1)F_{\ell_1} + G_{\ell_1}) \oplus \cdots \oplus ((t + \alpha_q)F_{\ell_q} + G_{\ell_q}),$$

$$\alpha_1, \ldots, \alpha_q \in \mathsf{C}, \tag{12.7.2}$$

be its canonical form under (C^T)*-congruence* (*cf. Theorem* 15.3.3). *Then a complex symmetric matrix pencil* $A'+tB'$ *is* ϕ*-congruent to* $A+tB$ *if and only if the canonical form of* $A'+tB'$ *under* (C^T)*-congruence is obtained from* (12.7.2) *by replacing some* (*or none*) *of the nonreal numbers* α_j *among* $\alpha_1, \ldots, \alpha_q$ *with their complex conjugates* $\overline{\alpha_j}$.

For the proof observe that the canonical form of $A + tB$ under ϕ-congruence is (12.7.2), up to permutation of blocks and replacement of each α_j with its complex conjugate (see Theorem 11.1.3).

Thus, a ϕ-congruence class of a complex symmetric matrix pencil $A+tB$ consists of exactly u classes of (C^T)-congruence, where the number u is computed as follows. For λ, a complex number with positive imaginary part, and a positive integer ℓ, let $\widetilde{s}(\lambda, \ell)$ be the total number of blocks of the form $(t + \alpha)F_\ell + G_\ell$ with $\alpha = \lambda$ or $\alpha = \overline{\lambda}$ in the canonical form (12.7.2) of $A + tB$ under (C^T)- congruence. (If there are no such blocks, we set $\widetilde{s}(\lambda, \ell) = 0$.) Then

$$u = \prod_{\ell > 0} \prod_{\{\lambda \in \mathsf{C} \,:\, \mathfrak{I}(\lambda) > 0\}} (\widetilde{s}(\lambda, \ell) + 1). \tag{12.7.3}$$

The proof of formula (12.7.3) is obtained in a manner similar to the proof of Theorem 9.7.5.

We leave it to the reader to formulate and prove results analogous to Theorem 12.7.5 for skewsymmetric complex matrix pencils and mixed matrix pencils—in other words, symmetric/skewsymmetric (Ex. 12.9.1).

12.8   PROOF OF THEOREM 12.7.4

Clearly, we need to prove only the "only if" part. In the case $\tau = \eta = 1$, this is clear in view of fact that ϕ-hermitian quaternionic pencils are H-strictly equivalent if and only if they are ϕ-congruent (Theorem 11.1.3).

Consider the case $\tau = \eta = -1$. Assume that $A+tB$ and $A'+tB'$ are H-strictly equivalent. Since by Theorem 15.3.2 the relation of C-strict equivalence of complex skewsymmetric pencils is the same as the relation of (C^T)-congruence, we may further replace $A + tB$ with its canonical form under (C^T)-congruence. In other words, we assume that $A + tB$ is given by

$$0_{u \times u} \;\oplus\; \oplus_{j=1}^p \left(t \begin{bmatrix} 0 & 0 & F_{\varepsilon_j} \\ 0 & 0_1 & 0 \\ -F_{\varepsilon_j} & 0 & 0 \end{bmatrix} + \begin{bmatrix} 0 & F_{\varepsilon_j} & 0 \\ -F_{\varepsilon_j} & 0 & 0 \\ 0 & 0 & 0 \end{bmatrix} \right)$$

$$\oplus\; \oplus_{j=1}^r \left(\begin{bmatrix} 0 & F_{k_j} \\ -F_{k_j} & 0 \end{bmatrix} + t \begin{bmatrix} 0 & G_{k_j} \\ -G_{k_j} & 0 \end{bmatrix} \right)$$

$$\oplus\; \oplus_{j=1}^q \left((t + \alpha_j) \begin{bmatrix} 0 & F_{\ell_j} \\ -F_{\ell_j} & 0 \end{bmatrix} + \begin{bmatrix} 0 & G_{\ell_j} \\ -G_{\ell_j} & 0 \end{bmatrix} \right), \tag{12.8.1}$$

where $\alpha_j \in \mathsf{C}$. Note the following ϕ-congruence relations (recall that we assume $\beta(\phi) \in \mathrm{Span}_\mathsf{R}(\mathsf{j}, \mathsf{k})$):

$$(S_{k_j})_\phi \begin{bmatrix} 0 & F_{k_j} + tG_{k_j} \\ -F_{k_j} - tG_{k_j} & 0 \end{bmatrix} S_{k_j}$$

$$= (-(\beta(\phi)F_{k_j} + t\beta(\phi)G_{k_j})) \oplus (\beta(\phi)F_{k_j} + t\beta(\phi)G_{k_j}); \tag{12.8.2}$$

$$(S_{\ell_j})_\phi \left((t + \alpha_j) \begin{bmatrix} 0 & F_{\ell_j} \\ -F_{\ell_j} & 0 \end{bmatrix} + \begin{bmatrix} 0 & G_{\ell_j} \\ -G_{\ell_j} & 0 \end{bmatrix} \right) S_{\ell_j}$$

$$= \left(-\left((t + \alpha_j)\beta(\phi)F_{\ell_j} + \beta(\phi)G_{\ell_j} \right) \right) \oplus \left((t + \alpha_j)\beta(\phi)F_{\ell_j} + \beta(\phi)G_{\ell_j} \right), \qquad (12.8.3)$$

$$\alpha_j \in \mathsf{R},$$

where

$$S_m = \frac{1}{\sqrt{2}} \begin{bmatrix} \beta(\phi)I_m & -\beta(\phi)I_m \\ I_m & I_m \end{bmatrix}, \qquad (S_m)_\phi = \frac{1}{\sqrt{2}} \begin{bmatrix} -\beta(\phi)I_m & I_m \\ \beta(\phi)I_m & I_m \end{bmatrix}.$$

Comparing with the canonical form under ϕ-congruence (Theorem 11.2.2), we see that the canonical form of $A + tB$ under ϕ-congruence is given by (12.8.1), where each block

$$\begin{bmatrix} 0 & F_{k_j} \\ -F_{k_j} & 0 \end{bmatrix} + t \begin{bmatrix} 0 & G_{k_j} \\ -G_{k_j} & 0 \end{bmatrix}$$

is replaced with the right-hand side of (12.8.2), and each block

$$(t + \alpha_j) \begin{bmatrix} 0 & F_{\ell_j} \\ -F_{\ell_j} & 0 \end{bmatrix} + \begin{bmatrix} 0 & G_{\ell_j} \\ -G_{\ell_j} & 0 \end{bmatrix}$$

is replaced with the right-hand side of (12.8.3). In other words, the blocks in the canonical form of $A+tB$ under ϕ-congruence that correspond to the real eigenvalues and to infinity appear in pairs, and in each such pair the two blocks have opposite signs. Of course, the same property is valid also for the canonical form of $A' + tB'$ under ϕ-congruence. Notice that $A + tB$ and $A' + tB'$ have the same canonical form under H-strict equivalence and that the canonical forms of quaternion ϕ-skewhermitian (with respect to a nonstandard involution) matrix pencils under ϕ-congruence and under H-strict equivalence can differ only in the signs associated with blocks corresponding to the real eigenvalues and to the eigenvalue at infinity (see Theorem 11.2.2). We obtain, therefore, that $A + tB$ and $A' + tB'$ have the same canonical form under ϕ-congruence. In other words, $A + tB$ and $A' + tB'$ are ϕ-congruent, as claimed.

Finally, consider the case $\eta = 1$, $\tau = -1$ (the remaining case $\eta = -1$, $\tau = 1$ can be easily reduced to the case under consideration by interchanging the roles of A and B). First of all, we will transform the primitive blocks (q-ϕ-h-sk2)–(q-ϕ-h-sk6) into different forms using ϕ-congruence, so that the obtained forms are easily comparable to the canonical form of quaternion ϕ-hermitian/skewhermitian matrix pencils under ϕ-congruence.

Claim 12.8.1. *The block (q-ϕ-h-sk5) is ϕ-congruent to*

$$(G + t\beta(\phi)F) \oplus (-(G + t\beta(\phi)F)), \qquad (12.8.4)$$

where we let $G = G_{\ell/2}$, $F = F_{\ell/2}$, and recall that $\ell/2$ is odd.

For the proof of the claim, consider the matrix pencil

$$\begin{bmatrix} 0 & G \\ G & 0 \end{bmatrix} + t \begin{bmatrix} 0 & -\beta(\phi)F \\ \beta(\phi)F & 0 \end{bmatrix}. \qquad (12.8.5)$$

The matrix pencil (12.8.5) may be considered as a pencil of complex hermitian matrices under the real linear map Ψ of $\mathrm{Span}_{\mathsf{R}}\{1, \beta(\phi)\}$ onto C via $1 \mapsto 1$ and $\beta(\phi) \mapsto i$. Transformations of the matrix pencil (12.8.5) of the form

$$\begin{bmatrix} 0 & G \\ G & 0 \end{bmatrix} + t \begin{bmatrix} 0 & -\beta(\phi)F \\ \beta(\phi)F & 0 \end{bmatrix}$$

$$\mapsto S_\phi \left(\begin{bmatrix} 0 & G \\ G & 0 \end{bmatrix} + t \begin{bmatrix} 0 & -\beta(\phi)F \\ \beta(\phi)F & 0 \end{bmatrix} \right) S,$$

where S is an invertible matrix with entries in $\mathrm{Span}_{\mathsf{R}}\{1, \beta(\phi)\}$, amount to C-congruences under the map Ψ. Next, we verify that the canonical form under C-congruence of (12.8.5), understood as a pencil of complex hermitian matrices, is equal to

$$(G + tF) \oplus (-(G + tF)). \tag{12.8.6}$$

Indeed, since the C-Kronecker form of (12.8.5) (again, under the map Ψ) is $(tI + J_{\ell/2}(0)) \oplus (tI + J_{\ell/2}(0))$, it follows from Theorem 15.3.1 that the canonical form of (12.8.5) under C-congruence is

$$\eta_1(G + tF) \oplus \eta_2(G + tF),$$

where η_1, η_2 are signs ± 1. However, the case $\eta_1 \eta_2 = 1$ is impossible, because if $\eta_1 \eta_2 = 1$ holds, then for large real values of t, the signature (the difference between the number of positive eigenvalues, counted with multiplicities, and the number of negative eigenvalues, counted with multiplicities) of the hermitian matrix $\eta_1(G + tF) \oplus \eta_2(G + tF)$ is not zero (this is where the hypothesis that $\ell/2$ is odd is used), whereas for the hermitian matrix (12.8.5) the signature is equal to zero for all real t, a contradiction with the inertia theorem for hermitian matrices. Thus, the canonical form of (12.8.5) under C-congruence must be (12.8.6). In particular,

$$S_\phi \begin{bmatrix} 0 & G \\ G & 0 \end{bmatrix} S = \begin{bmatrix} G & 0 \\ 0 & -G \end{bmatrix},$$

$$S_\phi \begin{bmatrix} 0 & -\beta(\phi)F \\ \beta(\phi)F & 0 \end{bmatrix} S = \begin{bmatrix} F & 0 \\ 0 & -F \end{bmatrix}$$

for some invertible matrix S with entries in $\mathrm{Span}_{\mathsf{R}}\{1, \beta(\phi)\}$. Thus,

$$S_\phi \begin{bmatrix} 0 & F \\ -F & 0 \end{bmatrix} S = \begin{bmatrix} \beta(\phi)F & 0 \\ 0 & -\beta(\phi)F \end{bmatrix},$$

and the claim follows.   □

In a completely analogous way, the next claim is verified:

Claim 12.8.2. (a) *The block* (q-ϕ-h-sk2) *is ϕ-congruent to* $F_k + t\beta(\phi)G_k$; *recall that k is odd.*

(b) *The block* (q-ϕ-h-sk3) *is ϕ-congruent to*

$$(F_{\frac{k}{2}} + t\beta(\phi)G_{\frac{k}{2}}) \oplus (-(F_{\frac{k}{2}} + t\beta(\phi)G_{\frac{k}{2}})); \tag{12.8.7}$$

recall that $k/2$ is even.

(c) *The block (q-ϕ-h-sk4) is ϕ-congruent to $G_\ell + t\beta(\phi)F_\ell$; recall that ℓ is even.*

Our final claim concerns (q-ϕ-h-sk6) with $\alpha \in \mathsf{C} \setminus \{0\}$ having zero real part.

Claim 12.8.3. *The block*

$$\begin{bmatrix} 0 & \alpha F_q + G_q \\ \alpha F_q + G_q & 0 \end{bmatrix} + t \begin{bmatrix} 0 & F_q \\ -F_q & 0 \end{bmatrix},$$

where $\alpha \in \mathsf{C} \setminus \{0\}$ has zero real part, is ϕ-congruent to a block of the form

$$(A_0 + tB_0) \oplus (-(A_0 + tB_0)),$$

where $A_0 = (A_0)_\phi \in \mathsf{H}^{q \times q}$ and $B_0 = -(B_0)_\phi \in \mathsf{H}^{q \times q}$. Moreover, the H-Kronecker form of the quaternion pencil $A_0 + tB_0$ consists of only one Jordan block of size $q \times q$ with eigenvalue α (or any similar eigenvalue).

Proof of Claim 12.8.3. First of all, notice the equality (recall that $\beta(\phi) \in \mathrm{Span}_{\mathsf{R}} \{\mathsf{j}, \mathsf{k}\}$, $\alpha \in \mathrm{Span}_{\mathsf{R}} \{\mathsf{i}\}$)

$$\alpha\beta(\phi) = -\beta(\phi)\alpha. \tag{12.8.8}$$

Define the matrix

$$Z = \mathrm{diag}\,(1, -1, \ldots, (-1)^{q-1}). \tag{12.8.9}$$

We obviously have $Z^{-1} = Z^T = Z$.

Assume first that q is odd. Then

$$Z(F_q G_q)Z = -F_q G_q, \quad Z = F_q Z F_q, \quad Z G_q Z = -G_q.$$

One now verifies that

$$\begin{bmatrix} \beta(\phi)I_q & -\beta(\phi)Z \\ Z & I_q \end{bmatrix}_\phi \left(\begin{bmatrix} 0 & \alpha F_q + G_q \\ \alpha F_q + G_q & 0 \end{bmatrix} + t \begin{bmatrix} 0 & F_q \\ -F_q & 0 \end{bmatrix} \right)$$

$$\cdot \begin{bmatrix} \beta(\phi)I_q & -\beta(\phi)Z \\ Z & I_q \end{bmatrix}$$

$$= 2(\alpha\beta Z F_q + Z G_q \beta(\phi) - (\beta(\phi)Z F_q)t)$$

$$\oplus 2(-\alpha\beta(\phi)Z F_q + Z G_q \beta(\phi) + (\beta(\phi)Z F_q)t),$$

and

$$Z(-\alpha\beta(\phi)Z F_q + Z G_q \beta(\phi) + (\beta(\phi)Z F_q)t)Z$$

$$= -(\alpha\beta(\phi)Z F_q + Z G_q \beta(\phi) - (\beta(\phi)Z F_q)t).$$

Assume now that q is even. Then

$$Z F_q Z = -F_q, \quad Z G_q Z = G_q,$$

where Z is defined by (12.8.9). Since ϕ is a nonstandard involution, it is easy to see that there exists $\gamma \in \mathsf{H}$ such that $\gamma^2 = -1$, $\alpha\gamma = -\gamma\alpha$, and $\phi(\gamma) = \gamma$. Now, a straightforward verification shows that

$$\begin{bmatrix} I_q & \gamma Z \\ \gamma Z & I_q \end{bmatrix}_\phi \left(\begin{bmatrix} 0 & \alpha F_q + G_q \\ \alpha F_q + G_q & 0 \end{bmatrix} + t \begin{bmatrix} 0 & F_q \\ -F_q & 0 \end{bmatrix} \right) \begin{bmatrix} I_q & \gamma Z \\ \gamma Z & I_q \end{bmatrix}$$

$$= (2(\gamma\alpha Z F_Q + \gamma Z G_q - \gamma t Z F_q)) \oplus (2(\gamma\alpha Z F_q + \gamma Z G_q + \gamma t Z F_q)),$$

and, furthermore (note that $(\gamma Z)_\phi = \gamma Z$),

$$(\gamma Z)(\gamma \alpha Z F_q + \gamma Z G_q + \gamma t Z F_q)(\gamma Z) = -(\gamma \alpha Z F_q + \gamma Z G_q - \gamma t Z F_q).$$

This completes verification of Claim 12.8.3.   □

Assume now that the complex pencils $A + tB$ and $A' + tB'$, where $A = A^T$, $B = -B^T$, $A' = A'^T$, and $B' = -B'^T$, are H-strictly equivalent. Since, by Theorem 15.3.2, the relation of C-strict equivalence of complex symmetric-skewsymmetric pencils is the same as the relation of $(C,{}^T)$-congruence, we may replace $A + tB$ with its canonical form under $(C,{}^T)$-congruence; in other words, we may assume that $A + tB$ is a direct sum of primitive blocks of types (q-ϕ-h-sk0)–(q-ϕ-h-sk6). In view of Claims 12.8.1, 12.8.2, and 12.8.3, under the ϕ-congruence, the blocks with eigenvalue zero having odd sizes, the blocks with eigenvalue at infinity having even sizes, and the blocks with nonzero complex eigenvalues having zero real parts appear in pairs with opposite signs for each of the two blocks in every such pair. The same statement holds for $A' + tB'$ as well.

Now observe that the canonical form under ϕ-congruence and the canonical form under H-strict equivalence of a quaternionic matrix pencil

$$X + tY, \qquad X = X_\phi \in \mathsf{H}^{m \times m}, \quad Y = -Y_\phi \in \mathsf{H}^{m \times m}, \qquad (12.8.10)$$

$$\phi \text{ nonstandard involution}$$

may differ, apart from a permutation of blocks, only in signs ± 1 that attached precisely to the blocks with eigenvalue zero having odd sizes, the blocks with eigenvalue at infinity having even sizes, and the blocks with nonzero eigenvalues having zero real parts (Theorems 12.1.2 and 12.3.1). In view of the statement in the preceding paragraph, we are done.   □

12.9   EXERCISES

Ex. 12.9.1. State and prove the analogues of Theorem 12.7.5 for skewsymmetric complex matrix pencils and symmetric/skewsymmetric complex matrix pencils.

Ex. 12.9.2. Consider the following property of a ϕ-hsk matrix pencil $A + tB$:

(A)   *If $A' + tB'$ is a ϕ-hsk matrix pencil which is strictly equivalent to $A + tB$, then $A' + tB'$ is ϕ-congruent to $A + tB$.*

Prove that if $A + tB$ has property (A) and if both A and B are invertible, then there is $\varepsilon > 0$ such that every ϕ-hsk matrix pencil $A' + tB'$ also has property (A), provided $\|A' - A\| + \|B' - B\| < \varepsilon$.

Ex. 12.9.3. Give examples of ϕ-hsk matrix pencils $A + tB$ for which the result of Ex. 12.9.2 fails if:

(a) only A is assumed invertible;

(b) only B is assumed invertible.

Ex. 12.9.4. Find all possible canonical forms (12.3.1) (or (12.3.5)) under each of the following conditions:

(1) $\operatorname{rank}(xA + yB) = 3$ for all real x, y not both zero;

(2) rank $(xA + yB) \leq 2$ for all real x and y;

(3) rank $B = 2$.

Ex. 12.9.5. Find all possible canonical forms (12.3.1) (or (12.3.5)) of the ϕ-hsk $n \times n$ matrix pencil $A + tB$ under each of the following conditions:

(1) B is $\beta(\phi)$-positive definite.

(2) B is $\beta(\phi)$-positive semidefinite.

(3) The $\beta(\phi)$-inertia of B is $(n - 2, 1, 1)$.

Ex. 12.9.6. Let $A + tB$ be a ϕ-hsk matrix pencil and assume that B is invertible and the sum of algebraic multiplicities of $A + tB$ corresponding to the eigenvalues with zero real parts of $A + tB$ does not exceed 2.

(a) Show that there exist not more than 4 mutually pairwise non-ϕ-congruent ϕ-hsk matrix pencils such that each of them strictly equivalent to $A + tB$.

(b) Prove that there is $\varepsilon > 0$ such that every ϕ-hsk matrix pencil $A' + tB'$ also has the propery described in Part (a), provided $\|A' - A\| + \|B' - B\| < \varepsilon$.

Ex. 12.9.7. State and prove generalization of Ex. 12.9.6 for ϕ-hsk matrix pencils $A + tB$ with invertible B and whose sum of algebraic multiplicities corresponding to the eigenvalues with zero real parts of does not exceed k, for a fixed integer $k \geq 2$.

Ex. 12.9.8. Find the canonical forms of the following ϕ-hsk matrix pencils $A + tB$ under strict equivalence and under ϕ-congruence:

(a) $\begin{bmatrix} t\beta(\phi) & \beta(\phi) + taq \\ -\beta(\phi) - taq & t\beta(\phi) \end{bmatrix}$;

(b) $\begin{bmatrix} q + t\beta(\phi) & q & 0 \\ q & q + t\beta(\phi) & 0 \\ 0 & 0 & t\beta(\phi) \end{bmatrix}$.

Here a is a real parameter and the nonzero quaternion q is such that $\beta(\phi)q = -q\beta(\phi)$.

Ex. 12.9.9. (a) Find the canonical form of the matrix $\begin{bmatrix} i & j \\ 0 & k \end{bmatrix}$ under ϕ-congruence, where the nonstandard involution ϕ is such that $\phi(i) = -i$, $\phi(j) = j$, $\phi(k) = k$.

(b) Repeat (a), but now, with ϕ such that $\phi(i) = i$, $\phi(j) = -j$, $\phi(k) = k$.

12.10   NOTES

The contents of this chapter are taken from Rodman [133] and [134]. Exposition and presentation in the chapter, as well as many proofs, follow the two papers.

The results of Theorems 12.1.2 and 12.3.1 were known before; see, e.g., Djoković et al. [34] (for the case when at least one of the matrices A or B is invertible).

Chapter Thirteen

Indefinite inner products: Nonstandard involution

In this chapter we fix a nonstandard involution ϕ.

In parallel with Chapter 10, we introduce indefinite inner products defined on $\mathsf{H}^{n \times 1}$ of the symmetric and skewsymmetric types associated with ϕ and matrices having symmetry properties with respect to one of these indefinite inner products. The symmetric-type inner product is a function

$$[\cdot, \cdot]^{(\phi)} : \mathsf{H}^{n \times 1} \times \mathsf{H}^{n \times 1} \longrightarrow \mathsf{H}$$

(the superscript $^{(\phi)}$ indicates that the inner product is associated with ϕ, in contrast with the inner product of Chapter 10) with the following properties:

(1′) Linearity in the first argument:

$$[x_1 \alpha_1 + x_2 \alpha_2, y]^{(\phi)} = [x_1, y]^{(\phi)} \alpha_1 + [x_2, y]^{(\phi)} \alpha_2$$

for all $x_1, x_2, y \in \mathsf{H}^{n \times 1}$ and all $\alpha_1, \alpha_2 \in \mathsf{H}$.

(2′) Symmetry: $[x, y]^{(\phi)} = \phi([y, x]^{(\phi)})$ for all $x, y \in \mathsf{H}^{n \times 1}$.

(3′) Nondegeneracy: if $x_0 \in \mathsf{H}^{n \times 1}$ is such that $[x_0, y]^{(\phi)} = 0$ for all $y \in \mathsf{H}^{n \times 1}$, then $x_0 = 0$.

The skewsymmetric-type inner product $[\cdot, \cdot]^{(\phi)}$ is defined by the properties (1′), (3′), and

(2″) antisymmetry: $[x, y]^{(\phi)} = -\phi([y, x]^{(\phi)})$ for all $x, y \in \mathsf{H}^{n \times 1}$.

It follows from (1′) and (2′), or from (1′) and (2″), that

$$[x, y_1 \alpha_1 + y_2 \alpha_2]^{(\phi)} = \phi(\alpha_1)[x, y_1] + \phi(\alpha_2)[x, y_2]$$

for all $x, y_1, y_2 \in \mathsf{H}^{n \times 1}$ and all $\alpha_1, \alpha_2 \in \mathsf{H}$.

In complete analogy with Proposition 10.0.1 we have the following.

Proposition 13.0.1. $[\cdot, \cdot]^{(\phi)}$ *is an inner product on* $\mathsf{H}^{n \times 1}$ *of symmetric-, resp. skewsymmetric-, type if and only if there exists a ϕ-hermitian, resp. ϕ-skewhermitian, invertible $n \times n$ matrix H such that*

$$[x, y]^{(\phi)} = y_\phi H x, \quad \text{for all} \quad x, y \in \mathsf{H}^{n \times 1}.$$

Such a matrix H is uniquely determined by the inner product.

The proof is essentially the same as that of Proposition 10.0.1 and is omitted (Ex. 13.7.1).

In this chapter we study indefinite inner products associated with ϕ of the symmetric- and skewsymmetric-type and matrices having symmetry properties with respect to one of these indefinite inner products. As in Chapter 10, we will often work with matrices H rather than directly with indefinite inner products.

The development in this chapter is often parallel to that of Chapter 10, but here the indefinite inner products are with respect to a nonstandard involution, rather with respect to the conjugation as in Chapter 10. We develop canonical forms for (H, ϕ)-symmetric and (H, ϕ)-skewsymmetric matrices (when the inner product is of the symmetric-type), and canonical forms of (H, ϕ)-Hamiltonian and (H, ϕ)-skew-Hamiltonian matrices (when the inner product is of the skewsymmetric-type). Applications include invariant Lagrangian subspaces and systems of differential equations with symmetries.

13.1   CANONICAL FORMS: SYMMETRIC INNER PRODUCTS

Let $H \in \mathsf{H}^{n \times n}$ be an invertible ϕ-hermitian matrix. Recall that a matrix $X \in \mathsf{H}^{n \times n}$ is called (H, ϕ)-symmetric if the equality $[x, Xy]_{H,\phi} = [Xx, y]_{H,\phi}$ holds for all $x, y \in \mathsf{H}^{n \times 1}$, where $[\cdot, \cdot]_{H,\phi}$ stands for the inner product of symmetric-type induced by H: $[x, y]_{H,\phi} = y_\phi H x$, $x, y \in \mathsf{H}^{n \times 1}$. It is easy to see that X is (H, ϕ)-symmetric if and only if the equation $HX = X_\phi H$ holds true—in other words, HX is ϕ-hermitian. A matrix $A \in \mathsf{H}^{n \times n}$ is said to be (H, ϕ)-skewsymmetric if the equality $[x, Ay]_{H,\phi} = -[Ax, y]_{H,\phi}$ holds true for all $x, y \in \mathsf{H}^{n \times 1}$ or, equivalently, if HA is ϕ-skewhermitian.

In this chapter we write H-symmetric and H-skewsymmetric, short for (H, ϕ)-symmetric and (H, ϕ)-skewsymmetric, respectively.

The next theorem describes the canonical form for pairs (H, X), where H is invertible ϕ-hermitian and X is H-symmetric. Recall that Inv (ϕ) stands or the set (real vector space) of quaternions that are fixed by ϕ.

Theorem 13.1.1. *Let $H \in \mathsf{H}^{n \times n}$ be an invertible matrix such that $H = H_\phi$, and let A be H-symmetric. Then there exists an invertible matrix $S \in \mathsf{H}^{n \times n}$ such that the matrices $S^{-1}AS$ and $S_\phi HS$ have the form*

$$S_\phi HS = F_{\ell_1} \oplus \cdots \oplus F_{\ell_q}, \quad S^{-1}AS = J_{\ell_1}(\alpha_1) \oplus \cdots \oplus J_{\ell_q}(\alpha_q), \quad (13.1.1)$$

where $\alpha_1, \ldots, \alpha_q \in$ Inv (ϕ). Moreover, the form (13.1.1) is unique up to an arbitrary permutation of the diagonal blocks and up to replacement of α_j in each block $J_{\ell_j}(\alpha_j)$ with a similar quaternion $\beta_j \in$ Inv (ϕ).

Conversely, if H and A have the form (13.1.1), then H is ϕ-hermitian and invertible and A is H-symmetric.

Proof. The converse part is checked by a straightforward verification. For the direct part we apply Theorem 11.1.3 to the ϕ-pencil $HA + tH$. Because H is invertible, only blocks $(t + \alpha_j)F_{\ell_j} + G_{\ell_j}$ can appear in the canonical form (11.1.1) of $HA + tH$. Since $A = H^{-1}(HA)$, this leads to formula (13.1.1), but with $J_{\ell_j}(\alpha_j)$ replaced with $J_{\ell_j}(\alpha_j)^T$. Then the transformation

$$F_{\ell_j} \mapsto (F_{\ell_j})_\phi F_{\ell_j} F_{\ell_j} = F_{\ell_j}, \quad J_{\ell_j}(\alpha_j)^T \mapsto F_{\ell_j}^{-1} J_{\ell_j}(\alpha_j)^T F_{\ell_j} = J_{\ell_j}(\alpha_j)$$

puts the pair (H, A) in the required form (13.1.1). The uniqueness statement in Theorem 13.1.1 follows from that of Theorem 11.1.3 applied to the pencil $HA + tH$. $\square$

Next, we present the canonical form for H-skewsymmetric matrices. It will be convenient to identify the primitive forms first. As before, we fix $\beta(\phi) \in \mathsf{H}$, such that $\phi(\beta(\phi)) = -\beta(\phi)$ and $|\beta(\phi)| = 1$.

(q-ϕ-H-sk2) $H = \varepsilon F_k$, $A = \beta(\phi)J_k(0)$, where $\varepsilon = 1$ if k is odd and $\varepsilon = \pm 1$ if k is even.

(q-ϕ-H-sk4)

$$H = \begin{bmatrix} 0 & F_\ell \\ F_\ell & 0 \end{bmatrix}, \quad A = \begin{bmatrix} J_\ell(\alpha) & 0 \\ 0 & -J_\ell(\alpha) \end{bmatrix},$$

where $\alpha \in \mathsf{H}$ satisfies the conditions $\phi(\alpha) = \alpha$ and $\mathfrak{R}(\alpha) > 0$.

(q-ϕ-H-sk5) $H = \delta F_s$, $A = \beta(\phi)J_s(\tau)$, where τ is positive real and $\delta = \pm 1$.

Alternatively, one can assume that τ is negative real, rather than positive real, in (q-ϕ-H-sk5). Indeed, if $\gamma \in \mathsf{H}$ is such that $\gamma^{-1}\beta(\phi)\gamma = -\beta(\phi)$ and $|\gamma| = 1$, then

$$\left(\operatorname{diag}\left(\gamma^{-1}, -\gamma^{-1}, \ldots, (-1)^{s-1}\gamma^{-1} \right) \right) \beta(\phi)J_s(\tau)$$

$$\cdot \left(\operatorname{diag}\left(\gamma, -\gamma, \ldots, (-1)^{s-1}\gamma \right) \right) = \beta(\phi)J_s(-\tau), \tag{13.1.2}$$

and

$$\left(\operatorname{diag}\left(\phi(\gamma), -\phi(\gamma), \ldots, (-1)^{s-1}\phi(\gamma) \right) \right) F_s$$

$$\cdot \left(\operatorname{diag}\left(\gamma, -\gamma, \ldots, (-1)^s\gamma \right) \right) = (-1)^{s-1}F_s,$$

where $\tau > 0$. Note that $\gamma^2 = -1$ and $\phi(\gamma) = \gamma$. Thus, the replacement of τ by $-\tau$ in (q-ϕ-H-sk5) will reverse the sign δ if s is odd and leave the sign δ invariant if s is even.

Also, the condition $\mathfrak{R}(\alpha) > 0$ in (q-ϕ-H-sk4) may be replaced by $\mathfrak{R}(\alpha) < 0$. This follows from the formulas

$$\begin{bmatrix} 0_\ell & \operatorname{diag}\left(1, -1, \ldots, (-1)^{\ell-1}\right) \\ \operatorname{diag}\left(1, -1, \ldots, (-1)^{\ell-1}\right) & 0_\ell \end{bmatrix}$$

$$\cdot \begin{bmatrix} J_\ell(-\alpha) & 0 \\ 0 & -J_\ell(-\alpha) \end{bmatrix}$$

$$\cdot \begin{bmatrix} 0_\ell & \operatorname{diag}\left(1, -1, \ldots, (-1)^{\ell-1}\right) \\ \operatorname{diag}\left(1, -1, \ldots, (-1)^{\ell-1}\right) & 0_\ell \end{bmatrix}$$

$$= \begin{bmatrix} J_\ell(\alpha) & 0 \\ 0 & -J_\ell(\alpha) \end{bmatrix},$$

$$\begin{bmatrix} 0_\ell & \operatorname{diag}\left(1, -1, \ldots, (-1)^{\ell-1}\right) \\ \operatorname{diag}\left(1, -1, \ldots, (-1)^{\ell-1}\right) & 0_\ell \end{bmatrix} \begin{bmatrix} 0 & F_\ell \\ F_\ell & 0 \end{bmatrix}$$

$$\cdot \begin{bmatrix} 0_\ell & \operatorname{diag}\left(1, -1, \ldots, (-1)^{\ell-1}\right) \\ \operatorname{diag}\left(1, -1, \ldots, (-1)^{\ell-1}\right) & 0_\ell \end{bmatrix}$$

$$= (-1)^{\ell-1} \begin{bmatrix} 0 & F_\ell \\ F_\ell & 0 \end{bmatrix},$$

and for ℓ even we apply further transformation

$$\begin{bmatrix} I & 0 \\ 0 & -I \end{bmatrix} \left(- \begin{bmatrix} 0 & F_\ell \\ F_\ell & 0 \end{bmatrix} \right) \begin{bmatrix} I & 0 \\ 0 & -I \end{bmatrix} = \begin{bmatrix} 0 & F_\ell \\ F_\ell & 0 \end{bmatrix},$$

$$\begin{bmatrix} I & 0 \\ 0 & -I \end{bmatrix} \begin{bmatrix} J_\ell(\alpha) & 0 \\ 0 & -J_\ell(\alpha) \end{bmatrix} \begin{bmatrix} I & 0 \\ 0 & -I \end{bmatrix} = \begin{bmatrix} J_\ell(\alpha) & 0 \\ 0 & -J_\ell(\alpha) \end{bmatrix}.$$

Theorem 13.1.2. *Let $A \in \mathsf{H}^{n \times n}$ be H-skewsymmetric. Then there exists an invertible quaternion matrix S such that $S_\phi H S$ and $S^{-1} A S$ have the following block diagonal form:*

$$S_\phi H S = H_1 \oplus H_2 \oplus \cdots \oplus H_m, \quad S^{-1} A S = A_1 \oplus A_2 \oplus \cdots \oplus A_m, \qquad (13.1.3)$$

where each pair (H_i, A_i) has one of the forms (q-ϕ-H-sk2), (q-ϕ-H-sk4), (q-ϕ-H-sk5). Moreover, the form (13.1.3) is uniquely determined by the pair (H, A), up to a permutation of blocks and up to a replacement in each block of the form (q-ϕ-H-sk4), the quaternion α with a similar quaternion α' such that $\phi(\alpha') = \alpha'$.

Conversely, if H and A are given by (13.1.3), then H is ϕ-hermitian and A is H-skewsymmetric.

Thus, the *sign characteristic* of a pair (H, A), where A is H-skewsymmetric, attaches a sign ± 1 to every partial multiplicity of A associated with nonzero eigenvalues having zero real parts (if any) and to every even partial multiplicity associated with the eigenvalue 0 (if A is not invertible).

A technical lemma will be needed for the proof of Theorem 13.1.2, as well as in later proofs.

Lemma 13.1.3. (a) *If $x \in \mathsf{C} \setminus \{0\}$, then there exists a (necessarily invertible) matrix $S \in \mathsf{C}^{m \times m}$ such that equations*

$$S^T (x F_m + G_m) S = F_m \qquad (13.1.4)$$

and

$$((J_m(x))^T)^{-1} S = S (J_m(x^{-1}))^T \qquad (13.1.5)$$

are satisfied.

(b) *If $x \in \mathsf{R} \setminus \{0\}$, then there exists a matrix $S \in \mathsf{R}^{n \times n}$ such that equations*

$$S^T (x F_m + G_m) S = \pm F_m \qquad (13.1.6)$$

and (13.1.5) are satisfied, with the sign $+1$ if $x > 0$ and m odd or if $x < 0$ and m even and with the sign -1 if $x > 0$ and m even or if $x < 0$ and m odd.

Proof. Proof of Part (a). We exhibit S in the following form: $S = [s_{j,k}]_{j,k=1}^m$ is lower triangular; $s_{j,k} = 0$ if $j < k$, and for $j \geq k$, the entries $s_{j,k}$ have the form

$$s_{j,k} = f_{j,k} x^{-j-k+2} s_{1,1}, \quad f_{j,k} \in \mathsf{C}.$$

The constants $f_{j,k}$ and $s_{1,1}$ are to be determined so that equalities (13.1.4) and (13.1.5) are satisfied. We rewrite (13.1.5) as follows:

$$\begin{bmatrix} 0 & 0 & \cdots & \cdots & 0 \\ -x^{-2} & 0 & \cdots & \cdots & 0 \\ x^{-3} & -x^{-2} & \cdots & \cdots & 0 \\ \vdots & \ddots & \ddots & \ddots & \vdots \\ (-1)^{m-1} x^{-m} & (-1)^{m-2} x^{-m+1} & \cdots & -x^{-2} & 0 \end{bmatrix} S$$

$$= S \begin{bmatrix} 0 & 0 & \cdots & \cdots & 0 \\ 1 & 0 & \cdots & \cdots & 0 \\ 0 & 1 & \cdots & \cdots & 0 \\ \vdots & \ddots & \ddots & \ddots & \vdots \\ 0 & 0 & \cdots & 1 & 0 \end{bmatrix} = \begin{bmatrix} 0 & 0 & \cdots & \cdots & 0 \\ s_{2,2} & 0 & \cdots & \cdots & 0 \\ s_{3,2} & s_{3,3} & \cdots & \cdots & 0 \\ \vdots & \ddots & \ddots & \ddots & \vdots \\ s_{m,2} & s_{m,3} & \cdots & s_{m,m} & 0 \end{bmatrix}. \quad (13.1.7)$$

To start with, we let

$$f_{j,j} = (-1)^{j-1}, \quad j = 1, 2, \ldots, m. \quad (13.1.8)$$

It is easy to see that (13.1.7) amounts, in view of (13.1.8), to the equalities

$$(-1)^{k-j} f_{j,j} + \cdots + (-1) f_{k-1,j} = f_{k,j+1}, \quad \text{for} \quad k > j+1, \quad (13.1.9)$$

which determine uniquely all elements $f_{k,j}$, provided $f_{1,1} = 1, f_{2,1}, \ldots, f_{m,1}$ are given.

To satisfy (13.1.4), we let $s_{1,1}$ be such that

$$s_{1,1}^2 = (-1)^{m-1} x^{2m-3}. \quad (13.1.10)$$

(Note that if x is real and $(-1)^{m-1} x$ is positive, then $s_{1,1}$ is real as well.) Then, for $j + k \geq m + 1$, the (j, k)th entry in the left-hand side of (13.1.4) coincides with the (j, k)th entry of F_m. Now, (13.1.4) is satisfied if and only if the following property holds true:

(A) *For all $j + k \leq m$, the (j, k)th entry of the matrix*

$$\begin{bmatrix} f_{m,1} + f_{m-1,1} & f_{m-1,1} + f_{m-2,1} & \cdots & f_{2,1} + f_{1,1} & f_{1,1} \\ f_{m,2} + f_{m-1,2} & f_{m-1,2} + f_{m-2,2} & \cdots & f_{2,2} & 0 \\ \vdots & \vdots & \vdots & \vdots & \vdots \\ f_{m,m-1} + f_{m-1,m-1} & f_{m-1,m-1} & \cdots & 0 & 0 \\ f_{m,m} & 0 & \cdots & 0 & 0 \end{bmatrix}$$

$$\cdot \begin{bmatrix} f_{1,1} & 0 & \cdots & 0 & 0 \\ f_{2,1} & f_{2,2} & \cdots & 0 & 0 \\ \vdots & \vdots & \vdots & \vdots & \vdots \\ f_{m-1,1} & f_{m-1,2} & \cdots & f_{m-1,m-1} & 0 \\ f_{m,1} & f_{m,2} & \cdots & f_{m,m-1} & f_{m,m} \end{bmatrix} \quad (13.1.11)$$

is equal to zero.

Note also that the matrix in (13.1.11) is symmetric. We prove by induction on m that one can chose $f_{2,1}, \ldots, f_{m,1}$ so that (13.1.8), (13.1.9), and (A) hold true. For $m = 2$, the choice $f_{2,1} = -\frac{1}{2}$ satisfies the required properties. Assume by induction that $f_{2,1}, \ldots, f_{m-1,1}$ have been already selected so that (13.1.8), (13.1.9), and (A) hold true (with m replaced by $m - 1$). We prove that $f_{m,1}$ can be selected so that (13.1.8), (13.1.9), and (A) hold true for m. The equality (13.1.9) implies that

$$f_{k,j} + f_{k-1,j} = -f_{k-1,j-1}, \quad \text{for} \quad k > j > 1;$$

therefore, by the induction hypotheses, the (j, k)th entries in (13.1.11) are zeros, provided $j + k \leq m$ and $j > 1$. By the symmetry of the matrix (13.1.11), the same holds true also for the (j,k)th entries, provided $j + k \leq m$ and $k > 1$. It remains to fix $f_{m,1}$ so that

$$(f_{m,1} + f_{m-1,1})f_{1,1} + (f_{m-1,1} + f_{m-2,1})f_{2,1}$$
$$+ \cdots + (f_{2,1} + f_{1,1})f_{m-1,1} + f_{1,1}f_{m,1} = 0,$$

and we are done with the proof of Part (a).

Part (b). If $(-1)^{m-1}x$ is positive, the result of Part (b) is obtained as a by-product of the proof of Part (a) (see the remark after (13.1.10)). Suppose now that $(-1)^{m-1}x$ is negative. Then choose $s_{1,1} \in \mathsf{R}$ in the proof of Part (a) so that $s_{1,1}^2 = (-1)^m x^{2m-3}$ to obtain the desired result. □

Proof of Theorem 13.1.2. The converse part is easily verified.

We prove the direct part. Note that A is H-skewsymmetric if and only the matrix pencil $H + tHA$ is ϕ-hermitian/skewhermitian; in this case, the transformation

$$H \mapsto S_\phi HS, \quad A \mapsto S^{-1}AS, \quad S \in \mathsf{H}^{\times n}, \qquad (13.1.12)$$

where S is invertible but otherwise arbitrary, is equivalent to ϕ-congruence of the pencil $H + tHA$. Thus, we can use Theorem 12.3.5. It follows that the pair (H, A) can be brought by a transformation of type (13.1.12) to a direct sum of the forms:

(a) $H = \varepsilon F_k$, $HA = \varepsilon\beta(\phi)G_k$, where $\varepsilon = 1$ if k is odd and $\varepsilon = \pm 1$ if k is even;

(b)

$$H = \begin{bmatrix} 0 & \alpha F_\ell + G_\ell \\ \alpha F_\ell + G_\ell & 0 \end{bmatrix}, \quad HA = \begin{bmatrix} 0 & F_\ell \\ -F_\ell & 0 \end{bmatrix}, \qquad (13.1.13)$$

where $\alpha \in \mathrm{Inv}\,(\phi)$, $\Re(\alpha) > 0$;

(c) $H = \delta F_s$, $HA = \delta\beta(\phi)F_s J_s(\tau)$, where $\delta = \pm 1$ and $\tau > 0$.

Forms (a) and (c) clearly lead to (q-ϕ-H-sk2) and (q-ϕ-H-sk5), respectively.

We conclude the proof by showing that the pair (H, A) given by (13.1.13) can be brought by a transformation of type (13.1.12) to the pair $(\widetilde{H}, \widetilde{A})$, where

$$\widetilde{H} = \begin{bmatrix} 0 & F_\ell \\ F_\ell & 0 \end{bmatrix}, \quad \widetilde{A} = \begin{bmatrix} J_\ell(\alpha^{-1}) & 0 \\ 0 & -J_\ell(\alpha^{-1}) \end{bmatrix}.$$

It will be convenient to do this in two steps. First, we show that (13.1.13) can be brought to the form

$$\left(\widetilde{H}, A_1 := \begin{bmatrix} -J_\ell(\alpha^{-1})^T & 0 \\ 0 & J_\ell(\alpha^{-1})^T \end{bmatrix} \right).$$

Let $q \in \mathsf{H}$ be a square root of -1 such that $\alpha \in \mathrm{Span}_{\mathsf{R}}\,\{1, q\}$ and $\phi(q) = q$. Identifying $\mathrm{Span}_{\mathsf{R}}\,\{1, q\}$ with C and using Lemma 13.1.3, we find $S \in \mathsf{H}^{\ell \times \ell}$ such that

$$S_\phi(\alpha F_\ell + G_\ell)S = F_\ell \quad \text{and} \quad S^{-1}(\alpha I + F_\ell G_\ell)^{-1}S = J_\ell(\alpha^{-1})^T.$$

Then

$$\widetilde{H} = \begin{bmatrix} S_\phi & 0 \\ 0 & S_\phi \end{bmatrix} H \begin{bmatrix} S & 0 \\ 0 & S \end{bmatrix} \quad \text{and} \quad A_1 = \begin{bmatrix} S^{-1} & 0 \\ 0 & S^{-1} \end{bmatrix} A \begin{bmatrix} S & 0 \\ 0 & S \end{bmatrix}.$$

Finally, observe that $\widetilde{H} = F_{2\ell}\widetilde{H}F_{2\ell}, \quad \widetilde{A} = F_{2\ell}A_1F_{2\ell}.$ □

Part (a) and the equivalence of (1) and (2) in Part (b) of the following corollary are immediate from Theorems 13.1.1 and 13.1.2. The equivalence of (2) and (3) in Part (b) follows by consideration of the Jordan form of A.

Corollary 13.1.4. (a) *Every matrix in* $\mathsf{H}^{n \times n}$ *is* H*-symmetric for some invertible* ϕ*-hermitian* $H \in \mathsf{H}^{n \times n}$.

(b) *The following statements are equivalent for a matrix* $A \in \mathsf{H}^{n \times n}$:

(1) A *is* H*-skewsymmetric for some* ϕ*-hermitian invertible* $H \in \mathsf{H}^{n \times n}$.

(2) *The partial multiplicities of eigenvalues* α *and* $-\alpha$ *of* A, *where* $\Re(\alpha) \neq 0$, *are the same; more precisely, for every positive integer* k *and every* $\alpha \in \mathsf{H}^{n \times n}$ *with* $\Re(\alpha) \neq 0$, *the number of partial multiplicities of* A *at* α *is equal to the number of partial multiplicities of* A *at* $-\alpha$, *and we set this number to be zero for all* k *if* α *(and then also* $-\alpha$*) is not an eigenvalue of* A.

(3) A *is similar to* $-A$.

13.2   CANONICAL FORMS: SKEWSYMMETRIC INNER PRODUCTS

In this section we assume that H is ϕ-skewhermitian and invertible, i.e., we are working with an indefinite inner product of skewsymmetric-type. Recall that a matrix $X \in \mathsf{H}^{n \times n}$ is called (H, ϕ)-*Hamiltonian* if the equality $[x, Xy]_{H,\phi} = -[Xx, y]_{H,\phi}$ holds true for all $x, y \in \mathsf{H}^{n \times 1}$, where $[\cdot, \cdot]_{H,\phi}$ stands for the inner product of skewsymmetric-type induced by H: $[x, y]_{H,\phi} = y_\phi Hx$, $x, y \in \mathsf{H}^{n \times 1}$. It is easy to see that X is (H, ϕ)-Hamiltonian if and only if the equation $HX = -X_\phi H$ holds true; in other words, HX is ϕ-hermitian.

Definition 13.2.1. A matrix $A \in \mathsf{H}^{n \times n}$ is said to be (H, ϕ)-*skew-Hamiltonian* if the equality $[x, Ay]_{H,\phi} = [Ax, y]_{H,\phi}$ holds true for all $x, y \in \mathsf{H}^{n \times 1}$ or, equivalently, if HA is ϕ-skewhermitian.

In this chapter we will write H-Hamiltonian and H-skew-Hamiltonian, short for (H, ϕ)-Hamiltonian and (H, ϕ)-skew-Hamiltonian, respectively.

In the next theorem we state the canonical form for Hamiltonian matrices. Again, we describe first the primitive forms.

(q-ϕ-H-H3) $L = \kappa\beta(\phi)F_k$, $A = \beta(\phi)J_k(0)$, where $\kappa = 1$ if k is even, and $\kappa = \pm 1$ if k is odd.

(q-ϕ-H-H4)

$$L = \begin{bmatrix} 0 & F_\ell \\ -F_\ell & 0 \end{bmatrix}, \quad A = \begin{bmatrix} -J_\ell(\alpha) & 0 \\ 0 & J_\ell(\alpha) \end{bmatrix},$$

where $\alpha \in \mathrm{Inv}\,(\phi)$, $\Re(\alpha) > 0$.

(q-ϕ-H-H5) $L = \delta\beta(\phi)F_s$, $A = \beta(\phi)J_s(\tau)$, where $\delta = \pm 1$ and τ is a negative real number.

Remark 13.2.2. One can replace "negative" by "positive" in (q-ϕ-H-$H5$), as can be seen from formulas (13.1.2) and

$$\left(\operatorname{diag}\left(\phi(\gamma), -\phi(\gamma), \ldots, (-1)^{s-1}\phi(\gamma)\right)\right) (\beta(\phi)F_s) \left(\operatorname{diag}\left(\gamma, -\gamma, \ldots, (-1)^{s-1}\gamma\right)\right)$$

$$= (-1)^{s-1}(\beta(\phi)F_s),$$

where $\gamma \in \mathsf{H}$ is such that $\gamma^{-1}\beta(\phi)\gamma = -\beta(\phi)$ and $|\gamma| = 1$. Note that the replacement of τ by its negative in (q-ϕ-H-$H5$) will reverse the sign δ if s is even and will leave the sign invariant if s is odd.

Theorem 13.2.3. *Let $A \in \mathsf{H}^{n\times n}$ be H-Hamiltonian, where $H \in \mathsf{H}^{n\times n}$ is invertible and ϕ-skewhermitian. Then there exists an invertible quaternion matrix S such that $S_\phi H S$ and $S^{-1}AS$ have the following block diagonal form:*

$$S_\phi H S = L_1 \oplus L_2 \oplus \cdots \oplus L_m, \quad S^{-1}AS = A_1 \oplus A_2 \oplus \cdots \oplus A_m, \qquad (13.2.1)$$

where each pair (L_i, A_i) has one of the forms (q-ϕ-H-$H3$), (q-ϕ-H-$H4$), (q-ϕ-H-$H5$). Moreover, the form (13.2.1) is uniquely determined by the pair (H, A), up to an arbitrary simultaneous permutation of blocks and up to a replacement of α in each block of the form (q-ϕ-H-$H4$) with a similar quaternion α' such that $\phi(\alpha') = \alpha'$.

Conversely, if H and A are given as in formula (13.2.1), then H is invertible ϕ-skewhermitian and A is H-Hamiltonian.

The *sign characteristic* of an H-Hamiltonian matrix A assigns a sign ± 1 to every partial multiplicity corresponding to a nonzero eigenvalue of A with zero real part (if any) and to every odd partial multiplicity corresponding to the eigenvalue zero of A (if A is not invertible).

Proof. The converse part being easily verified by a straightforward computation, we focus on the direct part.

Observe that A is H-Hamiltonian if and only if the matrix pencil $HA + tH$ is ϕ-hermitian/skewhermitian, and use the canonical form of Theorem 12.3.5. As a result we see that there is an invertible matrix $S \in \mathsf{H}^{n\times n}$ such that $S_\phi H S$ and $S^{-1}XS$ have the block diagonal form (13.2.1) with each pair (L_i, A_i) having one of the following three forms:

(a) $L = \kappa\beta(\phi)F_k$, $A = -\beta(\phi)J_k(0)^T$, where $\kappa = 1$ if k is even and $\kappa = \pm 1$ if k is odd;

(b) the form (q-ϕ-H-$H4$);

(c) $L = \delta\beta(\phi)F_s J_s(\tau)$, $A = -\beta(\phi)(J_s(\tau))^{-1}$, where $\tau > 0$.

Form (a) can be easily transformed to (q-ϕ-H-$H3$) by a transformation $(L, A) \mapsto (S_\phi L S, S^{-1}AS)$ for some invertible S. Indeed, we have

$$(\Xi_k(\beta(\phi)))_\phi (\beta(\phi)F_k) \Xi_k(\beta(\phi)) = \beta(\phi)F_k,$$

$$(\Xi_k(\beta(\phi)))^{-1} (-\beta(\phi)J_k(0)^T) \Xi_k(\beta(\phi)) = (-1)^{k-1}\beta(\phi)J_k(0),$$

which for k odd is in the form (q-ϕ-H-$H3$) and for k even can be transformed to that form using Remark 12.3.3.

As for (c), we will find an invertible matrix $S \in \mathsf{H}^{s \times s}$ such that

$$S_\phi \cdot \beta(\phi) F_s J_s(\tau) \cdot S = \widetilde{\delta} \beta(\phi) F_s, \tag{13.2.2}$$

$$S^{-1} \cdot (-\beta(\phi)(J_s(\tau))^{-1}) \cdot S = \beta(\phi) J_s(\tau^{-1}), \qquad \tau \in \mathsf{R} \setminus \{0\}, \tag{13.2.3}$$

where $\widetilde{\delta} = \pm 1$ (in fact, $\widetilde{\delta}$ coincides with the sign of τ). This will complete the proof of formula (13.2.1).

Let $(\beta(\phi), q_1, q_2)$ be a units triple. We seek S with entries in $\mathrm{Span}_{\mathsf{R}} \{1, q_1\}$, and identify $\mathrm{Span}_{\mathsf{R}} \{1, q_1\}$ with C. Then

$$\beta(\phi)S = \overline{S}\beta(\phi), \quad S^T \beta(\phi) = \beta(\phi)S^*, \quad S_\phi = S^T, \quad S^{-1}\beta(\phi) = \beta(\phi)\overline{S}^{-1},$$

and equations (13.2.2) and (13.2.3) read

$$S^*(F_s J_s(\tau))S = \widetilde{\delta} F_s, \quad \overline{S}^{-1}(J_s(\tau))^{-1}S = -J_s(\tau^{-1}), \qquad \tau < 0. \tag{13.2.4}$$

Thus, we seek invertible $S \in \mathsf{C}^{s \times s}$ that satisfies (13.2.4). In turn, setting $S = iS_0$, where $S_0 \in \mathsf{R}^{s \times s}$, (13.2.4) boils down to

$$S_0^T(\tau F_s + \widetilde{G}_s)S_0 = \pm F_s, \qquad (J_s(\tau))^{-1}S_0 = S_0 J_s(\tau^{-1}). \tag{13.2.5}$$

We now use Lemma 13.1.3(b), with $x = \tau$ and $m = s$. If $S \in \mathsf{R}^{s \times s}$ is the matrix satisfying (13.1.6) and (13.1.5), then it is easy to see that $S_0 := F_s S F_s$ satisfies (13.2.5); in verifying this, use the equalities

$$F_s(J_s(y))^T F_s = J_s(y), \quad F_s((J_s(y))^T)^{-1}F_s = (J_s(y))^{-1},$$

where $y \in \mathsf{H}$.

The uniqueness statement in Theorem 13.2.3 follows from that of Theorem 12.3.5. $\qquad \square$

Next, we present the canonical form for H-skew-Hamiltonian matrices.

Theorem 13.2.4. *Fix $\beta(\phi) \in \mathsf{H}$ such that $\phi(\beta(\phi)) = -\beta(\phi)$ and $|\beta(\phi))| = 1$. Let $H = -H_\phi \in \mathsf{H}^{n \times n}$ be an invertible matrix, and let A be H-skew-Hamiltonian. Then there exists an invertible matrix S such that the matrices $S^{-1}AS$ and $S_\phi HS$ have the form*

$$S_\phi HS = \eta_1 \beta(\phi) F_{m_1} \oplus \cdots \oplus \eta_p \beta(\phi) F_{m_p}$$

$$\oplus \begin{bmatrix} 0 & F_{\ell_1} \\ -F_{\ell_1} & 0 \end{bmatrix} \oplus \cdots \oplus \begin{bmatrix} 0 & F_{\ell_q} \\ -F_{\ell_q} & 0 \end{bmatrix},$$

$$S^{-1}AS = J_{m_1}(\gamma_1)^T \oplus \cdots \oplus J_{m_p}(\gamma_p)^T \oplus \begin{bmatrix} J_{\ell_1}(\alpha_1)^T & 0 \\ 0 & J_{\ell_1}(\alpha_1)^T \end{bmatrix}$$

$$\oplus \cdots \oplus \begin{bmatrix} J_{\ell_q}(\alpha_q)^T & 0 \\ 0 & J_{\ell_q}(\alpha_q)^T \end{bmatrix}, \tag{13.2.6}$$

where $\eta_1, \ldots, \eta_p$ are signs ± 1, the quaternions $\alpha_1, \ldots, \alpha_q \in \mathrm{Inv}\,(\phi) \setminus \mathsf{R}$, and $\gamma_1, \ldots, \gamma_p$ are real.

Moreover, the form (13.2.6) is unique up to an arbitrary simultaneous permutation of the diagonal blocks in each of the parts

$$\left(\oplus_{j=1}^{p} \left(\eta_j \beta(\phi) F_{m_j} \right), \quad \oplus_{j=1}^{p} J_{m_j}(\gamma_j)^T \right)$$

and

$$\left(\oplus_{j=1}^{q} \begin{bmatrix} 0 & F_{\ell_j} \\ -F_{\ell_j} & 0 \end{bmatrix}, \quad \oplus_{j=1}^{q} \begin{bmatrix} J_{\ell_j}(\alpha_j)^T & 0 \\ 0 & J_{\ell_j}(\alpha_j)^T \end{bmatrix} \right)$$

and up to replacements of α_j in each block $\begin{bmatrix} J_{\ell_j}(\alpha_j)^T & 0 \\ 0 & J_{\ell_j}(\alpha_j)^T \end{bmatrix}$ *by a similar quaternion $\lambda_j \in \mathrm{Inv}\,(\phi)$.*

Conversely, if H and A are given by formula (13.2.6), then H is ϕ-skewhermitian and A is H-skew-Hamiltonian.

Remark 13.2.5. In formula (13.2.6), any of the $J_{m_j}(\gamma_j)^T$'s can be replaced by $J_{m_j}(\gamma_j)$, and any of the $(J_{\ell_j}(\alpha_j)^T) \oplus (J_{\ell_j}(\alpha_j)^T)$'s can be replaced by $J_{\ell_j}(\lambda_j) \oplus J_{\ell_j}(\lambda_j)$, where $\lambda_j \in \mathrm{Inv}\,(\phi)$ is any quaternion similar to α. Use the equalities

$$F_{m_j} J_{m_j}(\gamma_j)^T F_{m_j} = J_{m_j}(\gamma_j), \quad j = 1, 2, \ldots, p,$$

and similar equalities for the blocks $J_{\ell_j}(\alpha_j)^T$ to verify this claim.

Proof. A is H-skew-Hamiltonian if and only if the matrix pencil $HA + tH$ is ϕ-skewhermitian. Now use the canonical form for $HA + tH$ of Theorem 11.2.2. □

Note that for H-skew-Hamiltonian matrices, the partial multiplicities corresponding to nonreal eigenvalues come in pairs.

Characterizations of matrices that are H-Hamiltonian or H-skew-Hamiltonian in some indefinite inner product of skewsymmetric-type, are given in the next corollary. Its proof is immediate from Theorems 13.2.3 and 13.2.4.

Corollary 13.2.6. (a) *A matrix $A \in \mathsf{H}^{n \times n}$ is H-skew-Hamiltonian for some ϕ-skewhermitian invertible H if and only if the partial multiplicities corresponding to nonreal eigenvalues of A come in pairs. More precisely, for all positive integers k and for all nonreal $\alpha \in \mathsf{H}$, the total number of partial multiplicities of A that are equal to k at eigenvalues similar to α is even, and we set this number to be zero for all k if α is not an eigenvalue of A.*

(b) *A matrix $A \in \mathsf{H}^{n \times n}$ is H-Hamiltonian for some ϕ-skewhermitian invertible H if and only if the partial multiplicities of eigenvalues α and $-\alpha$ of A, where $\Re(\alpha) \neq 0$, are the same. More precisely, for all positive integers k and for all α with $\Re(\alpha) \neq 0$, the total number of partial multiplicities of A that are equal to k at eigenvalues similar to α coincides with the total number of partial multiplicities of A that are equal to k at eigenvalues similar to $-\alpha$.*

As we have seen in Corollary 13.1.4, the condition in (b) is equivalent to A being similar to $-A$.

13.3 EXTENSION OF INVARIANT SEMIDEFINITE SUBSPACES

Let $G \in \mathsf{H}^{n \times n}$ be ϕ-skewhermitian. Recall that the $\beta(\phi)$-inertia of G are defined by

$$(\mathrm{In}_+(G), \mathrm{In}_-(G), \mathrm{In}_0(G)) = (p, q, n - p - q),$$

where p and q are taken from the canonical form (4.1.2) for G.

Theorem 13.3.1. *Let H be ϕ-skewhermitian and invertible, and let $A \in \mathsf{H}^{n \times n}$ be (H, ϕ)-Hamiltonian or (H, ϕ)-skew-Hamiltonian. Suppose subspace $\mathcal{M} \subseteq \mathsf{H}^{n \times 1}$ is (H, ϕ)-nonnegative, resp. (H, ϕ)-nonpositive, and A-invariant. Then there is an A-invariant (H, ϕ)-nonnegative, resp. (H, ϕ)-nonpositive, subspace $\mathcal{N} \subseteq \mathsf{H}^{n \times 1}$ that contains $\mathcal{M}$ and has dimension*

$$\dim_{\mathsf{H}} \mathcal{N} = \mathrm{In}_+ (H), \qquad resp. \quad \dim_{\mathsf{H}} \mathcal{N} = \mathrm{In}_- (H).$$

Proof. We consider the case of (H, ϕ)-nonnegative subspaces (for the (H, ϕ)-nonpositive subspaces apply the result for $-H$ in place of H). The proof is modeled after that of Theorem 10.2.1.

We use induction on the size n of the matrices H and A and on the dimension of $\mathcal{M}$. The case $n = 1$ is trivial, and the case $\mathcal{M} = \{0\}$ follows by a construction of an A-invariant H-nonnegative subspace in Step 1 below.

Step 1. The case $\mathcal{M} = \{0\}$. Assume first A is (H, ϕ)-Hamiltonian. We may suppose that H and A are given by the right-hand sides of (13.2.1). Note that

$$\mathrm{In}_+ \left(\begin{bmatrix} 0 & F_\ell \\ -F_\ell & 0 \end{bmatrix} \right) = \mathrm{In}_- \left(\begin{bmatrix} 0 & F_\ell \\ -F_\ell & 0 \end{bmatrix} \right) = \ell, \qquad (13.3.1)$$

$$\mathrm{In}_+ (\pm \beta(\phi) F_k) = \begin{cases} \dfrac{k}{2} & \text{if } k \text{ is even}; \\ \dfrac{k \pm 1}{2} & \text{if } k \text{ is odd}, \end{cases} \qquad (13.3.2)$$

and

$$\mathrm{In}_- (\pm \beta(\phi) F_k) = \begin{cases} \dfrac{k}{2} & \text{if } k \text{ is even}; \\ \dfrac{k \mp 1}{2} & \text{if } k \text{ is odd}. \end{cases} \qquad (13.3.3)$$

Therefore, to obtain a spanning set for an A-invariant (H, ϕ)-nonnegative subspace, we select vectors as follows (cf. selection of vectors in the proof of Theorem 10.2.1): for each pair of blocks

$$(L_i = \kappa \beta(\phi) F_k, \ A_i = \beta(\phi) J_k(0))$$

in (13.2.1), select $e_1, \ldots, e_k$ if k is even, select $e_1, \ldots, e_{(k-1)/2}$ if k is odd and $\kappa = -1$, and select $e_1, \ldots, e_{(k+1)/2}$ if k is odd and $\kappa = 1$. For each pair of blocks in (13.2.1),

$$\left(L_i = \begin{bmatrix} 0 & F_\ell \\ -F_\ell & 0 \end{bmatrix}, \ A_i = \begin{bmatrix} -J_\ell(\alpha) & 0 \\ 0 & J_\ell(\alpha) \end{bmatrix} \right), \qquad \alpha \in \mathrm{Inv} (\phi), \ \Re(\alpha) > 0,$$

select $e_1, \ldots, e_\ell$. For each pair of blocks in (13.2.1),

$$(L_i = \delta \beta(\phi) F_s, \ A_i = \beta(\phi) J_s(\tau)), \qquad \delta = \pm 1, \ \tau < 0,$$

select $e_1, \ldots, e_s$ if s is even, select $e_1, \ldots, e_{(s-1)/2}$ if s is odd and $\delta = -1$, and select $e_1, \ldots, e_{(s+1)/2}$ if s is odd and $\delta = 1$.

Next, assume A is (H, ϕ)-skew-Hamiltonian. We take A and H in the following form (Theorem 13.2.4 and Remark 13.2.5):

$$H = \eta_1 \beta(\phi) F_{m_1} \oplus \cdots \oplus \eta_p \beta(\phi) F_{m_p}$$

$$\oplus \begin{bmatrix} 0 & F_{\ell_1} \\ -F_{\ell_1} & 0 \end{bmatrix} \oplus \cdots \oplus \begin{bmatrix} 0 & F_{\ell_q} \\ -F_{\ell_q} & 0 \end{bmatrix},$$

$$A = J_{m_1}(\gamma_1) \oplus \cdots \oplus J_{m_p}(\gamma_p)$$

$$\oplus \begin{bmatrix} J_{\ell_1}(\alpha_1) & 0 \\ 0 & J_{\ell_1}(\alpha_1) \end{bmatrix} \oplus \cdots \oplus \begin{bmatrix} J_{\ell_q}(\alpha_q) & 0 \\ 0 & J_{\ell_q}(\alpha_q) \end{bmatrix},$$

where $\eta_1, \ldots, \eta_p \in \{1, -1\}$, the quaternions $\alpha_1, \ldots, \alpha_q \in \mathrm{Inv}\,(\phi) \backslash \mathsf{R}$, and $\gamma_1, \ldots, \gamma_p \in \mathsf{R}$. We assume also that $m_1, \ldots, m_r$ are odd and $m_{r+1} \ldots, m_p$ are even. Using (13.3.1), (13.3.2), and (13.3.3), one verifies that the following vectors span an A-invariant (H, ϕ)-nonnegative subspace, where we set $w = m_1 + \cdots + m_r$ and $u = m_1 + \cdots + m_p$:

$$e_1, e_2, \ldots, e_{(m_1+\eta_1)/2}, e_{m_1+1}, \ldots, e_{m_1+(m_2+\eta_2)/2}, \ldots,$$

$$e_{m_1+\cdots+m_{r-1}+1}, \ldots, e_{m_1+\cdots+m_{r-1}+(m_r+\eta_r)/2}, e_{w+1}, \ldots, e_{w+m_{r+1}/2},$$

$$e_{w+m_{r+1}+1}, \ldots, e_{w+m_{r+1}+m_{r+2}/2}, \ldots,$$

$$e_{w+m_{r+1}+\cdots+m_{p-1}+1}, \ldots, e_{w+m_{r+1}+\cdots+m_{p-1}+m_p/2},$$

$$e_{u+1}, \ldots, e_{u+\ell_1}, e_{u+2\ell_1+1}, \ldots, e_{u+2\ell_1+\ell_2}, \ldots,$$

$$e_{u+2\ell_1+\cdots+2\ell_{q-1}+1}, e_{u+2\ell_1+\cdots+2\ell_{q-1}+\ell_q}.$$

Step 2. The general case. We proceed as in the proof of Theorem 10.2.1. Using Lemma 4.5.1, we may assume that $\mathcal{M} = \mathrm{Span}_{\mathsf{H}} \{e_1, \ldots, e_{n_2+n_3}\}$,

$$H = \begin{bmatrix} 0 & 0 & I_{n_2} & 0 \\ 0 & \beta(\phi)I_{n_3} & 0 & 0 \\ -I_{n_2} & 0 & 0 & 0 \\ 0 & 0 & 0 & H_0 \end{bmatrix}, \quad \text{and}$$

$$A = \begin{bmatrix} A_{22} & A_{23} & A_{24} & A_{25} \\ A_{32} & A_{33} & A_{34} & A_{35} \\ 0 & 0 & A_{44} & A_{45} \\ 0 & 0 & A_{54} & A_{55} \end{bmatrix}, \tag{13.3.4}$$

where $H_0 = -(H_0)_\phi$, $A_{pp} \in \mathsf{H}^{n_p \times n_p}$ for $p = 2, 3, 4, 5$, and $n_2 + n_3 + n_4 + n_5 = n$. The condition $(HA)_\phi = \pm(HA)$ gives

$$A_{21} = 0, \quad A_{31} = 0, \quad A_{32} = 0, \quad A_{35} = 0, \quad A_{45} = 0,$$

and $(H_0 A_{55})_\phi = \pm H_0 A_{55}$. Letting $\beta(\phi)I_{p_0} \oplus -\beta(\phi)I_{q_0}$ be the canonical form of H_0, we obtain by the induction hypothesis that there is an A_{55}-invariant (H_0, ϕ)-nonnegative subspace $\mathcal{M}_0$ of dimension p_0. Then the subspace

$$\begin{bmatrix} \mathsf{H}^{n_2} \\ 0 \\ 0 \\ 0 \end{bmatrix} \dot{+} \begin{bmatrix} 0 \\ \mathsf{H}^{n_3} \\ 0 \\ 0 \end{bmatrix} \dot{+} \begin{bmatrix} 0 \\ 0 \\ 0 \\ \mathcal{M}_0 \end{bmatrix}$$

is clearly A-invariant (H, ϕ)-nonnegative and has the required dimension. □

Remark 13.3.2. Under the hypotheses of Theorem 10.2.1, consider the case when A is (H, ϕ)-Hamiltonian, and let Λ be a set of eigenvalues of A subject to conditions (10.2.5) and (10.2.6). As in Remark 10.2.2, one can prove that there is an A-invariant (H, ϕ)-nonnegative subspace $\mathcal{N}$ with the additional property that it contains the sums of the root subspaces of A corresponding to the eigenvalues in Λ. Likewise, there exists an A-invariant (H, ϕ)-nonpositive subspace $\mathcal{N}$ with this property.

For indefinite inner products of symmetric-type, i.e., those generated by invertible ϕ-hermitian matrices, we have the following result on invariant semidefinite extensions of invariant neutral subspaces.

In the next two theorems we assume that $H = H_\phi$ is invertible and $A \in \mathsf{H}^{n \times n}$ is (H, ϕ)-skewsymmetric. Note that HA is ϕ-skewhermitian.

Theorem 13.3.3. *Let $\mathcal{N} \subseteq \mathsf{H}^{n \times 1}$ be an A-invariant (H, ϕ)-neutral subspace. Then there exist A-invariant subspaces $\mathcal{L}_+$ and $\mathcal{L}_-$ such that each of them contains $\mathcal{L}$, and $\mathcal{L}_+$ is maximal (HA, ϕ)-nonnegative, whereas $\mathcal{L}_-$ is maximal (HA, ϕ)-nonpositive.*

As in Remark 10.7.1, one shows that, under the hypotheses of Theorem 13.3.3, if an A-invariant subspace $\mathcal{M}$ is (H, ϕ)-neutral, then $\mathcal{M}$ is also (HA, ϕ)-neutral; the converse holds, provided it is assumed in addition that A is invertible.

An analogue of Theorem 10.7.5 is valid as well. Recall the definition of a c-set given in Section 10.7.

Theorem 13.3.4. *Under the hypotheses of Theorem 13.3.3, assume in addition that the set $\mathcal{S}_0$ of eigenvalues with nonzero real parts of the restriction $A|_\mathcal{N}$ is such that*

$$\lambda_0 \in \mathcal{S}_0 \implies -\lambda_0 \notin \mathcal{S}_0.$$

Then for every c-set $\mathcal{S}$ containing $\mathcal{S}_0$ there exist subspaces $\mathcal{L}_\pm$ as in Theorem 13.3.3 with the additional property that $\mathcal{S}$ is the set of eigenvalues with nonzero real parts of $A|_{\mathcal{L}_\pm}$.

As in Open Problem 10.7.6, the question is open whether or not the results of Theorems 13.3.3 and 13.3.4 remain valid under the weaker assumption that $\mathcal{N}$ is A-invariant (HA, ϕ)-neutral.

The proofs of Theorems 13.3.3 and 13.3.4 are relegated to the next section.

We conclude this section with a formula for the dimensions of maximal (HA, ϕ)-nonnegative and maximal (HA, ϕ)-nonpositive subspaces in terms of the canonical form. Let $H = H_\phi \in \mathsf{H}^{n \times n}$ be invertible, and let $A \in \mathsf{H}^{n \times n}$ be (H, ϕ)-skewsymmetric. Then HA is ϕ-skewhermitian and its $\beta(\phi)$-inertia is given as follows, using the canonical form (13.1.3) of the pair (H, A):

(1) $\mathrm{In}_0 (HA)$ is equal to the number of pairs of blocks (H_i, A_i) of type (q-ϕ-H-sk2) in (13.1.3).

(2)

$$\mathrm{In}_+ (HA) - \mathrm{In}_+ (HA) = \sum{}' \varepsilon_i + \sum{}'' \delta_j, \qquad (13.3.5)$$

where the sum $\sum'$ is taken over all pairs (H_i, A_i) of type (q-ϕ-H-sk2) with even k and the sum $\sum''$ is taken over all pairs (H_j, A_j) of type (q-ϕ-H-sk5) with odd s. In (13.3.5), ε_i is the sign attached to the pair (H_i, A_i), and δ_j is the sign attached to the pair (H_j, A_j).

Formulas (1) and (2) are based on the following formulas for the $\beta(\phi)$-inertia for ϕ-skewhermitian matrices:

$$\text{In}_0\left(\varepsilon\beta(\phi)F_k J_k(0)\right) = 1; \tag{13.3.6}$$

$$\text{In}_\pm\left(\varepsilon\beta(\phi)F_k J_k(0)\right) = \begin{cases} \dfrac{k-1}{2} & \text{if } k \text{ is odd,} \\[2mm] \dfrac{k-1\pm\varepsilon}{2} & \text{if } k \text{ is even,} \end{cases} \tag{13.3.7}$$

where $\varepsilon = \pm 1$;

$$\text{In}_+\begin{bmatrix} 0 & -F_\ell J_\ell(\alpha) \\ F_\ell J_\ell(\alpha) & 0 \end{bmatrix} = \text{In}_-\begin{bmatrix} 0 & -F_\ell J_\ell(\alpha) \\ F_\ell J_\ell(\alpha) & 0 \end{bmatrix} = \ell, \tag{13.3.8}$$

where $\alpha \in \text{Inv}\,(\phi)$, $\Re(\alpha) > 0$; and

$$\text{In}_\pm\left(\delta\beta(\phi)F_s J_s(\tau)\right) = \begin{cases} \dfrac{s}{2} & \text{if } s \text{ is even,} \\[2mm] \dfrac{s+\delta}{2} & \text{if } s \text{ is odd,} \end{cases} \tag{13.3.9}$$

where τ is real and positive and $\delta = \pm 1$.

Verification of (13.3.6)–(13.3.9) is left as an exercise (Ex. 13.7.9).

Denoting by Υ the right-hand side of (13.3.5) and by t the number of pairs of blocks (H_i, A_i) of type (q-ϕ-H-sk2) in the canonical form of (H, A), it follows that the dimension of maximal (HA, ϕ)-nonnegative subspaces is

$$\text{In}_0\,(HA) + \text{In}_+\,(HA) = t + \frac{n + \Upsilon}{2}$$

and that of maximal (HA, ϕ)-nonpositive subspaces is $t + (n - \Upsilon)/2$.

13.4   PROOFS OF THEOREMS 13.3.3 AND 13.3.4

The proof is modeled (with necessary changes having been made) after the proofs of Theorems 10.7.2 and 10.7.5. Therefore, we omit some details in the proof.

Lemma 13.4.1. *Let $Z \in \mathsf{H}^{m \times m}$ be a ϕ-skewhermitian matrix partitioned as follows:*

$$Z = \begin{bmatrix} 0 & 0 & -(Q_1)_\phi \\ 0 & Q_2 & -(K_1)_\phi \\ Q_1 & K_1 & K_2 \end{bmatrix},$$

where the $p \times q$ block Q_1 is right-invertible (thus $p \le q$). Then

$$\text{In}_\pm(Z) + \text{In}_0(Z) = q + \text{In}_\pm(Q_2) + \text{In}_0(Q_2). \tag{13.4.1}$$

(The inertia here is the $\beta(\phi)$-inertia of ϕ-skewhermitian matrices.)

The proof is similar to that of Lemma 10.8.1 and is omitted.

We prove Theorems 13.3.3 and 13.3.4 only for HA-nonnegative subspaces (for nonpositive subspaces the proof is analogous, or else use $-H$ in place of H). The canonical form of the pair of matrices (A, H) (with invertible H) given in Theorem 13.1.2 allows us to reduce the proofs to separate consideration of two cases: (1) A is invertible; (2) A is nilpotent.

Assume first that A is invertible. Let $\widehat{H} = HA$. Then $\widehat{H}$ is ϕ-skewhermitian and invertible, and A is $(\widehat{H}, \phi)$-skewsymmetric: $\widehat{H}A = -A_\phi \widehat{H}$. The subspace $\mathcal{N}$ is easily seen to be $\widehat{H}$-neutral. By Theorem 13.3.1, there exist $(\widehat{H}, \phi)$-nonnegative, resp. $(\widehat{H}, \phi)$-nonpositive, subspaces which are A-invariant and have the properties of dimension and location of the spectrum, as required in Theorems 13.3.3 and 13.3.4.

Thus, it remains to prove these theorems for nilpotent A. We prove Theorem 13.3.3 assuming A is nilpotent.

Consider the (H, ϕ)-orthogonal companion of $\mathcal{N}$:

$$\mathcal{N}^{[\perp, \phi]} := \{x \in \mathsf{H}^{n \times 1} \mid x_\phi H y = 0 \quad \text{for all} \quad y \in \mathcal{N}\},$$

As $\mathcal{N}$ is H-neutral, we have $\mathcal{N} \subseteq \mathcal{N}^{[\perp, \phi]}$. Since A is (H, ϕ)-skewsymmetric and $\mathcal{N}$ is A-invariant, the subspace $\mathcal{N}^{[\perp, \phi]}$ is easily seen to be A-invariant as well. Note also that

$$\dim \mathcal{N} = n - \dim \mathcal{N}^{[\perp, \phi]}. \tag{13.4.2}$$

To verify (13.4.2), represent $\mathcal{N}^{[\perp, \phi]}$ as the solution set of the system of linear equations

$$(z_j)_\phi H x = 0, \quad j = 1, 2, \ldots, k,$$

where $z_1, \ldots, z_k$ is a basis for $\mathcal{N}$.

Assuming $\mathcal{N} \neq \mathcal{N}^{[\perp, \phi]}$, choose an (ordered) basis

$$(y_1, \ldots, y_n) \tag{13.4.3}$$

in $\mathsf{H}^{n \times 1}$ so that the first vectors in (13.4.3) form a basis of $\mathcal{N}$, the next vectors in (13.4.3) form a basis in some complement of $\mathcal{N}$ in $\mathcal{N}^{[\perp, \phi]}$, and the remaining vectors in (13.4.3) form a basis in some complement of $\mathcal{N}^{[\perp, \phi]}$ in $\mathsf{H}^{n \times 1}$. Then, as in the proof of Theorems 10.7.2 and 10.7.5, the proof reduces to the case when

$$\mathcal{N} = \mathcal{N}^{[\perp, \phi]} \quad \text{or} \quad \dim_\mathsf{H} \mathcal{N} = \dim_\mathsf{H} \mathcal{N}^{[\perp, \phi]} - 1. \tag{13.4.4}$$

Case 13.4.2. *Assume* $\mathcal{N} = \mathcal{N}^{[\perp, \phi]}$.

In view of (13.4.2) we have

$$\dim \mathcal{N} = \dim \mathcal{N}^{[\perp, \phi]} = \frac{n}{2}, \tag{13.4.5}$$

and n is necessarily even. Choosing a basis $(y_1', \ldots, y_n')$ in $\mathsf{H}^{n \times 1}$ such that the first part of its elements form a basis in $\mathcal{N}$, we represent A and H in the form

$$A' = \begin{bmatrix} B_{11} & B_{12} \\ 0 & B_{22} \end{bmatrix}, \quad \text{where} \quad B_{ij} \in \mathsf{H}^{n/2 \times n/2},$$

$$H' = ((y_i')_\phi H y_j')_{i,j=1}^n = \begin{bmatrix} 0 & H_1 \\ (H_1)_\phi & H_2 \end{bmatrix}.$$

Here

$$\mathcal{N} = \mathrm{Span}_{\mathsf{H}}\{e_1, \ldots, e_{n/2}\},$$

the matrix H_1 is invertible, and H_2 is ϕ-hermitian. Applying a transformation

$$A, H \quad \mapsto \quad S^{-1}AS, \; S_\phi HS \tag{13.4.6}$$

with $S = \begin{bmatrix} I & W_1 \\ 0 & W_2 \end{bmatrix}$ for suitable W_1 and W_2, we may (and do) assume that, in fact,

$$H' = \begin{bmatrix} 0 & I \\ I & 0 \end{bmatrix}. \tag{13.4.7}$$

Then, since A' is (H', ϕ)-skewsymmetric, we have

$$A' = \begin{bmatrix} B_{11} & B_{12} \\ 0 & -(B_{11})_\phi \end{bmatrix}, \qquad B_{12} \quad \text{is } \phi\text{-skewhermitian.} \tag{13.4.8}$$

Now argue as in the proof of Case 10.8.2, using Lemma 13.4.1 in place of Lemma 10.8.1.

Case 13.4.3. *Assume* $\dim \mathcal{N} = \dim \mathcal{N}^{[\perp, \phi]} - 1$.

Arguing as in the proof of Case 10.8.3, we represent A and H in the following form, conformally partitioned:

$$A = \begin{bmatrix} 0 & 0 & D_1 & D_2 & D_3 \\ C_1 & C_2 & D_4 & D_5 & D_6 \\ 0 & 0 & 0_{1\times 1} & D_7 & D_8 \\ 0 & 0 & 0 & 0 & -(C_1)_\phi \\ 0 & 0 & 0 & 0 & -(C_2)_\phi \end{bmatrix},$$

$$H = \begin{bmatrix} 0 & 0 & 0 & I & 0 \\ 0 & 0 & 0 & 0 & I \\ 0 & 0 & H_{22} & H_{231} & H_{232} \\ I & 0 & (H_{231})_\phi & 0 & 0 \\ 0 & I & (H_{232})_\phi & 0 & 0 \end{bmatrix}.$$

Here

$$\mathcal{N} = \mathrm{Span}_{\mathsf{H}}\{e_1, \ldots, e_{(n-1)/2}\}, \qquad \mathcal{N}^{[\perp, \phi]} = \mathrm{Span}_{\mathsf{H}}\{e_1, \ldots, e_{(n+1)/2}\},$$

$$\begin{bmatrix} D_2 & D_3 \\ D_5 & D_6 \end{bmatrix} \in \mathsf{H}^{(n-1)/2 \times (n-1)/2}, \qquad H_{22} = (H_{22})_\phi \in \mathsf{H} \setminus \{0\},$$

and $[C_1 \ C_2] \in \mathsf{H}^{q \times (n-1)/2}$ is right-invertible. Due to HA being ϕ-skewhermitian, we have $D_1 = -(H_{22}D_7)_\phi$, and

$$HA = \begin{bmatrix} 0 & 0 & 0 & 0 & -(C_1)_\phi \\ 0 & 0 & 0 & 0 & -(C_2)_\phi \\ 0 & 0 & 0 & H_{22}D_7 & -(D_4)_\phi \\ 0 & 0 & -(H_{22}D_7)_\phi & G & \star \\ C_1 & C_2 & D_4 & \star & \star \end{bmatrix},$$

where $G = -G_\phi$ and where we denote by $\star$ block entries of no immediate interest. Let

$$\widetilde{A} = \begin{bmatrix} 0_{1\times 1} & D_7 \\ 0 & 0 \end{bmatrix}, \qquad \widetilde{H} = \begin{bmatrix} 0_{1\times 1} & H_{22}D_7 \\ -(H_{22}D_7)_\phi & G \end{bmatrix} = -\widetilde{H}_\phi.$$

If $\mathcal{M}$ is an $\widetilde{A}$-invariant maximal $(\widetilde{H}, \phi)$-nonnegative subspace, then, in view of

Lemma 13.4.1, the subspace $\mathcal{N} + \begin{bmatrix} 0_{(n-1)/2\times 1} \\ \mathcal{M} \\ 0_{q\times 1} \end{bmatrix}$ is A-invariant and maximal

(HA, ϕ)-nonnegative, and we are done.

It remains, therefore, to prove existence of an $\widetilde{A}$-invariant subspace which is maximal $(\widetilde{H}, \phi)$-nonnegative. This can be done as in the proof of case 10.8.3, using the $\beta(\phi)$-inertia of ϕ-skewhermitian matrices rather than inertia of hermitian matrices; to verify that

$$\text{In}_+ \left(\begin{bmatrix} 0 & H_{22} \\ -H_{22} & G_{11} \end{bmatrix} \right) = 1, \quad \text{In}_0 \left(\begin{bmatrix} 0 & H_{22} \\ -H_{22} & G_{11} \end{bmatrix} \right) = 0,$$

where $H_{22} = (H_{22})_\phi \in \mathsf{H} \setminus \{0\}$, $G_{11} = -(G_{11})_\phi \in \mathsf{H}$, use the equality

$$\begin{bmatrix} a_\phi & 1 \\ b_\phi & 1 \end{bmatrix} \begin{bmatrix} 0 & H_{22} \\ -H_{22} & G_{11} \end{bmatrix} \begin{bmatrix} a & b \\ 1 & 1 \end{bmatrix} = \begin{bmatrix} \beta(\phi) & 0 \\ 0 & -\beta(\phi) \end{bmatrix},$$

where $a, b \in \mathsf{H}$ are found from the equations

$$-H_{22}b + b_\phi H_{22} = -\beta(\phi) - G_{11}, \quad a_\phi H_{22} = -G_{11} + H_{22}b.$$

(Note that the matrix $\begin{bmatrix} a & b \\ 1 & 1 \end{bmatrix}$ is necessarily invertible.)

The proofs of Theorems 13.3.3 and 13.3.4 are complete. $\qquad\square$

13.5 INVARIANT LAGRANGIAN SUBSPACES

Of special interest are invariant Lagrangian subspaces, which will be studied in this section. The exposition here is largely parallel to Sections 10.3 and 10.6, so some details will be omitted.

As everywhere in this chapter, fix a nonstandard involution ϕ.

Definition 13.5.1. Given a ϕ-hermitian or ϕ-skewhermitian matrix $G \in \mathsf{H}^{n\times n}$, a subspace $\mathcal{M} \subseteq \mathsf{H}^{n\times 1}$ is said to be (G, ϕ)-*Lagrangian* if it is (G, ϕ)-neutral and has dimension $n/2$.

Clearly, this definition makes sense only if n is even. Also, a subspace $\mathcal{M}$ is (G, ϕ)-Lagrangian if and only if the subspace $S^{-1}(\mathcal{M})$ is $(S_\phi GS, \phi)$-Lagrangian for all invertible matrices $S \in \mathsf{H}^{n\times n}$. As before, in Section 4.1, we denote by

$$(\text{In}_+(G), \ \text{In}_-(G), \ \text{In}_0(G))$$

the $\beta(\phi)$-inertia of a ϕ-skewhermitian matrix G. As follows from Theorems 4.2.6 and 4.2.7, and assuming n is even, (G, ϕ)-Lagrangian subspaces exist if and only if

$$\min\{\text{In}_+(G), \text{In}_-(G)\} + \text{In}_0(G) \geq \frac{n}{2} \tag{13.5.1}$$

if G is ϕ-skewhermitian, and

$$\lfloor \frac{\operatorname{rank} G + 1}{2} \rfloor \le \frac{n}{2} \qquad (13.5.2)$$

if G is ϕ-hermitian. We will use the notion of (G, ϕ)-Lagrangian subspaces for invertible G (and even n); in this case, (13.5.2) is always satisfied, and (13.5.1) becomes

$$\operatorname{In}_+ (G) = \operatorname{In}_- (G) = \frac{n}{2}. \qquad (13.5.3)$$

The following property of Lagrangian subspaces will be very useful.

Proposition 13.5.2. *Let*

$$G = \pm G_\phi = G_1 \oplus G_2 \oplus \cdots \oplus G_k \in \mathsf{H}^{n \times n},$$

where $G_j \in \mathsf{H}^{n_j \times n_j}$, $j = 1, 2, \ldots, k$, *and where* n *is even. Assume* G *is invertible. Then a subspace* $\mathcal{M} \subseteq \mathsf{H}^{n \times 1}$ *of the form*

$$\mathcal{M} = \mathcal{M}_1 \oplus \mathcal{M}_2 \oplus \cdots \oplus \mathcal{M}_k := \left\{ \begin{bmatrix} x_1 \\ x_2 \\ \vdots \\ x_k \end{bmatrix} : x_1 \in \mathcal{M}_1, \ldots, x_k \in \mathcal{M}_k \right\}$$

is (G, ϕ)-*Lagrangian if and only if each* $\mathcal{M}_j$ *is* (G_j, ϕ)-*Lagrangian, for* $j = 1, 2, \ldots, k$ *(in particular, each* n_j *must be even).*

The proof is analogous to that of Proposition 13.5.2 (Ex. 13.7.7).

Next, we consider existence of invariant Lagrangian subpaces. A necessary and sufficient condition for this to happen is given in the next theorem for the case of inner products of symmetric-type. As it turns out, existence of Lagrangian subspaces is generally not sufficient.

Theorem 13.5.3. *Let* $H \in \mathsf{H}^{n \times n}$ *be* ϕ-*hermitian and invertible.*

(a) *If* $A \in \mathsf{H}^{n \times n}$ *is* (H, ϕ)-*symmetric, then there exists an* A-*invariant* (H, ϕ)-*Lagrangian subspace if and only if every root subspace of* A *is even dimensional.*

(b) *If* $A \in \mathsf{H}^{n \times n}$ *is* (H, ϕ)-*skewsymmetric, then there exists an* A-*invariant* (H, ϕ)-*Lagrangian subspace if and only if the following two conditions are satisfied:*

(1) *The root subspace of* A *corresponding to the zero eigenvalue (if* A *if not invertible) is even dimensional.*

(2) *For every nonzero eigenvalue* λ *of* A *with zero real part (if any), the number of odd partial multiplicities corresponding to* λ *is even and exactly half of them have sign* -1 *(the other half having the sign* 1*) in the sign characteristic of* (A, H).

Proof. Note that for any invertible $S \in \mathsf{H}^{\times n}$, the subspace $\mathcal{M}$ is A-invariant (H, ϕ)-Lagrangian if and only if $S^{-1}(\mathcal{M})$ is $S^{-1}AS$-invariant $(S^* HS, \phi)$-Lagrangian. Therefore, without loss of generality, we may (and do) assume that the pair (A, H) is given by the canonical form of Theorem 13.1.1 (if A is (H, ϕ)-symmetric) or Theorem 13.1.2 (if A is (H, ϕ)-skewsymmetric).

Part (a). In view of Proposition 13.5.2 and the fact that every A-invariant subspace is the sum of its intersections with the root subspaces of A (Proposition 5.1.4), we need only to consider the situation when $\mathsf{H}^{n\times 1}$ is a root subspace for A. The condition that n is even is obviously necessary. Assuming n is even, we will exhibit an A-invariant (H, ϕ)-Lagrangian subspace. Rearranging (if necessary) the blocks in the canonical form, we let

$$H = F_{\ell_1} \oplus \cdots \oplus F_{\ell_{2u}} \oplus F_{\ell_{2u+1}} \oplus \cdots \oplus F_{\ell_q},$$
$$A = J_{\ell_1}(\alpha) \oplus \cdots \oplus J_{\ell_q}(\alpha), \qquad \alpha \in \mathsf{H},$$

where $\ell_1, \ldots, \ell_{2u}$ are odd and $\ell_{2u+1}, \ldots, \ell_q$ are even. Select the vectors:

$$e_1, \ldots, e_{(\ell_1-1)/2}, e_{(\ell_1+1)/2} + \beta(\phi)e_{\ell_1+(\ell_2+1)/2}, e_{\ell_1+1}, \ldots, e_{\ell_1+(\ell_2-1)/2}, \qquad (13.5.4)$$

$$e_{\ell_1+\cdots+\ell_{2j}+1}, \ldots, e_{\ell_1+\cdots+\ell_{2j}+\cdots+(\ell_{2j+1}-1)/2}, \qquad (13.5.5)$$

$$e_{\ell_1+\cdots+\ell_{2j}+\cdots+(\ell_{2j+1}+1)/2} + \beta(\phi)e_{\ell_1+\cdots+\ell_{2j}+\ell_{2j+1}+(\ell_{2j+2}+1)/2}, \qquad (13.5.6)$$

$$e_{\ell_1+\cdots+\ell_{2j}+\ell_{2j+1}+1}, \ldots, e_{\ell_1+\cdots+\ell_{2j}+\ell_{2j+1}+(\ell_{2j+2}-1)/2} \qquad (13.5.7)$$

for $j = 1, 2, \ldots, u - 1$, and, setting $w = \ell_1 + \cdots + \ell_{2u}$,

$$e_{w+1}, \ldots, e_{w+\ell_{2u+1}/2}, e_{w+\ell_{2u+1}+1}, \ldots, e_{w+\ell_{2u+1}+\ell_{2u+2}/2}, \ldots, \qquad (13.5.8)$$

$$e_{w+\ell_{2u+1}+\cdots+\ell_{q-1}+1}, e_{w+\ell_{2u+1}+\cdots+\ell_{q-1}+2}, \ldots, e_{w+\ell_{2u+1}+\cdots+\ell_{q-1}+\ell_q/2}. \qquad (13.5.9)$$

The selected vectors span an A-invariant (H, ϕ)-Lagrangian subspace in $\mathsf{H}^{n\times 1}$.

Part (b). In view of the canonical form (13.1.3), we can (and do) assume that $\mathsf{H}^{n\times 1}$ is either a root subspace of A corresponding to an eigenvalue with zero real part or the sum of two root subspaces of A corresponding to eigenvalues α and $-\alpha$ with nonzero real part. Three cases can occur:

(i)

$$H = F_{\ell_1} \oplus \cdots \oplus F_{\ell_v} \oplus \varepsilon_{v+1} F_{\ell_{v+1}} \oplus \cdots \oplus \varepsilon_q F_{\ell_q},$$
$$A = \beta(\phi) J_{\ell_1}(0) \oplus \cdots \oplus \beta(\phi) J_{\ell_q}(0),$$

where $\ell_1, \ldots, \ell_v$ are odd, $\ell_{v+1}, \ldots, \ell_q$ are even, and $\varepsilon_j = \pm 1$, $j = v + 1, \ldots, q$;

(ii)

$$H = \oplus_{j=1}^q (\delta_j F_{p_j}), \qquad A = \oplus_{j=1}^q (\beta(\phi) J_{p_j}(\tau)), \qquad (13.5.10)$$

where $\tau > 0$ and the δ_j's are signs ± 1;

(iii)

$$H = \oplus_{j=1}^s \begin{bmatrix} 0 & F_{k_j} \\ F_{k_j} & 0 \end{bmatrix}, \qquad A = \oplus_{j=1}^s \begin{bmatrix} J_{k_j}(\alpha) & 0 \\ 0 & -J_{k_j}(\alpha) \end{bmatrix},$$

where $\alpha \in \mathrm{Inv}\,(\phi)$ has positive real part.

In Case (i), condition (1) is obviously necessary, and assuming it holds true, we have that $v = 2u$ is even. We make the selection of vectors as in (13.5.4)–(13.5.9). Then the selected vectors span an A-invariant (H, ϕ)-Lagrangian subspace. In Case (iii), select

$$e_1, e_2, \ldots, e_{k_1}, e_{2k_1+1}, \ldots, e_{2k_1+k_2}, e_{2k_1+2k_2+1}, \ldots,$$

$$e_{2k_1+2k_2+k_3}, \ldots, e_{2k_1+\cdots+2k_{s-1}+k_s}.$$

Again, the selected vectors span an A-invariant (H, ϕ)-Lagrangian subspace.

Finally, consider Case (ii). Note that HA is ϕ-skewhermitian. Its $\beta(\phi)$-inertia is found by the formula

$$\text{In}_+ (HA) + \text{In}_- (HA) = n, \quad \text{In}_+ (HA) - \text{In}_- (HA) = \sum_{\{j \,:\, p_j \text{ is odd}\}} \delta_j.$$

Observe that all A-invariant (H, ϕ)-Lagrangian subspaces are, clearly, also (HA, ϕ)-Lagrangian. Thus, the necessary condition $\text{In}_+ (HA) = \text{In}_- (HA)$ for existence of an A-invariant (H, ϕ)-Lagrangian subspace boils down to item (2). Conversely, suppose (2) holds true. We will construct an A-invariant (H, ϕ)-Lagrangian subspace. Simultaneously rearranging blocks in the formulas for H and A in (13.5.10) (if necessary), we assume

$$H = (F_{p_1} \oplus -F_{p_2}) \oplus \cdots \oplus (F_{p_{2u-1}} \oplus -F_{p_{2u}}) \oplus \delta_{2u+1} F_{p_{2u+1}} \oplus \cdots \oplus \delta_q F_{p_q},$$

where $p_1, \ldots, p_{2u}$ are odd and $p_{2u+1}, \ldots, p_q$ are even. Select the vectors in $\mathsf{H}^{n \times n}$, as in (10.3.5)–(10.3.10). The selected vectors span an A-invariant (H, ϕ)-Lagrangian subspace. $\qquad\square$

Remark 13.5.4. For an (H, ϕ)-skewsymmetric matrix A, if there exists an A-invariant (H, ϕ)-Lagrangian subspace, then there exists such a subspace with additional properties regarding location of the spectrum, as in Remark 10.2.2. Namely, let Λ be a set of eigenvalues of A with the properties

$$\lambda \in \Lambda, \quad \mu \in \mathsf{H}, \quad \mu \text{ similar to } \lambda \quad \Longrightarrow \quad \mu \in \Lambda \qquad (13.5.11)$$

and

$$\lambda \in \Lambda \quad \Longrightarrow \quad -\lambda \notin \Lambda. \qquad (13.5.12)$$

Then there exists an A-invariant (H, ϕ)-Lagrangian subspace $\mathcal{M}$ that contains the sum of root subspaces for A corresponding to the eigenvalues in Λ. We leave it as an exercise to appropriately select vectors to span such a subspace $\mathcal{M}$ (Ex. 13.7.8).

Next, we consider invariant Lagrangian subspaces in the context of indefinite inner products of skewsymmetric-type. Thus, it will be assumed from now on in this section that $H \in \mathsf{H}^{n \times n}$ is ϕ-skewhermitian and invertible and (as before) that n is even. Then existence of (H, ϕ)-Lagrangian subspaces is guaranteed.

Theorem 13.5.5. (a) *Let $X \in \mathsf{H}^{n \times n}$ be (H, ϕ)-Hamiltonian. Then there exists an X-invariant (H, ϕ)-Lagrangian subspace if and only if every root subspace of X corresponding to eigenvalues with zero real part (including the zero eigenvalue if X is not invertible) has even (quaternion) dimension, and for each such eigenvalue, the signs in the sign characteristic corresponding to the odd multiplicities (if any) associated with that eigenvalue, sum up to zero.*

(b) *Let $X \in \mathsf{H}^{n \times n}$ be (H, ϕ)-skew-Hamiltonian. Then there exists an X-invariant (H, ϕ)-Lagrangian subspace if and only if every root subspace for X is even dimensional.*

Proof. Part (b). The "only if" part is clear. For the "if" part, in view of the canonical form (13.2.6) for the pair of matrices (H, X), the proof reduces to two cases:

(i)

$$H = \eta_1 \beta(\phi) F_{m_1} \oplus \cdots \oplus \eta_p \beta(\phi) F_{m_p}, \quad X = J_{m_1}(\gamma)^T \oplus \cdots \oplus J_{m_p}(\gamma)^T,$$

where γ is real, the η_j's are signs ± 1, and $m_1 + \cdots + m_p$ is even;

(ii)

$$H = \begin{bmatrix} 0 & F_{\ell_1} \\ -F_{\ell_1} & 0 \end{bmatrix} \oplus \cdots \oplus \begin{bmatrix} 0 & F_{\ell_q} \\ -F_{\ell_q} & 0 \end{bmatrix},$$

$$X = \begin{bmatrix} J_{\ell_1}(\alpha)^T & 0 \\ 0 & J_{\ell_1}(\alpha)^T \end{bmatrix} \oplus \cdots \oplus \begin{bmatrix} J_{\ell_q}(\alpha)^T & 0 \\ 0 & J_{\ell_q}(\alpha)^T \end{bmatrix},$$

where $\alpha \in \mathrm{Inv}\,(\phi) \setminus \mathsf{R}$.

In Case (i), argue as in the proof of Part (b) of Theorem 10.6.1. In Case (ii), existence of an X-invariant (H, ϕ)-Lagrangian subspace is evident: one such subspace is spanned by the following vectors:

$$e_1, e_2, \ldots, e_{\ell_1}, e_{2\ell_1+1}, \ldots, e_{2\ell_1+\ell_2}, \ldots,$$

$$e_{2\ell_1+2\ell_2+\cdots+2\ell_{q-1}+1}, \ldots, e_{2\ell_1+\cdots+2\ell_{q-1}+\ell_q}. \tag{13.5.13}$$

Part (a). The canonical form of Theorem 13.2.3 for the pair (H, X) will be used. Thus, we need to consider only the following three cases:

(iii)

$$H = \begin{bmatrix} 0 & F_{\ell_1} \\ -F_{\ell_1} & 0 \end{bmatrix} \oplus \cdots \oplus \begin{bmatrix} 0 & F_{\ell_q} \\ -F_{\ell_q} & 0 \end{bmatrix},$$

$$X = \begin{bmatrix} -J_{\ell_1}(\alpha) & 0 \\ 0 & J_{\ell_1}(\alpha) \end{bmatrix} \oplus \cdots \oplus \begin{bmatrix} -J_{\ell_q}(\alpha) & 0 \\ 0 & J_{\ell_q}(\alpha) \end{bmatrix},$$

where $\alpha \in \mathrm{Inv}\,(\phi)$, $\Re(\alpha) > 0$;

(iv)

$$H = \oplus_{j=1}^s \left(\kappa_j \beta(\phi) F_{p_j} \right) \oplus \oplus_{j=s+1}^q \beta(\phi) F_{p_j}, \quad X = \oplus_{j=1}^q \beta(\phi) J_{p_j}(0),$$

where $p_1, \ldots, p_s$ are odd, $p_{s+1}, \ldots, p_q$ are even, and the κ_j's are signs ± 1 $(j = 1, 2, \ldots, s)$;

(v)

$$H = \oplus_{j=1}^v \left(\delta_j \beta(\phi) F_{\ell_j} \right) \oplus \oplus_{j=v+1}^q \left(\delta_j \beta(\phi) F_{\ell_j} \right),$$

$$X = \oplus_{j=1}^q \beta(\phi) J_{\ell_j}(\tau), \quad \tau < 0,$$

where $\ell_1, \ldots, \ell_v$ are odd, $\ell_{v+1}, \ldots, \ell_q$ are even, and the δ_j's are signs ± 1 $(j = 1, 2, \ldots, q)$.

In case (iii) holds true, the condition in Theorem 13.5.5(a) is vacuous, and indeed vectors (13.5.13) span an X-invariant (H, ϕ)-Lagrangian subspace.

Suppose H and X are given as in (iv). For an (H, ϕ)-Lagrangian subspace to exist, it is obviously necessary that $s = 2u$ be even. Moreover, for the $\beta(\phi)$-inertia $\mathrm{In}_\pm(H)$ of H we have

$$\mathrm{In}_+(H) + \mathrm{In}_-(H) = n, \quad \mathrm{In}_+(H) - \mathrm{In}_-(H) = \sum_{j=1}^{v} \kappa_j,$$

so the necessary condition (13.5.3) holds true if and only if $\kappa_1 + \cdots + \kappa_s = 0$. This proved the "only if" direction of Theorem 13.5.5(a). For the "if" direction, we rearrange (if necessary) the blocks in (iv) so that $\kappa_j = (-1)^{j-1}$, for $j = 1, 2, \ldots, s = 2u$. Select the vectors in $\mathsf{H}^{n \times 1}$ as in (10.3.5)–(10.3.10). By inspection, the selected vectors span an X-invariant (H, ϕ)-Lagrangian subspace.

Finally, suppose H and X are given as in (v). As in Case (iv), we verify that the conditions of Theorem 13.5.5(a) are necessary for existence of an (H, ϕ)-Lagrangian subspace. The converse is proved as in Case (iv), by selecting the vectors (10.3.5)–(10.3.10). $\qquad\square$

A statement analogous to Remark 13.5.4 holds true also for X-invariant (H, ϕ)-Lagrangian subspaces, in case H is ϕ-skewhermitian and invertible and X is (H, ϕ)-Hamiltonian.

13.6 BOUNDEDNESS OF SOLUTIONS OF DIFFERENTIAL EQUATIONS

Consider the system of differential equations with constant coefficients

$$A_\ell x^{(\ell)}(t) + A_{\ell-1} x^{(\ell-1)}(t) + \cdots + A_1 x'(t) + A_0 x(t) = 0, \quad t \in \mathsf{R}, \qquad (13.6.1)$$

where $A_\ell, \ldots, A_1, A_0 \in \mathsf{H}^{n \times n}$, and $x(t)$ is an unknown ℓ times continuously differentiable $\mathsf{H}^{n \times 1}$-valued function of the real independent variable t. It will be assumed in addition that A_k is ϕ-hermitian if k is odd, A_k is ϕ-skewhermitian if k is even, and A_ℓ is invertible.

In this section we study boundedness properties of solutions of (13.6.1). The development here is parallel to that of Section 10.4. In particular, we will apply to (13.6.1) the definitions of bounded, forward bounded, backward bounded, and stably bounded systems given in Section 10.4.

As in Section 10.4, consider the companion matrix associated with system (13.6.1):

$$C = \begin{bmatrix} 0 & I_n & 0 & \cdots & 0 \\ 0 & 0 & I_n & \cdots & 0 \\ \vdots & \vdots & \vdots & \ddots & \vdots \\ 0 & 0 & 0 & \cdots & I_n \\ -A_\ell^{-1} A_0 & -A_\ell^{-1} A_1 & -A_\ell^{-1} A_2 & \cdots & -A_\ell^{-1} A_{\ell-1} \end{bmatrix} \in \mathsf{H}^{\ell n \times \ell n}, \qquad (13.6.2)$$

and define

$$G := \begin{bmatrix} A_1 & A_2 & A_3 & \ldots & A_\ell \\ -A_2 & -A_3 & \ldots & -A_\ell & 0_n \\ A_3 & A_4 & \ldots & 0_n & 0_n \\ -A_4 & \ldots & \ldots & 0_n & 0_n \\ \vdots & \ddots & \ddots & \ddots & \vdots \\ (-1)^{\ell-1} A_\ell & 0_n & 0_n & \ldots & 0_n \end{bmatrix} \in \mathsf{H}^{n\ell \times n\ell}. \qquad (13.6.3)$$

Clearly, G is ϕ-hermitian and invertible.

A computation shows that that the matrix GC is ϕ-skewhermitian; in other words, C is (G, ϕ)-skewsymmetric. This is a key fact that allows us to use the canonical form (13.1.3) of the pair of matrices (G, C) in the study of (13.6.1).

Theorem 13.6.1. *The following four statements are equivalent for the system* (13.6.1), *where* A_k *is* ϕ-*hermitian if* k *is odd,* A_k *is* ϕ-*skewhermitian if* k *is even, and* A_ℓ *is invertible*:

(a) *The system is forward bounded.*

(b) *The system is backward bounded.*

(c) *The system is bounded.*

(d) *All eigenvalues of* C *have zero real parts, and for every eigenvalue the geometric multiplicity coincides with the algebraic multiplicity.*

The proof is analogous to the proof of Theorem 10.4.2, and is omitted.

At present, we do not have results concerning stable boundedness of (13.6.1). Thus, we formulate an open problem.

Open Problem 13.6.2. *Develop criteria for stable boundedness of system of the type* (13.6.1), *under the hypotheses that* A_k *is* ϕ-*hermitian if* k *is odd,* A_k *is* ϕ-*skewhermitian if* k *is even, and* A_ℓ *is invertible.*

In particular, is the following statement true?

(A) *Assume that the system* (13.6.1) *is bounded, and for every nonzero eigenvalue* α *of* C *with zero real part (if any), the signs in the sign characteristic of* (G, C) *corresponding to the eigenvalues similar to* α *are all equal. Then* (13.6.1) *is stably bounded.*

In a similar vein the system of differential equations (13.6.1) is studied, where now the following is assumed.

(B) A_k *is* ϕ-*hermitian if* k *is even,* A_k *is* ϕ-*skewhermitian if* k *is odd, and* A_ℓ *is invertible.*

In this case, the matrix G defined by (13.6.3) is ϕ-skewhermitian and C is (G, ϕ)-Hamiltonian. The result of Theorem 13.6.1 remains valid, with a proof similar to that of Theorem 10.4.2.

To state and prove our result on stable boundedness under hypothesis (B), it will be convenient to introduce the following concept.

Definition 13.6.3. Given a matrix $Y \in \mathsf{H}^{n \times n}$ and a subspace $\mathcal{M} \subseteq \mathsf{H}^{n \times 1}$, we say that $\mathcal{M}$ is (Y, ϕ)-definite if

$$\min_{\{x \in \mathcal{M} \,:\, \|x\|=1\}} |x_\phi Y x| > 0.$$

Some properties of (Y, ϕ)-definite subspaces are collected in the next proposition.

Proposition 13.6.4. *Let* $Y \in \mathsf{H}^{n \times n}$ *and* $\mathcal{M} \subseteq \mathsf{H}^{n \times 1}$. *Then the following statements are equivalent:*

(i) $\mathcal{M}$ *is* (Y, ϕ)*-definite.*

(ii) $x_\phi Y x \neq 0$ *for every nonzero* $x \in \mathcal{M}$.

(iii) $\min_{\{x \in \mathcal{M} \,:\, \|x\| \geq \varepsilon\}} |x_\phi Y x| > 0$ *for some (equivalently, all) positive* ε.

(iv) *The subspace* $S^{-1}(\mathcal{M})$ *is* $(S_\phi Y S, \phi)$*-definite for some (equivalently, all) invertible matrices* $S \in \mathsf{H}^{n \times n}$.

(v) *All nearby subspaces, including* $\mathcal{M}$ *itself, are* (Y, ϕ)*-definite; in other words, there exists* $\delta > 0$*, which depends on* Y *and* $\mathcal{M}$ *only, for fixed* ϕ*, such that every subspace* $\mathcal{N} \subseteq \mathsf{H}^{n \times n}$ *satisfying* $\theta(\mathcal{M}, \mathcal{N}) < \delta$ *is* (Y, ϕ)*-definite.*

Proof. The equivalence of (i) and (iii) and implication (i) $\Rightarrow$ (ii) are easily seen by scaling the vector $x \in \mathcal{M}$. To show (ii) $\Rightarrow$ (i), note that by continuity of the function $|x_\phi Y x|$, $x \in \mathcal{M} \setminus \{0\}$, and by compactness of the unit sphere $\{x \in \mathcal{M} \,:\, \|x\| = 1\}$ in $\mathcal{M}$, the infimum

$$\inf_{\{x \in \mathcal{M} \,:\, \|x\|=1\}} |x_\phi Y x|$$

is attained (Theorem 3.10.5); therefore, the infimum cannot be zero in view of (ii). Hence, $\mathcal{M}$ is (Y, ϕ)-definite. We leave it to the reader to prove the equivalence of (i) and (iv).

Clearly (v) implies (i). The opposite implication can be proved as in the proof of Theorem 10.4.4, Step 2. □

It turns out that existence of definite subspaces imposes some restrictions on the matrix.

Proposition 13.6.5. *If* $\mathcal{M}$ *is a* (Y, ϕ)*-definite subspace, then*

$$\dim_{\mathsf{H}} (\mathrm{Ker}\,(Y - Y_\phi) \cap \mathcal{M}) \leq 1.$$

Proof. Assuming the contrary, let $\mathcal{M}_0$ be a two-dimensional subspace in $\mathcal{M}$ such that $(Y - Y_\phi)x = 0$ for every $x \in \mathcal{M}_0$. Then

$$
\begin{aligned}
x_\phi Y x &= x_\phi \left(\frac{1}{2}(Y + Y_\phi) \right) x + x_\phi \left(\frac{1}{2}(Y - Y_\phi) \right) x \\
&= x_\phi \left(\frac{1}{2}(Y + Y_\phi) \right) x, \qquad \text{for all } x \in \mathcal{M}_0. \qquad (13.6.4)
\end{aligned}
$$

By Theorem 4.1.2(a), in a suitable basis for $\mathcal{M}_0$, the ϕ-hermitian matrix $\frac{1}{2}(Y + Y_\phi)$ restricted to $\mathcal{M}_0$ has one of the three forms: I_2, 0_2, or $\begin{bmatrix} 1 & 0 \\ 0 & 0 \end{bmatrix}$. In all three cases, in view of (13.6.4), there exists a nonzero vector $x_0 \in \mathcal{M}_0$ such that $(x_0)_\phi Y x_0 = 0$. Indeed, this is obvious if $\frac{1}{2}(Y + Y_\phi)$ restricted to $\mathcal{M}_0$ is 0_2 or $\begin{bmatrix} 1 & 0 \\ 0 & 0 \end{bmatrix}$, and if $\frac{1}{2}(Y + Y_\phi)$ restricted to $\mathcal{M}_0$ is I_2, this follows from Lemma 4.4.1. Existence of such x_0 contradicts the hypothesis that $\mathcal{M}$ is (Y, ϕ)-definite. $\square$

We now return to system (13.6.1). Under hypothesis (B), sufficient conditions for stable boundedness are given in the next theorem.

Theorem 13.6.6. *Consider system (13.6.1) under hypothesis (B), and let C and G be defined by (13.6.2) and (13.6.3), respectively. Then the following statements are equivalent:*

(a) *Every root subspace of C is (G, ϕ)-definite.*

(b) *System (13.6.1) is bounded, and for every eigenvalue α of C with zero real part, the signs in the sign characteristic of (G, C) corresponding to the eigenvalues similar to α are all equal.*

Also, each condition (a) or (b) implies:

(c) *System (13.6.1) is stably bounded.*

If $\ell = 1$, then the conditions (a), (b), and (c) are equivalent.

Proof. Proof of (b) $\Rightarrow$ (a). By Theorem 13.6.1 (which holds true under hypothesis (B) as well), all eigenvalues of C have zero real parts, and for every eigenvalue of C, the geometric and algebraic multiplicities coincide. In view of the canonical form of the (G, ϕ)-Hamiltonian matrix C (Theorem 13.2.3), we need only to verify that for $G_0 = \beta(\phi)I_k$ the space $\mathsf{H}^{k \times 1}$ is (G_0, ϕ)-definite. To this end, note the equality $\alpha_\phi \beta(\phi)\alpha = \beta(\phi)\alpha^*\alpha$ for all $\alpha \in \mathsf{H}$, which can be verified by direct computation. Therefore, $x_\phi G_0 x = \beta(\phi)\|x\|^2$ for all $x \in \mathsf{H}^{k \times k}$, and the (G_0, ϕ)-definiteness of $\mathsf{H}^{k \times 1}$ follows.

Proof of (a) $\Rightarrow$ (b). The canonical form of (G, C) (Theorem 13.2.3) shows that all eigenvalues of C have zero real parts, and for every eigenvalue of C, the geometric and algebraic multiplicities coincide (otherwise, there exists an eigenvector x of C such that $x_\phi G x = 0$, a contradiction with (a)), so by Theorem 13.6.1 system (13.6.1) is bounded. Arguing by contradiction, suppose the condition on signs in (b) is not satisfied; we then produce an eigenvector x_0 of C such that $(x_0)_\phi G x_0 = 0$, a contradiction with (a). Indeed, in view of the canonical form, we need to consider only

$$G_0 = \beta(\phi)I_p \oplus -\beta(\phi)I_q, \qquad C_0 = -\beta(\phi)\tau I_{p+q},$$

in place of G, C, respectively, where p, q are positive integers and τ is a nonpositive real number. Then set $x_0 = e_1 + e_{p+1}$.

The proof that (b) and/or (a) imply (c) is completely analogous to that of Theorem 10.4.4, Step 2, and, therefore, is omitted.

Finally, assume $\ell = 1$, so (13.6.1) takes the form $A_1 x' + A_0 x = 0$, where $A_1 \in \mathsf{H}^{n \times n}$ is ϕ-skewhermitian invertible and $A_0 \in \mathsf{H}^{n \times n}$ is ϕ-hermitian, with $C = A_1^{-1} A_0$, $G = A_1$. Suppose (c) holds true. In particular, (13.6.1) is bounded, so by Theorem 13.6.1 all eigenvalues of C have zero real parts, and for every eigenvalue the geometric and algebraic multiplicities coincide. Arguing by contradiction, suppose that there is an eigenvalue α_0 of C such that the signs in the sign characteristic of (C, G) corresponding to α_0 are not all equal. We will construct a (G, ϕ)-Hamiltonian matrix $\widetilde{C}$ that is arbitrarily close to C and has eigenvalues with nonzero real parts, thereby contradicting the hypothesis that (13.6.1) is stably bounded. Using the canonical form for (C, G) (Theorem 13.2.3), it suffices to consider two cases:

(1) $\quad G = \begin{bmatrix} \beta(\phi) & 0 \\ 0 & -\beta(\phi) \end{bmatrix}, \qquad C = 0_2;$

(2) $\quad G = \begin{bmatrix} \beta(\phi) & 0 \\ 0 & -\beta(\phi) \end{bmatrix}, \qquad C = \begin{bmatrix} -\beta(\phi)\tau & 0 \\ 0 & -\beta(\phi)\tau \end{bmatrix},$ where τ is real and negative.

These two cases can be combined as one by assuming $\tau \leq 0$. Set

$$\widetilde{C} = \begin{bmatrix} -\beta(\phi)\tau & x \\ x & -\beta(\phi)\tau \end{bmatrix},$$

where $x > 0$ is close to zero. Then $\widetilde{C}$ is (G, ϕ)-Hamiltonian and has eigenvalues with real parts $\pm x$, as required. $\qquad\square$

Open Problem 13.6.7. *Are* (a), (b), *and* (c) *of Theorem* 13.6.6 *equivalent if* $\ell > 1$?

13.7   EXERCISES

In all exercises in this chapter, ϕ is a nonstandard involution.

Ex. 13.7.1. Prove Proposition 13.0.1.

Ex. 13.7.2. Show that system (13.6.1), under the hypotheses that A_k is ϕ-hermitian if k is odd, A_k is ϕ-skewhermitian if k is even, and A_ℓ is invertible, is stably bounded if it is bounded and all root subspaces of C corresponding to eigenvalues with zero real parts (if any) are 1-dimensional.

Ex. 13.7.3. Let H be ϕ-skewhermitian and A be (H, ϕ)-Hamiltonian. Prove that a root subspace $\mathcal{M}$ of A is (H, ϕ)-definite if and only if $x_\phi H x \neq 0$ for every eigenvector $x \in \mathcal{M}$ of A. Hint: Use the canonical form of Theorem 13.2.3.

Ex. 13.7.4. Consider system (13.6.1), and assume hypothesis (B). Show that if $x_\phi G x \neq 0$ for every eigenvector x of C, then the system (13.6.1) is stably bounded. (G and C are given by (13.6.3) and (13.6.2), respectively.)

Ex. 13.7.5. Consider system (13.6.1), and assume hypothesis (B). Prove that if every root subspace of C is G-definite, then every system sufficiently close to (13.6.1) and satisfying hypothesis (B) is stably bounded; in other words, there exists $\varepsilon > 0$ such that the system

$$A'_\ell x^{(\ell)}(t) + A'_{\ell-1} x^{(\ell-1)}(t) + \cdots + A'_1 x'(t) + A'_0 x(t) = 0,$$

$$A'_0, \ldots, A'_\ell \in \mathsf{H}^{n \times n},$$

is stably bounded as soon as A'_k is ϕ-hermitian if k is even, A'_k is ϕ-skewhermitian if k is odd, and inequalities $\|A'_k - A_k\| < \varepsilon$ for $k = 1, 2, \ldots, \ell$ hold true. (Note that $\|A'_\ell - A_\ell\| < \varepsilon$ for sufficiently small ε guarantees invertibility of A'_ℓ.)

Ex. 13.7.6. Is the result of Ex. 13.7.5 valid if it assumed that $x_\phi G x \neq 0$ for every eigenvector x of C, instead of assuming that every root subspace of C is G-definite?

Ex. 13.7.7. Prove Proposition 13.5.2.

Ex. 13.7.8. Complete the details in the proof of Remark 13.5.4.

Ex. 13.7.9. Verify the formulas (13.3.6), (13.3.7), (13.3.8), and (13.3.9).

Ex. 13.7.10. Let $H \in \mathsf{H}^{n \times n}$ be an invertible ϕ-hermitian matrix. Prove that if two (H, ϕ)-symmetric matrices A and A' are similar, then $A' = T^{-1}AT$ for some (H, ϕ)-orthogonal matrix T.

Ex. 13.7.11. Determine whether or not the result analogous to Ex. 13.7.10 holds true in each of the following three contexts:

(1) H is an invertible ϕ-hermitian matrix, and A, A' are (H, ϕ)-skewsymmetric similar matrices.

(2) H is an invertible ϕ-skewhermitian matrix, and A, A' are similar (H, ϕ)-Hamiltonian matrices.

(3) H is an invertible ϕ-skewhermitian matrix, and A, A' are similar (H, ϕ)-skew-Hamiltonian matrices.

Ex. 13.7.12. If the result analogous to Ex. 13.7.10 does not generally hold in one of the situations (1), (2), or (3) of Ex. 13.7.11, then find conditions on the Jordan structure of an (H, ϕ)-skewsymmetric, resp. (H, ϕ)-Hamiltonian or (H, ϕ)-skew-Hamiltonian, matrix A that would guarantee the following property:

(a) If A' is an (H, ϕ)-skewsymmetric, resp. (H, ϕ)-Hamiltonian or (H, ϕ)-skew-Hamiltonian, matrix that is similar to A, then there exists an (H, ϕ)-orthogonal, resp. (H, ϕ)-symplectic, matrix T such that $A' = T^{-1}AT$.

Ex. 13.7.13. Determine the possible canonical forms (13.2.1) and (13.2.6) under each of the following four conditions:

(a) H is $\beta(\phi)$-positive definite.

(b) The $\beta(\phi)$-inertia of H is $(n - 1, 1, 0)$.

(c) The rank of A is at most 2.

(d) $\operatorname{Rank}(A) = 3$.

Ex. 13.7.14. Find all A-invariant maximal (H, ϕ)-nonnegative subspaces and all A-invariant maximal (H, ϕ)-nonpositive subspaces for each of the following pairs of matrices (A_j, H_j), $j = 1, 2$, where $\beta = \beta(\phi)$ and the nonzero quaternion q is such that $q\beta = -\beta q$:

(a) $\quad H_1 = \begin{bmatrix} 0 & 0 & \beta \\ 0 & \beta & 0 \\ \beta & 0 & 0 \end{bmatrix}, \quad A_1 = \begin{bmatrix} 0 & 0 & 0 \\ 0 & q & q \\ 0 & q & q \end{bmatrix}.$

(b) $\quad H_2 = H_1, \quad A_2 = \begin{bmatrix} 0 & 0 & 1 \\ 0 & 0 & 0 \\ 0 & 0 & 0 \end{bmatrix}.$

Ex. 13.7.15. Suppose that either $H \in \mathsf{H}^{n \times n}$ is ϕ-skewhermitian and invertible and A is (H, ϕ)-skew-Hamiltonian, or H is ϕ-hermitian and A is (H, ϕ)-symmetric.

Show that if there exists an A-invariant H-Lagrangian subspace, then for every $\varepsilon > 0$ there is an (H, ϕ)-skew-Hamiltonian, resp. (H, ϕ)-symmetric, matrix A' for which no A'-invariant H-Lagrangian subspace exists and such that $\|A' - A\| < \varepsilon$.

Hint: Assume H is ϕ-hermitian, and let H and A be given by the right-hand sides of the canonical form (13.1.1). Let

$$A' = \widetilde{J}_{\ell_1}(\alpha_1) \oplus J_{\ell_2}(\alpha_2) \oplus \cdots \oplus J_{\ell_q}(\alpha_q),$$

where

$$\widetilde{J}_{\ell_1}(\alpha_1) := J_{\ell_1}(\alpha_1) + \begin{bmatrix} 0_{(\ell_1-1) \times 1} & 0_{(\ell_1-1) \times (\ell_1-1)} \\ \varepsilon & 0_{1 \times (\ell_1-1)} \end{bmatrix}, \quad \varepsilon > 0.$$

What are the eigenvalues of $\widetilde{J}_{\ell_1}(\alpha_1)$?

Ex. 13.7.16. Prove that each of the following four sets, for a fixed invertible $H \in \mathsf{H}^{n \times n}$, is closed:

$\{A \in \mathsf{H}^{n \times n} \; : \; H = H_\phi, \; A \text{ is } (H, \phi)\text{-symmetric},$
          and there exist A-invariant H-Lagrangian subspaces$\}$.

$\{A \in \mathsf{H}^{n \times n} \; : \; H = H_\phi, \; A \text{ is } (H, \phi)\text{-skewsymmetric},$
          and there exist A-invariant H-Lagrangian subspaces$\}$.

$\{A \in \mathsf{H}^{n \times n} \; : \; H = -H_\phi, \; A \text{ is } (H, \phi)\text{-Hamiltonian},$
          and there exist A-invariant H-Lagrangian subspaces$\}$.

$\{A \in \mathsf{H}^{n \times n} \; : \; H = -H_\phi, \; A \text{ is } (H, \phi)\text{-skew-Hamiltonian},$
          and there exist A-invariant H-Lagrangian subspaces$\}$.

Hint: Show that if a sequence $\{A_m\}_{m=1}^{\infty}$ belongs to one of those sets and if $\widetilde{A} = \lim_{m \to \infty} A_m$, then $\widetilde{A}$ also belongs to the set.

Ex. 13.7.17. Are the sets of Ex. 13.7.16 closed if H is not assumed to be fixed? In other words, determine if the following set is closed:

$\{A \in \mathsf{H}^{n \times n} \; : \; \text{for some invertible } H = H_\phi, \; A \text{ is } (H, \phi)\text{-symmetric},$
          and there exist A-invariant H-Lagrangian subspaces$\}$.

Do the same for three other sets in Ex. 13.7.16.

13.8 NOTES

The material and presentation in this chapter is based on Rodman [132, 133]. In particular, the proofs are taken from these two papers.

The results of Theorems 13.1.2, 13.1.1, 13.2.4, and 13.2.3 can be found, for example, in Djoković et al. [34]. Theorem 13.1.1 is a particular case of Theorem 7.5 of Rodman [132]. (The setting of Rodman [132, Theorem 7.5] is more general in that the symmetry is understood with respect to several nonstandard involutions.)

Theorems 13.1.2 and 13.2.3 are found in Rodman [133, Theorems 6.2 and 6.4, respectively].

The canonical form for H-skew-Hamiltonian matrices of Theorem 13.2.4 is taken from Rodman [132, Theorem 8.9].

Chapter Fourteen

Matrix equations

Here, we present applications to polynomial matrix equations, algebraic Riccati equations, and linear quadratic regulators. Without attempting to develop in-depth exposition of the topics (this would take us too far afield), we present these applications in basic forms. Maximal invariant semidefinite or neutral subspaces will play a key role.

14.1 POLYNOMIAL EQUATIONS

The approach to studying polynomial equations using companion matrices of Section 5.12 extends to polynomial matrix equations. Consider the matrix equation

$$Z^n + A_{n-1} Z^{n-1} + \cdots + A_1 Z + A_0 = 0, \tag{14.1.1}$$

where $A_0, \ldots, A_{n-1} \in \mathsf{H}^{m \times m}$ are given and $Z \in \mathsf{H}^{m \times m}$ is the unknown matrix. Let C be the *block companion matrix* corresponding to equation (14.1.1):

$$C = \begin{bmatrix} 0 & I_m & 0 & 0 & \cdots & 0 \\ 0 & 0 & I_m & 0 & \cdots & 0 \\ \vdots & \vdots & \vdots & \ddots & \cdots & \vdots \\ 0 & 0 & 0 & 0 & \cdots & I_m \\ -A_0 & -A_1 & -A_2 & -A_3 & \cdots & -A_{n-1} \end{bmatrix} \in \mathsf{H}^{(mn) \times (mn)}.$$

Theorem 14.1.1. *There exists a one-to-one correspondence between solutions Z of (14.1.1) and C-invariant subspaces $\mathcal{M}$ which are direct complements to the subspace $\mathrm{Span}_{\mathsf{H}} \{e_{m+1}, e_{m+2}, \ldots, e_{mn}\}$ in $\mathsf{H}^{(mn) \times (mn)}$. The correspondence is given by the formula*

$$\mathcal{M} = \mathrm{Ran} \left(\begin{bmatrix} I_m \\ Z \\ \vdots \\ Z^{n-1} \end{bmatrix} \right). \tag{14.1.2}$$

Proof. If Z is a solution of (14.1.1), then we have

$$C \operatorname{col} (Z^j)_{j=0}^{n-1} = \operatorname{col} (Z^j)_{j=0}^{n-1} Z, \tag{14.1.3}$$

which implies that the subspace $\mathrm{Ran} \operatorname{col} (Z^j)_{j=0}^{n-1}$ is C-invariant. (Recall that we use $\operatorname{col} (Z^j)_{j=0}^{n-1}$ for the matrix in the right-hand side of (14.1.2).) Since the matrix

$$\begin{bmatrix} I_m & 0 \\ \operatorname{col} (Z^j)_{j=1}^{n-1} & I_{mn-m} \end{bmatrix}$$

is invertible, it follows that the subspaces $\operatorname{Ran}\operatorname{col}(Z^j)_{j=0}^{n-1}$ and

$$\operatorname{Span}_{\mathsf{H}}\{e_{m+1}, e_{m+2}, \ldots, e_{mn}\} = \operatorname{Ran}\begin{bmatrix} 0_{m\times m(n-1)} \\ I_{m(n-1)\times m(n-1)} \end{bmatrix} \qquad (14.1.4)$$

are direct complements to each other.

Conversely, if $\mathcal{M} \subseteq \mathsf{H}^{mn\times 1}$ is a subspace as in Theorem 14.1.1, then the condition that $\mathcal{M}$ is a direct complement to (14.1.4) means that

$$\mathcal{M} = \operatorname{Ran}\left(\operatorname{col}(X_j)_{j=1}^n\right)$$

for some matrices $X_1, \ldots, X_n \in \mathsf{H}^{m\times m}$ with invertible X_1. Multiplying $\operatorname{col}(X_j)_{j=1}^n$ on the right by X_1^{-1}, we may assume that, in fact, $X_1 = I_m$. Now the condition that $\mathcal{M}$ is C-invariant implies that $X_j = X_{j-1}X_2$ for $j = 2, 3, \ldots, n$, so $\mathcal{M}$ has the form (14.1.2). Now the C-invariance of $\mathcal{M}$ means that equality (14.1.3) holds, and equating the block bottom row in the both sides of (14.1.3), we see that Z is a solution of (14.1.1). $\qquad\square$

In contrast with the scalar polynomial equation (5.12.3), the matrix equation (14.1.1) is not guaranteed a solution.

Example 14.1.2. The equation

$$Z^2 = \begin{bmatrix} 0 & 1 \\ 0 & 0 \end{bmatrix}, \qquad Z \in \mathsf{H}^{2\times 2}, \qquad (14.1.5)$$

has no solutions Z. Indeed, if such Z existed, then it must have the only eigenvalue zero. However, both possibilities for the Jordan form of Z—namely, $\begin{bmatrix} 0 & 1 \\ 0 & 0 \end{bmatrix}$ and $0_{2\times 2}$—lead to a contradiction with (14.1.2). $\qquad\square$

In view of Example 14.1.2, it it of interest to develop criteria, or sufficient conditions, for a matrix polynomial equation (14.1.1) to have solutions. Theorems 14.1.3 and 14.1.5 below provide such results.

Theorem 14.1.3. *Assume that the companion matrix C has nm mutually non-similar eigenvalues. Then equation (14.1.1) has solutions.*

The proof of Theorem 14.1.3 is obtained at once by combining Theorem 14.1.1 and the following general result concerning direct complements (take $v_1, \ldots, v_{mn}$ of Proposition 14.1.4 to be the eigenvectors of C that correspond to the nm mutually nonsimilar eigenvalues; see Proposition 5.3.9).

Proposition 14.1.4. *Let $v_1, \ldots, v_{nm}$ be a basis for $\mathsf{H}^{nm\times nm}$. Then for every subspace $\mathcal{M} \subseteq \mathsf{H}^{nm\times nm}$, $\mathcal{M} \neq \mathsf{H}^{nm\times nm}$, there are vectors $v_{i_1}, \ldots, v_{i_{nm-s}}$ among $v_1, \ldots, v_{nm}$ such that $\operatorname{Span}_{\mathsf{H}}\{v_{i_1}, \ldots, v_{i_{nm-s}}\}$ and $\mathcal{M}$ are direct complements to each other; here $s = \dim\mathcal{M}$.*

For a proof of Proposition 14.1.4, apply the replacement theorem 3.1.1, with $p = mn$ and $u_1, \ldots, u_s$ a basis for $\mathcal{M}$.

Theorem 14.1.5. *(a) If $A_0, A_1 \in \mathsf{H}^{m\times m}$ are hermitian, then the equation*

$$Z^2 + A_1 Z + A_0 = 0, \qquad Z \in \mathsf{H}^{m\times m},$$

has a solution.

(b) *If* $A_0, A_1, A_2 \in \mathsf{H}^{m \times m}$ *are hermitian, then the equation*

$$Z^3 + A_2 Z^2 + A_1 Z + A_0 = 0, \qquad Z \in \mathsf{H}^{m \times m},$$

has a solution.

Proof. We start with Part (b). Let

$$G = \begin{bmatrix} -A_1 & -A_2 & -I_m \\ -A_2 & -I_m & 0 \\ -I_m & 0 & 0 \end{bmatrix}, \qquad C = \begin{bmatrix} 0 & I_m & 0 \\ 0 & 0 & I_m \\ -A_0 & -A_1 & -A_2 \end{bmatrix}.$$

Then $G = G^*$ and $(GC)^* = GC$. Also,

$$\mathrm{In}_+ G = m, \qquad \mathrm{In}_- G = 2m,$$

in view of Lemma 10.8.1. By Theorem 10.2.1, there exists an m-dimensional G-nonnegative C-invariant subspace $\mathcal{M}$.

We claim that $\mathcal{M}$ is a direct complement to $\mathrm{Span}\,\{e_{m+1}, e_{m+2}, \ldots, e_{3m}\}$ in $\mathsf{H}^{3m \times 1}$. Suppose not; then there exists a *nonzero* vector

$$y = \begin{bmatrix} 0 \\ y_1 \\ y_2 \end{bmatrix} \in \mathcal{M}, \qquad y_1, y_2 \in \mathsf{H}^{m \times 1}.$$

A computation shows that $y^* G y = -y_1^* y_1$, and since $\mathcal{M}$ is G-nonnegative, we must have $y_1 = 0$. Since $\mathcal{M}$ is C-invariant, we now have

$$Cy = \begin{bmatrix} 0 \\ y_2 \\ -A_2 y_2 \end{bmatrix} \in \mathcal{M},$$

and analogous computation shows $y_2 = 0$, a contradiction. Now an application of Theorem 14.1.1 yields part (b).

Part (a). Here we let

$$G = \begin{bmatrix} A_1 & I_m \\ I_m & 0 \end{bmatrix}, \qquad C = \begin{bmatrix} 0 & I_m \\ -A_0 & -A_1 \end{bmatrix}.$$

We have $G = G^*$, $(GC)^* = GC$, and $\mathrm{In}_+ G = \mathrm{In}_- G = m$. By Theorem 10.2.1, there exists an m-dimensional G-nonnegative C-invariant subspace $\mathcal{N}$. The subspace $\mathcal{N}$ is a direct complement to $\mathrm{Span}\,\{e_{m+1}, \ldots, e_{2m}\}$ in $\mathsf{H}^{2m \times 1}$. Indeed, otherwise we would have a vector

$$y = \begin{bmatrix} 0 \\ y_1 \end{bmatrix} \in \mathcal{N}, \qquad \text{where } y_1 \in \mathsf{H}^{m \times 1} \setminus \{0\}.$$

Then

$$Cy = \begin{bmatrix} y_1 \\ -A_1 y_1 \end{bmatrix} \in \mathcal{N},$$

and a computation shows that

$$\begin{bmatrix} y^* \\ (Cy)^* \end{bmatrix} G \begin{bmatrix} y & Cy \end{bmatrix} = \begin{bmatrix} 0 & y_1^* y_1 \\ y_1^* y_1 & \star \end{bmatrix}, \tag{14.1.6}$$

where $\star$ stands for an entry of no immediate interest. Since $y_1 \neq 0$, the right-hand side of (14.1.6) is an indefinite 2×2 hermitian matrix, a contradiction with $\mathcal{N}$ being G-nonnegative. As in the proof of part (b), it remains to apply Theorem 14.1.1. $\quad\square$

14.2   BILATERAL QUADRATIC EQUATIONS

In the rest of this chapter, we study quadratic matrix equations of the form

$$ZBZ + ZA - DZ - C = 0, \tag{14.2.1}$$

where $A \in \mathsf{H}^{n \times n}$, $B \in \mathsf{H}^{n \times m}$, $C \in \mathsf{H}^{m \times n}$, $D \in \mathsf{H}^{m \times m}$ are given matrices and solutions $Z \in \mathsf{H}^{m \times n}$ are to be found. Equation (14.2.1) is termed *bilateral* because the unknown matrix Z appears on both sides in the expression in the left-hand side of (14.2.1), in contrast with the unilateral equation (14.1.1), where the unknown matrix appears only on the right side of the expression in (14.1.1).

Definition 14.2.1. For any $Z \in \mathsf{H}^{m \times n}$, we call the n-dimensional subspace

$$G(Z) := \operatorname{Ran} \begin{bmatrix} I_n \\ Z \end{bmatrix} \subseteq \mathsf{H}^{(m+n) \times 1}$$

the *graph* of Z.

We connect solutions of (14.2.1) with invariant subspaces of the $(m+n) \times (m+n)$ matrix

$$T = \begin{bmatrix} A & B \\ C & D \end{bmatrix}. \tag{14.2.2}$$

Proposition 14.2.2. *For any $Z \in \mathsf{H}^{m \times n}$, the graph of Z is T-invariant if and only if Z is a solution of* (14.2.1).

Proof. If the graph of Z is T-invariant, then

$$\begin{bmatrix} A & B \\ C & D \end{bmatrix} \begin{bmatrix} I_n \\ Z \end{bmatrix} = \begin{bmatrix} I_n \\ Z \end{bmatrix} X \tag{14.2.3}$$

for some matrix X. The first block row in this equality gives $X = A + BZ$, and substituting into the second block row gives $C + DZ = Z(A + BZ)$; i.e., Z is a solution of (14.2.1). Conversely, if Z is a solution of (14.2.1), then (14.2.3) holds with $X = A + BZ$. $\square$

Clearly, $G(Z_1) = G(Z_2)$ holds for $Z_1, Z_2 \in \mathsf{H}^{m \times n}$ if and only if $Z_1 = Z_2$. Thus, Proposition 14.2.2 establishes a one-to-one correspondence between solutions of (14.2.1) and certain T-invariant subspace—namely, those T-invariant subspaces that are graph subspaces or, in other words, T-invariant subspaces that are direct complements to $\operatorname{Span}_\mathsf{H} \{e_{n+1}, \dots, e_{m+n}\}$ in $\mathsf{H}^{(m+n) \times 1}$. Although T has invariant subspaces of every dimension from 0 to $m + n$, T-invariant subspaces of dimension n that are graph subspaces need not exist.

Example 14.2.3. Equation (14.1.5) of Example 14.1.2 can be recast in the form (14.2.1) with

$$B = I, \quad A = D = 0, \quad C = \begin{bmatrix} 0 & 1 \\ 0 & 0 \end{bmatrix}.$$

Then

$$T = \begin{bmatrix} 0 & 0 & 1 & 0 \\ 0 & 0 & 0 & 1 \\ 0 & 1 & 0 & 0 \\ 0 & 0 & 0 & 0 \end{bmatrix}.$$

It is easy to see that the Jordan from of T is $J_4(0)$, and the unique 2-dimensional T-invariant subspace is $\mathrm{Span}_H \{e_1, e_3\}$, which is not a graph subspace. Thus, (14.1.5) has no solutions. $\qquad\Box$

Note that the set of $(m + n) \times (m + n)$-matrices that have $m + n$ pairwise nonsimilar eigenvalues is open and dense in $H^{(m+n)\times(m+n)}$. Indeed, the openness of this set follows from the continuous dependence of eigenvalues on the matrix (Theorem 5.2.5), and the denseness can be easily obtained using the Jordan form of a matrix $T \in H^{(m+n)\times(m+n)}$ (perturb the diagonal entries in the Jordan form so that the resulting matrix has $m + n$ pairwise nonsimilar eigenvalues). For such matrices there exists a basis of eigenvectors in $H^{(m+n)\times 1}$ (Proposition 5.3.9). It follows that generically equation (14.2.1) has a finite number of solutions, and the number of solutions does not exceed $(m + n)!/(m!\, n!)$, the number of ways a set of n vectors can be chosen from a basis of $m + n$ eigenvectors of T in $H^{(m+n)\times 1}$.

14.3 ALGEBRAIC RICCATI EQUATIONS

Here we specialize equation (14.2.1) by introducing certain symmetries and changing the notation somewhat. The equation we consider now has the form

$$ZDZ + ZA + A^*Z - C = 0, \qquad (14.3.1)$$

where all matrices are in $H^{n\times n}$ and D and C are assumed to be hermitian. Equation (14.3.1) is known as *algebraic Riccati equation*.

Introduce the matrices

$$T = \begin{bmatrix} A & D \\ C & -A^* \end{bmatrix}, \quad K = \begin{bmatrix} 0 & I_n \\ -I_n & 0 \end{bmatrix}, \quad H = \begin{bmatrix} -C & A^* \\ A & D \end{bmatrix}. \qquad (14.3.2)$$

Then KT is hermitian and HT is skewhermitian, so T is K-Hamiltonian and (assuming H is invertible) H-skewhermitian.

Because of the symmetry of equation (14.3.1), we can say more about T-invariant subspaces.

Proposition 14.3.1. *Let Z be a solution of (14.3.1), with the corresponding graph subspace $G(Z)$. Then:*

(1) *Z is hermitian if and only if $G(Z)$ is K-neutral;*

(2) *the graph $G(Z)$ is H-nonpositive, resp. H-nonnegative, if and only if*

$$(Z^* - Z)(A + DZ) \leq 0,$$

resp. $(Z^* - Z)(A + DZ) \geq 0$;

(3) *the graph $G(Z)$ is H-neutral if and only if*

$$(Z^* - Z)(A + DZ) = 0.$$

Note that the matrix $(Z^* - Z)(A + DZ)$ in (3) is hermitian.

The proof of Proposition 14.3.1 is by straightforward computation; in particular, the following equality is used:

$$\begin{bmatrix} I \\ Z \end{bmatrix}^* H \begin{bmatrix} I \\ Z \end{bmatrix} = (Z^* - Z)(A + DZ) + ZDZ + ZA + A^*Z - C.$$

In view of Proposition 14.2.2, we would like to identify situations when an n-dimensional T-invariant subspace is a graph subspace. This is the subject matter of the next lemma.

Definition 14.3.2. A pair of matrices (A, B), where $A \in \mathsf{H}^{p \times p}$ and $B \in \mathsf{H}^{p \times q}$, is said to be *controllable* if

$$\mathsf{H}^{p \times 1} = \sum_{j=0}^{\infty} \mathrm{Ran}\,(A^j B). \tag{14.3.3}$$

This terminology comes from consideration of *linear time-invariant control systems*

$$\frac{dx}{dt} = Ax(t) + Bu(t), \qquad t \geq 0 \tag{14.3.4}$$

(cf. (14.3.16)). One can show that (A, B) is controllable if and only if the state of system (14.3.4), represented by the vector $x(t)$, can be driven from any position to any other position in a prescribed period of time by a suitable choice of the continuous control function $u(t)$. This is a standard fact in the theory of linear control systems; for a proof in the context of complex matrices; see, e.g., Gohberg et al. [54, Section 8.2].

Since the degree of the minimal polynomial for A does not exceed $2p$ (Ex. 5.16.23), condition (14.3.3) is equivalent to

$$\mathsf{H}^{p \times 1} = \sum_{j=0}^{2p-1} \mathrm{Ran}\,(A^j B). \tag{14.3.5}$$

We leave the proof as Ex. 14.4.3. (For a stronger result, see Ex. 14.4.4.)

Lemma 14.3.3. *Assume that D is positive semidefinite and the pair (A, D) is controllable. Let $\mathcal{L}$ be an n-dimensional T-invariant H-nonpositive subspace of $\mathsf{H}^{2n \times 1}$. Then $\mathcal{L}$ is a graph subspace.*

Proof. For the subspace $\mathcal{L}$, write

$$\mathcal{L} = \mathrm{Ran} \begin{bmatrix} X_1 \\ X_2 \end{bmatrix}, \quad \text{where } X_1, X_2 \in \mathsf{H}^{n \times n}.$$

We are going to prove that X_1 is invertible.

Observe that T-invariance of $\mathcal{L}$ means

$$\begin{bmatrix} A & D \\ C & -A^* \end{bmatrix} \begin{bmatrix} X_1 \\ X_2 \end{bmatrix} = \begin{bmatrix} X_1 \\ X_2 \end{bmatrix} Q$$

for some $Q \in \mathsf{H}^{n \times n}$, i.e.,

$$AX_1 + DX_2 = X_1 Q, \tag{14.3.6}$$

$$CX_1 - A^* X_2 = X_2 Q. \tag{14.3.7}$$

The H-nonpositivity of $\mathcal{L}$ means that the matrix

$$[X_1^* \quad X_2^*] \begin{bmatrix} -C & A^* \\ A & D \end{bmatrix} \begin{bmatrix} X_1 \\ X_2 \end{bmatrix}$$

$$= X_2^* D X_2 + X_1^* A^* X_2 + X_2^* A X_1 - X_1^* C X_1 \tag{14.3.8}$$

is negative semidefinite.

Let $\mathcal{K} = \operatorname{Ker} X_1$. We have for every $x \in \mathcal{K}$:

$$x^* X_2^* D X_2 x + x^* X_1^* A^* X_2 x + x^* X_2^* A X_1 x - x^* X_1^* C X_1 x = x^* X_2^* D X_2 x,$$

which is real nonpositive in view of the negative semidefiniteness of (14.3.8). But $D \geq 0$, so we must have $X_2 x \in \operatorname{Ker} D$—in other words,

$$X_2(\mathcal{K}) \subseteq \operatorname{Ker} D. \tag{14.3.9}$$

Further, equation (14.3.6) implies that $Q(\mathcal{K}) \subseteq \mathcal{K}$. Now equation (14.3.7) gives, for every $x \in \mathcal{K}$,

$$A^* X_2 x = -C X_1 x + A^* X_2 x = -X_2 Q x \in X_2(\mathcal{K}).$$

Thus,

$$A^* X_2(\mathcal{K}) \subseteq X_2(\mathcal{K}) \subseteq \operatorname{Ker} D. \tag{14.3.10}$$

We claim that, more generally,

$$(A^*)^r X_2(\mathcal{K}) \subseteq \operatorname{Ker} D, \qquad r = 0, 1, \ldots. \tag{14.3.11}$$

Indeed, we have already proved (14.3.11) for $r = 0$ and $r = 1$. Arguing by induction, assume (14.3.11) holds true for $r - 1$. Using (14.3.10), it is found that

$$(A^*)^r (X_2(\mathcal{K})) = (A^*)^{r-1}(A^* X_2(\mathcal{K})) \subseteq (A^*)^{r-1}(X_2(\mathcal{K})) \subseteq \operatorname{Ker} D,$$

and (14.3.11) follows. Now, for every $x \in \mathcal{K}$, we have

$$\begin{bmatrix} D \\ DA^* \\ \vdots \\ D(A^*)^{n-1} \end{bmatrix} X_2 x = 0,$$

or

$$(X_2 x)^* [D \quad AD \quad A^2 D \quad \ldots \quad A^{n-1} D] = 0,$$

and in view of controllability of (A, D), the equality $X_2 x = 0$ is obtained. But the only vector x for which $X_1 x = X_2 x = 0$ is the zero vector, because otherwise the dimension of $\mathcal{L}$ would be smaller than n, a contradiction with our assumption. So $\mathcal{K} = \{0\}$, and X_1 is invertible. Now

$$\mathcal{L} = \operatorname{Ran} \begin{bmatrix} I \\ X_2 X_1^{-1} \end{bmatrix}$$

is indeed a graph subspace. $\qquad \square$

Recall that T is H-skewhermitian (assuming that H is invertible). Thus, existence of T-invariant maximal H-nonnegative (or maximal H-nonpositive) subspaces is ensured by Theorem 10.2.1. Combining this observation, with Proposition 14.3.1 and Lemma 14.3.3, we arrive at a complete description of a class of solutions of (14.3.1) in terms of invariant subspaces.

Theorem 14.3.4. *Assume that D is positive semidefinite, the pair (A, D) is controllable, and*

$$\text{In}_+ H = \text{In}_- H = n, \tag{14.3.12}$$

where $H = \begin{bmatrix} -C & A^* \\ A & D \end{bmatrix}$ *(in particular H is invertible). Then (14.3.1) admits solutions Z such that*

$$(Z^* - Z)(A + DZ) \leq 0. \tag{14.3.13}$$

Moreover, the formula

$$\mathcal{M} = \text{Ran} \begin{bmatrix} I \\ Z \end{bmatrix} \tag{14.3.14}$$

establishes a one-to-one correspondence between solutions Z of (14.3.1) with the property (14.3.13) and the set of T-invariant maximal H-nonpositive subspaces $\mathcal{M}$.

Specializing to hermitian solutions, we obtain the following.

Theorem 14.3.5. *Under the hypotheses of Theorem 14.3.4, equation (14.3.1) admits hermitian solutions if and only if for every eigenvalue λ of T with zero real part (if any), the number of odd partial multiplicities corresponding to λ is even, and exactly half of them have sign -1 in in the sign characteristic of the pair (T, H). Moreover, formula (14.3.14) establishes a one-to-one correspondence between hermitian solutions Z of (14.3.1) and the set of T-invariant H-Lagrangian subspaces $\mathcal{M}$.*

Proof. We verify first that a solution Z of (14.3.1) satisfies

$$(Z^* - Z)(A + DZ) = 0 \tag{14.3.15}$$

if and only if Z is hermitian. Indeed, the "if" part is trivial. Conversely, suppose (14.3.15) holds true. By Proposition 14.3.1, $G(Z)$ is H-neutral. Observe the equality $H = -KT$. In particular, it follows that T is invertible, and $G(Z)$, being T-invariant, is also $-T^{-1}$-invariant. Now, for $x, y \in G(Z)$, we have

$$x^* K y = x^* H(-T^{-1})y = 0,$$

because $G(Z)$ is H-neutral. Thus, $G(Z)$ is K-neutral, and by Proposition 14.3.1, Z is hermitian.

By the same Proposition 14.3.1, solutions Z with the property (14.3.15), or what is the same, hermitian solutions, are characterized by the property that their graph subspaces $G(Z)$ are T-invariant H-Lagrangian. Now apply Theorem 10.3.3(b). $\square$

Algebraic Riccati equations are ubiquitous in many applications—in particular, in control systems. Several of these applications are outlined in Lancaster and Rodman [91]. Here we present just one application—linear quadratic regulator, in a basic form.

Consider a primitive time-invariant linear control system

$$\frac{dx}{dt} = Ax(t) + Bu(t), \quad x(0) = x^0, \quad t \geq 0, \tag{14.3.16}$$

where $A \in \mathsf{H}^{n \times n}$, $B \in \mathsf{H}^{n \times m}$, and $x(t) \in \mathsf{H}^{n \times 1}$ and $u(t) \in \mathsf{H}^{m \times 1}$ are vector valued functions known as *state* and *input* (or *control*) vectors, respectively. Observe that

$x(t)$ is uniquely defined by a given integrable control vector function $u(t)$ and the initial value x^0:

$$x(t) = e^{At}x^0 + \int_0^t e^{A(t-s)}Bu(s)\,ds, \qquad t \ge 0. \qquad (14.3.17)$$

It will be assumed throughout that control functions $u(t)$ are defined for $0 \le t < \infty$ and have the property that $u \in L^2_m(0,T)$ for all $T > 0$. Here the space $L^2_m(0,T)$ is the quaternion Hilbert space of square integrable functions on $(0,T)$ with values in $\mathsf{H}^{m\times 1}$ equipped with the quaternion-valued inner product

$$\langle y(t), z(t) \rangle_{L^2_m(0,T)} = \int_0^T (z(t))^* y(t)\,dt.$$

Fix two matrices: $Q \in \mathsf{H}^{n\times n}$ positive semidefinite and $R \in \mathsf{H}^{m\times m}$ positive definite. Define the *quadratic cost functional* by

$$J^u(x^0) := \int_0^\infty [x(t)^* Q x(t) + u(t)^* R u(t)]\,dt.$$

Here $x(t)$ is given by formula (14.3.17). Thus, $J^u(x^0)$ is a function of the control vector $u(t)$ and the initial value x^0, and $0 \le J^u(x^0) \le \infty$. The *optimal cost* at x^0 is defined by

$$\widehat{J}(x^0) := \inf_u J^u(x^0),$$

and an *optimal control* is then a control vector function $u(t)$ for which the infimum is attained.

It turns out that, under appropriate hypotheses, the optimal control exists and can be given by a *feedback* mechanism; i.e., the optimal control is coupled to the state vector by an equation of the form $u(t) = -Fx(t)$, for some fixed (i.e., independent of t) matrix $F \in \mathsf{H}^{m\times n}$. For such $u(t)$ the initial value problem (14.3.16) takes the form

$$\frac{dx}{dt} = (A - BF)x(t), \qquad x(0) = x^0,$$

with solution

$$x(t) = e^{(A-BF)t}x^0,$$

and so

$$u(t) = -Fe^{(A-BF)t}x^0, \qquad t \ge 0.$$

In connection with the linear quadratic regulator problem, introduce the following algebraic Riccati equation:

$$ZBR^{-1}B^*Z - ZA - A^*Z - Q = 0. \qquad (14.3.18)$$

Here is the main result concerning the linear quadratic regulator.

Theorem 14.3.6. *Assume (in addition to the assumptions made above) that there exists $Y \in \mathsf{H}^{m\times n}$ such that all eigenvalues of $A + BY$ have negative real parts, and assume that the pair (A^*, Q) is controllable. Then there is a unique positive semidefinite solution $Z_0 \in \mathsf{H}^{n\times n}$ of (14.3.18). Moreover:*

(a) *$\widehat{J}(x^0) = (x^0)^* Z_0 x^0$ for all $x^0 \in \mathsf{H}^{n\times 1}$;*

(b) *for each x^0 there is a unique optimal control $u(t)$; this control is determined by the feedback matrix $F = R^{-1}B^*Z_0$, and then*

$$u(t) = -R^{-1}B^*Z_0 e^{(A-BR^{-1}B^*Z_0)t}x^0, \qquad t \geq 0;$$

(c) *all eigenvalues of $A - BR^{-1}B^*Z_0$ have negative real parts;*

(d) *Z_0 is positive definite;*

(e) *Z_0 is the maximal hermitian solution of (14.3.18): if Z is any hermitian solution of (14.3.18), then $Z_0 - Z$ is positive semidefinite.*

Theorem 14.3.6 can be proved in the same way as the corresponding result for complex matrices (see, e.g., Theorem 16.3.3 of Lancaster and Rodman [91]). A complete proof of the theorem would take us too far afield; therefore, we omit details and refer the reader to Lancaster and Rodman [91].

14.4 EXERCISES

Ex. 14.4.1. Show that the solution set of every matrix polynomial equation of the form (14.1.1) is compact (or empty).

Ex. 14.4.2. Let $A \in \mathsf{H}^{p \times p}$, $B \in \mathsf{H}^{p \times q}$, and define the subspace

$$\mathcal{C}(A, B) := \sum_{j=0}^{\infty} \mathrm{Ran}\,(A^j B) \subseteq \mathsf{H}^{p \times 1}.$$

Show that if

$$\mathrm{Ran}\,(A^k B) \subseteq \sum_{j=0}^{k-1} \mathrm{Ran}\,(A^j B)$$

for some integer $k \geq 1$, then $\mathcal{C}(A, B) = \sum_{j=0}^{k-1} \mathrm{Ran}\,(A^j B)$.

Ex. 14.4.3. Prove that (14.3.5) is equivalent to controllability of the pair (A, B). Hint: Use Ex. 14.4.2.

Ex. 14.4.4. Let $A \in \mathsf{H}^{p \times p}$, $B \in \mathsf{H}^{p \times q}$. Prove that the pair (A, B) is controllable if and only if

$$\mathsf{H}^{p \times 1} = \sum_{j=0}^{p-1} \mathrm{Ran}\,(A^j B).$$

Hint: Prove that $\mathrm{Ran}\,(A^p B)$ is contained in $\sum_{j=0}^{p-1} \mathrm{Ran}\,(A^j B)$. To this end, use the Jordan form of A to reduce the proof to the case when A is a complex matrix; then use the fact that a complex $p \times p$ matrix is a root of the characteristic polynomial of the matrix which is of degree p.

Ex. 14.4.5. For all pairs of matrices $A \in \mathsf{H}^{p \times p}$, $B \in \mathsf{H}^{p \times q}$, prove that the subspace $\sum_{j=0}^{\infty} \mathrm{Ran}\,(A^j B)$ is A-invariant.

Ex. 14.4.6. Show that (14.3.12) is satisfied if both matrices C and D are positive definite.

14.5 NOTES

Various types of quadratic equations with quaternion coefficients have been treated extensively in literature; see, e.g., Porter [121], Huang and So [68], and Janovska and Opfer [73].

Lemma 14.3.3 was proved for complex matrices in Čurilov [31] and independently by Lancaster and Rodman [88].

Exposition in Section 14.3 (except Theorem 14.3.5) is adapted from Lancaster and Rodman [91].

Pairs of real matrices (A, H), where H is skewsymmetric and A is $(H,^T)$-Hamiltonian, play a key role in several important problems of applied analysis—in particular, Riccati equations (which are ubiquitous in systems and control); see, e.g., Abou-Kandil et al. [1], Lancaster and Rodman [91], Mehrmann [112], gyroscopic vibrating systems in Lancaster et al. [86], Hamiltonian systems, and transfer functions with symmetries and their factorizations (see, e.g., Alpay et al. [2], Furhmann [47], Lancaster and Rodman [90], and Ran and Rodman [126]). In the framework of real matrices, invariant subspaces of H-Hamiltonian and symplectic matrices that have neutrality or definitiveness properties with respect to the skew symmetric inner product induced by H have been studied in Rodman [129, 130], Lancaster and Rodman [89], Freiling et al. [46], and Mehl et al. [107]. Many of these studies were motivated largely by applications to algebraic Riccati equations. Although of most importance are hermitian solutions of the algebraic Riccati equations, nonhermitian solutions are also of interest (see, e.g., Abou-Kandil et al. [1]).

Linear control systems with quaternion coefficients have been studied in Pereira et al. [117] and Pereira and Vettori [118].

The linear quadratic regulator problem (in the setting of real and complex matrices) was treated first in Kalman [76], and since then the topic has generated an immense amount of literature.

Chapter Fifteen

Appendix: Real and complex canonical forms

For the reader's convenience, we state here (without proof, but with references) canonical forms for real and complex matrices and for pairs of real and complex matrices, or matrix pencils, with symmetries. All these forms are known, and most are well-known. Our main sources for the material in this chapter are Gantmacher [48, 49] and expository papers by Thompson [151] and Lancaster and Rodman [93, 92]; Thompson [151] also contains an extensive bibliography.

15.1 JORDAN AND KRONECKER CANONICAL FORMS

For Jordan forms, we use the complex and real Jordan blocks (1.2.1) and (1.2.2).

Theorem 15.1.1. *Let* $\mathsf{F} = \mathsf{R}$ *or* $\mathsf{F} = \mathsf{C}$. *Let* $A \in \mathsf{F}^{n \times n}$. *Then there exists an invertible* $S \in \mathsf{F}^{n \times n}$ *such that* $S^{-1}AS$ *has the form*

$$S^{-1}AS = J_{m_1}(\lambda_1) \oplus \cdots \oplus J_{m_p}(\lambda_p), \quad \lambda_1, \ldots, \lambda_p \in \mathsf{C} \tag{15.1.1}$$

if $\mathsf{F} = \mathsf{C}$ *or the form*

$$\begin{aligned} S^{-1}AS = {} & J_{m_1}(\lambda_1) \oplus \cdots \oplus J_{m_p}(\lambda_p) \\ & \oplus J_{2m_{p+1}}(\mu_{p+1} \pm i\nu_{p+1}) \oplus \cdots \oplus J_{2m_q}(\mu_q \pm i\nu_q) \end{aligned} \tag{15.1.2}$$

if $\mathsf{F} = \mathsf{R}$, *where*

$$\lambda_1, \ldots, \lambda_p, \mu_{p+1}, \nu_{p+1}, \ldots, \mu_q, \nu_q \in \mathsf{R}$$

and $\nu_{p+1}, \ldots, \nu_q$ *are positive.*

The forms (15.1.1) *and* (15.1.2) *are uniquely determined by* A, *up to an arbitrary permutation of the constituent Jordan blocks in the complex case, or up to an arbitrary permutation of blocks in each of the two parts*

$$\oplus_{i=1}^{p} J_{m_i}(\lambda_i) \quad \text{and} \quad \oplus_{j=p+1}^{q} J_{2m_j}(\mu_j \pm i\nu_j)$$

in the real case.

In (15.1.2) the cases when $p = 0$, resp. $q = p$, are not excluded; in these cases A has no real eigenvalues, resp. has no nonreal eigenvalues.

Proofs of Theorem 15.1.1 in the complex case can be found in many textbooks on linear algebra, such as Finkbeiner [42], Smith [147], Horn and Johnson [62], Lancaster and Tismenetsky [94], Gohberg et al. [55], Gantmacher [48], and Meyer [114]. In the real case, a complete proof is given in Gohberg et al. [54, Chapter 12], Shilov [145], and Lancaster and Tismenetsky [94].

Let $\mathsf{F} = \mathsf{R}$ or $\mathsf{F} = \mathsf{C}$.

Definition 15.1.2. Two matrix pencils $A + tB, A' + tB'$, where $A, B, A', B' \in \mathsf{F}^{m \times n}$, are said to be F-*strictly equivalent* if $PAQ = A'$, $PBQ = B'$ for some invertible matrices $P \in \mathsf{F}^{m \times m}$, $Q \in \mathsf{F}^{n \times n}$.

For the Kronecker form, besides Jordan blocks, we also need singular blocks, given by (1.2.10). A compete proof of the following theorem is given, e.g., in Gantmacher [48, 49] and in the Appendix of Gohberg et al. [54] for the complex case and in Gantmacher [48, Chapter XII] for the real case.

Theorem 15.1.3. *Let* $\mathsf{F} = \mathsf{R}$ *or* $\mathsf{F} = \mathsf{C}$. *Every pencil* $A + tB \in \mathsf{F}(t)^{m \times n}$ *is* F-*strictly equivalent to a matrix pencil in the block diagonal form*

$$
\begin{aligned}
0_{u \times v} \quad &\oplus \quad L_{\varepsilon_1 \times (\varepsilon_1 + 1)} \oplus \cdots \oplus L_{\varepsilon_p \times (\varepsilon_p + 1)} \oplus L^T_{\eta_1 \times (\eta_1 + 1)} \oplus L^T_{\eta_q \times (\eta_q + 1)} \\
&\oplus \quad (I_{k_1} + t J_{k_1}(0)) \oplus \cdots \oplus (I_{k_r} + t J_{k_r}(0)) \\
&\oplus \quad (t I_{\ell_1} + J_{\ell_1}(\alpha_1)) \oplus \cdots \oplus (t I_{\ell_s} + J_{\ell_s}(\alpha_s))
\end{aligned}
\tag{15.1.3}
$$

if $\mathsf{F} = \mathsf{C}$ *and in the block diagonal form*

$$
\begin{aligned}
0_{u \times v} \quad &\oplus \quad L_{\varepsilon_1 \times (\varepsilon_1 + 1)} \oplus \cdots \oplus L_{\varepsilon_p \times (\varepsilon_p + 1)} \oplus L^T_{\eta_1 \times (\eta_1 + 1)} \oplus L^T_{\eta_q \times (\eta_q + 1)} \\
&\oplus \quad (I_{k_1} + t J_{k_1}(0)) \oplus \cdots \oplus (I_{k_r} + t J_{k_r}(0)) \\
&\oplus \quad (t I_{\ell_1} + J_{\ell_1}(\alpha_1)) \oplus \cdots \oplus (t I_{\ell_w} + J_{\ell_w}(\alpha_w)) \\
&\oplus \quad (t I_{2\ell_{w+1}} + J_{2\ell_{w+1}}(\alpha_{w+1} \pm i \beta_{w+1})) \\
&\oplus \quad \cdots \oplus (t I_{2\ell_s} + J_{2\ell_s}(\alpha_s \pm i \beta_s))
\end{aligned}
\tag{15.1.4}
$$

if $\mathsf{F} = \mathsf{R}$.

In (15.1.3) *and* (15.1.4) *we have*

$$
\varepsilon_1 \leq \cdots \leq \varepsilon_p; \quad \eta_1 \leq \cdots \leq \eta_q; \quad k_1 \leq \cdots \leq k_r
$$

are positive integers, $\alpha_1, \ldots, \alpha_s \in \mathsf{C}$ *if* $\mathsf{F} = \mathsf{C}$ *and*

$$
\alpha_1, \ldots, \alpha_w, \alpha_{w+1}, \ldots, \alpha_s \in \mathsf{R}; \quad \beta_{w+1}, \ldots, \beta_s > 0
$$

if $\mathsf{F} = \mathsf{R}$.

Moreover, the integers u, v, and $\{\varepsilon_i\}_{i=1}^p$, $\{\eta_j\}_{j=1}^q$, $\{k_y\}_{y=1}^r$ are uniquely determined by the pair A, B, and the part

$$
(t I_{\ell_1} + J_{\ell_1}(\alpha_1)) \oplus \cdots \oplus (t I_{\ell_s} + J_{\ell_s}(\alpha_s))
$$

if $\mathsf{F} = \mathsf{C}$, *or the part*

$$
(t I_{\ell_1} + J_{\ell_1}(\alpha_1)) \oplus \cdots \oplus (t I_{\ell_w} + J_{\ell_w}(\alpha_w))
$$

$$
\oplus (t I_{2\ell_{w+1}} + J_{2\ell_{w+1}}(\alpha_{w+1} \pm i \beta_{w+1})) \oplus \cdots \oplus (t I_{2\ell_s} + J_{2\ell_s}(\alpha_s \pm i \beta_s))
$$

if $\mathsf{F} = \mathsf{R}$, *is uniquely determined by* A *and* B *up to an arbitrary permutation of the diagonal blocks.*

The cases when some of the parts in (15.1.3) and (15.1.4) are empty are not excluded. A similar remark applies to other canonical forms in this chapter.

Two corollaries are noteworthy and will be used extensively in the text.

Corollary 15.1.4. *Let* $\mathsf{F} = \mathsf{R}$ *or* $\mathsf{F} = \mathsf{C}$. *If* $A, B \in \mathsf{F}^{n \times n}$ *are such that* $A^{\oplus s}$ *is similar (over* F) *to* $B^{\oplus s}$ *for some positive integer* s, *then* A *is similar to* B (*over* F).

Corollary 15.1.5. *Let* $\mathsf{F} = \mathsf{R}$ *or* $\mathsf{F} = \mathsf{C}$. *If* $A + tB$ *and* $A' + tB'$ *are matrix pencils, where* $A, B, A', B' \in \mathsf{F}^{m \times n}$, *are such that* $(A + tB)^{\oplus s}$ *and* $(A' + tB')^{\oplus s}$ *are strictly equivalent* (*over* F) *for some positive integer* s, *then* $A + tB$ *and* $A' + tB'$ *are strictly equivalent* (*over* F) *as well.*

Proof. The proofs of Corollaries 15.1.4 and 15.1.5 are based on the uniqueness parts in Theorems 15.1.1 and 15.1.3. We provide details for the case $\mathsf{F} = \mathsf{C}$ in Corollary 15.1.4 only; all other cases in Corollaries 15.1.4 and 15.1.5 are treated analogously. For a fixed $\lambda \in \mathsf{C}$ and fixed positive integer k, let $q_{\lambda,k} \geq 0$, resp. $q'_{\lambda,k} \geq 0$, be the number of times the Jordan block $J_k(\lambda)$ appears in the complex Jordan form of A, resp. B. The condition that $A^{\oplus s}$ is similar to $B^{\oplus s}$ implies that $A^{\oplus s}$ and $B^{\oplus s}$ have the same Jordan form, and so

$$s q_{\lambda,k} = s q'_{\lambda,k}.$$

Thus, $q_{\lambda,k} = q'_{\lambda,k}$. Consequently, A and B have the same complex Jordan form, and their similarity (over C) follows. $\square$

15.2   REAL MATRIX PENCILS WITH SYMMETRIES

In this section we provide canonical forms for pairs of real matrices, either one of which is symmetric or skewsymmetric, or what is the same, corresponding matrix pencils. In addition to the standard matrices F_m, G_m, the following standard matrices and matrix pencils will be needed here:

$$Y_{2m} = \begin{bmatrix} 0 & & & & 1 & 0 \\ & & & & 0 & -1 \\ & & 1 & 0 & & \\ & & 0 & -1 & & \\ & & \cdots & & & \\ 1 & 0 & & & & \\ 0 & -1 & & & & 0 \end{bmatrix}; \tag{15.2.1}$$

$$Z_{2m}(t, \mu, \nu) := (t + \mu)F_{2m} + \nu Y_{2m} + \begin{bmatrix} F_{2m-2} & 0 \\ 0 & 0_2 \end{bmatrix},$$

where $\mu \in \mathsf{R}$, $\nu \in \mathsf{R} \setminus \{0\}$.

We start with the case of two real symmetric matrices.

Theorem 15.2.1. (a) *Every matrix pencil* $A + tB$, *where* $A = A^T \in \mathsf{R}^{n \times n}$, $B = B^T \in \mathsf{R}^{n \times n}$, *is* R-*congruent to a real symmetric matrix pencil of the form*

$$0_u \quad \oplus \quad \left(t \begin{bmatrix} 0 & 0 & F_{\varepsilon_1} \\ 0 & 0 & 0 \\ F_{\varepsilon_1} & 0 & 0 \end{bmatrix} + G_{2\varepsilon_1 + 1} \right)$$

$$\oplus \quad \cdots \quad \oplus \left(t \begin{bmatrix} 0 & 0 & F_{\varepsilon_p} \\ 0 & 0 & 0 \\ F_{\varepsilon_p} & 0 & 0 \end{bmatrix} + G_{2\varepsilon_p + 1} \right)$$

$$\oplus \quad \delta_1 \left(F_{k_1} + tG_{k_1} \right) \oplus \cdots \oplus \delta_r \left(F_{k_r} + tG_{k_r} \right)$$

$$\oplus \quad \eta_1 \left((t + \alpha_1) F_{\ell_1} + G_{\ell_1} \right) \oplus \cdots \oplus \eta_q \left((t + \alpha_q) F_{\ell_q} + G_{\ell_q} \right)$$

$$\oplus \quad Z_{2m_1}(t, \mu_1, \nu_1) \oplus \cdots \oplus Z_{2m_s}(t, \mu_s, \nu_s). \tag{15.2.2}$$

Here, $\varepsilon_1 \leq \cdots \leq \varepsilon_p$ and $k_1, \ldots, k_r$ are positive integers, α_j's and μ_j's are real numbers, $\nu_1, \ldots, \nu_s$ are positive numbers, and $\delta_1, \ldots, \delta_r, \eta_1, \ldots, \eta_q$ are signs, each equal to $+1$ or -1.

The form (15.2.2) is uniquely determined by $A + tB$ up to an arbitrary permutation of the blocks in each of the three parts

$$Z_{2m_1}(t, \mu_1, \nu_1) \oplus \cdots \oplus Z_{2m_s}(t, \mu_s, \nu_s),$$

$$\eta_1 \left((t + \alpha_1) F_{\ell_1} + G_{\ell_1}\right) \oplus \cdots \oplus \eta_q \left((t + \alpha_q) F_{\ell_q} + G_{\ell_q}\right),$$

and

$$\delta_1 \left(F_{k_1} + tG_{k_1}\right) \oplus \cdots \oplus \delta_r \left(F_{k_r} + tG_{k_r}\right).$$

(b) *Every matrix pencil $A + tB$, where $A = A^T \in \mathsf{R}^{n \times n}$, $B = B^T \in \mathsf{R}^{n \times n}$, is R-strictly equivalent to a unique (up to a permutation of constituent blocks as in Part (a)) real symmetric matrix pencil of the form (15.2.2), with all signs taken to be $+1$.*

The signs $\delta_1, \ldots, \delta_r, \eta_1, \ldots, \eta_q$ form the *sign characteristic* of the real symmetric pencil $A + tB$. Thus, the sign characteristic associates a sign ± 1 to every partial multiplicity of a real eigenvalue and of the eigenvalue at infinity of $A + tB$.

The result of Theorem 15.2.1 has a long history. We refer the reader to Thompson [151] and Lancaster and Rodman [92] for complete proofs and to Thompson [151] for an extensive bibliography.

Next, we consider pairs of real skewsymmetric matrices.

Theorem 15.2.2. (i) *Every matrix pencil $A + tB$, where A and B are real skewsymmetric matrices, is R-strictly equivalent to a pencil of skewsymmetric matrices of the form*

$$
0_{u \times u} \;\oplus\; \oplus_{j=1}^{p} \left(t \begin{bmatrix} 0 & 0 & F_{\varepsilon_j} \\ 0 & 0_1 & 0 \\ -F_{\varepsilon_j} & 0 & 0 \end{bmatrix} + \begin{bmatrix} 0 & F_{\varepsilon_j} & 0 \\ -F_{\varepsilon_j} & 0 & 0 \\ 0 & 0 & 0 \end{bmatrix} \right)
$$

$$
\oplus \;\; \oplus_{j=1}^{r} \left(\begin{bmatrix} 0 & F_{k_j} \\ -F_{k_j} & 0 \end{bmatrix} + t \begin{bmatrix} 0 & G_{k_j} \\ -G_{k_j} & 0 \end{bmatrix} \right)
$$

$$
\oplus \;\; \oplus_{j=1}^{q} \left((t + \alpha_j) \begin{bmatrix} 0 & F_{\ell_j} \\ -F_{\ell_j} & 0 \end{bmatrix} + \begin{bmatrix} 0 & G_{\ell_j} \\ -G_{\ell_j} & 0 \end{bmatrix} \right)
$$

$$
\oplus \;\; \oplus_{j=1}^{s} \begin{bmatrix} 0 & Z_{2m_j}(t, \mu_j, \nu_j) \\ -Z_{2m_j}(t, \mu_j, \nu_j) & 0 \end{bmatrix}, \tag{15.2.3}
$$

where the positive integers ε_j's satisfy $\varepsilon_1 \leq \cdots \leq \varepsilon_p$, the numbers α_j, μ_j, ν_j are all real, and $\nu_1, \ldots, \nu_s$ are positive.

The form (15.2.3) is uniquely determined by $A + tB$ up to arbitrary permutations of diagonal blocks in each of the three parts

$$\oplus_{j=1}^{r} \left(\begin{bmatrix} 0 & F_{k_j} \\ -F_{k_j} & 0 \end{bmatrix} + t \begin{bmatrix} 0 & G_{k_j} \\ -G_{k_j} & 0 \end{bmatrix} \right),$$

$$\oplus_{j=1}^{q} \left((t + \alpha_j) \begin{bmatrix} 0 & F_{\ell_j} \\ -F_{\ell_j} & 0 \end{bmatrix} + \begin{bmatrix} 0 & G_{\ell_j} \\ -G_{\ell_j} & 0 \end{bmatrix} \right),$$

and

$$\oplus_{j=1}^{s} \begin{bmatrix} 0 & Z_{2m_j}(t,\,\mu_j,\nu_j) \\ -Z_{2m_j}(t,\,\mu_j,\nu_j) & 0 \end{bmatrix}.$$

(ii) *Two matrix pencils $A_1 + tB_1$ and $A_2 + tB_2$ with skewsymmetric A_1, B_1, A_2 and B_2 are R-congruent if and only if they are R-strictly equivalent.*

For a complete proof, as well as bibliography, we refer to Thompson [151] and Lancaster and Rodman [93]. The particular form (15.2.3) is taken from Lancaster and Rodman [93].

We observe that *in the blocks*

$$\begin{bmatrix} 0 & Z_{2m_j}(t,\,\mu_j,\nu_j) \\ -Z_{2m_j}(t,\,\mu_j,\nu_j) & 0 \end{bmatrix}$$

in (15.2.3), ν_j may be replaced by $-\nu_j$, the blocks F_{2m_j} may be replaced by $-F_{2m_j}$, the blocks H_{2m_j} may be replaced by $-H_{2m_j}$, and the blocks

$$\begin{bmatrix} F_{2m_j-2} & 0 \\ 0 & 0_2 \end{bmatrix}$$

may be replaced by

$$\begin{bmatrix} -F_{2m_j-2} & 0 \\ 0 & 0_2 \end{bmatrix} \quad \text{or by} \quad \begin{bmatrix} 0_2 & 0 \\ 0 & \pm F_{2m_j-2} \end{bmatrix}.$$

To make this statement explicit, set (we write m instead of m_j for brevity)

$$W_{2m}(x,y,z) = xF_{2m} + yY_{2m} + z\begin{bmatrix} F_{2m-2} & 0 \\ 0 & 0 \end{bmatrix},$$

$$\widetilde{W}_{2m}(x,y,z) = xF_{2m} + yH_{2m} + z\begin{bmatrix} 0 & 0 \\ 0 & F_{2m-2} \end{bmatrix}$$

(Y_{2m} is given by (15.2.1)), where x, y, z are independent real variables, and introduce the skewsymmetric matrix

$$Q_{2m} = \mathrm{diag}\left(\begin{bmatrix} 0 & 1 \\ -1 & 0 \end{bmatrix}, \begin{bmatrix} 0 & -1 \\ 1 & 0 \end{bmatrix}, \dots, (-1)^{m+1}\begin{bmatrix} 0 & 1 \\ -1 & 0 \end{bmatrix}\right) = -Q_{2m}^T,$$

and the symmetric diagonal matrix

$$D_{4m} = \begin{bmatrix} \widetilde{D}_{2m} & 0 \\ 0 & -\widetilde{D}_{2m} \end{bmatrix}, \quad \text{where} \quad \widetilde{D}_{2m} = \mathrm{diag}\,(1,-1,\dots,1,-1).$$

Then we have the following formulas:

$$D_{4m}\begin{bmatrix} 0 & W_{2m}(x,y,z) \\ -W_{2m}(x,y,z) & 0 \end{bmatrix}D_{4m}$$

$$= \begin{bmatrix} 0 & W_{2m}(x,-y,z) \\ -W_{2m}(x,-y,z) & 0 \end{bmatrix}; \qquad (15.2.4)$$

$$
\begin{bmatrix} F_{2m} & 0 \\ 0 & F_{2m} \end{bmatrix}
\begin{bmatrix} 0 & W_{2m}(x,y,z) \\ W_{2m}(-x,-y,-z) & 0 \end{bmatrix}
\begin{bmatrix} F_{2m} & 0 \\ 0 & F_{2m} \end{bmatrix}
$$
$$
= \begin{bmatrix} 0 & \widetilde{W}_{2m}(x,-y,z) \\ \widetilde{W}_{2m}(-x,y,-z) & 0 \end{bmatrix};
$$

and

$$
\begin{bmatrix} Q_{2m} & 0 \\ 0 & (-1)^{m+1}Q_{2m} \end{bmatrix}
\begin{bmatrix} 0 & W_{2m}(x,y,z) \\ W_{2m}(-x,-y,-z) & 0 \end{bmatrix}
$$
$$
\cdot \begin{bmatrix} -Q_{2m} & 0 \\ 0 & (-1)^{m}Q_{2m} \end{bmatrix}
= \begin{bmatrix} 0 & W_{2m}(-x,-y,z) \\ W_{2m}(x,y,-z) & 0 \end{bmatrix}.
$$

$$(15.2.5)$$

Formulas (15.2.4) and (15.2.5) are taken from Lancaster and Rodman [93]. Other variations of the canonical from of Theorem 15.2.2 are given in Lancaster and Rodman [93] as well.

To state the canonical form of real matrix pencils $A + tB$, where one of the matrices A, B is symmetric and the other is skewsymmetric, it will be convenient to identify the primitive canonical real matrix pencils with this symmetry first. It will be assumed that A is symmetric and B is skewsymmetric (the situation when A is skewsymmetric and B is symmetric can be dealt with by simply interchanging the roles of A and B). The standard matrix

$$
\Xi_k := \Xi_k(1) = \begin{bmatrix}
0 & 0 & 0 & \cdots & 0 & 0 & 1 \\
0 & 0 & 0 & \cdots & 0 & -1 & 0 \\
0 & 0 & 0 & \cdots & 1 & 0 & 0 \\
\vdots & \vdots & \ddots & & \vdots & \vdots & \vdots \\
0 & 1 & 0 & \cdots & 0 & 0 & 0 \\
(-1)^{k-1} & 0 & 0 & \cdots & 0 & 0 & 0
\end{bmatrix} = (-1)^{k-1}\Xi_k^T \quad (15.2.6)
$$

will be used (cf. (1.2.6)). Thus, Ξ_k is symmetric if k is odd and skewsymmetric if k is even.

The primitive canonical real matrix pencils of symmetric/skewsymmetric-type are as follows; the prefix r-sy-sk stands for real, symmetric, skewsymmetric.

(r-sy-sk0)   a square-size zero matrix.

(r-sy-sk1)

$$
\begin{bmatrix} F_{2\varepsilon} & 0 \\ 0 & 0_1 \end{bmatrix}
+ t \begin{bmatrix} 0 & 0 & F_\varepsilon \\ 0 & 0_1 & 0 \\ -F_\varepsilon & 0 & 0 \end{bmatrix}.
$$

(r-sy-sk2)

$$
F_k + t \begin{bmatrix} 0_1 & 0 & 0 \\ 0 & 0 & F_{\frac{k-1}{2}} \\ 0 & -F_{\frac{k-1}{2}} & 0 \end{bmatrix}, \qquad k \text{ odd.}
$$

(r-sy-sk3)

$$
F_k + t \begin{bmatrix} 0_1 & 0 & 0 & 0 \\ 0 & 0 & 0 & F_{\frac{k-2}{2}} \\ 0 & 0 & 0_1 & 0 \\ 0 & -F_{\frac{k-2}{2}} & 0 & 0 \end{bmatrix}, \qquad k \text{ even and } \frac{k}{2} \text{ even.}
$$

(r-sy-sk4)

$$
G_\ell + t \begin{bmatrix} 0 & F_{\frac{\ell}{2}} \\ -F_{\frac{\ell}{2}} & 0 \end{bmatrix}, \qquad \ell \text{ even.}
$$

The matrix $\widetilde{G}_\ell$ of (1.2.5) can be used in (r-sy-sk4) in place of G_ℓ.

(r-sy-sk5)

$$
\begin{bmatrix} 0 & G_{\frac{\ell}{2}} \\ G_{\frac{\ell}{2}} & 0 \end{bmatrix} + t \begin{bmatrix} 0 & F_{\frac{\ell}{2}} \\ -F_{\frac{\ell}{2}} & 0 \end{bmatrix}, \qquad \ell \text{ even and } \frac{\ell}{2} \text{ odd.}
$$

(r-sy-sk6)

$$
\begin{bmatrix} 0 & \alpha F_{\frac{\ell}{2}} + G_{\frac{\ell}{2}} \\ \alpha F_{\frac{\ell}{2}} + G_{\frac{\ell}{2}} & 0 \end{bmatrix} + t \begin{bmatrix} 0 & F_{\ell/2} \\ -F_{\ell/2} & 0 \end{bmatrix},
$$

where ℓ is even and $\alpha \in \mathsf{R} \setminus \{0\}$.

The matrix $\widetilde{G}_{\ell/2}$ can be used in (r-sy-sk5) and in (r-sy-sk6) in place of $G_{\ell/2}$.

(r-sy-sk7)

$$
\begin{bmatrix} 0 & 0 & \cdots & 0 & 0 & \nu \Xi_2^{m+1} \\ 0 & 0 & \cdots & 0 & -\nu \Xi_2^{m+1} & -I_2 \\ 0 & 0 & \cdots & \nu \Xi_2^{m+1} & -I_2 & 0 \\ \vdots & \vdots & \cdot^{\cdot^{\cdot}} & \vdots & \vdots & \vdots \\ (-1)^{m-1}\nu \Xi_2^{m+1} & -I_2 & 0 & \cdots & 0 & 0 \end{bmatrix}
$$

$$
+ t \begin{bmatrix} 0 & 0 & \cdots & 0 & \Xi_2^m \\ 0 & 0 & \cdots & -\Xi_2^m & 0 \\ \vdots & & \vdots & \cdot^{\cdot^{\cdot}} & \vdots & \vdots \\ 0 & (-1)^{m-2}\Xi_2^m & \cdots & 0 & 0 \\ (-1)^{m-1}\Xi_2^m & 0 & \cdots & 0 & 0 \end{bmatrix}, \qquad \nu > 0.
$$

Here $\Xi_2 = \begin{bmatrix} 0 & 1 \\ -1 & 0 \end{bmatrix}$. The pencil in (r-sy-sk7) is $2m \times 2m$, where m is a positive integer. Note that the size $2m \times 2m$ matrices

$$
\begin{bmatrix} 0 & 0 & \cdots & 0 & 0 & \nu \Xi_2^{m+1} \\ 0 & 0 & \cdots & 0 & -\nu \Xi_2^{m+1} & -I_2 \\ 0 & 0 & \cdots & \nu \Xi_2^{m+1} & -I_2 & 0 \\ \vdots & \vdots & \cdot^{\cdot^{\cdot}} & \vdots & \vdots & \vdots \\ (-1)^{m-1}\nu \Xi_2^{m+1} & -I_2 & 0 & \cdots & 0 & 0 \end{bmatrix}
$$

and

$$\begin{bmatrix} 0 & 0 & \cdots & 0 & \Xi_2^m \\ 0 & 0 & \cdots & -\Xi_2^m & 0 \\ \vdots & \vdots & \cdot^{\cdot^{\cdot}} & \vdots & \vdots \\ 0 & (-1)^{m-2}\Xi_2^m & \cdots & 0 & 0 \\ (-1)^{m-1}\Xi_2^m & 0 & \cdots & 0 & 0 \end{bmatrix}$$

are symmetric and skewsymmetric, respectively, for every positive integer m and every real ν.

(r-sy-sk8)

$$\begin{bmatrix} 0 & J_{2m}(a \pm ib)^T \\ J_{2m}(a \pm ib) & 0 \end{bmatrix} + t \begin{bmatrix} 0 & I_{2m} \\ -I_{2m} & 0 \end{bmatrix},$$

where $a, b > 0$. The matrix pencil here is $4m \times 4m$.

We remark that in (r-sy-sk8), any of the following seven matrices can be used in place of $J_{2m}(a \pm ib)$. Of course, the transposed matrix should then be used simultaneously for $J_{2m}(a \pm ib)^T$:

$$J_{2m}(a \pm i(-b)), \quad J_{2m}(-a \pm ib), \quad J_{2m}(-a \pm i(-b)), \quad (J_{2m}(a \pm ib))^T,$$

$$(J_{2m}(a \pm i(-b)))^T, \quad (J_{2m}(-a \pm ib))^T, \quad (J_{2m}(-a \pm i(-b)))^T.$$

To verify this, use: (1) the property that every regular real matrix pencil is R-strictly equivalent to its transpose (see Corollary 7.3.6 and Theorem 7.6.3, or Corollary 3.4 in Lancaster and Rodman [92] for complex matrix pencils); and (2) the equalities

$$\begin{bmatrix} X & 0 \\ 0 & (X^{-1})^T \end{bmatrix} \left(\begin{bmatrix} 0 & J^T \\ J & 0 \end{bmatrix} + t \begin{bmatrix} 0 & I \\ -I & 0 \end{bmatrix} \right) \begin{bmatrix} X^T & 0 \\ 0 & X^{-1} \end{bmatrix}$$

$$= \begin{bmatrix} 0 & J \\ J^T & 0 \end{bmatrix} + t \begin{bmatrix} 0 & I \\ -I & 0 \end{bmatrix},$$

where J is one of the matrices $J_{2m}(a \pm i(-b))$, $J_{2m}(-a \pm ib)$, $J_{2m}(-a \pm i(-b))$ and where the invertible real matrix X is such that $X J^T X^{-1} = J$. The existence of X can be established directly, and it also follows from the well-known fact that every real or complex $n \times n$ matrix is similar to its transpose; see, e.g., Horn and Johnson [62].

One easily verifies that each of the pencils (r-sy-sk0)–(r-sy-sk8) is of mixed type (symmetric/skewsymmetric).

Theorem 15.2.3. *Let $A+tB$ be a real symmetric/skewsymmetric matrix pencil.*

(a) *$A + tB$ is R-strictly equivalent to a direct sum of blocks of types (r-sy-sk0)–(r-sy-sk8). In this direct sum, several blocks of the same type and of different sizes may be present. The direct sum is uniquely determined by A and B, up to an arbitrary permutation of blocks.*

(b) *$A+tB$ is R-congruent to a real symmetric/skewsymmetric pencil of the form*

$$(A_0 + tB_0) \oplus \oplus_{j=1}^r \delta_j \left(F_{k_j} + t \begin{bmatrix} 0_1 & 0 & 0 \\ 0 & 0 & F_{\frac{k_j-1}{2}} \\ 0 & -F_{\frac{k_j-1}{2}} & 0 \end{bmatrix} \right) \tag{15.2.7}$$

$$\oplus\ \oplus_{w=1}^{p}\ \eta_w \left(G_{\ell_w} + t \begin{bmatrix} 0 & F_{\frac{\ell_w}{2}} \\ -F_{\frac{\ell_w}{2}} & 0 \end{bmatrix} \right) \oplus \oplus_{u=1}^{q} \zeta_u \tag{15.2.8}$$

$$\left(\begin{bmatrix} 0 & 0 & \cdots & 0 & 0 & \nu_u \Xi_2^{m_u+1} \\ 0 & 0 & \cdots & 0 & -\nu_u \Xi_2^{m_u+1} & -I_2 \\ 0 & 0 & \cdots & \nu_u \Xi_2^{m_u+1} & -I_2 & 0 \\ \vdots & \vdots & \cdots & \vdots & \vdots & \vdots \\ (-1)^{m_u-1}\nu_u\Xi_2^{m_u+1} & -I_2 & 0 & \cdots & 0 & 0 \end{bmatrix} \right.$$

$$\left. +t \begin{bmatrix} 0 & 0 & \cdots & 0 & \Xi_2^{m_u} \\ 0 & 0 & \cdots & -\Xi_2^{m_u} & 0 \\ \vdots & \vdots & \cdots & \vdots & \vdots \\ 0 & (-1)^{m_u-2}\Xi_2^{m_u} & \cdots & 0 & 0 \\ (-1)^{m_u-1}\Xi_2^{m_u} & 0 & \cdots & 0 & 0 \end{bmatrix} \right). \tag{15.2.9}$$

Here, $A_0 + tB_0$ is a direct sum of blocks of types (r-sy-sk0), (r-sy-sk1), (r-sy-sk3), (r-sy-sk5), (r-sy-sk6), and (r-sy-sk8), in which several blocks of the same type and of different and/or the same sizes may be present, and the k_j's are odd positive integers, the ℓ_w's are even positive integers, the ν_y's are positive real numbers, and $\delta_j, \eta_w, \zeta_y$ are signs ± 1.

The blocks in (15.2.7) and (15.2.9) are uniquely determined by $A + tB$ up to an arbitrary permutation of constituent matrix pencil blocks in $A_0 + \lambda B_0$ and an arbitrary permutation of blocks in each of the three parts

$$\oplus_{j=1}^{r} \delta_j \left(F_{k_j} + t \begin{bmatrix} 0_1 & 0 & 0 \\ 0 & 0 & F_{\frac{k_j-1}{2}} \\ 0 & -F_{\frac{k_j-1}{2}} & 0 \end{bmatrix} \right),$$

(15.2.8), and (15.2.9).

The special properties of the Kronecker form for real symmetric/skewsymmetric matrices have been known since the paper by Kronecker [82], and, later, by Williamson [161]; see also Mal'cev [105] for the case when the skewsymmetric matrix is invertible. The result of part (b) was first proved in Williamson [161]; more detailed expositions are found in Thompson [151] and Lancaster and Rodman [93], where complete proofs are given. The particular form (15.2.7), (15.2.8), (15.2.9) is taken from Lancaster and Rodman [93].

The signs $\delta_j, \eta_w, \zeta_y$ form the *sign characteristic* of the symmetric/skewsymmetric real matrix pencil $A + tB$. It is easy to see that the block

$$F_{k_j} + t \begin{bmatrix} 0_1 & 0 & 0 \\ 0 & 0 & F_{\frac{k_j-1}{2}} \\ 0 & -F_{\frac{k_j-1}{2}} & 0 \end{bmatrix}, \quad k_j \text{ odd},$$

is R-strictly equivalent to $I + tJ_{k_j}(0)$, and the block

$$G_{\ell_w} + t \begin{bmatrix} 0 & F_{\frac{\ell_w}{2}} \\ -F_{\frac{\ell_w}{2}} & 0 \end{bmatrix}, \quad \ell_w \text{ even},$$

is R-strictly equivalent to $tI + J_{\ell_w}(0)$. Moreover, a computation shows that

$$
\begin{bmatrix}
0 & 0 & \cdots & 0 & 0 & \nu_u \Xi_2^{m_u+1} \\
0 & 0 & \cdots & 0 & -\nu_u \Xi_2^{m_u+1} & -I_2 \\
0 & 0 & \cdots & \nu_u \Xi_2^{m_u+1} & -I_2 & 0 \\
\vdots & \vdots & \cdots & \vdots & \vdots & \vdots \\
(-1)^{m_u-1}\nu_u \Xi_2^{m_u+1} & -I_2 & 0 & \cdots & 0 & 0
\end{bmatrix}
$$

$$
\cdot
\begin{bmatrix}
0 & 0 & \cdots & 0 & \Xi_2^{m_u} \\
0 & 0 & \cdots & -\Xi_2^{m_u} & 0 \\
\vdots & \vdots & \cdots & \vdots & \vdots \\
0 & (-1)^{m_u-2}\Xi_2^{m_u} & \cdots & 0 & 0 \\
(-1)^{m_u-1}\Xi_2^{m_u} & 0 & \cdots & 0 & 0
\end{bmatrix}^{-1}
$$

$$
=
\begin{bmatrix}
\nu_u \Xi_2 & 0 & 0 & \cdots & 0 \\
(-1)^{m_u-1}\Xi_2^{m_u} & \nu_u \Xi_2 & 0 & \cdots & 0 \\
0 & (-1)^{m_u-2}\Xi_2^{m_u} & \nu_u \Xi_2 & \cdots & 0 \\
\vdots & \vdots & \vdots & \ddots & \vdots \\
0 & 0 & 0 & \cdots (-1)\Xi_2^{m_u} & \nu_u \Xi_2
\end{bmatrix},
$$

which is easily seen to have the real Jordan form $J_{2m_u}(\pm \nu_u i)$. Thus, the sign characteristic of a symmetric/skewsymmetric real matrix pencil associates a sign ± 1 to every odd partial multiplicity at infinity, to every even partial multiplicity at zero, and to every partial multiplicity corresponding to a pure imaginary eigenvalue with positive (or negative) imaginary part.

15.3   COMPLEX MATRIX PENCILS WITH SYMMETRIES

In this section we present canonical forms of complex matrix pencils with various symmetries. We start with the hermitian symmetry.

Theorem 15.3.1. (a) *Every matrix pencil $A + tB$, where $A = A^* \in \mathsf{C}^{n \times n}$, $B = B^* \in \mathsf{C}^{n \times n}$, is C-congruent to a complex hermitian matrix pencil of the form*

$$
0 \; \oplus \; \left(t \begin{bmatrix} 0 & 0 & F_{\varepsilon_1} \\ 0 & 0 & 0 \\ F_{\varepsilon_1} & 0 & 0 \end{bmatrix} + G_{2\varepsilon_1+1} \right)
$$

$$
\oplus \; \cdots \; \oplus \left(t \begin{bmatrix} 0 & 0 & F_{\varepsilon_p} \\ 0 & 0 & 0 \\ F_{\varepsilon_p} & 0 & 0 \end{bmatrix} + G_{2\varepsilon_p+1} \right)
$$

$$
\oplus \; \delta_1 \left(F_{k_1} + tG_{k_1} \right) \oplus \cdots \oplus \delta_r \left(F_{k_r} + tG_{k_r} \right)
$$

$$
\oplus \; \eta_1 \left((t+\alpha_1) F_{\ell_1} + G_{\ell_1} \right) \oplus \cdots \oplus \eta_q \left((t+\alpha_q) F_{\ell_q} + G_{\ell_q} \right)
$$

$$
\oplus \; \left(\begin{bmatrix} 0 & (t+\beta_1)F_{m_1} \\ (t+\overline{\beta}_1)F_{m_1} & 0 \end{bmatrix} + \begin{bmatrix} 0 & G_{m_1} \\ G_{m_1} & 0 \end{bmatrix} \right)
$$

$$
\oplus \; \cdots \; \oplus \left(\begin{bmatrix} 0 & (t+\beta_s)F_{m_s} \\ (t+\overline{\beta}_s)F_{m_s} & 0 \end{bmatrix} + \begin{bmatrix} 0 & G_{m_s} \\ G_{m_s} & 0 \end{bmatrix} \right). \quad (15.3.1)
$$

Here, $\varepsilon_1 \leq \cdots \leq \varepsilon_p$ and $k_1, \ldots, k_r$ are positive integers, α_j are real numbers, β_j are complex nonreal numbers with positive imaginary parts, and $\delta_1, \ldots, \delta_r, \eta_1, \ldots, \eta_q$ are signs, each equal to $+1$ or -1.

The form (15.3.1) *is uniquely determined by* $A + tB$ *up to an arbitrary permutation of blocks in each of the three parts*

$$\oplus_{j=1}^{r} \left(\delta_j \left(F_{k_j} + tG_{k_j}\right)\right), \quad \oplus_{j=1}^{q} \left(\eta_j \left((t + \alpha_j) F_{\ell_j} + G_{\ell_j}\right)\right),$$

and

$$\oplus_{j=1}^{s} \left(\begin{bmatrix} 0 & (t + \beta_j)F_{m_j} \\ (t + \overline{\beta}_j)F_{m_j} & 0 \end{bmatrix} + \begin{bmatrix} 0 & G_{m_j} \\ G_{m_j} & 0 \end{bmatrix}\right).$$

(b) *Every matrix pencil* $A + tB$, *where* $A = A^* \in \mathsf{C}^{n \times n}$, $B = B^* \in \mathsf{C}^{n \times n}$, *is* C-*strictly equivalent to a unique* (*up to a permutation, as in Part* (a)) *complex hermitian matrix pencil of the form* (15.3.1), *with all signs taken to be* +1.

The signs $\delta_1, \ldots, \delta_r, \eta_1, \ldots, \eta_q$ form the *sign characteristic* of the complex hermitian pencil $A + tB$. Thus, the sign characteristic associates a sign ± 1 to every index of a real eigenvalue and at infinity of $A + tB$.

We will not consider separately complex skewhermitian matrices because one can multiply a skewhermitian complex matrix by i to obtain a hermitian matrix and then apply the results of Theorem 15.3.1.

We study next complex matrix pencils with symmetries with respect to transposition. Recall that two complex matrix pencils $A + tB$ and $A' + tB'$ are $(\mathsf{C},^T)$-*congruent* if $A + tB = S^T(A' + tB')S$ for some invertible complex matrix S. It turns out that for complex pencils with symmetries respecting transposition, $(\mathsf{C},^T)$-congruence is the same as C-strict equivalence:

Theorem 15.3.2. *Let* $A_1, B_1, A_2, B_2 \in \mathsf{C}^{n \times n}$, *and assume that* $A_j^T = \eta_1 A_j$, $B_j^T = \eta_2 B_j$ *for* $j = 1, 2$, *where* η_1, η_2 *are signs* ± 1. *Then the pencils* $A_1 + tB_1$ *and* $A_2 + tB_2$ *are* $(\mathsf{C},^T)$-*congruent if and only if they are* C-*strictly equivalent.*

The detailed proof of this theorem, as well as that of Theorems 15.3.3, 15.3.5, and 15.3.6 below, is found in Thompson [151].

Theorem 15.3.3. *Every matrix pencil* $A + tB$, *where* A *and* B *are complex symmetric matrices, is* $(\mathsf{C},^T)$-*congruent to a pencil of symmetric matrices of the form*

$$0_{u \times u} \quad \oplus \quad \oplus_{i=1}^{s} \begin{bmatrix} 0 & L_{\varepsilon_i \times (\varepsilon_i+1)}^{T} \\ L_{\varepsilon_i \times (\varepsilon_i+1)} & 0 \end{bmatrix} \oplus \oplus_{j=1}^{r} \left(F_{k_j} + tG_{k_j}\right)$$

$$\oplus \quad \oplus_{j=1}^{q} \left((t + \alpha_j) F_{\ell_j} + G_{\ell_j}\right), \tag{15.3.2}$$

where $\varepsilon_1 \leq \cdots \leq \varepsilon_s$ *and* $k_1 \leq \cdots \leq k_r$ *are positive integers, and* $\alpha_1, \ldots, \alpha_q \in \mathsf{C}$. *The form* (15.3.2) *is uniquely determined by* $A + tB$ *up to arbitrary permutation of blocks in the part*

$$\oplus_{j=1}^{q} \left((t + \alpha_j) F_{\ell_j} + G_{\ell_j}\right).$$

As a corollary we obtain the following.

Corollary 15.3.4. *A pencil of complex* $n \times n$ *matrices is* C-*strictly equivalent to a pencil of complex symmetric matrices if and only if its left indices, arranged in nondecreasing order, coincide with its right indices, also arranged in nondecreasing order.*

For complex matrices rather than pencils, this result is found in Gantmacher [48].

Theorem 15.3.5. *Every matrix pencil* $A + tB$, *where* A *and* B *are complex skewsymmetric matrices, is* $(\mathsf{C},^T)$-*congruent to a pencil of skewsymmetric matrices of the form*

$$0_u \quad \oplus \quad \oplus_{j=1}^{p} \left(t \begin{bmatrix} 0 & 0 & F_{\varepsilon_j} \\ 0 & 0_1 & 0 \\ -F_{\varepsilon_j} & 0 & 0 \end{bmatrix} + \begin{bmatrix} 0 & F_{\varepsilon_j} & 0 \\ -F_{\varepsilon_j} & 0 & 0 \\ 0 & 0 & 0 \end{bmatrix} \right)$$

$$\oplus \quad \oplus_{j=1}^{r} \left(\begin{bmatrix} 0 & F_{k_j} \\ -F_{k_j} & 0 \end{bmatrix} + t \begin{bmatrix} 0 & G_{k_j} \\ -G_{k_j} & 0 \end{bmatrix} \right)$$

$$\oplus \quad \oplus_{j=1}^{q} \left((t + \alpha_j) \begin{bmatrix} 0 & F_{\ell_j} \\ -F_{\ell_j} & 0 \end{bmatrix} + \begin{bmatrix} 0 & G_{\ell_j} \\ -G_{\ell_j} & 0 \end{bmatrix} \right), \qquad (15.3.3)$$

where the positive integers ε_j's *satisfy* $\varepsilon_1 \leq \cdots \leq \varepsilon_p$ *and* $\alpha_1, \dots, \alpha_q \in \mathsf{C}$.

The form (15.3.3) *is uniquely determined by* $A + tB$ *up to arbitrary permutations of diagonal blocks in each of the two parts*

$$\oplus_{j=1}^{r} \left(\begin{bmatrix} 0 & F_{k_j} \\ -F_{k_j} & 0 \end{bmatrix} + t \begin{bmatrix} 0 & G_{k_j} \\ -G_{k_j} & 0 \end{bmatrix} \right)$$

and

$$\oplus_{j=1}^{q} \left((t + \alpha_j) \begin{bmatrix} 0 & F_{\ell_j} \\ -F_{\ell_j} & 0 \end{bmatrix} + \begin{bmatrix} 0 & G_{\ell_j} \\ -G_{\ell_j} & 0 \end{bmatrix} \right).$$

One can replace G_k by $\widetilde{G}_k$ in Theorems 15.3.3 and 15.3.5. Note the general fact that if a field is algebraically closed with characteristic not equal to two, then two pencils of skewsymmetric matrices are congruent (with respect to transposition) over the field if and only if the pencils are strictly equivalent over the same field; see, e.g., MacDuffee [104], Mal'cev [105] (for the case when one matrix is invertible), Gantmacher [48], or Thompson [151].

To deal with the case of mixed pencils, where one matrix is symmetric and the other is skewsymmetric, we indicate the primitive forms first, with the prefix c-sy-sk (complex symmetric/skewsymmetric). The forms (c-sy-sk0), (c-sy-sk1), (c-sy-sk2), (c-sy-sk3), (c-sy-sk4), and (c-sy-sk5) are the same as (r-sy-sk0), (r-sy-sk1), (r-sy-sk2), (r-sy-sk3), (r-sy-sk4), and (r-sy-sk5), respectively. The form (c-sy-sk6) is

$$\begin{bmatrix} 0 & \alpha F_{\frac{\ell}{2}} + G_{\frac{\ell}{2}} \\ \alpha F_{\frac{\ell}{2}} + G_{\frac{\ell}{2}} & 0 \end{bmatrix} + t \begin{bmatrix} 0 & F_{\frac{\ell}{2}} \\ -F_{\frac{\ell}{2}} & 0 \end{bmatrix},$$

where ℓ is even and $\alpha \in \mathsf{C} \setminus \{0\}$. The matrices $\widetilde{G}_m$ may be used in place of G_m in (c-sy-sk4), (c-sy-sk5), and (c-sy-sk6). Note that the forms (c-sy-sk0)–(c-sy-sk6) are indeed complex symmetric/skewsymmetric pencils.

Theorem 15.3.6. *Let* $A + tB$ *be a complex symmetric/skewsymmetric matrix pencil:* $A = A^T$, $B = -B^T$. *Then* $A + tB$ *is* $(\mathsf{C},^T)$-*congruent to a direct sum of blocks of types* (c-sy-sk0)–(c-sy-sk6). *In this direct sum, several blocks of the same type and of different sizes may be present. The direct sum is uniquely determined by* A *and* B, *up to an arbitrary permutation of blocks.*

The following meta-corollary will be useful.

Corollary 15.3.7. *Let* $F = R$ *or* $F = C$. *Fix signs* $\xi = \pm 1$, $\eta = \pm 1$, *and fix an involutory transformation on matrices* $X \mapsto X^\star$, *where* $X^\star = X^T$ *or* $X^\star = X^*$ *is independent of the matrix* X (*in the real case* $X^T = X^*$). *Let there be given two matrix pencils* $A + tB$ *and* $A' + tB'$, *where* $A, B, A', B' \in F^{n \times n}$ *and*

$$A = \xi A^\star, \quad B = \eta B^\star, \quad A' = \xi (A')^\star, \quad B' = \eta (B')^\star.$$

If for some positive integer s, $(A + tB)^{\oplus s}$ *is* $(F, \star)$-*congruent to* $(A' + tB')^{\oplus s}$, *then* $A + tB$ *is* $(F, \star)$-*congruent to* $A' + tB'$

The proof of the meta-corollary follows from the uniqueness of the respective canonical form for matrix pencils with symmetries, analogously to the proof of the case $F = C$ in Corollary 15.1.4. We omit details.

Bibliography

[1] H. Abou-Kandil, G. Freiling, V. Ionescu, and G. Jank. *Matrix Riccati Equations in Control and Systems Theory*, Birkhäuser Verlag, Basel, 2003.

[2] D. Alpay, J. A. Ball, I. Gohberg, and L. Rodman. Realization and factorization for rational matrix functions with symmetries. *Operator Theory: Advances and Appl.* **47**, 1–60, Birkhäuser, Basel, 1990.

[3] D. Alpay and H. Dym. Structured invariant spaces of vector valued rational functions, Hermitian matrices, and a generalization of the Iohvidov laws, *Linear Algebra and Appl.* **137/138** (1990), 137–181.

[4] D. Alpay, A. C. M. Ran, and L. Rodman. Basic classes of matrices with respect to quaternionic indefinite inner product spaces, *Linear Algebra and Appl.* **416** (2006), 242–269.

[5] E. Artin. *Geometric Algebra*, Princeton University Press, Princeton, NJ, 1957.

[6] M. Artin. *Algebra*, Prentice Hall, Englewood Cliffs, NJ, 1991.

[7] Y.-H. Au-Yeung. A theorem on a mapping from a sphere to the circle and the simultaneous diagonalization of two hermitian matrices, *Proc. Amer. Math. Soc.* **20** (1969), 545–548.

[8] ———. A simple proof of convexity of the field of values defined by two hermitian forms, *Aequationes Math* **12** (1975), 82–83.

[9] ———. On the convexity of the numerical range in quaternionic Hilbert space, *Linear and Multilinear Algebra* **16** (1984), 93–100.

[10] Y.-H. Au-Yeung and Y. T. Poon. A remark on the convexity and positive definiteness concerning Hermitian matrices, *Southeast Asian Bull. Math.* **3** (1979), 85–92.

[11] Y. H. Au-Yeung and N. K. Tsing. An extension of the Hausdorff-Toeplitz theorem on the numerical range, *Proc. Amer. Math. Soc.* **89** (1983), 215–218.

[12] ———. Some theorems on the generalized numerical ranges, *Linear and Multilinear Algebra* **15** (1984), 3–11.

[13] T. Ya. Azizov and I. S. Iohvidov. *Linear Operators in Spaces with an Indefinite Metric*, John Wiley, Chichester, 1989 (translation from Russian).

[14] G. Berhuy and F. Oggier. *An Introduction to Central Simple Algebras and Their Applications to Wireless Communication*, Amer. Math. Soc., Providence, RI, 2013.

[15] V. Bolotnikov. Private communication.

[16] Y. Bolshakov, C. V. M. van der Mee, A. C. M. Ran, B. Reichstein, and L. Rodman. Extensions of isometries in finite dimensional indefinite scalar product spaces and polar decompositions, *SIAM J. of Matrix Analysis and Appl.* **18** (1997), 752–774.

[17] K. C. Border. *Fixed Point Theorems with Applications to Economics and Game Theory*, Cambridge University Press, Cambridge, 1985.

[18]   J. L. Brenner. Matrices of quaternions. *Pacific J. Math.* **1** (1951), 329–335.

[19]   L. Brickman. On the field of values of a matrix, *Proc. Amer. Math. Soc.* 12 (1961), 61–66.

[20]   E. Brieskorn. *Lineare Algebra und Analytische Geometrie*, Volume 1 and 2, Vieweg Verlag, Braunschweig, 1985 and 1993.

[21]   A. M. Bruckner, J. B. Bruckner, and B. S. Thomson. *Real Analysis*, Prentice Hall, Upper Saddle River, NJ, 1997.

[22]   A. Bunse-Gerstner, R. Byers, and V. Mehrmann. A quaternion QR algorithm, *Numer. Math.* **55** (1989), 83–95.

[23]   N. Burgoyne and R. Cushman. Normal forms for real linear Hamiltonian systems, *Lie Groups: History Frontiers and Applications* **VII** (1977), 483–529, Math Sci. Press, Brookline, MA.

[24]   L. X. Chen. Inverse matrix and properties of double determinant over quaternion field, *Sci. China Ser. A* **34** (1991), 528–540.

[25]   ———. Definition of determinant and Cramer solutions over the quaternion field, *Acta Mathematics Sinica, New Series* **7** (1991), 171–180.

[26]   W.-S. Cheung, C.-K. Li, and L. Rodman. Operators with numerical range in a closed subspace, *Taiwanese J. of Math.* **11** (2007), 471–481.

[27]   R. V. Churchill and J. W. Brown. *Complex Variables and Applications*, 5th ed., McGraw-Hill, New York, 1990.

[28]   N. Cohen and S. De Leo. The quaternionic determinant, *Electronic J. of Linear Algebra*, **7** (2000), 100–111.

[29]   P. M. Cohn. *Skew Fields: Theory of General Division Rings*, Cambridge University Press, New York, 1995.

[30]   J. H. Conway and D. A. Smith. *On Quaternions and Octonions: Their Geometry, Arithmetic, and Symmetry*, A. K. Peters, Natick, MA, 2003.

[31]   A. N. Čurilov. On the solutions of quadratic matrix equations, *Non-linear Vibrations and Control Theory* (Udmurt State University, Izhevsk) **2** (1978), 24–33 (Russian).

[32]   J. Dieudonné. Les determinants sur un corp non-commutatif, *Bull. Soc. Math. France* **71** (1943), 27–45.

[33]   D. Z. Djoković. Classification of pairs consisting of a linear and a semilinear map, *Linear and Multilinear Algebra* **20** (1978), 147–165.

[34]   D. Z. Djoković, J. Patera, P. Winternitz, and H. Zassenhaus. Normal forms of elements of classical real and complex Lie and Jordan algebras, *J. Math. Phys.* **24** (1983), 1363–1374.

[35]   C. H. Edwards, Jr. *Advanced Calculus of Several Variables*. Academic Press, New York and London, 1973.

[36]   S. Eilenberg and I. Niven. The "fundamental theorem of algebra" for quaternions, *Bull. Amer. Math. Soc.* **50** (1944), 246–248.

[37]   B. Farb and R. K. Dennis. *Noncommutative Algebra*, Springer Verlag, New York, 1993.

[38]   D. R. Farenick and B. A. F. Pidkowich. The spectral theorem in quaternions, *Linear Algebra and Appl.* **371** (2003), 75–102.

[39] H. Faßbender and Kh. Ikramov. Several observations on symplectic, Hamiltonian, and skew-Hamiltonian matrices, *Linear Algebra and Appl.* **400** (2005), 15–29.

[40] H. Faßbender, D. S. Mackey, and N. Mackey. Hamilton and Jacobi come full circle: Jacobi algorithms for structured eigenproblems, *Linear Algebra and Appl.* **332–334** (2001), 37–80.

[41] H. Faßbender, D. S. Mackey, N. Mackey, and H. Xu. Hamiltonian square roots of skew-Hamiltonian matrices, *Linear Algebra and Appl.* **287** (1999), 125–159.

[42] D. Finkbeiner. *Introduction to Matrices and Linear Transformations*, W. H. Freeman, San Francisco, 1978.

[43] P. Finsler. Über das Vorkommen definiter und semidefiniter Formen in Scharen quadratischer Formen, *Comm. Math. Helv.* **9** (1937), 188–192.

[44] W. Fleming. *Functions of Several Variables*, Springer Verlag, New York, 1977.

[45] J. N. Franklin. *Matrix Theory*, Prentice Hall, Englewood Cliffs, NJ, 1968.

[46] G. Freiling, V. Mehrmann, and H. Xu. Existence, uniqueness, and parametrization of Lagrangian invariant subspaces, *SIAM J. of Matrix Analysis and Appl.* **23** (2002), 1045–1069.

[47] P. A. Fuhrmann. On Hamiltonian rational transfer functions, *Linear Algebra and Appl.* **63** (1984), 1–93.

[48] F. R. Gantmacher. *The Theory of Matrices*, Vols. 1 and 2, Chelsea, New York, 1959 (translation from Russian).

[49] ———. *Applications of the Theory of Matrices*, Interscience Publishers, New York, 1959 (translation of part II of the Russian original).

[50] ———. *The Theory of Matrices*, 3rd. ed., Nauka, Moscow, 1967. (Russian.)

[51] F. R. Gantmacher and M. G. Krein. *Oscillation Matrices and Kernels and Small Vibrations of Mechanical Systems*, rev. ed., AMS Chelsea Publishing, Providence, RI, 2002 (translation based on the 1941 original).

[52] I. Gohberg, P. Lancaster, and L. Rodman. *Matrices and Indefinite Scalar Products.*, Operator Theory: Advances and Appl. **8**, Birkhäuser, Basel and Boston, 1983.

[53] ———. *Indefinite Linear Algebra and Applications*, Birkhäuser, Boston, 2006.

[54] ———. *Invariant Subspaces of Matrices with Applications*, John Wiley, New York, 1986; republication SIAM, Philadelphia, 2006.

[55] ———. *Matrix Polynomials*, Academic Press, New York, London, 1982; republication SIAM, Philadelphia, 2009.

[56] G. H. Golub and C. F. Van Loan. *Matrix Computations*, 2nd. ed., The Johns Hopkins University Press, Baltimore and London, 1989.

[57] B. Gordon and T. S. Motzkin. On the zeros of polynomials over division rings, *Trans. Amer. Math. Soc.* **116** (1965), 218–226.

[58] R. M. Guralnick and L. S. Levy. Presentations of modules when ideals need not be principal, *Illinois J. Math.* **32** (1988), 593–653.

[59] R. M. Guralnick, L. S. Levy, and C. Odenthal. Elementary divisor theorem for noncommutative PIDs, *Proc. Amer. Math. Soc.* **103** (1988), 1003–1011.

[60] K. Gürlebeck and W. Sprössig. *Quaternionic and Clifford Calculus for Physicists and Engineers*, John Wiley, Chichester, 1997.

[61] I. N. Herstein. *Abstract Algebra*, 2nd ed., McMillan, New York, 1990.

[62] R. A. Horn and C. R. Johnson. *Matrix Analysis*, Cambridge University Press, Cambridge, 1985.

[63] ———. *Topics in Matrix Analysis*, Cambridge University Press, Cambridge, 1991.

[64] R. A. Horn and V. V. Sergeichuk. Canonical matrices of bilinear and sesquilinear forms, *Linear Algebra and Appl.* **428** (2008), 193–223.

[65] ———. Canonical forms for complex matrix congruence and *congruence, *Linear Algebra and Appl.* **416** (2006) 1010–1032.

[66] ———. Congruences of a square matrix and its transpose, *Linear Algebra and Appl.* **389** (2004), 347–353.

[67] L. K. Hua. On the theory of automorphic functions of a matrix variable, II. The classification of hypercircles under the symplectic group, *Amer. J. Math.* **66** (1944), 531–563.

[68] L. Huang and W. So. Quadratic formulas for quaternions, *Appl. Math. Letters* **15** (2002), 533–540.

[69] T. W. Hungerford. *Algebra*, Springer Verlag, New York. 1974.

[70] I. S. Iohvidov, M. G. Krein, and H. Langer. *Introduction to the Spectral Theory of Operators in Spaces with an Indefinite Metric*, Mathematical Research, vol. 9, Akademie-Verlag, Berlin, 1982.

[71] N. Jacobson. *The Theory of Rings*. American Mathematical Society, RI, 1943.

[72] D. Janovska and G. Opfer. A note on the computation of all zeros of simple quaternionic polynomial, *SIAM J. Numerical Analysis* **48** (2010), 244–256.

[73] ———. The algebraic Riccati equations for quaternions, *Advances in Applied Clifford Algebras*, to appear.

[74] R. E. Johnson. On the equation $\chi\alpha = \gamma\chi + \beta$ over an algebraic division ring, *Bull. Amer. Math. Soc.* **50** (1944), 202–207.

[75] M. A. Kaashoek, C. V. M. van der Mee, and L. Rodman. Analytic operator functions with compact spectrum, II. Spectra pairs and factorization, *Integral Equations and Operator Theory* **5** (1982), 791–827.

[76] R. E. Kalman. Contributions to the theory of optimal control, *Boletin Sociedad Matematica Mexicana* **5** (1960), 102–119.

[77] W. Kaplan. *Advanced Calculus*, 3rd ed., Addison - Wesley, Reading, MA, 1984.

[78] M. Karow. Self-adjoint operators and pairs of Hermitian forms over the quaternions, *Linear Algebra and Appl.* **299** (1999), 101–117.

[79] ———. Note on the equation $ax - xb$ over the quaternions, unpublished manuscript.

[80] ———. Private communication.

[81] M. Koecher and R. Remmert. Hamiltons' quaternions, in *Numbers* (J. Ewing, ed.), pp. 189–220, Springer Verlag, New York, 1991 (English translation).

[82] L. Kronecker. *Collected Works*, Chelsea, 1968.

[83] ———. Über die congruenten Transformationen der bilinearer Formen, *Monats. der Akademie der Wissenschaften, Berlin* (1874), 397–447.

[84] J. B. Kuipers. *Quaternions and Rotation Sequences*, Princeton University Press, Princeton, 2002.

[85] P. Lancaster, A. S. Markus, and Q. Ye. Low rank perturbations of strongly definitizable transformations and matrix polynomials, *Linear Algebra and Appl.* **197/198** (1994), 3–29.

[86] P. Lancaster, A. S. Markus, and F. Zhou. A wider class of stable gyroscopic systems. *Linear Algebra Appl.* 370 (2003), 257–267.

[87] P. Lancaster, A. S. Markus, and P. Zizler. The order of neutrality for linear operators on inner product spaces, *Linear Algebra and Appl.* **259** (1997), 25–29.

[88] P. Lancaster and L. Rodman. Existence and uniqueness theorems for the algebraic Riccati equations, *International J. of Control* **32** (1980), 285–309.

[89] ———. Invariant neutral subspaces for symmetric and skew real matrix pairs, *Canadian J. Math.* **46** (1994), 602–618.

[90] ———. Minimal symmetric factorizations of symmetric real and complex rational matrix functions, *Linear Algebra and Appl.* **220** (1995), 249–282.

[91] ———. *Algebraic Riccati Equations*, Oxford University Press, New York, 1995.

[92] ———. Canonical forms for hermitian matrix pairs under strict equivalence and congruence, *SIAM Review* **47** (2005), 407–443.

[93] ———. Canonical forms for symmetric/skew-symmetric real matrix pairs under strict equivalence and congruence, *Linear Algebra and Appl.* **406** (2005), 1–76.

[94] P. Lancaster and M. Tismenetsky. *The Theory of Matrices*, 2nd ed., Academic Press, Orlando, 1985.

[95] P. Lancaster and Q. Ye. Variational properties and Rayleigh quotient algorithms for symmetric matrix pencils, *Operator Theory: Advances and Appl.* **40** (1989), 247–278.

[96] S. R. Lay. *Convex Sets and Their Applications*, Dover, Mineola, NY, 2007.

[97] N. Le Bihan and J. Mars. Singular value decomposition for quaternion matrices: A new tool for vector-sensor signal processing, *Signal Processing* **84** (2004), 1177–1199.

[98] H. C. Lee. Eigenvalues of canonical forms of matrices with quaternion coefficients, *Proc. Royal Irish Acad.* **52** Sect. A (1949), 253–260.

[99] J. M. Lee and D. A. Weinberg. A note on canonical form for matrix congruence. *Linear Algebra and Appl.* **249** (1996), 207–215.

[100] L. S. Levy and J. C. Robson. Matrices and pairs of modules, *J. of Agebra* **29** (1974), 427–454.

[101] W. W. Lin, V. Mehrmann, and H. Xu. Canonical forms for Hamiltonian and symplectic matrices and pencils, *Linear Algebra and Appl.* **302/303** (1999), 469–533.

[102] T. A. Loring. Factorization of matrices of quaternions, *Expositiones Mathematicae*, **30** (2012), 250–264.

[103] P. Lounesto. *Clifford Algebras and Spinors*, 2nd ed., Cambridge University Press, Cambridge, 2001.

[104] C. C. MacDuffee. *The Theory of Matrices*, Berlin, 1933; Chelsea, New York, 1946.

[105] A. I. Mal'cev. *Foundations of Linear Algebra*, W. H. Freeman, San Francisco and London, 1963 (translation from Russian).

[106] C. V. M. van der Mee, A. C. M. Ran, and L. Rodman. Classes of plus-matrices in finite dimensional indefinite scalar product spaces, *Integral Equations and Operator Theory* **30** (1998), 432–451.

[107] C. Mehl, V. Mehrmann, A. C. M. Ran, and L. Rodman. Perturbation analysis of Lagrangian invariant subspaces of symplectic matrices, *Linear and Multilinear Algebra* **57** (2009), 141–184.

[108] ———. Eigenvalue perturbation theory of classes of structured matrices under generic sutructured rank one perturbations, *Linear Algebra and Appl.* **435** (2011), 687–716.

[109] C. Mehl, V. Mehrmann, and H. Xu. Canonical forms for double structured matrices and pencils, *Electronic J. of Linear Algebra* **7** (2000), 112–151.

[110] C. Mehl, A. C. M. Ran, and L. Rodman. Semidefinite invariant subspaces: degenerate inner products, *Operator Theory: Advances and Appl.*, Current trends in operator theory and its applications, **149** (2004), 467–486.

[111] C. Mehl and L. Rodman. Symmetric matrices with respect to sesquilinear forms, *Linear Algebra and Appl.* **349** (2002), 55–75.

[112] V. L. Mehrmann. *The Autonomous Linear Quadratic Control Problem,* Lecture Notes in Control and Information Sciences **163**, Springer-Verlag, Berlin, 1991.

[113] V. Mehrmann and H. Xu. Structured Jordan canonical forms for structured matrices that are Hermitian, skew-Hermitian, or unitary with respect to an indefinite inner product, *Electronic J. of Linear Algebra* **5** (1999), 67–103.

[114] C. D. Meyer. *Matrix Analysis and Applied Linear Algebra*, SIAM, Philadelphia, 2000.

[115] Y. Nakayama. A note on the elementary divisor theory in non-commutative domains, *Bull. Amer. Math. Soc.* **44** (1938), 719–723.

[116] R. Pereira. *Quaternionic Polynomials and Behavioral Systems*, Doctoral Thesis, Universidade de Aviero, 2001.

[117] R. Pereira, P. Rocha, and P. Vettori. Algebraic tools for the study of quaternionic behavioral systems, *Linear Algebra and Appl.* **400** (2005), 121–140.

[118] R. Pereira and P. Vettori. Stability of quaternionic linear systems, *IEEE Trans. on Automatic Control* **31** (2006), 518–523.

[119] A. Pogorui and M. Shapiro. On the structure of the set of zeros of quaternionic polynomials, *Complex Var. Elliptic Funct.* **49** (2004), 379–389.

[120] Y. T. Poon. Generalized numerical ranges, joint positive definiteness and multiple eigenvalues, *Proc. of Amer. Math. Soc.* **125** (1997), 1625–1634.

[121] R. M. Porter. Quaternionic linear and quadratic equations, *J. Natur. Geom.* **11** (1997), 101–106.

[122] A. C. M. Ran. Minimal factorizations of selfadjoint rational matrix functions, *Integral Equations and Operator Theory*, **6** (1982), 850–869.

[123] A. C. M. Ran and L. Rodman. Stability of invariant maximal semidefinite subspaces, I. *Linear Algebra and Appl.* **62** (1984), 51–86.

[124] ———. Stability of invariant maximal semidefinite subspaces, II. Applications: selfadjoint rational matrix functions, algebraic Riccati equations, *Linear Algebra and Appl.* **63** (1984), 133–173.

[125] ———. Stability of invariant Lagrangian subspaces. I. *Operator Theory: Advances and Appl.* **32** (1988), 181–218, Birkhäuser, Basel.

[126] ———. Stable invariant Lagrangian subspaces: factorization of symmetric rational matrix functions and other applications. *Linear Algebra and Appl.* **137/138** (1990), 575–620.

[127] R. von Randow. The involutory antiautomorphisms of the quaternion algebra, *American Mathematical Monthly*, **74** (1967), 699–700.

[128] R. T. Rockafellar. *Convex Analysis*, Princeton University Press, Princeton, NJ, 1970.

[129] L. Rodman. Non-Hermitian solutions of algebraic Riccati equations. *Canadian J. Math.* **49** (1997), 840–854.

[130] ———. Invariant subspaces of selfadjoint matrices in skew symmetric inner products, *SIAM J. on Matrix Analysis and Appl.* **26** (2005), 901–907.

[131] ———. Similarity vs unitary similarity and perturbation analysis of sign characteristics: Complex and real indefinite inner products, *Linear Algebra and Appl.* **416** (2006), 945–1009.

[132] ———. Canonical forms for symmetric and skew-symmetric quaternionic matrix pencils, *Operator Theory: Advances and Appl.* **176** (2007), 199–254.

[133] ———. Canonical forms for mixed symmetric-skewsymmetric quaternion matrix pencils, *Linear Algebra and Appl.* **424** (2007), 184–221.

[134] ———. Comparison of congruences and strict equivalences for real, complex, and quaternionic matrix pencils with symmetries, *Electronic J. of Linear Algebra* **16** (2007), 248–283.

[135] ———. Pairs of hermitian and skew-hermitian quaternionic matrices: canonical forms and their applications, *Linear Algebra and Appl.* **429** (2008), 981–1019.

[136] ———. Pairs of quaternionic selfadjoint operators with numerical range in a half-plane, *Linear and Multilinear Algebra* **56** (2008), 179–184.

[137] ———. Stability of invariant subspaces of quaternion matrices, *Complex Analysis and Operator Theory* **6** (2012), 1069–1119.

[138] ———. Strong stability of invariant subspaces of quaternion matrices, *Operator Theory: Advances and Appl.* **237** (2013).

[139] ———. Hamiltonian square roots of skew Hamiltonian quaternionic matrices, *Electronic J. of Linear Algebra* **17** (2008), 168–191.

[140] ———. Invariant neutral subspaces for Hamiltonian matrices, *Electronic J. of Linear Algebra* **27** (2014), 55–99.

[141] V. V. Sergeichuk. Classification problems for systems of forms and linear mappings, *Math. USSR-Izv.*, **31** (1988), 481–501 (translation from Russian).

[142] ———. Classification of sesquilinear forms, pairs of Hermitian forms, self-conjugate and isometric operators over the division ring of quaternions, *Math. Notes* **49** (1991), 409–414 (translation from Russian).

[143] ———. Canonical matrices of isometric operators on indefinite inner product spaces, *Linear Algebra and Appl.* **428** (2008), 154–192.

[144] M. A. Shayman. On the variety of invariant subspaces of a finite-dimensional linear operator, *Trans.Amer. Math. Soc* **274** (1982), 721–747.

[145] G. E. Shilov. *An Introduction to the Theory of Linear Spaces*, Prentice Hall, Englewood Cliffs, 1961 (translation from Russian).

[146] L.-s. Siu. *A Study of Polynomials, Determinants, Eigenvalues and Numerical Ranges over Real Quaternions*, M. Ph. Thesis, University of Hong Kong, 1997.

[147] L. Smith. *Linear Algebra*, Springer Verlag. New York, 1984.

[148] W. So and R. C. Thompson. Convexity of the upper complex plane part of the numerical range of a quaternionic matrix, *Linear and Multilinear Algebra* **41** (1996), 303–365.

[149] G. W. Stewart and J.-g. Sun. *Matrix Perturbation Theory*, Academic Press, Boston, 1990.

[150] E. Study. Zur Theorie der Linearen Gleichungen, *Acta Math.* **42** (1920), 1–61.

[151] R. C. Thompson. Pencils of complex and real symmetric and skew matrices, *Linear Algebra and Appl.* **147** (1991), 323–371.

[152] F. Uhlig. A recurring theorem about pairs of quadratic forms and extensions: a survey, *Linear Algebra and Appl.* **25** (1979), 219–237.

[153] J. Vince. *Quaternions for Computer Graphics*, Springer, London, 2011.

[154] W. R. Wade. *An Introduction to Analysis*, Prentice Hall, Upper Saddle River, NJ, 1995.

[155] C. T. C. Wall. Stability, pencils, and polytopes, *Bull. London Math. Soc.* **12** (1980), 401–421.

[156] Z.-X. Wan. *Geometry of Matrices*, World Scientific Publishing, River Edge, NJ, 1996.

[157] J. P. Ward. *Quaternions and Cayley Numbers*, Kluwer Academic Publishers, Dordrecht, 1997.

[158] R. Webster. *Convexity*, Oxford University Press, Oxford, New York Tokyo, 1994.

[159] B. Wie, H. Weiss, and A. Arapostathis. Quarternion feedback regulator for spacecraft eigenaxis rotations, *Journal of Guidance, Control, and Dynamics* **12** (1989), 375–380.

[160] N. A. Wiegmann. Some theorems on matrices with real quaternion entries, *Canadian J. of Math.* **7** (1955), 191–201.

[161] J. Williamson. On the algebraic problem concerning the normal forms of linear dynamical systems, *Amer. J. of Math.* **58** (1936), 141–163.

[162] L. A. Wolff. Similarity of matrices in which the elements are real quaternions, *Bull. Amer. Math. Soc.* **42** (1936), 737–743.

[163] R. M. W. Wood. Quaternionic eigenvalues. *Bull. London Math. Soc.* **17** (1985), 137–138.

[164] F. Zhang. Quaternions and matrices of quaternions, *Linear Algebra and Appl.* **251** (1997), 21–57.

[165] P. Zizler. Definitizable operators on a Krein space, *Bull. Canadian Math.Soc.* **38** (1995), 496–506.

Index

Princeton Series in Applied Mathematics